Vegetable Oil Fuels

Proceedings of the International Conference on Plant and Vegetable Oils as Fuels

August 2-4, 1982
Holiday Inn
Fargo, North Dakota

Published by
American Society of Agricultural Engineers
2950 Niles Road • St. Joseph, Michigan 49085

Library of Congress Card Number (LCCN) 82-72554
International Standard Book Number (ISBN) 0-916150-46-1

The American Society of Agricultural Engineers is not responsible for statements and opinions advanced in its meetings or printed in its publications. They represent the views of the individuals to whom they are credited and are not binding on the Society as a whole.

Table of Contents

PREFACE

Modern agriculture is dependent on the diesel engine. Traditional sources of diesel engine fuel for agriculture have been under threat of supply disruptions, quality deterioration and cost escalations. Are there alternatives?

Plant and vegetable oils, "Bio-oils", are the most promising candidates as substitutes or extenders for fossil-derived diesel fuels for agriculture. These oils have the potential to make farming communities self-sufficient in the fuel they need to continue their business.

The idea is not new. As early as 1900, a diesel-cycle engine was demonstrated running wholly on ground-nut oil at the Paris exposition. Interest has recently been rekindled in bio-oils as diesel fuels and researchers in many parts of the world are applying new technologies to the concept.

The objectives of this Conference are to serve to coordinate worldwide research effort on the subject by:

- promoting technical interaction and international co-operation
- providing a forum for personal contacts
- further promoting renewable fuels.

In November 1977 the first International Symposium on Alcohol Fuel Technology was held in Wolfsburg, West Germany. At that time, few of the participants could have anticipated the progress that has been made as fuel alcohols make another bid for a significant place in the liquid fuels market-place. Undoubtedly the first and subsequent International Alcohol Fuel Symposia helped galvanize this development. Could a similar thing happen for plant and vegetable oils?

Plans to hold this Conference were conceived in 1979 and formalized at a meeting in Fargo, North Dakota in March 1980. Mr. Charles Moses of Interstate Grain initiated the Fargo meeting. Since then Drs. Roald Lund, George Pratt, Ken Kaufman and members of the several committees listed have been diligent in making preparations for the Conference. Fargo was selected, and endorsed by ASAE, due to its location in one of the most intensive vegetable oil producing areas of the world.

Over seventy paper proposals were received for this conference. From these, the Program Committee selected for presentation fifty substantive contributions involving more than one hundred people worldwide.

(continued)

Three years after the idea to hold this Conference we find ourselves in a different energy scenario, one well summarised by Christopher Byron in TIME, April 19, 1982:

> *"The current fortuitious situation of declining oil prices provides an ideal opportunity to press ahead with greater efforts at conservation and the development of non-OPEC energy supplies. It was just such efforts in the past three years that helped bring about the present break in oil cartel prices."*

There are still unprecedented challenges.

GRAEME R. QUICK,
Glenfield, Australia.
April 23, 1982.

INTERNATIONAL CONFERENCE ON PLANT AND
VEGETABLE OILS AS FUELS

August 2-4, 1982
Holiday Inn
Fargo, ND

Sponsor:
American Society of Agricultural Engineers

Cosponsors (with liaisons):
American Agricultural Economics Association
(Sydney James)
American Association of Cereal Chemists
(Everett Pryde)
American Oil Chemists Society
(James Lyon)
American Society for Testing and Markets
(Bobby Hall)
American Society of Agronomy
(David Kral)
American Society of Mechanical Engineers, Diesel and Gas Engine Power Division
(Robert Parker)
American Soybean Association
(David Erickson)
Australian Sunflower Association
(Owen Duncan)
Agricultural Engineering Society, Australia
(Graeme Quick)
Canadian Society of Agricultural Engineering
(R. C. Strayer)
Canola Council of Canada
(Gerald Zoerb)
Commission Internationale du Genie Rural (CIGR)
(Lawrence Skromme)
Crop Science Society of America
(David Kral)
Farm and Industrial Equipment Institute
(James Ebbinghaus)
National Cottonseed Products Association
(Thomas Wedegaertner)
North Dakota Sunflower Council
(Larry Kleingartner)
South African Society of Agricultural Engineers
(J. J. Bruwer)
South Dakota Sunflower Council
(Samuel Heikes)
Sunflower Association of America
(Donald Lilleboe)
USDA-ARS
(Landy Altman)
US DOE
(Marilyn Ripin)

CONFERENCE COMMITTEES

Conference Chairman:

Dr. George L. Pratt
Chairman
Department of Agricultural Engineering
North Dakota State University
FARGO, NORTH DAKOTA 58107

Conference Co-ordinator:

Dr. Graeme R. Quick
Director of Agricultural Engineering
NSW Department of Agriculture
Agricultural Engineering Centre
GLENFIELD, NSW, 2167. AUSTRALIA

STEERING COMMITTEE

George L. Pratt, Chairman
North Dakota State University

Graeme Quick, Conference Coordinator
South Wales Government

Charles Moses
Interstate Seed Company

Ken Domier
University of Alberta

Gordon Lund
Farmer (ND)

Ralph Matthews
Allis-Chalmers

Stanley Clark
Kansas State University

Charles Sheets
J. I. Case Company

Kenneth L. McFate
University of Missouri

Sherwood S. DeForest
Montgomery Assoc.

John Bailey
Catepillar Tractor Company

Mark Purschwitz (Ex Officio)
American Society of Agr. Engineers

LOCAL ARRANGEMENTS COMMITTEE

Harvey J. Hirning, Chairman
North Dakota State University

Delmer Helgeson
North Dakota State University

Harold Klosterman
North Dakota State University

Vince Lindstrom
Fargo Convention Bureau

Charles Moses
Interstate Seed Company

Norm Robinson
Holiday Inn (Fargo)

Gordon Lund
Farmer (ND)

Don Lilleboe
Sunflower Association of America

Henry L. Kucera
North Dakota State University

George L. Pratt
North Dakota State University

PROGRAM COMMITTEE

Kenton R. Kaufman, Chairman
North Dakota State University

William E. Splinter
University of Nebraska

Wesley F. Buchele
Iowa State University

Everett Pryde
Northern Regional Research Lab.

James Butler
Georgia Coastal Experiment Station

Ralph Matthews
Allis-Chalmers

Norm Sauter
Deere and Company

John C. Basiletti
International Harvester Company

Dick Thow
The French Oil Mill Machinery Co.

Graeme Quick
New South Wales Government

Don Zimmerman
North Dakota State University

Joe Ward
Simon-Rosedown Limited

J. J. Bruwer
Republic of South Africa

Warren Roller
Ohio Agricultural R & D Center

Delmer Helgeson
North Dakota State University

William Larsen
Montana State University

John Bailey
Catepillar Tractor Company

Charles Peterson
University of Idaho

Allen A. Housh
Cargill Inc.

PUBLICITY COMMITTEE

Frank Buckingham, Chairman
Agri-Business Writing

Don Lilleboe
Sunflower Association of America

Cecil Hammond
University of Georgia

Charles Humphrey
North Dakota State University

Carroll Goering
University of Illinois

Vernon Miller
Progressive Farmer

Mark Purschwitz (Ex Officio)
American Society of Agr. Engineers

PROCEEDINGS COMMITTEE

Leslie F. Backer, Chairman
North Dakota State University

Graeme Quick
New South Wales Government

George L. Pratt
North Dakota State University

Kenton R. Kaufman
North Dakota State University

Don Zimmerman
North Dakota State University

VEGETABLE OILS AND ANIMAL FATS FOR DIESEL FUELS: A SYSTEMS STUDY

E. S. Lipinsky

S. Kresovich

C. K. Wagner

H. R. Appelbaum

T. A. McClure

J. L. Otis

D. A. Trayser

BATTELLE
Columbus Laboratories
505 King Avenue
Columbus, Ohio 43201

Despite the temporary glut of petroleum and the middle distillate fuels derived from this non-renewable resource, users of middle distillate fuels are threatened by potential shortages and/or inordinate price increases. Middle distillate fuel users are essential elements of not only the United States economic structure, but that of the entire world. For example, farmers use diesel fuel to power their agricultural equipment, long-haul truck transportation is predominantly operated with diesel fuel, railroads operate primarily on diesel fuel, and electric utilities use diesel electricity generators in remote locations and No. 2 fuel oil in gas turbines for peaking power. Therefore, a shortage of diesel fuel can shut down both U.S. agriculture and the transportation of many commodities to markets. Military uses of diesel fuel also are significant, but the United States Armed Forces are better prepared for diesel fuel shortages than are the agricultural and transportation sectors.

It is the purpose of this paper to provide some information on the possible use of vegetable oils and animal fats as substitute fuels and as emergency diesel fuels in the United States. A substitute fuel can replace a portion of demand for a conventional fuel over a relatively long time span, whereas an emergency fuel is one that is used in the event of a crisis. Although there is considerable interest in vegetable oils and animal fats as fuels for electric power and railroad transportation applications, the scope of this paper is confined to systems for making and using triglyceride fuels (a comprehensive term for both vegetable oils and animal fats) in agricultural, automotive, and highway transportation applications.

TRIGLYCERIDE SYSTEMS

Satisfactory substitution of petroleum-based diesel fuels with triglyceride-based fuels requires the development of an integrated system for the production, processing, and end use of the new fuels on a basis that is both technically attractive and economically rewarding to all of the elements of

The authors are: E. S. LIPINSKY, Senior Research Leader, T. A. MCCLURE, Senior Research Scientist, C. K. WAGNER, Research Scientist, S. KRESOVICH, Principal Research Scientist, J. L. OTIS, Principal Research Scientist, H. R. APPLEBAUM, Researcher, and D. A. TRAYSER, Principal Mechanical Engineer

the system. The overall system is composed of three subsystems. The first subsystem consists of farms that produce oilseed crops. The farms may be dedicated solely to oilseed production or may enter and leave this system or demand changes. The second subsystem involves production of triglycerides and protein. The processing units that manufacture the triglyceride fuels may be dedicated to fuel production or may be pressed into service in response to a crisis or in response to rising prices of alternative products. The end use subsystem involves both the manufacturers of diesel engines and the owners of the present stock of auto-ignition engines. Thus, the system consists of both the hardware (land, agricultural equipment, processing equipment, and diesel engines) and the owners of these capital goods (farmers and oilseed processors, truck and tractor owners).

Agricultural System Opportunities

A recent study conducted by the authors for the U.S. Department of Energy indicates that it is technically feasible to manufacture oilseeds on a sufficiently large scale to permit manufacture of an additional one billion to five billion gallons, over and above domestic U.S. vegetable oil requirements and traditional export requirements (Lipinsky, et al, 1981). This ambitious goal can be attained by means of the restructuring shown in Table 1. Most of the increase in oilseed production arises from new crops and conversion of land from pasture and other uses. These conversions require considerable time; therefore, they are not acceptable for emergency sources of diesel fuel. They constitute possible profit opportunities for permanent substitution of petroleum-derived diesel fuels.

The quick turnaround required for emergency diesel fuel production involves either diversion of a crop initially destined for another application (usually food or industrial chemicals) or adoption of modified crop rotations so that land is pressed into service for the production of diesel fuel raw material on short notice. The crop rotation option is potentially capable of producing approximately one billion gallons per year, as shown in Table 2.

Oilseed Processing and Conversion

The key issues in the processing of oilseeds for the production of diesel fuels are:

(1) Whether the processing should be conducted at the local level in relatively small units or whether the oilseeds should be processed in large-scale units resembling those now employed for the production of soybean oil

(2) Whether processing and conversion can be confined to simple removal of the oil or whether transesterification also must be conducted

(3) Whether the oilseed processing/conversion unit should be specialized for the use of a predetermined oilseed or whether it should be a general facility capable of processing whichever raw material is cheapest at the moment.

The investigations of the authors (Lipinsky, et al, 1981) indicate that conventional large-scale units produce cheaper vegetable oil than do small, on-farm facilities, even when some consideration is given to reduced raw material costs and labor cost savings on the farm. The bases for the preliminary evaluation of an on-farm unit, a cooperative-scale unit, and a small-scale commercial unit are presented in Table 3. Table 4 contains the results of the evaluation.

Table 1. U.S. Agricultural Potential for Additional Fats and Oils Production for Diesel Fuel Use

Source	Basis for Estimate	Estimate (Billion Gallons Per Year)[a] Near Term[b]	Long Term[c]
Animal fats	20 percent of 1980 production	0.1	0.2
Crop rotation	20 million acres at 60 gallons per acre, with 1/3 available the first year that sufficient processing equipment is available	0.4	1.2
Sequential cropping	25 million acres at 40 gallons per acre[d] with 20 percent available the second year that sufficient processing equipment is available	0.2	1.0
New cropland from pasture, range, and forest	20 million acres at 40 gallons per acre[e]	0.2	0.8
New oilseed crops (e.g., Chinese Tallow tree)	10 million acres at 170 gallons per acre		1.7
Total		0.9	4.9

[a] Agricultural potential for raw materials containing indicated amounts of fats and oils suitable for diesel fuel use. Time frame for construction of facilities to extract this much oil must be estimated separately.

[b] Near term is within one or two years of the onset of an emergency of sufficient magnitude to stimulate intensive implementation.

[c] Long term is 15 years after onset of an emergency of sufficient magnitude to stimulate intensive implementation.

[d] Yields 25 percent less than the U.S. average for the crop are assumed because the timing of planting and/or harvest usually will be suboptional.

[e] Yields 25 percent less than the U.S. average for the crop are assumed because less fertile land is used.

Source: Battelle-Columbus estimates

Table 2. Hypothetical Production of Vegetable Oils as Diesel Fuel Substitutes, Utilizing Potential Cropland in Seven Farm Production Regions*

Farm Production Region	Area with High and Medium Potential for Conversion to Cropland, 1977** (Million Acres)	Primary Oilseed Crop	Assume 20% of Potential New Cropland Converted to Primary Oilseed Crop (Million Acres)	Assumed Oil Yield (Gallons/Acre)			Total Oil Production (Million Gallons)		
				Low	Med.	High	Low	Med.	High
Northern Plains	17.7	Sunflower	3.54	46	61	76	163	216	269
Southern Plains	20.1	Sunflower	4.02	33	44	55	133	177	221
Lake States	8.5	Sunflower	1.70	46	61	76	78	104	129
Corn Belt	14.4	Soybeans	2.88	31	41	51	89	118	147
Delta States	10.1	Soybeans	2.02	24	32	40	48	65	81
Southeast	15.8	Peanuts	3.16	65	87	109	205	275	344
Appalachian	14.4	Soybeans	2.88	26	34	43	75	98	124
TOTAL	101.0		20.20	39	52	65	791	1053	1315

*Does not include potential contribution of oilseed plants that could be double-cropped or intercropped with other crops.

**Includes pasture, rangeland, forests and other lands including farmsteads, windbreaks, etc.; from U.S. Department of Agriculture, National Agricultural Lands Study. Final Report, 1981, Washington, D.C.

Table 3. Capital Investments

Capital Investment for On-Farm Unit

Basis $22000

Hourly feed rate:	88.2 pounds undehulled seeds per hour
Operation :	2,184 hours per year (156 days, per year, 14 hours per day)
Efficiency :	90 percent

Sunflower seeds required:	86.7 tons/year
Filtered oil produced :	6,670 gallons/year
Meal cake produced (14 percent oil) :	60.5 tons/year
Sludge produced :	136 gallons/year

Peanut seeds required :	86.7 tons/year
Filtered oil produced :	8,510 gallons/year
Meal cake produced (14 percent oil) :	53.4 tons/year
Sludge produced :	174 gallons/year

Capital Investment for Cooperative Unit

Basis $260000

Hourly feed rate:	458 pounds undehulled seeds per hour
Operation :	7,920 hours per year (330 days per year, 24 hours per day)
Efficiency :	90 percent

Sunflower seeds required:	1,632.3 tons/year
Filtered oil produced :	147,436 gallons/year
Meal cake produced (7 percent oil) :	1,053 tons/year
Sludge produced :	3,009 gallons/year

Peanut seeds required :	1,632.3 tons/year
Filtered oil produced :	179,640 gallons/year
Meal cake produced (7 percent oil) :	930.2 tons/year
Sludge produced :	3,667 gallons/year

Capital Investment for Commercial Unit

Basis $2800000.

Hourly feed rate:	16,667 pounds/hour seeds
Operation :	7,920 hours per year (330 days per year, 24 hours per day)
Efficiency of Operations :	95 percent

Sunflower seeds required:	62,701 tons/year
Filtered and winterized oil produced :	5,890,540 gallons/year
Meal cake produced (6 percent oil) :	40,020 tons/year

Peanut seeds required :	62,701 tons/year

Table 3. Capital Investments(Cont'd)

Filtered and winterized oil produced	: 7,533,780 gallons/year
Waste Hulls	: 10,659 tons/years
Meal cake produced (4 percent oil)	: 23,186 tons/year

Table 4. Summary of Economic Analysis

Zero Profit Emergency Economics Analysis

Sunflower	On-Farm	Cooperative	Commercial
Seeds processed, tons/year	86.7	1,632	62,700
Gallons oil produced per year	6,670	147,400	5,890,500
Gallons diesel fuel equivalent	6,350	140,400	5,610,000
Meal produced, tons/year	60.5	1,053	40,020
Conversion cost, dollars/year	8,321	200,700	1,825,000
Raw material cost, dollars/year	11,271[a]	244,800[b]	12,540,000[c]
Less by product meal credit[d]	7,865	136,900	5,203,000
Total cost, dollars/year	11,727	308,600	9,162,000
Cost per gallon diesel oil equivalent, dollars/gallon	1.85	2.20	1.63
Peanuts			
Seeds processed, tons/year	86.7	1,632	62,700
Gallons oil produced per year	8,510	179,600	7,533,800
Gallons diesel fuel equivalent	8,110	171,100	7,175,000
Meal produced, tons/year	53.4	930	23,190
Conversion cost, dollars/year	9,203	201,200	1,498,100
Raw material cost, dollars/year[e]	13,525	293,760	15,048,000
Less by product meal credit	8,010	139,600	4,521,300
Total cost, dollars/year	14,718	355,360	12,024,800
Cost per gallon diesel oil equivalent, dollars/gallon	1.81	2.08	1.68

[a] Assumes on-farm value of $130 per ton.

[b] Assumes local value of $150 per ton.

[c] Assumes commercial market value of $200 per ton.

[d] Assumes by-product credit of $130 per ton for sunflower meal, $150 per ton for peanut meal from the on-farm and cooperative units, and $195 per ton for peanut meal from the commercial unit.

[e] Peanut acreage allotments distort costs. Incremental production of less than food-grade peanuts assumed at 20 percent more than sunflowers.

Source: Lipinsky, et al, 1981

Although the advantage of the large central processing unit might be overcome for the production of unprocessed vegetable oil fuels, the economy of such advantage is likely to be insurmountable if the end use requirements of diesel engines necessitate the use of transesterified products. Whether transesterification is needed depends partly on the oil under consideration (Bruwer, et al, 1981). Beef tallow will require transesterification anywhere whereas a highly unsaturated oil may require no transesterification in semitropical areas. In northern Canada, probably every oil requires transesterification because climate is the major factor.

Diesel Engine Subsystem

Vegetable oils and animal fats have heats of combustion that are within about 10% that of petroleum-derived diesel fuel (Kaufman, et. al, 1981). Although the cetane numbers of vegetable oils and animal fats are less than those of diesel fuel, the knocking qualities seem to be adequate in preliminary tests. Compared with petroleum-derived fuels, the ash and sulfur contents of vegetable oils and animal fats are excellent.

In a critical emergency, two offsetting considerations must be taken into account:

(1) The critical emergency is likely to last for only a short time; therefore, immediate performance outweighs damage that might occur over the long term

(2) In a critical emergency, there is no opportunity for the engine manufacturer to adjust engine design.

The first consideration liberalizes the opportunity to use triglyceride fuels and the second one minimizes the opportunity. If the critical emergency does not arise for many years, engine design could be adjusted to take critical emergencies into account.

The major potential drawbacks in using unmodified vegetable oils as diesel fuel include:

(1) Fuel line and filter clogging due to fines, phosphatide gums, waxes, or high melting saturated fat

(2) Polymerization or partial oxidation during storage of the oil, which raises viscosity

(3) Polymerization in the combustion chamber due to heat or free radicals which leads to deposits

(4) Poor ignition and combustion characteristics, resulting from improper atomization of this high viscosity fuel

(5) Coking

(6) Polymerization of the triglycerides in the lube oil due to blowby of unsaturated fuels capable of autoxidation.

Although many diesel fuel-vegetable oil combinations may be acceptable for short-term emergencies involving hours or a few days, acceptability of vegetable oils in long-term usage requires that these potential drawbacks be more thoroughly defined and, if serious, reduced or eliminated.

Methyl esters with low viscosities approaching that of diesel fuel can be made from vegetable oils by reaction with methanol. The glycerin byproduct is valuable for chemical uses. Removal of this possible acrolein precursor is environmentally desirable as well. However, the methyl esters of saturated fatty acids still tend to solidify in cool weather conditions and methyl ester products are still subject to the free radical polymerization problems that are associated with buildup in crank case viscosity and coking.

Microemulsions of vegetable oil products with alcohols and water in diesel fuel are under development by several organizations (Georing, et al, 1982) (Anon. 1981). Experimental emulsions have been prepared that are low in viscosity, stable in storage and apparently have favorable performance characteristics. More data are needed regarding the long-term performance of these products.

RESEARCH AND DEVELOPMENT NEEDS

The fact that farmers, processors, and engine owners/manufacturers must all change their behavior for the triglyceride fuel system to emerge places the new system at a considerable disadvantage. A concerted research and development program will be needed to make possible the substantial use of vegetable oils and animal fats as diesel fuels. The program's most critical element is investigation of the effects of various fats and oils on the durability and performance of diesel engines. For each oil, it will be necessary to evaluate various purity levels that are associated with the extent of processing and conversion to establish which oils are acceptable and at which level of purity. The oils that fail in performance and/or engine durability evaluations may be modified to achieve acceptance through chemical conversion to esters or by microemulsification, etc.

Processing systems need to be developed to achieve the desired levels of purity at reasonable costs. Oilseed crop production needs to be enhanced by selection, breeding, and improved agronomic practices for those crops that appear to be attractive on both technical and economic grounds.

Research and development on conversion of fats and oils into diesel products through transesterification, microemulsification, and other processes involves performance trials, research to determine capital and operating costs, and process investigation. Such studies should be geared to overcoming serious technical and/or economic drawbacks of the fats and oils from which the derivatives are made.

Fuel performance evaluations should be phased so that in-depth evaluations are conducted only on fuels that pass screening criteria regarding physical properties and economic attractiveness. The following considerations are important in the in-depth studies:

- The types of engine should be those used by the customers.
- Durability tests should represent long runs using test cycles approved by the Engine Manufacturers Association.
- Adequate testing should take place with vehicles or engines under adverse weather conditions.
- Failure evaluations involving fuels that have become contaminated with fines or high viscosity ingredients should be conducted so that the sensitivity of engine durability and performance to fuel composition and purity can be assessed.
- Adequate experimental design with respect to replication and statistical rigor should be employed by the investigators.
- Complete vegetable oil characterization needs to be undertaken.

Information pertaining to crop production practices and production budgets for fuel-grade peanuts, rape, and Chinese tallow tree (Morgan and Schultz, 1981) needs to be developed and evaluated for various cultivars in various geographical areas. Even soybeans and sunflowers, for which more information is available, have serious agronomic uncertainties when they are to be used in sequential cropping systems.

The in-depth research and development on crop production should involve only those crops that pass both agronomic and economic screening criteria. Programs should use well-replicated experimental designs with field plots of adequate size. The emphasis should be on regional studies that are meaningful in geographical scope.

Emergency diesel fuel requires that the triglyceride fuel substitution be compatible with whatever diesel engines are used by farmers and truckers at the time that the crisis occurs. However, the greatest challenge is to design an engine that is tailored to vegetable oil fuel. Vegetable oils are more homogeneous than are petroleum-based fuels. The boiling point range is much more limited but opportunities for precombustion pyrolysis are abundant. However, an engine optimized for vegetable oils or animal fats would be suboptimal during times of petroleum abundance.

CONCLUSIONS

Vegetable oils and animal fats have substantial prospects as long-term substitutes for diesel fuels. If special auto-ignition engines were developed to handle vegetable oils, on-farm production and use might succeed. In the absence of such engine development, it is likely that large, centralized facilities to manufacture vegetable oils and their methyl esters will be the successful processing route. Vegetable oils are likely to succeed first in geographical areas with benign climates.

Vegetable oils and animal fats have limited prospects as diesel fuels for acute emergencies. The high viscosity of vegetable oils and the necessity to make substantial capital investments to obtain oils from oilseeds render the system relatively inflexible.

Acknowledgement

The authors gratefully acknowledge support from the Department of Energy's Biomass Energy Technology Division and the Office of Vehicle and Engine Research and Development of the research on which this paper is based. Dr. Marilyn Ripin and Dr. Eugene Ecklund in particular were supportive of this systems study. The editorial and secretarial efforts of I. R. Morgan and J. P. Carlyle are greatly appreciated by the project team.

REFERENCES

(1) Anon., 1981. Chem. Week 128: 34 (August 26).

(2) Bruwer, J. J., et al. 1981. The Utilization of Sunflower Seed Oil as a Renewable Fuel for Diesel Engines. Agricultural Energy 2, American Society of Engineers, St. Louis: 385-390.

(3) Georing, C. E., A. W. Schwab, and E. H. Pryde. 1982. Evaluations of Microemulsions as Engine Fuel. Paper presented at the ACS Great Lakes Regional Meeting, Normal, Illinois, June 7-9.

(4) Kaufman, K. R., M. Ziejewski, M. Marchl, H. Kucera, and A. E. Jones. 1981. Performance of Diesel Oil and Sunflower Oil Mixtures in Diesel Farm Tractors. Paper #81-1054. Presented at the ASAE Summer Meeting, Orlando, Florida, June 21-24.

(5) Lipinsky, E. S., T. A. McClure, S. Kresovich, J. L. Otis, C. K. Wagner, D. A. Trayser, and H. R. Appelbaum. 1981. Systems Study of Animal Fats and Vegetable Oils for Use as Substitute and Emergency Diesel Fuels. Phase II Final Report from Battelle to U.S. Department of Energy Biomass Energy Technology Division, Office of Vehicle and Engine Research and Development.

(6) Morgan, R. P. and E. B. Schultz, Jr. 1981. Fuels and Chemicals from Novel Seed Oils. Chemical & Engineering News, 59: 69-77 (September 7).

COMPARISON OF OILSEED YIELDS: A PRELIMINARY REVIEW[1/]

James A. Duke[2/] and Marvin O. Bagby[3/]

INTRODUCTION

In the April-May issue of "Crops and Soils" we read: "In order for agriculture to survive, a continuous, yet inexpensive, fuel supply is needed. Oilseed crops just might be able to supply some of this fuel...Only a small percentage (around 10 percent) of a farm's acreage is required to produce the fuel necessary for crop production." To address the implications of the last sentence, we have compiled this review. Taking the 10% "tithe" as a rule of thumb, we generalize that 90% of an oil yield might be considered as profit.

It is difficult to find realistic yield predictions for most oilseed crops. Experimental yields are often ten times higher than on-the-farm yields, herein called production yields. To give some idea of the range of variation we reproduce here Table 1, assembled for other reasons, hence not necessarily biased to show high or low oil yields (Duke, 1978). These yield data, tabulated elsewhere, may not always agree with the newer data summarized in the discussion. Table 1 shows seed yields ranging from 200 to 14,000 kg/ha, the low one too low to consider and the high one suspiciously high. The yield of 14,000 kg oil per hectare is equivalent to more than 30 barrels of oil per hectare, even after the "tithe." It was the only 5-digit oilseed yield.

To compare oil seeds, we reproduce pertinent portions of the yield and energy paragraphs from oilseed species information summaries being compiled by Dr. Duke for DOE Grant No. 59-2246-1-6-054-0 with Dr. Bagby as ADODR. Under that grant, Dr. Duke is preparing information summaries on 200 species that have been suggested as energy species. Interested readers are invited to correspond for further details, including the references from which these yield comparisons are drawn. We solicit further information or references to other published yield figures on these and other species that should be considered as energy candidates.

Following Table 1, which lists many oilseeds alphabetically by common name, we reroduce energy/yield paragraphs, alphabetically by scientific name. Excepting the peanut and soybean, we have intentionally omitted from further discussion the legume family (see Duke's 1981 Handbook of Legumes of World Economic Importance, Plenum Press, New York). The grass, carrot, pigweed, and lambsquarter families, some of which have high seed yields, were not included in Table 1 because generally their oil contents are not high.

1/ Because of relocations, this manuscript, submitted late, has not been through the normal review process.

2/ Economic Botany Laboratory, Building 001, Room 133, BARC-West, Beltsville, Maryland 20705.

3/ Northern Agricultural Energy Center, Northern Regional Research Center, 1815 North University, Peoria, Illinois 61604.

Table 1. Oil Seed Yields

Common Name	Scientific Name	Seed or Oil Yield (kg/ha)
Ajipo	Pachyrhizus tuberosus (Lam.) Spreng.	600 sd
Almond	Prunus dulcis (Mill.) D.A.Webb	3,000 sd
Bean, adzuki	Vigna angularis (Willd.) Ohwi & Ohashi	1,100 sd
Bean, African locust	Parkia filicoidea Welw. ex Oliv.	5,000 sd
Bean, broad	Vicia faba L.	6,600 sd
Bean, green	Phaseolus vulgaris L.	2,500 sd
Bean, lablab	Dolichos lablab L.	1,400 sd
Bean, lima	Phaseolus lunatus L.	2,500 sd
Bean, mung	Vigna radiata (L.) Wilczek var. radiata	1,100 sd
Bean, rice	Vigna umbellata (Thunb.) Ohwi & Ohashi	200 sd
Bean, scarlet runner	Phaseolus coccineus L.	1,700 sd
Bean, tepary	Phaseolus acutifolius A.Gray	1,700 sd
Cacao	Theobroma cacao L.	3,300 sd
Canicha	Sesbania bispinosa (Jacq.) W.F.Wight	1,000 sd
Cashew	Anacardium occidentale L.	1,000 sd
Castorbean	Ricinus communis L.	5,000 sd
Chickpea	Cicer arietinum L.	2,000 sd
Colocynth	Citrullus colocynthis (L.) Schrad.	6,700 sd
Crambe	Crambe abyssinica Hochst. ex R.E.Fries	5,000 sd
Croton, purging	Croton tiglium L.	900 sd
Crownvetch	Coronilla varia L.	500 sd
Fenugreek	Trigonella foenum-graecum L.	3,000 sd
Gourd, buffalo	Cucurbita foetidissima HBK	3,000 sd
Gram, black	Vigna mungo (L.) Hepper	800 sd
Groundnut, Bambarra	Voandzeia subterranea (L.) Thou.	4,200 sd
Guar	Cyamopsis tetragonoloba (L.) Taub.	2,000 sd
Jicama	Pachyrhizus erosus (L.) Urb.	600 sd
Lablab	Dolichos lablab L.	1,400 sd
Lentil	Lens culinaris Medik.	1,700 sd
Lupine, European blue	Lupinus angustifolius L.	1,000 sd
Lupine, European yellow	Lupinus luteus L.	1,000 sd
Lupine, white	Lupinus albus L.	1,000 sd
Marihuana	Cannabis sativa L.	1,500 sd
Meadowfoam, Baker's	Limnanthes bakeri J.T.Howell	400 sd
Meadowfoam, Douglas's	Limnanthes douglasii R.Br.	1,900 sd
Mu-oil tree	Aleurites montana (Lour.) Wils.	5,500 oil
Mustard, black	Brassica nigra (L.) Koch	1,100 sd
Mustard, white	Sinapis alba L.	8,000 sd
Niger seed	Guizotia abyssinica (L.f.) Cass	600 sd
Nut, macadamia	Macadamia spp.	4,000 sd
Oilvine, Zanzibar	Telfairia pedata (Sm. ex Sims) Hook.	2,000 sd
Palm, African Oil	Elaeis guineensis Jacq.	2,200 oil
Pea, cow	Vigna unguiculata (L.) Walp.	2,500 sd
Pea, garden	Pisum sativum L.	1,800 sd
Peanut	Arachis hypogaea L.	5,000 sd
Perilla	Perilla frutescens (L.) Britt.	1,500 sd
Poppy, opium	Papaver somniferum L.	900 sd
Rape	Brassica napus L.	3,000 sd
Safflower	Carthamus tinctorius L.	4,500 sd
Soybean	Glycine max (L.) Merr	3,100 sd
Stylo, Townsville	Stylosanthes humilis H.B.K.	1,200 sd
Sunflower	Helianthus annuus L.	3,700 sd
Tallow tree, Chinese	Sapium sebiferum (L.) Roxb.	14,000 sd
Trefoil, birdsfoot	Lotus corniculatus L.	600 sd
Velvetbean	Mucuna deeringiana (Bort) Merr.	2,000 sd
Walnut, black	Juglans nigra L.	7,500 sd
Walnut, Persian	Juglans regia L.	7,500 sd

DISCUSSION

ABLEMOSCHUS MOSCHATUS Medik. MALVACEAE AMBRETTE

Telek and Martin (1981) cite seed yields of 836-1,693 kg/ha in Louisiana, 1,880-3,330 in Puerto Rico, with the oil yield in Louisiana averaging 226 kg/ha, 433 kg/ha in Puerto Rico.

ALEURITES FORDII Hemsl. EUPHORBIACEAE TUNG-OIL TREE

Trees yield 4.5-5 MT (metric tons)/ha.

ALEURITES MONTANA (Lour.) Wils. ANACARDIACEAE WOOD-OIL TREE, MU-OIL TREE

A. montana is reported to give much higher yields of fruits than A. fordii. The percentage of kernels in the seeds is about 56%, and of oil in the kernels, about 59.3%. Yields of oil per tree in China is reckoned to be about 3.2 kg; in Florida, 4.5-9 kg. Trees yield about 45-68 kg nuts/year, these yielding about 35-40% oil. In one Malawi trial, N treatments gave an increase of 519 kg/ha dry seed over a trial mean of 1,070 kg/ha. With tung cake and ammonium sulphate, air-dry tung seed yields of 12-17 year old trees was 2,013-2,367 kg/ha, of 6-9 year olds 766-1,546 kg/ha (Spurling and Spurling, 1974).

ANACARDIUM OCCIDENTALE L. ANACARDIACEAE CASHEW

Yields are said to range from 0-48 kg/tree/year, with an average yield of 800-1,000 kg/ha. Heavy bearing trees often produce nuts considered too small for the trade. Indian field trials showed that fertilizers could increase yields of 15-year-old trees from less than 1 kg/tree to >4 and enabled 6-year olds to average 5.7. Regular applications of 250 g N, 150 g P_2O_5, and 150 g K_2O/tree resulted in average yield increases of 700-1600 kg/ha (Nambiar and Haridasan, 1979). In Pernambuco, trees produced 1.5-24.0 kg each/year, averaging 10.3 kg per tree (Johnson, 1973). At Pacajus (Ceara, Brazil) trees average 17.4 kg/yr with one tree bearing 48 kg/yr. A perennial species, the cashew has already, in the past, yielded alcohol from the "apple", oil from the nut, and charcoal from the wood.

ARACHIS HYPOGAEA L. FABACEAE PEANUT

The world low production yield was 400 kg/ha in Mozambique, an international production yield was 1,016 kg/ha, and the world high production yield was 3,783 in Malaysia. (FAO, 1980) Yields reported for India vary considerably from 580 kg/ha in Mysore to 714 kg/ha in the Central Provinces, to 833 in Hyderabad, 879 in Bombay to 1,089 in Madras. These are production figures, experimental figures as high as 1,918 kg/ha for rainfed and 3,660 kg/ha for irrigated are reported. Then the Wealth of India (C.S.I.R., 1948-1976) reports yields of 1,793 kg/ha in China, and 2,453 in Mauritius. Experimental yields of 6,000 kg/ha are reported, the 35-55% oil percentage suggesting potential oil yields of more than 3 MT ha. Peanut oil is the cheapest and most extensively used vegetable oil in India (C.S.I.R., 1948), and has also been used for fuel and illumination. However, if energy is the target, it seems highly improbable that peanut oil yields will ever compare with oil palm yields.

BRASSICA JUNCEA (L.) Czern. BRASSICACEAE MUSTARD GREENS

As much as 12 MT leaf is reported per hectare. In one Indian study, the highest seed yield was 1.13 MT/ha from crops sown October 15, with 200,000 plants/hectare. Seeds sown on November 4 yielded only 0.49 MT/ha (Patel et al, 1980). In another study (Maity et al, 1980), comparing irrigating and nitrogen levels, seed yields were 0.82-2.37 MT/ha, the maximum being irrigated and fertilized with 150 kg N/ha. Other Indian studies suggest 1,000-2,000 kg seed per hectare, with an oil content of 30-38%. In headon comparative trials in Davis, California, Knowles, Kearney, and Cohen (1981) reported seed yields of 1,737-2,549 kg/ha for B. juncea, 980-2,003 for B. campestris, 2,063-2,852 for B carinata, and 1,202-2,083 for B. napus. Pryde and Doty (1981) suggest average oil yields of 409 kg/ha from a seed yield of 1,166 as an overall Canadian rapeseed yield.

BRASSICA NAPUS L. BRASSICACEAE RAPE, COLZA

Seed yields vary in different regions from 900-1,500 kg/ha (rarely to 3,000 kg/ha); in North Africa it may be only 300-350 kg/ha. The oil content runs 35-45%, and oil yields of more than 1 MT/ha are reported. In the Palouse region of northern Idaho and eastern Washington, Peterson et al (VODF = (Vegetable Oils as Diesel Fuel) Seminar II, 1981), report winter rape yields of 2,200 to 2,750 kg/ha with 45% oil. In Canada, Finlayson, Krzymanski, and Downey (1973), report yields of only 718 kg/ha in the low glucosinolate cv 'Bronowski' compared to 1,304 for 'Target'. Yield data for their 1972 trials at Saskatoon were ca 2,960 kg/ha (42.7% oil) for 'Target', ca 2,560 (39.6%) for 'Zephyr', ca 3,010 (44.2%) for 'Midas', ca 2,630 (42.4%) for 'SZN71-1788', ca 2,500 (41.7%) for 'SZN71-1787', ca 2,720 (42.3%) for 'SZN71-1785', ca 2,550 kg/ha (41.1% oil) for 'SZN71-1784', nearly all yielding more than a metric ton oil per hectare (Finlayson, Krzymanski, and Downey, 1974). In three experiments comparing autumn- and spring-sown rape in Britain, seed yields for the spring sown ranged from 963-2,284 kg/ha, for autumn-sown, from 1,787-2,783 kg/ha. Not only did the autumn-sown crop have higher yields, it had a higher oil content (42.0-44.5%) than the spring sown (35.8-38.5%) (Scott et al, 1973). Scott et al, 1973, indicate aerial DM (= dry matter) yields of 1-2 MT with seed yields of about the same magnitude suggesting a straw factor of 1.

BRASSICA NIGRA (L.) Koch BRASSICACEAE BLACK or BROWN MUSTARD

Pure crop of mustard under good cultural practice and mechanized harvesting in the United States yields up to 1,100 kg seed/ha, compared to 400 kg/ha reported in Sri Lanka.

BRASSICA RAPA L. BRASSICACEAE TURNIP

Seed yields in Minnesota and Canada run over 1,000 kg/ha/yr, and the oil from such seeds is being considered for energy purposes.

CANNABIS SATIVA L. MORACEAE HEMP

In some areas of India, fiber yields of 850-1,700 kg/ha compared to 1,300-1,700 kg/ha seed. Plants remaining in the field after harvesting for fiber are allowed to set seed. They are cut after the fruits are ripened, then dried and threshed for seed collection. Grown solely for seeds, an average crop yields 1.3 to 1.6 MT/ha seed, with 25-30% oil content suggesting oil yields less than 500 kg/ha.

CARTHAMUS TINCTORIUS L. ASTERACEAE SAFFLOWER

Pryde and Doty (1981) suggest average US safflower oil yields of 762 kg/ha out of an average seed yield of 2,119 kg/ha. Telek and Martin (1981) suggest oil yields of 600 kg/ha. In California average yields are 1,900 kg/ha, but yields above 4,500 kg/ha are not uncommon; in the Great Plains yields run about 850 kg/ha. In Arizona 'Dart' yields 4,004 kg/ha or 1,622 kg oil/ha, 25% more than the open-headed Gila which is susceptible to birds. In India, when seed oil is the object, yields are about 90-130 kg florets and 440 to 660 kg seed/ha. Intercropped yields are only 110-275 kg/ha seed (WOI = Wealth of India). Under rainfed conditions, hybrid sorghum yielded 4.21 MT/ha in the kharif monsoon and the following safflower yielded 470 kg seed. Grown after fallow however, the safflower yielded 1,220 kg/ha (Veeranna, Jagannath, and Gidnavar, 1980). Raghu and Sharma (1978) report India yields of 1,350 kg rainfed, 1,940 kg irrigated. According to Khoshoo (1981), the Btu value per gallon of safflower oil is 130,730, cf. 130,450 for rapeseed, 130,900 for sunflower, and 138,252 for diesel. Viscosity is a main drawback, i.e. 32.7 for safflower, 35.3 for sunflower, 51.3 for rapeseed, and only 2.9 for diesel. Tractors were run on 100% safflower oil for over 90 hours to cut hay and cultivate in Australia, and diesel engines fueled with safflower oil were run more than 700 hours in Idaho with no obvious difference ascribed. Indians have "successfully cultivated safflower on sodic soil of Banthra with pH 8.5-9.0, EC value below one, and ESP between 25-40, and a yield of 0.36" tons/ha oil was obtained. In northern Idaho it requires about 25 gallons of liquid fuel to harvest 212-262 gallons of extractable oil per hectare from safflower and sunflower (Khoshoo, 1982). According to the Wealth of India, the oil content ranges from 20-30%, 92 kg yielding 16 kg oil, 28 kg oil cake, and 38 kg husk.

CITRULLUS COLOCYNTHIS (L.) Schrad. CUCURBITACEAE COLOCYNTH

Duke (1978) reported a seed yield of 6,700 kg/ha. If yields of 6,700 kg/ha are attainable with low energy inputs, and if oil yields are 47.2% as the Food Composition Tables suggest, oil yields might exceed 3,000 kg/ha placing this among the serious oilseed energy candidates, with medicinal byproducts. (cf. 3,000 kg seed/ha with 24-34% oil for the buffalo gourd, Cucurbita foetidissima.)

COCOS NUCIFERA L. ARECACEAE COCONUT

Pryde and Doty (1981) put the average oil yield at 1,050 kg/ha, Telek and Martin (1981), at 600 kg/ha. A well-managed plantation should yield 6,000 nuts per hectare, enough for one ton of copra. According to the Wealth of India, efficient pressing of 1,000 kg copra will yield 625 kg coconut oil and 350 kg coconut cake, the latter containing 7-10% oil. The factor 63% is generally used for converting copra to oil equivalent. Yields of copra as high as 5 MT/ha have been reported, but oil yields of 900-1,350 kg/ha. Australians respect the Iodine number as one chemical means to rate the propensity of a bio-oil to react. Quick (1981) tested linseed oil (Iodine number 180) which cokes up fuel injectors in less than 20 hours and rapeseed oil (Iodine number ca 100) which logs into the hundreds of hours before the onset of severe injector coking. Coconut oil (Iodine number 10) should be a very good candidate from this viewpoint. This could be very important in developing tropical countries where diesel fuel is scarce and often <u>more</u> expensive than coconut oil. One Australian patent suggests that distillation of coconuts at 550° gave 11.5% charcoal, 11% fuel gas, 37.5% copra spirit, 12.5% olein distillate, 12.5% black oil, 1% crude acetic acid, 0.15% glycerol, and 0.85% acetone + methanol. Coconut trees can be tapped to yield 250 liters of toddy (12-17.5% sucrose) which natural fermentation takes to 2.7-5.8% ethanol. Of course, you can't have your coconut toddy and eat or drink or burn it too. (Duke, 1977).

CRAMBE ABYSSINICA Hochst. ex R.E.Fries BRASSICACEAE CRAMBE

Yields vary widely from 1,125-1,624 kg/ha in Russia and 450-2,522 kg/ha in the United States, with yields highest in weed-free fields. In irrigated fields with additional nitrogen yields up to 5 T/ha have been attained. Buchanan and Duke (1981) take a conservative 1,120 kg/ha for their yield calculations. That is slightly less than is reported by a farmer in Kentucky who followed his crambe with soybean. Yields of more than 1,400 kg/ha are reported over a four-year period in Sweden (FCA 04534(035)). Test plantings in Russia, under a wide variety of ecological conditions, gave oil contents of 25-33% for seed with hulls (dehulled seeds reached 54%) (Cornelius and Simmons, 1969). Lessman and Anderson (1981) report yields of 169-840 kg/ha in Indiana, 2,000 in Missouri, suggesting oil yields of about 60-600 kg/ha.

ELAEIS GUINEENSIS Jacq. ARECACEAE AFRICAN OIL PALM

According to the Wealth of India, the oil yield of oil palm is "higher than that of any other oilseed crop", producing 2.5 MT oil per ha per year, with 5 MT recorded. Pryde and Doty (1981) put the average palm oil yield (for Malaysia) at 3,475 kg/ha, Telek and Martin at 2,200. Yields of semi-wild palms vary widely, usually ranging from 1.2 to 5 MT of bunches per hectare per year. One MT of bunches yields about 80 kg oil by local soft oil extraction, or 180 kg by hydraulic hand-press. Estate yields in Africa vary from 7.5-15 MT bunches per hectare per year; in Sumatra and Malaya, 15-25 MT, with some fields producing 30-38 MT. Bunch yields may attain 22,000 kg/ha, of which only about 10% is oil, indicating oil yields of only 2,200 kg/ha. Higher yields are attainable. Corley (1981) suggests plantation yields of 2-6 MT/ha mesocarp oil, experimentally up to 8.5 MT/ha. An energy evaluation of all the wastes from the palm oil fruit was made and it revealed that this can satisfy ca 17% of Malaysia's energy requirements. Palm oil could satisfy 20% more (Keong, 1981). An alcoholic wine can be made from the sap of the male spikes, 150 trees yielding about 4,000 liters of palm wine per hectare, per year.

GLYCINE MAX (L.) Merr. FABACEAE SOYBEAN

According to corrected Agricultural Statistics 1981, soybean yields are 2,510 kg/ha for Canada, 1,800 for the US, 1,870 for Mexico, 2,100 for Argentina, 1,720 for Brazil, 1,960 for Colombia, 1,350 for Paraguay, 917 for Romania, 1,570 for Yugoslavia, and only 616 kg/ha for USSR. Whigham (1981) suggests 5,000-6,000 kg/ha as yield with top management and optimum environmental conditions. With only 13 to 25% oil, this suggests a top oil yield of 1,500 kg/ha. Pryde and Doty (1981) suggest an average oil yield of only 319 kg/ha from 1,788 kg/ha seed. Telek and Martin (1981) suggest average oil yields of 300 kg/ha.

GOSSYPIUM HIRSUTUM L. MALVACEAE COTTON

Pryde and Doty (1981) put the average oil yield at only 140 kg/ha. According to Parnell (1981), annual US cottonseed yields are 800-950 kg/ha. About 6-8% is planted, with small quantities used for feed; the remainder is crushed for cottonseed oil (ca $0.45/kg in April 1982), with the byproduct being cottonseed meal (ca $145 ton). A metric ton of cottonseed yields ca 160 kg oil, 450 kg meal, 68 kg linters, 250 kg hulls and 72 kg trash, waste and invisible losses (Parnell, 1981). The world low production yield figure was 73 kg/ha in Grenada, the international production yield was 1,251 kg/ha, and the world high production yield was 3,306 in Guatemala (FAO, 1980).

GUIZOTIA ABYSSINICA (L.f.) Cass. ASTERACEAE NIGER

Average seed yields in India range from 100-200 kg/ha when grown with ragi, and 300-400 kg/ha when grown in pure stands. Similar yields obtained in Ethiopia when grown alone. In Kenya, monocultural yields average 600 kg/ha. Bhardway and Gupta (1977) report seed yields of 1,000 to 1,200 kg/ha on fertile Himalayan soils, 200-400 kg/ha in degraded habitats. Oil yields range about 235 kg/ha, especially in fields where KNP combination fertilizers have been used. Of three row spacings, 30 cm was reported to give the highest yields (552 kg/ha) at Jabalpur, India. Plants whose growing tips were nipped yielded 462 kg/ha compared with 489 for the controls (Singh et al, 1973). Since it is difficult to keep up with the birdseed market for this potential weed, there have been few suggestions of Guizotia as a diesel substitute. Yields are generally so low, that without marked improvement, oil yields of higher than 500 kg/ha are difficult to visualize. For example, Rao and Rao (1980), testing 8 cvs, reported the highest yield as only 205 kg/ha seed with the highest oil content of 41.9%. Oil recovery uses ranges from 25-35% in India.

HELIANTHUS ANNUUS L. ASTERACEAE SUNFLOWER

The world low production yield was 308 kg/ha in Algeria, the international production yield was 1,266 kg/ha, and the world high production yield was

2,420 kg/ha in Austria (FAO, 1980). With DM yields ranging from 4 to 9 MT/ha (in three months) and seed yields ranging from 300 to more than 3,000 kg/ha, a straw factor of 3 seems appropriate. According to USDA Secretary's Annual Report (1980), with an average yield of ca 1,500 kg/ha (North Dakota), a hectare would yield nearly 225 gallons of oil, 75% of which could be extracted on the farm. Twelve to 15 gallons are required to raise a hectare; hence the fuel from one hectare could produce 8-11 hectares of crop. In the US, the highest average commercial yields occurred in North Dakota and Minnesota, which averaged 1,170 and 1,267 kg/ha respectively, compared with 1,019 kg/ha for Texas. Pryde and Doty (1981) suggest average oil yields of 589 kg/ha from 1,469 kg/ha seed. Telek and Martin (1981) suggest oil yields of 450 kg/ha. Experimentally, at Davis, California, April plantings yielded 2,592-3,181 kg/ha (45.5-48.5% oil), May, 2,676-3,161 kg/ha (45.5-48.4% oil), June plantings 956-2,643 kg/ha (40.8-43.7% oil), and July plantings 702-2,447 kg/ha (40.2-42.6% oil). The lowest oil yield was 282 kg/ha, the highest, 1,543 kg/ha (Beard & Ingebretsen, 1980). In India, rainfed sunflower gave seed yields of 1,120 kg/ha in pure stands, 1,050-1,070 intercropped with cowpea, and 1,010-1,070 kg/ha intercropped with peanuts (Chandrasekar and Morachan, 1979). North Dakota researchers are testing a small auger press, operated on the farm, that can extract ca 75-80% of the oil in sunflower seeds, or ca 55 gallons (barely more than one 42-gallon barrel) from an average yield of 1,400 lbs/acre. According to S & E Newsmakers #4 (September 1981), to farm and produce 8 to 11 acres of crop, our usual 10:1 ratio. Pratt et al (VODF Seminar II, 1981) report an endurance test involving engines fueled with various mixtures of sunflower oil (25-50%) with diesel oil (75-50%). Two motors needed repair, ten were operating with no apparent difficulties, of which two were said to be doing even better. Ohio yields on poor soils (Wood County) were only 260 lb/acre (yielding 9.3 gallons of screw press oil); and on good soils (Champaign County), 1,680 lb/acre (yielding 69.1 gallons oil) cropped after wheat in a double cropping system (Ohio Report July-August 1981, p. 63). Sunflower oil should be dewaxed before being used as a diesel substitute. In Australia, sunflower first commercially planted in 1967, has great potential for expansion as a rainfed energy crop. Little water is required for processing oilseeds (unlike ethanol), and the seed coat can provide sufficient energy for heat and steam for oil extraction. Australians figure a net energy gain of 2 liters for every 3 liters produced (Quick, 1981). A hundred kg of dry seed will yield about 40 kg oil, 15-25 kg hulls, and 40 kg proteinaceous meal. Hulls have been pressed into fuel "logs". Threshed heads are ground and fed to cattle elsewhere. The heads are rich in pectin (Robinson, Ag. Ext. Service, Univ. of Minn.) Sheaffer et al (1976, Univ. Md. Ag. Exp. Station) report studies showing that sunflower yields 33.1 MT silage/ha compared to corn at 19.26 MT/ha. According to the phytomass files (Duke, 1981), annual DM productivity ranges from 3 to 15 MT/ha. DM yields averaged closer to 5 MT spaced at 43,000 plants/ha, 8 MT spaced at 172,000 plants/ha near Clarksville, Maryland. In these experiments, the sunflower followed barley. Jake Page's discussion in Science 81 (July-August 1981, 92-93) is picturesque, so we quote: "But I happen to like sunflowers... They can be grown almost anywhere in the country and you can grow between 500 and 3,000 pounds of sunflower seeds on an American acre in three months if you're clever. The soil can be lousy, the rainfall terrible...if the average American corn farmer put 10 percent of his land into sunflowers, he could become self-sufficient in fuel. It seems that using vegetable oil may be more efficient, in a net energy sense, than growing plants for conversion into alcohol (another nice alternative fuel) because the processing for alcohol is more elaborate, expensive, and energy intensive."

JUGLANS NIGRA L. JUGLANDACEAE EASTERN BLACK WALNUT

Although Duke (1978) reported yields of 7.5 MT seeds, this is probably optimistic. Elsewhere it is said that 95% of the wild black walnut seeds are empty or aborted. Perhaps yields could be as high as 2.5 MT/ha under intensive management, which is attainable in the commercial walnut, Juglans regia. 'Thomas', 'Ohio' and 'Myers' begin bearing nuts in second or third year after planting, while native trees usually do not begin to bear until about 10 years after planting. In 5-6 years, these three varieties bear about one-forth bushel of nuts; at 15-20 years of age, the first two bear 2 bu of nuts, 'Myers' about 1 bu, and native trees about one-fourth bu. For lumber, trees yield about 1,150 board feet at 76 years old. Trees may bear at rates of 7,500 seed/ha. Oil contents of the seeds run about 60%, suggesting that if the walnut yields of 7.5 MT/ha were attained, there might be as much as 4.5 MT/oil there. Hulls and exocarp might be used to fuel the processing, as the value of the timber improves with age (one tree commanded $35,000 at an Ohio auction). Prunings and culls, as well as fallen and dead limbs, might amount to 5 MT/ha/year.

JUGLANS REGIA L. JUGLANDACEAE ENGLISH WALNUT

Newer varieties begin producing nuts in 5-6 years; by 7-8 years they produce about 2.5 tons of nuts per hectare. Orchards on relatively poor, unirrigated mountain soil report 1.5-2.25 tons/ha; orchards in well cultivated valleys, 6.5-7.5 tons/ha. According to the Wealth of India, a grown individual can yield about 185 kg, but 37 kg is more likely. If the walnuts yields of 7,500 kg/ha (Duke, 1978) yielded all their 65% (63-67% oil) oil, there is a potential oil yield of nearly 5 MT per year, a very worthwhile target, if attainable. The green hulls have recoverable ascorbic acid (content 2.5-5% of dry weight). Hulls contain 12.2%, bark 7.5%, leaf blades 9-11% tannin. After extraction of the vitamin C and tannin, the residues might be used for fuel or ethanol. Prunings from the trees might contribute another 5 MT biomass per year.

LESQUERELLA spp. BRASSICACEAE

Buchanan and Duke (1981), anticipating a need for 90,700,000 kg lesquerella seed for hydroxy fatty acids, accept 1,121 kg/ha as a yield estimate for lesquerella. Under dry conditions in small experimental plots, Lesquerella fendleri has averaged 1,600 kg/ha. With yields ranging from 1,000-1,600 kg/ha, and oil percentage running about 30% (11-39%), oil yields as high as 1 MT/ha seem unlikely without more germplasm and selection.

LIMNANTHES DOUGLASII R.Br. LIMNANTHACEAE MEADOW FOAM

Yields of 400 kg/ha are reported for Limnanthes bakeri, 900 for L. alba, and 1,900 for L. douglasii, the most promising oilseed species in the genus. Researchers hope to double yields, but for now, with 24-30% oil content, we visualize oil yields of 500 kg/ha difficult to attain (cf Simmondsia, also considered a potential substitute for sperm whale oil). Oregon could already produce 1,100 kg seed/ha yielding 275 kg oil (Oregon's Agricultural Progress, Spring/Summer 1975).

LINUM USITATISSIMUM L. LINACEAE FLAX

Flax straw yields average 5-7 MT/ha. Fiber yields run 200-1,200 kg/ha. Seed production figures indicate seed yields of 220 to 2,820 kg/ha but, locally, yields can be much higher. Pryde and Doty (1981) cite average oil yields of only 230 kg/ha from a seed yield of 650 kg/ha. Telek and Martin (1981) put the oil yield at 225 kg/ha. Record seed yields in the US range from 2,460 kg/ha in the northern states to 4,390 kg/ha for winter sowings in Arizona/ California (Dybing and Lay, 1981). According to Agricultural Statistics 1981, yields of flaxseed per harvested acre ranged between 7.9-13.7 bushels per acre (roughly 25.4 kg/bushel) with the average farmer's price of $2.38 to $9.67 per bushel. Production figures for most countries fall under 1 MT/acre=2.5 MT/ha, but in 1980, Mexico, Uruguay, and Egypt exceeded this. With seed yields of 1,000-4,000 kg/ha quite feasible, and reported oil contents of 34.37%, we can see oil yields of 1,500 kg/ha with slight agronomic improvement. At VODF Seminar II, it was concluded that "linseed oil has caused so many problems in engine tests as to be totally rejected by several investigators as a candidate fuel oil."

MACADAMIA INTEGRIFOLIA Maiden & Betche PROTEACEAE MACADAMIA NUTS

In Australia, estimated yields are about 45 kg/tree annually; in Hawaii, average yields are 135 kg/tree. New cultivars are known to yield as much as 3.75 tons/ha, averaging 1 ton of kernels, which should contain more than 700 kg oil/ha renewably (oil makes up 65-75% of the kernel). According to Saleeb et al (1973) nuts of M. integrifolia and M. tetraphylla are equal in oil content with an iodine value of 75.4 and 71.8 respectively. They describe a method for partially extracting the oil (6-14% of the weight of intact oven dry kernels), rendering them more attractive, digestible, and less fattening, while diverting 14% of the weight to oil production.

PAPAVER SOMNIFERUM L. PAPAVERACEAE OPIUM POPPY

Yields of opium are about 0.4 to 0.9% with a dry morphine output equaling about 2 kg/ha. From field harvested opium this would yield the standing biomass as residue. In Tasmania, with 0 kg N and 0 kg P, Laughlin (1978) reported capsule yields of 1,197 kg/ha, seed yields of 1,712 and total plant yields of 7,948, with 0 P and 100 N, there were 1,220 kg capsule, 1,645 kg seed, and 8,273 kg total plant, with 0 N and 100 P there were 1,286 kg capsule, 1,829 kg seed, and 8,785 kg total plant, while with 100 kg N and 100 kg P/ha, there were 1,452 kg capsules, 1,760 kg seed, and 9,910 kg/ha total plant. German studies reported in Hortic. Abs. 01489(049) reported seed yields of 750-1,200 kg/ha and empty pod yields of 580-1,000 kg/ha. If his review, Loof (1966) suggests that average seed yields of 1,200-1,800 kg/ha would be realistic under European conditions with non-lodging cvs and adequate moisture. Experimental yields have attained 3,000 kg/ha (Loof, 1966). With the seeds running 45-50% oil, we can visualize oil yields of perhaps 1,500 kg/ha, with an edible 1,500 kg oilcake, and perhaps another 2-8,000 kg straw to fuel processing.

PERILLA FRUTESCENS (L.) Britt. LAMIACEAE PERILLA

Seed yields run from 500 to 1,700 kg/ha, with 35-45% oil, indicating oil potential of 500 kg/ha or more. Nagzo et al (1975) estimated dry leaf yield at only 1,000-1,300 kg/ha, which contains 0.41% perillaldehyde and 0.36% essential oil.

PRUNUS DULCIS (Mill.) D.A.Webb. ROSACEAE ALMOND

According to the Wealth of India, average California yields are ca 400 kg/ha, but they attain over 1,200 kg/ha. However, for Baluchistan they report 2,375 kg/ha, basing this on an optimistic yield of 7.3 kg for each of 325 trees/ha. Yields of 2-3 kg per tree seem more realistic. Duke (1978), however, reports seed yields of 3,000 kg/ha. With an oil yield of 50-55%, it is easy to project oil yields of 1,500 kg/ha.

RICINUS COMMUNIS L. EUPHORBIACEAE CASTORBEAN

Average seed yields range from 900-1,000 kg/ha under irrigation, and 300-400 kg/ha without adequate moisture. Some improved open-pollinated varieties in Brazil and the United States yield 1,300-1,500 kg/ha. In the United States, hybrid varieties produce up to 3,000 kg/ha, with exceptional yields up to 5,000 kg/ha. Compare this with the average world yield of 514 kg/ha (Core, 1981). It is not uncommon for the exceptional yields to be ten times the world average. And too often, papers such as this one draw on the exceptional rather than the average yield to make energy projections. Oil content of seeds varies from 35-55%, suggesting potential oil yields of 200-2,750 kg/ha. Telek and Martin (1981) put average oil yields at 485 kg/ha.

SAPIUM SEBIFERUM (L.) Roxb. EUPHORBIACEAE CHINESE TALLOW TREE

In plantations, trees should be planted one rod apart each way, giving 400 trees per hectare, and if trimmed to a convenient size for hand-harvesting, would yield 14 tons seed/ha, containing 2.6 tons of oil, 2.8 tons of tallow, 1.5 tons protein concentrate, 1.1 tons fibrous coat and 4.5 tons shell. Oil, tallow and protein meal would bring about $750 per hectare. This yield could increase with age. Scheld et al (1980) report yields of 4,000 to 10,000 kg/ha (in excess of 11,000 kg/ha in VODF Seminar II, 1981), and cite estimates of 25 barrels of oil per year as a sustained energy yield. Assuming an annual oil yield of 25 barrels per hectare, it would take only 24 million hectares of Sapium to replace the ca 8% of our petroleum which goes into chemical production.

SESAMUM INDICUM L. PEDALIACEAE SESAME

The world low production was 40 kg/ha in Bulgaria, the international production yield was 300 kg/ha, and the world high production yield was 1,000 kg/ha in Iran. Oil content runs about 50%, leaving an oilpress cake residue of ca 60% (Devendra & Raghavan, 1978). The dried stalks and leaves are said to be unfit for livestock feed and inferior for fiber, hence a logical energy candidate. Yields range up to 2,250 kg/ha of seed, but the world average was about 330 kg/ha. Indian production figures were as low as 157 kg/ha in 1967. Oil content of seed varies from 50-57%, depending on the variety. In United States, yields are up to 1,200 kg/ha with fertility and irrigation. Pryde and Doty (1981) suggest average oil yields at 220 kg/ha, Telek and Martin (1981) at 360. Intercropping with sesame has increased eggplant yields to 150-175% in nematode infested soil.

SIMMONDSIA CHINENSIS (Link) C.Schneid. BUXACEAE JOJOBA

Buchanan and Duke (1981) accept a figure near 2,250 kg/ha for yields of jojoba. Individual plants may yield 5 kg (dry weight) seeds and more, of which 50% (43-56%) by weight is a colorless, odorless liquid wax commonly called "jojoba oil" (N.A.S. 1975). Yermanos (1973) suggested that a 5 year old orchard should yield about 825 kg of nuts per hectare, increasing to 4,125 kg/ha in the 12th year, suggesting a renewable "oil" yield of ca 2 MT/year. Such yields may be optimistic, even for well managed plantations. However, Yermanos still says, "Extrapolating from data obtained from the harvest of wild plants and from limited yields of 8 years or older plantations might range from 1-2 tons of dry clean seed per acre." (2.5-5 MT/ha) possibly yielding 1-2.5 MT oil (from an undated article entitled Jojoba...).

SINAPIS ALBA L. BRASSICACEAE WHITE OR YELLOW MUSTARD

Under mechanization in the US, farmer yields of nearly 900 kg/ha have been obtained. If the experimental seed yields of 8,000 kg/ha (Duke, 1978) are correct, the 25-30% oil content could add up to nearly 2.5 MT oil per hectare.

STOKESIA LAEVIS (Hill) Green ASTERACEAE STOKES' ASTER

Estimating a current need for 136,100,000 kg Stokes' aster seed for epoxy fatty acids, Buchanan and Duke (1981) accept 1,121 as a rather conservative estimate of seed yields for Stokesia laevis. Gunn and White (1974) put preliminary seed yield estimates at 350-1,160 kg/ha. Oil yields for the achenes would be about 40% (27-44%) of the achene yields, assuming total extraction, suggesting oil yields of less than 500 kg/ha.

TELFAIRIA PEDATA (Sm. ex Sims) Hook. CUCURBITACEAE ZANZIBAR OILVINE

Average yields of the nuts are 1,000-2,000 kg/ha. The oil content of the seed is about 62% of its weight, or approximately 35% of the entire weight of the whole nut. This would suggest oil yields up to 700 kg/ha. Dr. T. W. Whitaker (personal communication, June 1982), suggests that this should be a promising species, but not so exciting as the Asian Hodgsonia of the Cucurbitaceae family. USDA germplasm teams to China should negotiate for some of this subtropical germplasm.

THEOBROMA CACAO L. STERCULIACEAE COCOA

The world low production yield is 29 kg/ha in American Samoa, an international production yield of 346 kg/ha, and a world high production yield of 2,000 kg/ha in Haiti. Yields of 3,375 kg/ha of dry beans are possible on good plantations. The oil content (35-50%) suggests potential oil yields of more than 1,750 kg/ha.

REFERENCES

Beard, B.H. and Ingebretsen, K.H. 1980. Spring planting is best for oilseed sunflower. Cal. agric. June(1980): 5-6.

Bhardwaj, S.P. and Gupta, R.K. 1977. Tilangi, a potential high yielding oil seed crop. Indian farming 27(6): 18-19.

Buchanan, R.A. and Duke, J.A. 1981. Botanochemical crops. pp. 157-179 in: McClure, T.A. and Lipinsky, E.S., eds. CRC handbook of biosolar resources. Vol. II. Resource materials. CRC Press, Inc. Boca Raton, FL. 575 pp.

Chandrasekar, V.P. and Morachan, Y.B. 1979. Effect of advanced sowing of intercrops and nitrogen levels on yield components of rainfed sunflower. Madras agric. j. 66(9): 578-581.

Core, E.L. 1981. Castor bean Ricinus communis. pp. 193-197 in: McClure, T.A. and Lipinsky, E.S., eds. 1981. CRC handbook of biosolar resources. Vol. II. Resource materials. CRC Press, Inc. Boca Raton, FL. 575 pp.

Corley, R.H.V. 1981. Oil palm. pp. 397-403 in McClure, T.A. and Lipinsky, ES., eds. CRC handbook of biosolar resources. Vol. II. Resource materials. CRC Press. Boca Raton, FL. 575 pp.

Cornelius, J.A. and Simmons, E.A. 1969. Crambe abyssinica - a new commercial oilseed. Trop. sci. 11(1): 17-23.

C.S.I.R. (Council of scientific and industrial research). 1948-1976. The Wealth of India. 11 volumes. New Delhi.

Dorrell, D.G. 1981. Sunflower Helianthus annuus. pp. 105-114 in: McClure, T.A. and Lipinsky, E.S., eds. CRC handbook of biosolar resources. Vol. II. Resource materials. CRC Press, Inc. Boca Raton, FL. 575 pp.

Duke, J.A. 1977. Palms as energy sources: a solicitation. Principes 21(2): 60-62.

Duke, J.A. 1978. The quest for tolerant germplasm, pp. 1-61 in: Crop tolerance to suboptimal land conditions. ASA special publ. No. 32. American Society for Agronomy. Madison. 343 pp.

Duke, J.A. 1981. Handbook of legumes of world economic importance. Plenum Press, New York. 345 pp.

Dybing, C.D. and Lay, C. 1981. Flax Linum usitatissimum. pp. 71-85 in: McClure, T.A. and Lipinsky, E.S., eds. CRC handbook of biosolar resources Vol. II. Resource materials. CRC Press, Inc. Boca Raton, FL. 575 pp.

Gunn, C.R. and White, G.M. 1974. Stokesia laevis, taxonomy and economic value. Econ. Bot. 28(2): 130-135.

Johnson, D. 1973. Cashew cultivation in Brazil. Agron. mocamb. lourenco marques 7(3): 119-129.

Keong, W.K. 1981. Safe energy from palm oil and its wastes. Agric. wastes 3(3): 191-200. Dept. Envir. Sci., Universiti Pertanian Malaysia, Serdang, Malaysia.

Knowles, P.F., Kearney, T.E., and Cohen, D.B. 1981. Species of rapeseed and mustard as oil crops in California. pp. 255-268 in: Pryde, E.H., Princen, L.H., and Mukherjee, K.D. 1981. New sources of fats and oils. AOCS monograph No. 9. American Oil Chemists' Society. Champaign, IL. 340 pp.

Laughlin, J.C. 1978. The effect of band placed nitrogen and phosphorus fertilizer on the yield of poppies (Papaver somniferum L.) grown on krasnozem soil. Acta horticulturae 73: 165-169.

Lessman, K.J. and Anderson, W.P. 1981. Crambe. pp. 223-246 in: Pryde, E.H., Princen, L.H., and Mukherjee, K.D. 1981. New sources of fats and oils. AOCS monograph No. 9. American Oil Chemists' Society. Champaign, IL. 340 pp.

Loof, B. 1966. Poppy cultivation (review article). Field crop abstracts 19(1): 1-5.

Maity, P.K., Sengupta, A.K., and Jana, P.K. 1980. Response of mustard variety varuna (Brassica juncea) to levels of irrigation and nitrogen. Indian agriculturist 24(1): 43-47. (FCAb. 00658 (035))

Nagao, Y., Fujioka, S., Takahashi, T., and Matsuoka, T. 1975. Studies on the quality of the Chinese drug "Soyo" and the cultivation of the original plant 2. Perilla frutescens. J. Takeda Res. Lab. 34(1): 33-42.

Nambiar, M.C. and Haridasan, M. 1979. Fertilizing cashew for higher yields. Indian farming 28(12): 16-17.

N.A.S. 1975. Products from jojoba: a promising new crop for arid lands. National Academy of Sciences, Washington. 30 pp.

Parnell, Jr., C.B. 1981. Cotton Gossypium hirsutum. pp. 115-122 in: McClure, T.A. and Lipinsky, E.S., eds. CRC handbook of biosolar resources. Vol. II. Resource materials. CRC Press, Inc. Boca Raton, FL. 575 pp.

Patel, J.R., Parmar, M.T. and Patel, J.C. 1980. Effect of different sowing dates, spacings and plant populations on yield of mustard. Indian j. agronomy 25(3): 526-7.

Pryde, E.H. and Doty, H.O. Jr. 1981. World fats and oils situation. pp. 3-14 in: Pryde, E.H., Princen, L.H., and Mukherjee, K.D. 1981. New sources of fats and oils. AOCS monograph No. 9. American Oil Chemists' Society. Champaign, IL. 340 pp.

Quick, G.R. 1981. Abstract. A summary of some current research in Australia on vegetable oils as candidate fuels for diesel engines. Seminar II, Peoria, USDA. 1981.

Raghu, J.S. and Sharma, S.R. 1978. Response to irrigation and fertility levels of safflower. Indian j. agronomy 23(2): 93-97.

Rao, T.S. and Rao, M.S.K. 1980. Performance of niger genotypes at Raichur. Current research 9(5): 78. 1980.

Saleeb, W.F., Yermanos, D.M., Huszar, C.K., Storey, W.B., and Labanauskas, C.K. 1973. The oil and protein in nuts of Macadamia tetraphylla L. Johnson, Macadamia integrifolia Maiden and Betche, and their F_1 hybrid. J. Amer. Soc. Hort. Sci. 98(5): 453-456.

Singh, P.P., Shrivastava, P.S. and Beohar, A.B.L. 1973. Effect of row and plant spacings and nipping of tops on the seed yield of niger (Guizotia abyssinica Cass). JNKVV 7(4): 258-260.

Spurling, A.T. and Spurling, D. 1974. Effect of various organic and inorganic fertilizers on the yield of Montana tung (Aleurites montana) in Malawi. Trop. agr. 51(1): 1-12.

Telek, L. and Martin, F.W. 1981. Okra seed: a potential source for oil and protein in the humid lowland tropics. pp. 37-53 in: Pryde, E.H., Princen, L.H., and Mukherjee, K.D. 1981. New sources of fats and oils. AOCS monograph No. 9. American Oil Chemists' Society. Champaign, IL. 340 pp.

Whigham, D.K. 1981. Soybeans. Glycine max. pp. 95-104 in: McClure, T.A. and Lipinsky, E.S., eds. CRC handbook of biosolar resources. Vol. II. Resource materials. CRC Press, Inc. Boca Raton., FL. 575 pp.

Yermanos, D.M. 1973. Jojoba ... a brief survey of the agronomic potential. California agriculture (Sept.).

ENERGY AND ECONOMIC EFFICIENCY FOR OFF-SITE PROCESSING OF SELECTED OILSEEDS

Josef M. Broder, Asst. Professor, Agricultural Economics
Univ. of Georgia, Athens, GA.
Kenneth C. Schneeberger, Professor, Agricultural Economics
Univ. of Missouri, Columbia, MO.
James G. Beierlein, Asst. Professor, Agricultural Economics
Penn. State Univ., University Park, PA.

ABSTRACT

This paper discusses the supply side of oilseed processing and measures the energy requirements involved in generating oilseed oil using current production technologies and existing large-scale-off-site oilseed processing facilities. An examination is made of the net energy gains associated with commercial oilseed processing under short-run conditions. Estimates of net energy gains from oilseed processing from various studies are examined along with current and future economic characteristics and potentials of selected oilseeds. From an energy balance standpoint, this study found that existing large scale commercial soybean processors can generate a positive energy balance. From an economic standpoint, none of the oilseed oils examined in this study were found to be economically competitive with diesel fuel under prevailing market prices and conditions. However, the competitive potential of selected oilseed oils may be enhanced by adverse economic conditions such as outright embargoes on crude oil imports or drastic increases in crude oil prices. Likewise, further cost reducing technological advancements in oilseed production and processing may also enhance the competitive position of selected oilseeds.

INTRODUCTION

Recent energy price and supply instabilities have brought about a renewed interest in alternative sources of fuel. Researchers from various disciplines are giving increased attention to renewable energy sources. Agricultural scientists share a general interest and expertise in renewable soil based energy sources and a particular interest in plant and vegetable oils.

While the process of extracting oils from plants is not new, the potential for wide scale adoption of such oils for energy is a relatively new phenomenon. This potential for wide scale adoption depends upon several technological and economic considerations. From an economic standpoint, this potential depends on the relative costs of producing and processing plant and vegetable oils and their ability to compete in conventional fuel markets. Until such oils can be purchased at competitive prices, their wide spread adoption under normal market conditions seem doubtful.

This paper will discuss the supply side of plant and vegetable potentials under existing production and processing methods. Particular attention will be given to the energy requirements involved in making these oils available, using current production technologies and existing off-site processing facilities which are currently engaged in processing such oilseeds. While the processing requirements of oil as an energy fuel may not be as stringent as that needed for feed or food, the capabilities of existing oil processing facilities with their potentials for generating alternative fuels cannot be ignored.

The objectives of this paper are to:

1. Describe the energy requirements associated with producing and processing selected oilseeds.

2. Estimate net energy gains associated with large-scale-off-site processing of oilseeds.

3. Examine existing and future economic characteristics and potentials of selected oilseeds.

Individual processor data from a recent survey of large-scale-off-site oilseed (LSOS) processors will be analyzed to determine the energy requirements associated with processing selected oil seeds. These findings will be compared to those found by other researchers in separate studies of LSOS processing. Next, an attempt is made to identify energy and economic differences related to plant size.

CURRENT STUDY

To learn more about energy use in food processing in general and oilseed processing in particular, a national energy survey was undertaken in the Fall of 1978 by the U.S. Department of Agriculture, ESCS in cooperation with four land-grant universities. A 16 percent response rate to the national survey yielded 24 responses from fats and oil products industry. Of these responses, nine responses were from the Standard Industrial Classification (SIC) group 2075-FOS soybean processors and eleven were from the SIC group 2074-FOC cottonseed processors. These oilseed processors, which were located in the South and Midwest, provided the data base for our analysis.

ENERGY BALANCE

In determining the technical and economic feasibility of oilseed products as fuel, Hammond et. al., have indicated the need to estimate an energy balance or some measure of net energy gains from oilseed processing and/or production. Kaufman and Pratt refer to this energy balance as an energy ratio. In general, a positive energy balance is an important consideration for economic feasibility. In computing an energy ratio for oilseeds, the analyst must designate which energy sources and outputs should be included in the equation. Previous researchers have used different methodologies under varying assumptions which will be discussed later.

For our study, processing energy inputs were estimated from and limited to, purchased energy sources reported in Table 1. Processing energy outputs were estimated by the energy potential of oil produced from oilseed processors. This particular methodology represents a short-run analysis, limited to readily available and variable energy sources. Other indirect processing energy inputs and outputs were omitted because they were not readily usable as a fuel source (meal, hulls, lint, etc.) or the energy embodied in the processing input was fixed in the short-run (processing machinery and equipment).

Table 2 illustrates the energy efficiency of selected oilseeds examined in our study. The average weekly capacities of the SIC groups were 1718 and 5835 metric tons for cottonseed and soybean processors, respectively. Cottonseed firms processed 73,569 metric tons during the year or about 30 percent of the volume handled by soybean processors (250,060 metric tons). The average cottonseed processors generated an average of 9,089 metric tons of oil for an annual total of 353,440 Giga-Joules in oil energy. The average soybean processor generated 47,465 metric tons of oil for an annual total of 1,868,504 Giga-Joules of oil energy.

Using these energy outputs, two separate energy ratios were computed for each processing group. The first ratio is limited to processing energy inputs and gives an indication of net energy gains through processing. Cottonseed firms used an average 97,065 GJ that year and generated a processing energy ratio of 3.62, meaning that 3.62 Kilo-Joule of oil energy were generated for each Kilo-Joule of processing energy input. That same ratio for soybean processors was 2.60.

A second energy ratio was estimated as a net energy return to both processing and production energy. USDA-FEA estimates shown in Table 2 were used to estimate oilseed production energy requirements.[1] All energy used in producing the raw oilseed product was included in computing production energy requirements on the assumption that the entire oilseed plant must be grown before oil can be extracted. A system of partitioning energy used in oilseed production in proportion to net oil weight was thought to be methodologically inappropriate because such a portion of the total production energy would have produced only a portion of the initial raw oilseed crop. Thus, the energy ratio for production and processing was found to be .17 for cottonseed oil and 1.13 for soybean oil. For cottonseed oil, this energy ratio of less than one implies that more total energy is needed to produce and process a unit of cottonseed oil than the energy generated by a unit of cottonseed oil. Estimates for soybean oil are positive and translate into 1,214 Giga-Joules or 198 equivalent barrels of crude petroleum oil per 1000 hectares, assuming an average yield of 263 kg of oil per hectare.

Because of differences in processing channels, the reader is cautioned when using these ratios for comparing the over-all processing efficiencies of cottonseed and soybeans. In particular, cottonseed processors in our survey purchased delinted cottonseed and thus, energy requirements for delinting cottonseed are excluded from Table 2. Because of this efficiency ratios for cottonseed processing in this particular study are thought to be biased upwards relative to soybean efficiency ratios.

OTHER STUDIES

In and of themselves these energy ratios for selected oilseed oils indicate that, in the case of soybeans, the current network of LSOS processors has the technical potential for producing fuel from oilseed. Before turning to an economic analysis of oilseeds a brief review of energy ratio's from other studies will be presented. Five studies were identified in the literature which either computed an energy ratio for a selected oilseed or provided the data from which ratios could be computed. A comparison of energy efficiency estimates among selected LSOS processors is shown in Table 3. These include a 1975 study Fritsh, et. al., at the Center for Science in the Public Interest; a study by the USDA reported in the Milk and Liquid Food Transporter; a 1980 study by Hammond, et.al., from the University of Georgia, a study by Kaufman and Pratt from North Dakota State.

Except for the study by Hammond, et. al., all of the energy ratios represent energy returns to both production and processing. In general, all of these other studies found positive energy ratios for cottonseed oil, soybean oil, peanut oil and sunflower oil. However, before forming conclusions about their findings, a brief critique of their methodology is in order.

Relative to our estimates for cottonseed oil, estimates by Fritsch, et. al., were disturbingly high. While not explicitly stated, we suspect that their estimates were derived through the partitioning method described earlier. More specifically, they estimated production and processing inputs by multiplying the total energy used for production and processing by the percentage of crop going to oilseed mills and that figure by the percentage of oil extracted from the oilseed itself.

The encouragingly high energy ratio estimates were found in the USDA study (energy ratio = 3.26) but a detailed description of the study methodology was not available and hence comparisons between their estimates for soybeans and our estimates for soybeans may not be valid.

Original estimates for soybeans by Hammond et. al., reflected returns to oil, meal and hulls. The estimates reported in Table 3 represent returns to oil only, consistent with the short-run methodology adopted for our study. The Hammond et. al., estimates of energy ratio exclude processing energy and tend to be biased upward. When processing energy estimates from our study are added, the Hammond, et. al. ratio drops to 0.9, indicating an unfavorable energy balance.

Estimates of a sunflower energy ratio made by Quick are reported by Kaufman and Pratt. While the exact methodology of these estimates were not reported, there appears room for optimism that LSOS processing of sunflowers appears to yield the highest energy ratio relative to energy ratios generated by other oilseeds reported in separate studies. The Kaufman and Pratt estimate translates into 18,993 Giga-Joules or 3105 equivalent barrels of crude petroleum oil per 1000 hectares of sunflowers (@ a yield of 588 kg of sunflower oil per hectare).

PROCESSING SCALE

Relationships between energy efficiency and scale of processing were analyzed using the sample of LSOS processors. The distribution-free Spearman rank correlation coefficient was used to identify relationships between scale and energy efficiency among oilseed processors surveyed in our study. Correlation coefficients shown in Table 4 were estimated for alternative measures of scale and energy efficiency. These data indicate that larger off-site processors appeared to be more energy efficient on a total processing basis. However, when scale and efficiency were defined in oil terms, there seemed to be greater energy efficiencies associated with smaller processors. However, the statistical strength of these coefficients tend to be weak as given by lower levels of statistical significance. More rigorous modeling is recommended to validate these findings.

ECONOMIC ANALYSIS

Despite the technical potential of producing and processing selected oilseeds for fuel their remains the question of economic feasibility. Since detailed production and processing costs were not available, our methodology for determining the economic potential for selected oilseeds was based on economic theory. We assumed that a high degree of competition exists in oilseed production and processing and that prevailing market prices for oilseed oil are a reasonable approximation of the economic costs of oilseed oil.[2] An assumed absence of long run excess industry profits suggest that current oilseed oil prices approximate their opportunity cost of production. On this assumption we adapted the methodology developed by Erickson to estimate competitive market price levels for selected oilseed oils. Shown in Table 5 are prices of selected oilseed oils which are equal (when adjusted for energy potential) to diesel fuel at $.26 per liter (or $1.00 per gallon). Also shown in Table 5 are price levels to which diesel fuel must increase before oilseed oils become competitive on a price per KJ basis. This analysis assumes that increases in the price of diesel fuel will not effect the price of oilseed oil. The economic gap between oilseed oil and diesel fuel might persist for diesel fuel prices above those reported in Table 5, if increases in diesel fuel prices impact upon the cost and/or price of oilseed oil. The economic senerio for lower oilseed oil prices may also be frustrated if lower oilseed oil prices stem from lower diesel fuel prices.

CONCLUSIONS

This analysis of the energy and economic efficiency of oilseed processing indicated that large scale processing operations can be energy efficient. From an energy balance standpoint, this study found that existing large scale commercial soybean processors can generate a positive energy balance.

From an economic standpoint, none of the oilseed oils examined in this study were found to be economically competitive with diesel fuel under prevailing market prices and conditions. Under adverse economic conditions such as outright embargoes on crude oil imports or drastic increases in crude oil prices, oilseed oil might serve as a short-run alternative to diesel fuel. Likewise, further cost reducing technological advancements in oilseed production and processing may also enhance the competitive position of oilseed oil.

This study attempted to document the technical and economic potentials of only a few energy producing plants. A longer run strategy for energy production from plants and vegetables should take into consideration the technical and economic potentials of a broad range of plants and vegetables, especially those which can be effectively produced on marginal land with a minimum of resources.

NOTES

[1] The reader should note the errors in the reported BTU/acre of the USDA-FEA study. Column 11 of Table 1 on page 3 of this study should be in thousands.

[2] Total economic efficiency losses due to imperfect competition were estimated at 6.2 percent of GNP (Scherer, p. 408). Hence some economic costs of oilseed oil may be inflated in concentrated processing industries.

REFERENCES

Anon. 1980-81. The Power of Vegetable Oils. Milk and Liquid Food Transporter. Nov. 1980, Jan. 1981.

Erickson, David. 1980. Soy Oil as Diesel Fuel: Economic and Technical Perspectives. Fact sheet. American Soybean Association.

Fritsch, A.J., L.W. Dujack and D.A. Jimerson. 1975. Energy and Food: Energy Used in Production, Processing, Delivery and Marketing of Selected Food Items. Center for Science in the Public Interest.

Hammond, C., L. Samples and B. Tyson. Energy Balance for Some Vegetable Oil Fuels Grown in the Southeast. University of Georgia, Agricultural Engineering, Athens.

Kaufman, K.R. and G.L. Pratt. 1981. Sunflower Oil Production/Utilization. Alcohol and Vegetable Oil as Alternative Fuels, Purdue University, Agricultural Engineering Department, W. Lafayette.

Pryde, E.H. 1981. Vegetable Oil vs Diesel Fuel: Chemistry and Availability of Vegetable Oils. Alcohol and Vegetable Oil as Alternative Fuels, Purdue University, Agricultural Engineering Department, W. Lafayette.

Scherer, F.M. 1973. Industrial Market Structure and Economic Performance. Chicago, Rand McNally. p. 408.

USDA-ERS. 1982. Fats and Oils Outlook and Situation, February.

USDA-FEA. 1977. A Guide to Energy Savings For the Field Crops Producer. USDA-ERS, Washington, D.C.

Table 1. Distribution of Energy Used by Oilseed Processors in Joule Equivalents, 1977-78

Energy Source	SIC GROUP: Cottonseed Processors (2074-FOC)	SIC GROUP: Soybean Processors (2075-FOS)
	percent	percent
Fuel Oil	5.0	32.7
Natural Gas	45.2	53.5
Electricity	38.1	12.3
Coal	0	0
LP Gas	< 0.1	< 0.1
Gasoline	1.8	0.2
Diesel	9.8	0.6
Number of Observations	11	9

Table 2. A Comparison of Energy Efficiencies from Processing and Production of Selected Oilseeds, 1977-78

	SIC GROUP AVERAGE	
	Cottonseed Processors[a] (2074-FOC)	Soybean Processors (2075-FOS)
Number of firms sampled	11	9
Processing Output		
Weekly capacity (tons)[b]	1,718	5,835
Annual total tons (all products)	73,569	250,060
Annual total oil tons	9,089	47,465
Annual total oil energy (G Joule)	353,444	1,868,504
K Joule per kg oil	38,847	39,356
Processing Input		
Total annual energy (G Joule)	97,065	716,383
K Joule per kg (all products)	1,323	2,863
K Joule per kg oil	10,725	15,084
Production Input[c]		
K Joule per kg (all products)	24,918	3,731
K Joule per kg oil	201,599	19,659
Energy Efficiency		
Processing:		
Net KJ gain/kg (all products)	3,475	4,605
Net KJ gain/kg oil	28,121	24,272
Total kg per kg oil	8.09	5.27
RATIO: KJ output/KJ input	3.62	2.60
Processing and Production:		
Net KJ gain/kg (all products)	-20,118	875
Net KJ gain/kg oil	-173,478	4,612
RATIO: KJ output/KJ input	0.18	1.13
Avg. kg oil per hectare[d]	140	263
Net gain from 1000 hectares:		
-Giga-Joules	-24,287	1,214
-Equivalent barrels petroleum	-3,968	198

[a]delinting energy excluded from some observations

[b]metric tons

[c]USDA-FEA, p. 3. Estimates include total production energy

[d]E.H. Pryde, p. 228

[e]One barrel petroleum crude oil = 6.12 Giga-Joules

Table 3. A Comparison of Energy Efficiency Estimates for Oilseeds From Selected Studies

	ENERGY EFFICIENCY[a]						
						Net gain per 1000 hectares	
Studies	Input KJ per Kg oil	Output KJ per Kg oil	Total Kg per Kg oil	RATIO: KJ output per KJ input	Avg. Kg oil per hect.[d]	Giga Joules	Equivalent barrels petroleum[e]
Broder, et. al.							
Cottonseed	212,296	38,847	8.1	0.17	140	-24,287	-3,968
Soybeans	34,750	39,356	5.3	1.13	263	1,214	198
Fritsch et. al[b]							
Cottonseed	16,398	38,847	6.1	2.36	140	3,143	514
Soybeans	26,865	39,356	4.5	1.46	263	3,285	538
USDA							
Soybeans	12,140	39,356	5.6	3.26	263	7,178	1,173
Hammond et. al.[c]							
Soybeans	27,140	39,356	5.6	1.45	263	3,212	525
Peanuts	19,876	40,147	2.0	2.02	790	16,014	2,617
Kaufman & Pratt							
Sunflowers	6,334	38,635	2.2	6.10	588	18,993	3,105

[a]includes production and processing energy

[b]energy requirements prorated on basis of oil content percentage

[c]excludes processing energy

[d]E.H. Pryde, p. 228

[e]One barrel petroleum crude oil = 6.12 Giga-Joules

Table 4. Spearman Rank Correlation Coefficients Between Energy Efficiency and Scale of Oilseed Processing Plant

RELATIVE RANKINGS	SIC GROUP[a] Cottonseed Processors	Soybean Processors	Total
Number of observations	11	9	20
Annual total output[b] vs kg/KJ	0.1818 (.5926)	0.000 (1.000)	0.0692 (.7720)
Annual oil output vs kg oil/KJ	-0.0363 (.9155)	-0.5667 (0.1116)	-0.1278 (0.5912)

[a]significance level in parentheses

[b]includes all products

Table 5. A Comparison of Selected Oilseed Energy and Prices With Diesel Fuel, 1979

Output Ratios	Oilseed: Soybean Oil	Cottonseed Oil	Peanut Oil	Sunflower Oil	Diesel Fuel
KJ/kg oil	39,356	38,847	40,147	38,635	45,590
KJ/liter[a]	36,326	35,856	37,056	35,660	42,080
Kg/liter Diesel Fuel	1.07	1.08	1.05	1.09	.923
$/kg = $.26/liter Diesel Fuel	.242	.241	248	.239	—
$/kg (1981)[b]	.539	.530	.546	.526	—
Equivalent Diesel fuel $/liter	.577	.572	.573	.573	—

[a]excludes additional energy potential from mixtures of diesel fuel and oilseed oil

[b]December, 1981 prices, "Fats and Oils Outlook and Situation"

SUNFLOWER PRODUCTION, HARVESTING,
DRYING AND STORAGE

Vernon Hofman   Duane Berglund   Kenneth Hellevang

Production

Sunflower, produced for its edible oil, has recently evolved as an important cash crop for the Dakotas and Minnesota. This oilseed crop has increased from a mere 81,000 hectares in the mid-1960's to over 1,620,000 hectares in 1981. The greatest proportion of the crop planted in 1981 was located in North Dakota. Over 90 percent of the sunflower crop planted in the United States is of oilseed varieties. Total production has decreased slightly in the past two years (Figure 1). Highest crop area and production occurred in 1979 with over 2 million hectares of sunflower. Approximately 3.2 billion kilograms of sunflower seed was produced in 1979.

Figure 1. SUNFLOWER PRODUCTION

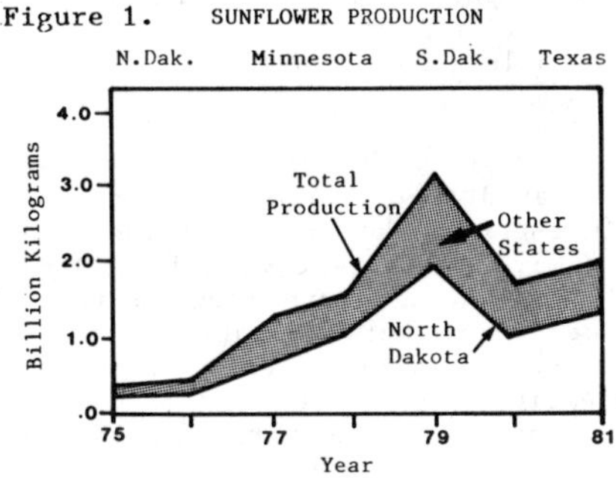

Increasing yields and favorable prices have made sunflower an attractive cash crop in the Dakotas and western Minnesota. Sunflower tends to fit well in small grain cropping rotation. Many producers find that by placing sunflower in a small grain rotation that a biological break is developed to various disease and insects which affect small grains. Sunflower are very tolerant to many of these pest problems. Also, producing sunflower is an efficient use of both labor and field machinery since it tends to spread out the cropping season compared to only producing small grain. Usually sunflower is planted after small grains in the spring and harvested late in the fall, following small grain harvest.

Figure 2. A sunflower crop will mature in 100 to 110 days after date of emergence.

The authors are Extension Agricultural Engineer, Extension Agronomist, Extension Agricultural Engineer, North Dakota State University, Fargo, North Dakota 58105

North Dakota's average oilseed sunflower yield has continued to increase steadily during the late 1970's. Average sunflower yields during the past five years has been approximately 1350 kilograms per hectare. The drouth conditions in 1980 caused an estimated 18 percent yield reduction in the North Dakota sunflower production compared to the 5 year average yield. The wheat crop, however, was reduced an estimated 33 percent compared to the previous wheat crops. Sunflower yields were not reduced as drastically since sunflower appeared to make efficient use of mid to late summer rainfall. The steady yield incease, with exception of 1980, can be attributed not only to higher producing, better quality hybrid seed, but also to better cultural management practices. These include: weed and insect control, improved fertility management, planting and harvesting practices, all enhanced by grower experience.

Selection of a sunflower hybrid variety is the initial management decision a sunflower producer must make. Most of the hybrids currently being grown are related to USDA developed parent material. The objective of the USDA-ARS breeding program at North Dakota State University involves developing and releasing of inbred lines for sunflower hybrid seed production to the private seed industry. This involves release of both the restorer lines and the female cytoplasmic male sterile lines. Criteria one should consider for selection of sunflower hybrids are as follows:

- Yield potential
- Disease resistance
- Maturity and dry down
- Oil percentage
- Stalk strength & lodging resistance
- Seed size, uniformity of seed
- Percent hybrid seed
- Self-compatability (fertilization)
- Germination percent
- Reputable seed company

Planting of sunflower is recommended from May 20 to May 31. Date of planting studies in North Dakota have indicated that extreme early plantings may increase insect problems later in the growing season. Yields have generally been higher in mid-May to late May plantings compared to June planting dates in North Dakota. Later June plantings have shown a trend towards lower percent oil and kilograms per cubic meter. Sunflower seedlings less than 15 centimeters tall are fairly tolerant to frost. However, once beyond this stage, they are susceptible to frost. Approximately 100-110 frost-free days from date of emergence are required to safely produce current maturing hybrids.

Soil temperature should be between $4^{o}C$ and $10^{o}C$ for germination. Soil temperatures below this level delays germination and extends the period of susceptibility to seedling diseases.

Hybrid sunflower seed is sold in various numbered sizes ranging from 1 through 5. The smaller the number such as 1 or 2, the larger the seed size, and the larger the number such as 5, the smaller the seed size. For example, a seed lot of size 1 seed would contain approximately 11,000 seeds per kilogram whereas a lot of size 5 seed would contain 22,000-24,000 seeds per kilogram. Hybrid sunflower seed is sold both by the kilogram and the number of seeds per bag. Seed sizes most available for the producers are 3 and 4. Seeding depth and emergence studies in sunflower have been conducted by the North Dakota State University Experiment Station. Research has indicated that seed size has not caused great fluctuations in seedling emergence or had any great effect on yields. Emergence has been good even at the 7.5 cm depth with most seed sizes. It is recommended, however, that under dry planting conditions, when deep seed placement is required that larger sized seed be utilized. The larger seeds may have a slight edge in seedling vigor.

Sunflower is usually considered a row crop with varying row width, depending upon the equipment available. Performance has been better when the supplemental nitrogen applied via fertilizer. Little sunflower yield response has been found to both phosphorus and potassium in most North Dakota trials. However, under soil tests which are extremely low in phosphorus (P_2O_5), slight yield response has been reported.

Insect damage has been a major concern of many sunflower producers in North Dakota. The major problem insects include: cutworms (*Euxoa sp.*), stem weevils (*Cylindrocopturus adspersus*), seed weevil (*Smicronyx fulvus*), sunflower moth (*Homoesoma electellum*) and sunflower midge (*Contarinia schulzi*). Cutworms can be a problem early in the season before or immediately after sunflower emergence. Sunflower midge causes major damage to developing sunflower buds and heads. The midge is confined to a localized area 65 km either side of the Red River which separates North Dakota and Minnesota. Delayed planting practices has lessened damage caused by the midge. Later plantings have also reduced problems associated with sunflower moth and stem weevil. There are a number of insecticides presently cleared for control of these insect problems. Many are short term insecticides having little if any systemic activity or lasting residual effect.

Diseases occurring in sunflower also can greatly reduce yield and hinder harvest operations. Sunflower diseases such as rust (*Puccinia helianthi*), downy mildew (*Plasmopara halstedi*), and verticillium wilt (*Verticillium dahliae*) can greatly reduce yields. Most sunflower hybrids currently planted have moderate to excellent resistance to these three diseases. Diseases such as sclerotinia stalk rot (white mold), phoma black stem, fusarium stalk rot and alternaria are presently the main disease threat to future sunflower production. Little if any hybrid tolerance or resistance to these diseases have been reported. Crop rotations of three to four years between planting sunflower or other susceptible crops such as dry beans, soybeans, sugarbeets, mustard or safflower is highly recommended. Small grains and corn should be grown in rotations with sunflower to minimize potential of the diseases commonly associated with sunflower and several other host row corps.

Sunflower, like any other crop, can have its yield severely reduced by weeds throughout the growing season. Sunflower, being planted as a row crop, can be kept somewhat weed free through tillage and cultivation practices. Pre-cultivation practices may include cross harrowing when the sunflower seedlings are 7.5 to 15 cm tall or using a rotary hoe when the plants are small. Research by the North Dakota State University Experiment Station has been conducted from the late 1960's to present observing and screening herbicides for weed control in sunflower. A number of herbicides are cleared at the present time for weed control in sunflower. These include Eptam (EPTC), Treflan (trifluralin), Tolban (profluralin), and Prowl (pendimethalin). These herbicides must be applied preplant incorporated and give excellent control of grassy weeds such as green foxtail (*Setaria viridis* L. Beauv), yellow foxtail (*Setaria lutescens* Weigel, Hubb.) and wild oats (*Avena fatua* L.), plus some broadleaf weeds. Amiben (chloramben) is also recommended for use on sunflower when there is a possibility of certain broadleaf weeds such as wild mustard (*Brassica kaber* D. C. Wheeler). Wild oat control can be obtained with Carbyne 2EC (barban) applied early postemergence on sunflower when the wild oat is in the 1½ to 2-leaf seedling stage. We look in the future to see numerous other herbicides cleared for use on sunflower.

Volunteer sunflower in crops following sunflower can be difficult weeds to control. In small grains adequate control can be obtained using 2,4-D, MCPA, bromoxynil and various mixtures of bromoxynil plus MCPA (Bronate or Brominal Plus) or Banvel (dicamba). These herbicides applied during the stage of early crop development can result in excellent control of volunteer sunflower and other broadleaf weeds which will compete within the crop. In corn, volunteer sunflower can be controlled adequately with

width of the row has been between 50 and 75 centimeters; however, widths as wide as 100 centimeters and as narrow as 35 centimeters have produced good yields. A difference of a few centimeters in row width would not justify the investment for different equipment.

Sunflower harvest populations in North Dakota for oilseed range from 40,000 plants per hectare in the western, drier regions to 55,000 plants per hectare in the eastern, higher rainfall region. Under irrigation, the recommended harvest population ranges from 60,000 to 65,000 plants per hectare. Confectionary sunflower should be planted at a population between 30,000 and 45,000 plants per hectare. Planting sunflower seeds at a rate 15 percent greater than the desired harvest population is a good practice. This overseeding will compensate for seedling losses due to soil crusting, poor germination, insect and disease problems. Table 1 shows the seed spacing for various row widths assuming a 15 percent stand reduction.

Table 1. Seed Spacing Required for Various Plant Populations Assuming a 15 Percent Loss in Germination and Stand.

Plants per	Row Width Centimeters					
Hectare	100	90	75	65	55	50
30,000	28.3	31.4	37.7	43.6	51.5	56.6
33,000	25.7	28.6	34.3	39.6	46.8	51.5
36,000	23.6	26.2	31.5	36.3	42.9	47.2
39,000	21.8	24.2	29.0	33.5	39.6	43.6
42,000	20.2	22.5	26.9	31.1	36.8	40.5
45,000	18.8	20.9	25.2	29.0	34.3	37.7
48,000	17.7	19.6	23.6	27.2	32.2	35.4
51,000	16.6	18.5	22.2	25.6	30.3	33.3
54,000	15.7	17.5	20.9	24.2	28.6	31.5
57,000	14.9	16.5	19.9	22.9	27.1	29.8
60,000	14.1	15.7	18.9	21.8	25.7	28.3
63,000	13.5	15.0	18.0	20.7	24.5	27.0
66,000	12.9	14.3	17.1	19.8	23.4	25.7

Proper adjustment and operation of planting equipment is one of the most important operations in sunflower production. Plateless and cyclo air-planters have been used effectively to get good seed distribution. However, conventional planters will provide good seed distribution by using correct planter plates, properly sized seed, keeping seed knockouts in good condition and traveling at a reasonable speed. Several commercial seed companies have proper plates for different seed sizes available. Grain drills are sometimes used on farms where row crop equipment is not available.

Sunflower will compensate for stand variation by producing larger seed heads when planted in low populations. Also, stalk diameter and stalk strength increases with low populations. Dry down period and maturity, however, are lengthened when sunflower are planted in low populations. High population sunflower are not always the highest yielding. Lodging, small heads and small seed size can be detrimental when sunflower are planted at high populations. Water availability becomes the limiting factor. Sunflower, however, is an efficient plant concerning water use. Research has been reported that sunflower uses approximately 35.5 to 38 cm of water during the growing season to produce 1680 kg per ha.

Sunflower has responded very well in fertility trials to various levels of nitrogen. Research indicates that approximately 2.25 kg of actual N are required per 45 kg of oilseed produced. This nitrogen includes both the nitrate nitrogen available in the soil prior to planting time and the

the use of either 2,4-D, Banvel or atrazine. In crops such as soybean or dry beans, volunteer sunflower can be adequately controlled with the use of Basagran (bentazon). In soybeans, Blazer (acifluorfen) and Dyanap (naptalam + dinoseb) can also be used for volunteer sunflower control.

Harvesting

A sunflower crop is normally ready for harvest about 120 days after planting. High summer temperatures can reduce the time from planting to harvest. Sunflower is physiologically mature when the back of the head has turned from green to yellow and the bracts are turning brown. This is about 30 to 45 days after bloom, depending on temperature during maturation. Application of preharvest drying aid chemicals (desiccants) may permit earlier harvest in some years. Desiccants, if used, are applied at about 25 to 30 percent seed moisture or when physiologically mature. The head, stalk and leaves dry more quickly with a desiccant but the seed drydown rate may be only slightly affected.

In the northern region, a killing frost will speed drying and harvest. If frost is too early, yield and oil losses are higher. Many northern growers harvest sunflower at high moisture levels and then dry to reduce shatter loss and exposure to birds.

Combines suitable for threshing small grains can be adapted to harvest sunflower. A variety of header attachments are available. Most attachments operate on a head-stripper principle.

Long seed gathering pans extending ahead of the cutter bar are used to collect shattered seed. The width of gathering pans varies. Narrow 23 cm pans spaced on 30 cm centers can operate on any desired row spacing (Figure 3) while wider and more efficient 75 to 100 cm spaced pans are limited to a fixed row spacing (Figure 4).

Figure 3. Narrow header pans mounted on 30 cm centers can operate on any row spacing.

Figure 4. Wide seed pans spaced 75 to 100 cm are limited to a particular row width but are more efficient in saving seed.

Machines with a small 40 to 50 centimeter diameter shielded reel can handle tall plants while taking in only a minimum length of stalk with the head (Figure 5). This permits threshing when the seed is dry but stalk moisture may remain above 50 percent. Cleaner threshing is also accomplished when only the head enters the machine. Optional forward rotating stalk-walker shafts, introduced from Argentina, can be mounted under the cutter bar to reduce hangup of stalks between pans.

Combines equipped with axial flow cylinders have been effective in obtaining clean seed, particularly under wet conditions. A recently introduced multirow crop header is effective in fields with lodged sunflower (Figure 6).

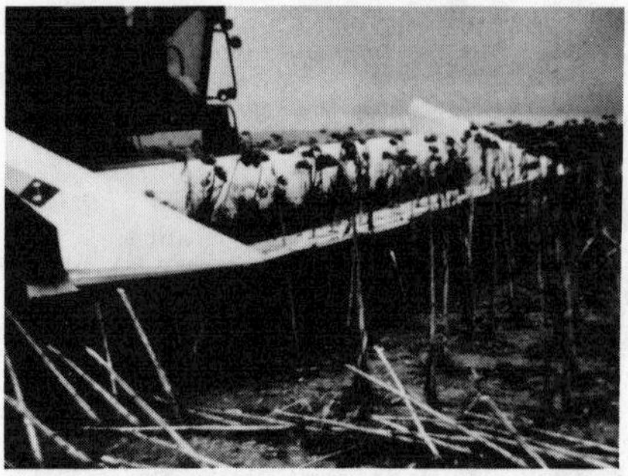

Figure 5. A header with a shielded reel eliminates tall stalks from entering the combine. This allows for cleaner threshing as less material enters the machine.

Figure 6. This multi-row crop header is very effective in fields with lodged sunflower.

Cylinder Speed: Sunflower threshes relatively easy. The cylinder speed should be as slow as possible, generally 300 to 450 rpm, and still thresh the seeds out of the head. Excessive cylinder speed causes excessive dehulling of seeds, and the chaffer, sieve, and tailings-return may become overloaded with small pieces of broken heads.

Concave Adjustment: When the crop is dry (10 percent moisture or less), the concave should be set wide open. Use a closer concave clearance only if some seed is left in the head. Under most conditions, it is better to decrease concave clearance rather than increase cylinder speed to get more complete threshing. It is usually possible to set cylinder concaves and cleaning shoe to obtain less than 5 percent dockage. Under abnormally wet conditions, it may be necessary to use higher cylinder speeds and closer concave settings even though trash in the seed increases.

Fan Setting and Sieves: Sunflower seed is easily blown over the chaffer and sieve. Seed forced into the tailings auger returns through the cylinder and may be dehulled. Only enough air should be used to keep the trash floating on the sieve. Adjust the chaffer and sieve for minimum tailings return.

When the combine is adjusted correctly for threshing sunflower, the threshed heads will come through only slightly broken and with only blank seed remaining in the head. Improper combine settings may crush the kernel, but leave the hull intact. This is critical for confectionary sunflower used for the human food market. Dockage should be less than 5 percent with little or no dehulled seed. Under ideal harvest conditions and with proper machine adjustment, harvest losses can be reduced to less than 5 percent and dockage less than 2 percent.

Drying

Sunflower is normally harvested at moisture contents above that recommended for storage. This is done to reduce field and harvest losses. Also the sunflower may not dry down in the field forcing mechanical drying.

Sunflower can be dried in conventional crop dryers; bin, batch and continuous flow dryers have been used successfully. Sunflower dries easily due to the relatively small amount of water removed. Since sunflower weighs about half as much as corn per cubic meter, only half as much water is removed during drying. Drying corn from 25 percent to 15 percent removes 85 kg of moisture per cubic meter. Drying sunflower from 20 pecent to 10 percent removes only about 40 kg of moisture per cubic meter.

The resistance to air flow, static pressure, of sunflower is comparable to barley. This is less than for wheat but more than the static pressure of corn. Fines and foreign material increase the static pressure, so it is important to clean the sunflower. Also some increase in static pressure is caused because sunflower settles during storage. Consequently, be sure to include a compaction factor and a factor for uncleaned seeds in static pressure calculations.

Drying temperatures up to 104°C do not have an adverse affect on the oil percentage or fatty acid composition of oil type sunflower. Drying confectionary sunflower at high temperatures may cause the kernels to be wrinkled or even scorched. Industry dries confectionary sunflower at 82°C without harm to the sunflower meats. Both temperature and time are factors so the temperature for bin drying is about 43°C. Low temperature-natural air drying in bin should only have enough heat added to control the equilibrium moisture content, usually less than 6°C. The kernel temperature should not exceed 43°C if the sunflower is to be used for seed.

Confectionary sunflower should be dried to 10 percent wet basis and oil sunflower to 8 percent. An electronic meter used to measure the moisture immediately after removal from a high temperature dryer will indicate an erroneous value. The meter will indicate the seed is dryer than the actual moisture content of the kernel since the hull will be drier than the kernel and the meter reading is based on the surface condition. The moisture will migrate from the kernel to the hull in a period of hours causing the moisture reading to increase. Consequently, the meter reading should be adjusted to account for this and the sunflower checked at least 12 hours later to confirm the sunflower is at the desired moisture content.

Fire Hazard

A serious fire hazard exists when drying sunflower. Fine hairs or fibers from the sunflower and a dirty dryer are the major causes of fires. During handling, the fibers from the sunflower hulls come loose. These fibers and other foreign material are ignited if drawn by the fan through the burner. Then the ignited materials go into the dryer plenum carrying the fire to the sunflower unless the fire burns out before contacting the sunflower. By facing a portable dryer into the wind, most of the material will be blown away from the burner intake. A permanent dryer should be faced into the prevailing wind. Screens can be used to filter the air or a tube may be placed on the burner intake to draw clean air away from the dryer, but care must be taken not to restrict the airflow (Figure 7). Partial air recirculating dryers may increase the fire hazard if combustible material is in the air that is reheated and comes in contact with the sunflower.

Thoroughly cleaning the dryer daily reduces the fire hazard. Operators have indicated they became lax in cleaning the dryer prior to a fire. If seeds are caught in the drying column, that spot will overdry increasing the fire hazard. Multistage dryers are frequently set with high temperatures on the wet sunflower and lower heat on the dryer seeds because the fire hazard increases as the sunflower seed becomes dryer.

A dryer should be monitored constantly when drying sunflower due to the fire hazard. Operators can usually control the fire if noticed early. Many fires are extinguished by shutting off the fan because without the air flow the fire smolders and dies. Water should be available to apply to the fire spot. A limited amount of water wets the hulls and may control the fire if applied early. Once the fire is established, the fire is an oil fire and is hard to control with water. The dryer should be unloaded if the fire progresses to that stage.

Storage

The storage of sunflower is similar to any other crop. Sunflower weighs between 325 and 400 kilograms per cubic meter. So a structure for most other crops will hold sunflower. Sunflower should be cleaned because foreign material inhibits good aeration and is usually wetter than the sunflower seed. The recommended storage moisture content is 8 percent for oil seeds and 10 percent for confectionary.

Cooling the sunflower seed greatly increases the storability and decreases insect damage. Sunflower should be cooled to about $0^{\circ}C$ which nearly stops microbial activity. Insects become dormant below about $10^{\circ}C$ so cooling controls insects too. If sunflower are not uniformaly cooled, moisture migration will cause wetting of the sunflower in the top center of the storage (Figure 8). As the outdoor temperature cools, the air inside the storage wall also cools sinking to the bottom. This forces the warm air in the center up. The air deposits moisture on the sunflower in the top center as it cools. Cooling the entire storage limits the moisture migration. Aeration fans sized to deliver 2 $L/s \cdot m^3$ will adequately cool the sunflower to the outdoor temperature in about 1 week of continuous operation (Figure 9). The cooling cycles should start as soon as the outdoor temperature is about $7^{\circ}C$ cooler than the sunflower and continue until the sunflower has been cooled to about $0^{\circ}C$. The sunflower should not be peaked. Utilizing the bin roof area for storage increases storage problems.

The sunflower should be checked at least weekly. An increase in temperature of the stored sunflower may indicate mold or insect activity. Checking the moisture content will indicate if moisture migration is occurring. Once a problem starts it grows very rapidly.

Figure 7. A portable batch dryer used for drying high-moisture sunflower. Note the long, flexible snorkel used to draw clean air to the fan.

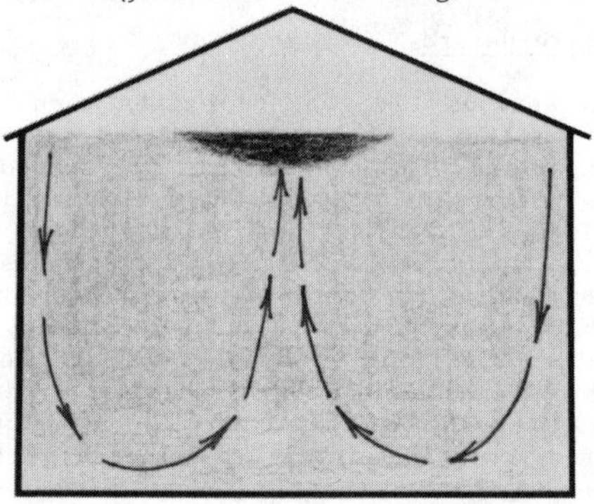

Figure 8. The normal pattern for moisture migration during the fall and winter. High moisture contents and crusting usually occurs at the top center of the bin.

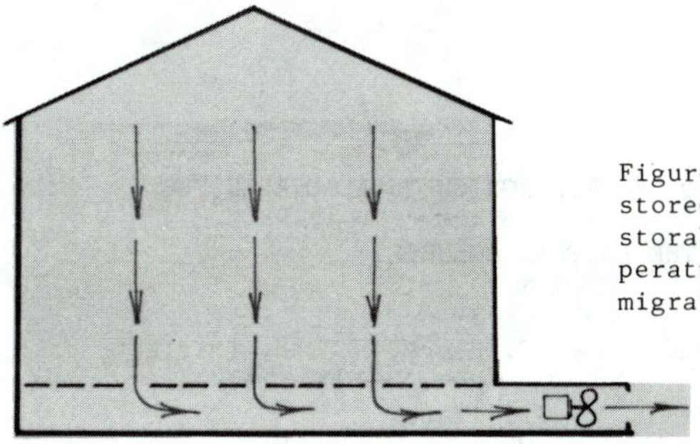

Figure 9. Aeration cools the stored sunflower increasing the storability and equalizes the temperature which limits moisture migration.

- aerate with 0.1 cfm/bu in storage
- start aeration when air temp. falls 20° below seed temp

DOUBLE-CROP SUNFLOWERS FOR AGRICULTURAL DIESEL FUEL IN THE EASTERN CORNBELT

T. L. Glenn, H. M. Keener, J. E. Henry, G. B. Triplett, Jr.

Assoc. Member ASAE   Affiliate Member ASAE

ABSTRACT

Agronomic and engineering information on double-crop sunflower production, processing and utilization is presented. This and other available information is used to assess feasibility and future directions in the use of sunflower oil for agricultural diesel fuel in the U.S. Eastern Corn Belt area. Double-cropping yields varied considerably due to precipitation extremes, plus different soil characteristics and management practices. Average expeller oil yields of 0.344 kg of oil per kg of moisture free seed were achieved with a feed rate of 125 kg per hour for a range in seed and processing conditions. Results from feasibility analyses suggest that sunflower oil can be grown in Ohio and processed in a community cooperative plant with a favorable energy ratio and marginal profitability.

Agriculture and the farming sectors within this nation and throughout the world are looking for ways to control their own liquid fuel supplies. International tensions and threats existing in the world today will greatly influence the long-term availability of OPEC oil. On a national and societal basis, where liquid, mobile fuels are concerned, overall fuel production energy efficiencies are vital, overall fuel economics are crucial, and liquid fuel allocation procedures may be required. But all of these are totally inconsequential to the large grain producer when the total fuel supply system has misfired to the extent that his diesel fuel supply has been delayed by 30 days. If it is the wrong 30 days in the seasonal cycle he has lost a crop, and likely is bankrupt. Multiply this by the productive capacity of even a small percentage of our farms, and society has a food shortage problem.

While ethanol production has received considerable attention during the past several years, from many standpoints, vegetable oils are a better alternative liquid fuel for agriculture. This is because oil expelling is much simpler, the fuel is better suited for diesel engines, the by-products are more easily handled and utilized, and the energy return ratios are much better - compared to alcohol. Because of these reasons, viable systems for producing and processing sunflowers as a source of liquid fuel for agriculture in Ohio and the Eastern Corn Belt are being investigated.

The authors are: T. L. GLENN, Research Scientist, H. M. KEENER, Associate Professor, J. E. HENRY, Assistant Professor, Agricultural Engineering Dept., and G. B. TRIPLETT, JR., Professor, Agronomy Dept., Ohio Agricultural Research and Development Center, Wooster, Ohio.

DOUBLE-CROP SUNFLOWER PRODUCTION

It is estimated that 65 liters of sunflower oil per hectare of cropped land in Ohio would be needed to replace diesel fuel now being used. For Ohio, current estimates of cropped land is 4.5 million hectares, of which approximately 34 percent is normally planted to corn for grain, while 32 percent is soybeans and 15 percent is wheat (Carter, 1978; Carter, 1982). The remainder is other grains, forages and vegetable crops (Fig. 1). This translates into a total fuel need of 293 million liters of sunoil.

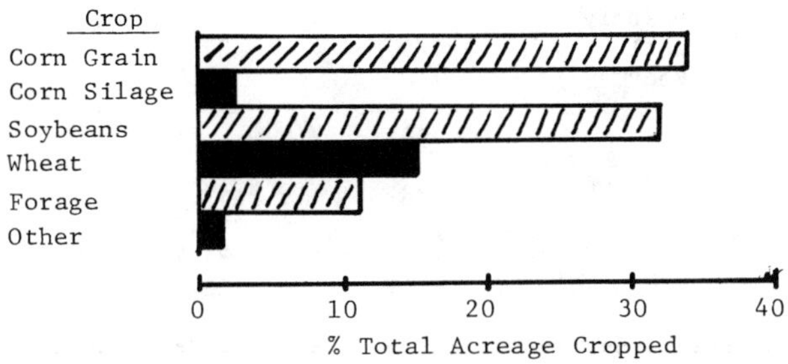

Fig. 1. 1981 Cropping Pattern of Ohio Agriculture

Field Studies

Yield potential of double-cropping sunflowers after wheat was studied in 1980 and 1981. In July, sunflowers were planted using no-tillage methods after wheat harvest at four locations in Ohio. A herbicide application of glyphosate, linuron and alachlor was made following planting. Harvest dates, moisture and yield are shown in Table 1. Plot sizes were 0.4 hectares.

In 1980, yields ranged from 0.29 to 1.88 tonne per hectare, the highest yield occurring on well drained soil receiving 84 kg N per hectare. The other plots received less than 59 kg N per hectare.

In 1981, average yield from three sites was 0.78 tonne per hectare. A fourth plot suffered from bird damage (estimated at 75 percent loss) and yielded 0.21 tonne per hectare. Volunteer wheat was a problem at two of the four locations.

It is of interest to note, in 1980 rainfall received at the Wayne County plot was 33.8 cm during a 32 day period beginning the last 11 days of July (more than three times the normal rainfall for that period). In 1981, rainfall patterns were reversed with plots receiving 1/3 to 1/2 the normal rainfall amounts in July and August.

Ohio's single-crop sunflower yields have been evaluated by Schmidt and Kroetz during 1979, 1980 and 1981. Average results for four hybrids versus planting date at Hoytville, Ohio in 1979 were 3.07 t/ha for May 18, 2.16 t/ha for June 18, and 1.34 t/ha for July 16. Yield values appear lower in 1980 and 1981, but direct comparisons are impossible because of location and hybrid changes and limited planting dates. For purpose of feasibility analysis, the single-crop yield was assumed to be 2.0 tonne per hectare, while the double-crop yield was assumed to be 1.2 tonne per hectare (10 percent moisture).

Table 1. 1980 and 1981 Ohio Double-Crop Sunflower Yields

Plot location (county)	Soil drainage classification	Date planted	Harvest moisture (% w.b.)	Yield (kg/ha at 10% m.c.)
		1980		
Champagin	Well	7/11	20.1	1,880
Darke	Poor	7/15	19.3	330
Wayne	Moderately well	7/17	32.7	570
Wood	Poor	7/23	18.5	290
		1981		
Pickaway		7/8	13.0	680
Darke		7/10	10.5	1,040
Wood		7/14	31.7	620
Wayne		7/17	22.5	210[a]

[a] This figure represents an estimated 25 percent of the crop. The remainder was lost to birds.

Pressing Trials and Oil Yields

1980 crop sunflower seeds were screw-pressed for oil using an industrial type screw press at the test laboratories of French Oil Mill Machinery Company, Piqua, Ohio. The press used is a geometrically scaled down version (half size) of their F44 machine.

Whole sunflower seeds were processed in their laboratory on three different days. Results from two days processing in May 1981 are shown in Table 2. Because of FOM's tight schedule and the variety of products run through their press, there was no serious attempt to optimize the processing results reported here.

Table 2. Oil Processing Results of Ohio Double-Crop Sunflower Seeds

Plot Location (Ohio County)	Screw Press Processing Results[a]				
	Feed Rate	Input Moisture Content	Oil Output Rate		Output Oil Ratio
	(kg/hr)	(% w.b.)	(kg/hr)	(L/hr)	(kg oil/kg moisture free seed)
Champaign	128	7.8	41.8	45.3	.353
Champaign[b]	124	2.9	45.4	49.2	.376
Darke	136	6.2	41.5	45.0	.326
Wayne	121	6.4	39.9	43.3	.354
Wood	116	10.2	32.2	34.9	.310

[a] Results are from processing whole sunflower seeds (1980 crop) at French Oil Mill Machinery Company, Piqua, Ohio, May 4 and 5, 1981. The press used had an 89 mm inside cage diameter with a 559 mm long main shaft and a 140 mm long feed screw.

[b] Preprocessed by cooking 4 hours at 85 deg. C.

The average oil output from the press was 43.5 liters per hour for 6.7 percent moisture content whole seed processed at an input rate of 125 kg per hour. A 7.1 percent increase in oil output was observed after the input seeds were cooked at 85 deg. C for 4 hours. The cooking produced a 4.4 percent moisture content reduction. Resultant average oil yield was 0.344 kg oil per kg of moisture-free seed. The screw-press oil yields ranged up to 689 liters per hectare based upon the production land area used for growing the seeds.

Land Requirements to Supply Oil

Using this information, it is possible to estimate the amount of land needed to be diverted to sunflower production to obtain liquid fuel (oil) for Ohio's agriculture. Figure 2a is a solution for the case of single-crop sunflowers replacing corn acreage and double-cropped sunflowers after wheat. Solutions are given for a use level of 47, 65 (current estimate to produce and harvest agricultural crops in Ohio, ignoring transportation and processing requirements) and 94 liters oil per hectare. The basic equation that was used in this analysis is:

$$0.9\,\varepsilon\left[a_3\, y_1\, f_3 + a_2\, y_{d1}\, f_{2d}\right] \geq \rho\, q\, [1 + a_2\, f_{2d}]$$

where
- ρ = sunflower oil density kg/liter, 0.92
- a_2 = fraction of acreage in wheat, 0.15
- a_3 = fraction of past acreage in corn, 0.34
- f_{2d} = fraction of wheat double cropped to sunflowers
- f_3 = fraction of past corn acreage now planted to sunflowers
- y_1 = single crop yield for sunflowers (2000 kg/hectare, 10% w.b.)
- y_{d1} = double-crop yield for sunflowers (1200 kg/hectare, 10% w.b.)
- q = average fuel used per hectare, 47, 65, or 94 liters
- ε = conversion factor of moisture free whole sunflower seed to oil, .344

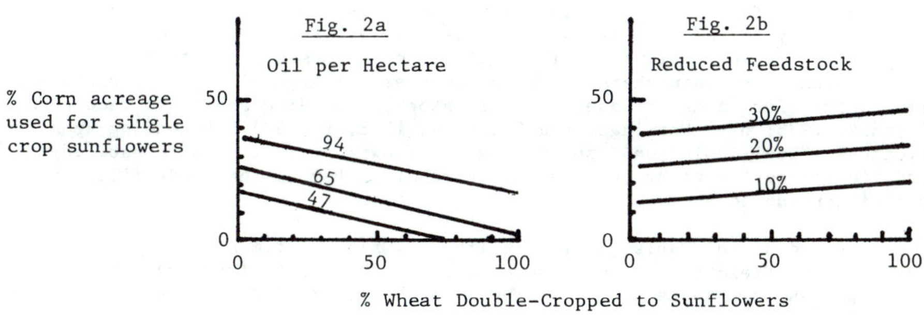

Figure 2. (a) Revised cropping patterns in Ohio that will produce 47,65 and 94 liters of sunflower oil per acre of land formed.

(b) Reduction in feedstock as a percent of corn production before diverting to sunflowers (weight basis).

If crop production can be carried out at the 65 liters per hectare level, then sunflowers replacing approximately 14.5 percent of the corn acreage and 50 percent of the wheat acreage, double-cropped, would meet Ohio's agricultural needs. This would be reduced if yields can be improved.

In addition to the oil produced, a feed by-product, approximately 65 percent of the original seed dry matter, is produced. Figure 2b is a graph showing the percent reduction in feed stock as sunflowers are substituted for corn and are double-cropped after wheat. For the case of 14.5 percent corn acreage reduction and 50 percent of the wheat acreage being double cropped, the percent reduction in feeds is estimated to be 10 percent.

Proximate analyses of sunflower meal (without hulls) mechanically extracted via a screwpress are given by Ensminger and Olentine (1978). Because of its relatively high protein content, it will be used as both a protein and energy source. Also, the mechanically produced meal's high oil content could lead to it being used as a speciality feed product for livestock such as dairy cattle and horses.

Anticipated energy inputs and outputs for Ohio sunflower production (single- and double-crop) and processing for agricultural diesel engine fuel are shown in Table 3. The "fossil energy" input designation refers to energy calculation procedures used to obtain drying and oil recovery figures (Keener et al., 1981). The actual electrical energy required has been divided by 0.33 to convert to a fossil energy input at the generating plant. Similarly, the steam input energy listed has been adjusted to obtain a fossil fuel input after dividing vaporization energy (estimate based upon information from Olsson, 1980) by 0.60.

Pimmentel (1980) conversion factors were used to estimate energy inputs for pesticide application rates, while Goering and Daugherty (1981) was used as the primary reference regarding seed, machinery and labor inputs for conventionally tilled sunflowers. Their labor and machinery figures have been reduced because no-till production has been assumed.

Diesel fuel estimates are based on published average fuel requirements for various no-till field operations (Deere & Company, 1980).

Mechanically expelled oil is considered here as the sole output oil source. Large solvent extraction plants might be willing to purchase output meal from a group of farmer-owned cooperative plants, in which case their output oil will probably be used solely in the food-grade oil industry. The farmer-owned cooperative plant output meal might also be used in the specialty feed industry where certain characteristics such as its high oil content would make it useful for some classes of livestock, including show and race horses (Tyznik, 1975; plus Ensminger and Olentine, 1978, p. 928). Depending upon the eventual "best solution" to mating sunflower oil to the diesel engine, additional sunflower processing may be required beyond the crude filtered oil output considered here.

Also included in Table 3 are results of Ohio experimental work. The Ohio double-crop field test results were based upon average 1980 and 1981 yields, plus processing tests run at the French Oil Mill Machinery Co. on 1980 harvested crop.

Average fertilizer usage per hectare for the Ohio field studies was 73 kg of N as ammonium nitrate, 55 kg of P_2O_5 as triple superphosphate, and 53 kg of K_2O as muriate of potash. The energy conversion factors used were 61.5, 12.56 and 6.70 MJ per kg, respectively. They were developed from information presented by Lockeretz (1980) based upon data from the Tennessee Valley Authority for producing and transporting commercial fertilizers. The nitrogen fertilizer

Table 3. Fossil Energy Inputs and Outputs for Agricultural Diesel Engine Fuel from Sunflowers in Ohio, No-Till

INPUTS (MJ of fossil energy)/ha	Ohio field studies (double crop)	Expected feasible goals in Ohio (double crop)	Expected feasible goals in Ohio (single crop)	Remarks
Labor		9	9	75% of conventional till
Machinery		649	649	80% " " "
Diesel fuel		984	984	No till
Fertilizer	5535	6561	7255	
Seed		915	915	
Pesticides	2945	1746	1746	Includes formulation, packaging & transport
Drying[c]	578	899	615	Low-temp, in-bin
Oil Recovery				
steam	0	1116	1861	60% fuel burning efficiency
electricity	371	130	216	33% fossil energy conversion
Total	(11986)[a]	13009	14250	
% charged to expelled oil	50.1	56.5	56.5	
OUTPUTS				
Seed yield (kg/ha)	771[b]	1200	2000	10% moisture content (w.b.)
Expelled oil, %	31	35	35	Depends on seed oil content
Energy, MJ/ha	9444	16,617	27,696	39.565 MJ/Kg oil
OUTPUT/INPUT	(1.57)[a]	2.26	3.44	

[a] Figured by using labor, machinery, diesel fuel and seed energy input estimates from other columns.

[b] Neglecting results from plot with 75% bird loss.

[c] Using 21% and 15% initial moistures for double and single crop respectively.

energy included in the Ohio goal columns is based upon application of anhydrous ammonia with a conversion factor of 50.2 MJ per kg.

Results showed sunflowers could be grown for an input of 6.76 MJ per kilogram of dry sunflower seed produced. This results in a favorable energy ratio of 3.44 energy units of expelled oil per unit of input fossil energy. Dollar cost of producing sunflowers (Schmidt, 1980) is estimated at $0.22/kg for double-crop and $0.26/kg for single-crop sunflowers if yields are 1.2 and 2.0 tonne per hectare, respectively. The lower cost for double crop arises from assigning lower input cost (land and machinery) in producing the crop.

PRESSING FACILITIES

On-farm liquid fuel production requires a considerable investment in capital and management. The time devoted to on-farm fuel production, while being a considerable drain on the farmer's efforts, is generally not enough to make an adequate return on his investment in buildings and equipment. This conflict in available time and return can probably best be resolved by going to larger off-farm plants operating full time. And yet the need remains for farmer controlled liquid fuel supplies.

In order to achieve rurally-controlled diesel fuel supplies, some type of cooperative processing plant using mechanical oil extraction is likely.

Figure 3 is a schematic of such a plant. This plant is typical of oil seed processing plants (Anderson International Corporation and French Oil Mill Machinery Company). Data supplied by Anderson (Olsson, 1980; Crane, 1982) for equipment suggested a cooperative size plant, capable of producing 4.2 million liters of oil per year (approximatey 16,700 liters per day for 250 days a year) would cost 1.2 million dollars for land, buildings, equipment and investment in utilities. This plant would require 9 1/2 persons (part-time office staff) to operate on a continous (24 hour) basis when in operation. Table 4 gives a cost analysis for the facility. Another much more elaborate plant, per French Oil Mill Machinery Company's information (Thow, 1981), is estimated to have a capacity of 23,000 liters per day (24 hour) when processing sunflower seed containing 8 percent moisture. Produced meal would contain a residual oil content of 4 percent or less. Total system cost for FOM's plant would be approximately $1.9 million. These estimates give an investment cost of $0.29-$0.34 per liter. Increasing operational time to 300 days a year reduces investment cost to $0.25-$0.28 per liter. To meet Ohio's needs of 293 million liters, total investment would approach 70-80 million dollars with about 65 plants strategically located among Ohio's 88 counties.

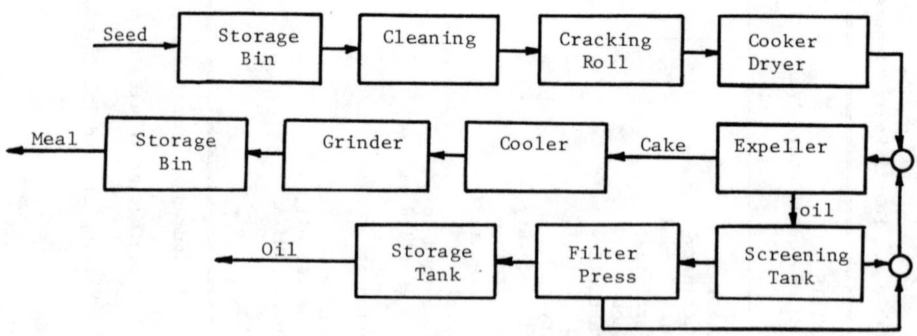

Fig. 3. Elements of Cooperative Processing Plant for Mechanical Extraction of Oil from Whole Sunflower Seed

Table 4. Budgeted Cost of 4.2 Million Liters per Year Facility

	Fixed Cost		
Land (site preparation)	$ 50,000		
90 x 120 ft building	60,000		
Equipment (press room)	650,000		
Utilities, installation	100,000		
Storage	150,000		
Other equipment	150,000		
Total			$1,160,000
	Variable Cost		
Labor			
Wages	$187,000		
Fringe benefits	28,000	$215,000	
Utilities			
Steam (20,000 lb per day @ 150 psi)	92,600		
Electrical (450 kWh per day)	11,250		
Other	7,000		
		110,850	
Total			$325,850

Cost per liter = (0.33 * 1.16 + 0.33)/4.2 = $0.169

Annual fixed cost is assumed to be .33 times total fixed cost.

Estimates of labor and utility cost for operation of 4.2 million liters per year facility is $325,850 per year for 250 days. Going to 300 days would increase labor cost to $252,000 per year and utility cost to $133,000 per year. This, combined with increased capacity to 5.0 million liters per year, lowers the estimated total processing cost to $0.15 per liter.

If the sunflower crop cost was zero, oil cost would look very good in today's market place. However, the cost of oil seed ranges from $0.22-$0.26/kg ($0.64-$0.76/liter of oil). This seed cost is a conservative estimate based upon our actual output oil data and not the Anderson International Corporation figures used in Table 3 under feasible goals. Adding this to the processing cost gives a total cost of from $0.79-$0.92 per liter of sun oil made locally in a cooperative plant. Subtracting off a credit for the meal by-product of $0.29 per liter ($0.15 per kg of meal) leaves a cost of $0.50-$0.60 per liter produced. This cost is double ($0.26 per liter) March 1982 diesel prices but less than average crude vegetable oil price of $0.64 per liter in 1980.

Processing Notes

Oil processing of seeds in a screw press of French Oil Mill Machinery Company has been done by Glenn et al. (1981). Results of those tests indicated oil yields are affected by preprocessing with highest yields being obtained with the driest seeds. Maximum oil yield was 38 percent during the test. One aspect of the test was that the unheated seed, although yielding less oil, gave a very bright clean looking oil. Comparison of its performance in a Ford 3000 diesel engine indicated decreased power output and fuel efficiency when using oil from cooked seed. This could mean less preprocessing of seed produces a better "fuel" oil. Other engine test results showed that using a 1:1 volumetric mixture of No. 1 diesel fuel and uncooked sunflower oil gave a 5 percent increase in maximum horsepower output and similar thermal efficiencies when compared to results using 100 percent No. 1 diesel fuel (Martin and Reilly). After one year ambient temperature storage in plastic containers, oil from uncooked seeds was mixed in a 1 to 1 volumetric ratio with No. 2 diesel fuel and performance wise compared in short term diesel engine tests with 100 percent No. 2 diesel fuel. An approximate 5 percent increase in maximum horsepower with similar thermal efficiencies was again observed for the sunflower oil-diesel fuel mixture when compared to the 100 percent No. 2 diesel fuel results. The one-year storage did not appear to adversely affect short-term engine tests results.

DISCUSSION OF RESULTS

This paper has looked at two important aspects of sunflowers for diesel fuel in the Eastern Corn Belt (Ohio).

1) Expected yields and cost of double-cropping sunflowers.

2) Processing needs and cost in Ohio for fuel self-sufficiency.

Results show favorable energy ratios for production and processing of sunflowers to oil. However, cost analysis indicates cost of seed production even for double-cropped sunflowers is high because of low yields. Comparing cost of production for a community cooperative plant with market prices received for unrefined food grade oil at $0.70 per kg, suggest the cooperative sized plant is marginally economical in years the crop is not needed for fuel. If such systems are to be practical, consistently high yields need to be achieved. This is one area where major research emphasis is needed in the future.

Results on processing of oil indicates that heating and lower moistures can improve overall yield. However, short-term engine tests indicate some unfavorable results when using oil from heated seeds.

REFERENCES

1. Carter, H. L. 1982. Ohio crop production annual summary, 1981. Published by the Ohio Crop Reporting Service, Columbus, Ohio, January.

2. Crane, D. W. 1982. Personal communication. Anderson International Corp., 6200 Harvard Ave., Cleveland, Ohio.

3. Deere & Company. 1980 Farm Management Notebook. Moline, Illinois.

4. Ensminger, M. E. and C. G. Olentine, Jr. 1978. Feeds and nutrition - complete. The Ensminger Publishing Co., Clovis, California. 1417 p.

5. Glenn, T. L., W. L. Roller, J. E. Henry and G. B. Triplett, Jr. 1981. Sunflower oil for fuel. Ohio Report 66(4): 63--July-August. Ohio Agricultural Research and Development Center, Wooster, Ohio.

6. Goering, C. E. and M. J. Daugherty. 1981. Energy inputs and outputs of eleven vegetable oil fuels. ASAE Paper No. 81-3586. American Society of Agricultural Engineers, St. Joseph, Michigan.

7. Helgeson, D. L., et al. 1977. The economic feasibility of establishing oil sunflower processing plants in North Dakota. Bulletin 503, North Dakota Agricultural Experiment Station, Fargo, North Dakota. 98 p.

8. Keener, H. M., T. L. Glenn and R. N. Misra. 1981. Minimizing fossil fuel energy in corn drying systems. TRANS. of the ASAE 24(5): 1357-1362, 1366.

9. Lockeretz, W. 1980. Energy inputs for nitrogen, phosphorus and potash fertilizers. Chapter in Handbook of Energy Utilization in Agriculture, CRC Press, Boca Raton, Florida.

10. Martin, D. J. and J. E. Reilly. 1981. Engine performance tests - sunflower oil fuel. Term project report for Agricultural Engineering Course Number 535, The Ohio State University, Columbus, Ohio. June.

11. Olsson, R. W. 1980. Personal communication. Anderson IBEC (Anderson International Corp., 6200 Harvard Ave., Cleveland, Ohio.

12. Pimmentel, D. 1980. Energy inputs for the production, formulation, packaging and transport of various pesticides. Chapter in Handbook of Energy Utilization in Agriculture, CRC Press, Boca Raton, Florida.

13. Pimmentel, D. 1980. Handbook of Energy Utilization in Agriculture. CRC Press, Boca Raton, Florida.

14. Schmidt, W. H. and M. E. Kroetz. 1979 (1980, 1982). Ohio sunflower variety performance trial 1979 (1980, 1981). Agronomy Series 223, Ohio Agricultural Research and Development Center and Ohio Cooperative Extension Service, Ohio State University, Columbus, Ohio.

15. Thow, R. L. 1981. Personal communication. The French Oil Mill Machinery Co., Piqua, Ohio.

16. Tyznik, W. J. 1975. Energy for horses. Paper presented at California Livestock Symposium. Dept. of Animal Science, Ohio Agricultural Research and Development Center, Ohio State University, Columbus, Ohio.

ANALYSIS OF VEGETABLE OIL PRODUCTION IN CENTRAL IOWA*

P. W. Claar II

T. S. Colvin
Member ASAE

S. J. Marley
Member ASAE

ABSTRACT

Vegetable oil can be used as an emergency substitute for diesel fuel for farming applications. This paper is an economic and energy analysis for vegetable oil production on a 180-ha (450-acre) central Iowa farm. The following data are presented as the basis for the economic analysis: (1) The yields of four varieties of sunflowers at three planting dates, (2) the measured sunflower harvesting losses--preharvest, header, threshing, and separating and cleaning for each variety and date of planting; and (3) the quantities of sunflower oil yielded from the pressing operation. Based on the data presented, it was concluded that even though a farmer could satisfactorily produce sunflowers, the on-farm processed sunflower oil does not compete with current diesel fuel prices. On-farm processed soybean oil has more potential as a substitute fuel from an economic standpoint in central Iowa.

INTRODUCTION

Current technology enables the American farmer to be highly efficient in the production of crops. Agricultural crop production is the business of using energy to capture energy in the form of food, feed, and fiber. For maximum economic returns and crop yields, the production operations--particularly planting and harvesting--must be performed at the proper time. Thus, dependable sources of energy must be available at critical times to keep production operations on track.

Crop production is highly dependent on petroleum-based fuels and chemicals. Crop production operations use an estimated 89% of the energy expenditures for production agriculture (11% for livestock production) (Machine Design Staff 1981). Due to dwindling petroleum reserves and increasing prices, production agriculture is expected to become nearly energy self-sufficient in the next decade.

The utilization of agricultural biomass is a means of making American production agriculture less dependent on fossil fuels. Farm crops and their residues are net energy producers--they produce five to ten times as much energy as is expended in growing the crops (Green 1975, and Nelson et al. 1975). Field crops and their residues, including oil seed crops, provide a renewable energy source that can be substituted for both liquid and solid fuels on the farm.

Sunflowers produce more oil per unit of land than other widely grown vegetable oil crops, such as soybeans. The agronomic aspects of sunflower production

The authors are: P. W. CLAAR II, Instructor; T. S. COLVIN, Agricultural Engineer, USDA-ARS; S. J. MARLEY, Professor, Iowa State University, Ames.

*Journal Paper No. J-10649 of the Iowa Agriculture and Home Economics Experiment Station, Ames, Iowa. Project No. 2460.

are discussed by Beard (1981). Research into alternative forms of energy that can be substituted for diesel fuel has indicated that vegetable oils compare satisfactorily with diesel fuel and do not significantly reduce engine power and performance (Quick 1980 and Bruwer et al. 1980).

The objectives of this paper are:

1. Determine the yields and harvesting losses of four sunflower varieties grown in central Iowa.
2. Determine the diesel fuel and vegetable oil requirements for conventional and no-till systems of corn, soybeans, winter wheat, and sunflower production.
3. Estimate the expected profit for several potential crop production systems on a 180-ha (450 acre) cash grain farm in central Iowa.
4. Estimate vegetable oil production costs using conventional and no-till crop production systems.

MEASUREMENT OF SUNFLOWER YIELDS AND HARVEST LOSSES ON A CENTRAL IOWA FARM

This study was conducted on two fields during the seasons of 1980 and 1981 to determine the yield and harvest losses of sunflowers grown in central Iowa.

The treatments and sunflower seed yields for 1980 and 1981 are shown in Table 1. In 1980 the experiment was designed as a randomized complete block with eight treatments and three replications. The treatments shown in Table 1 were four varieties and two planting dates. The first step was to measure the preharvest losses that were on the ground on each of the 24 plots. Each plot had an area of 0.19 ha (0.46 acre) [12 rows 0.76 m (30.0 in.) wide]. The next step was harvesting. During the harvesting process, material being discharged from the rear of the combine was caught to measure the threshing and separating losses, and the sunflower seed yield was weighed and recorded. The final step was to measure the header losses on each plot.

Table 1. Sunflower Seed Yields for 1980 and 1981 at Ames, Iowa

Planting date	Variety	Seed yield t/ha	Seed yield (ton/acre)
9 June 1980	A - 448 Sigco	3.52	(1.57)
	B - USDA 894	3.17	(1.38)
	C - 454 Sigco	3.24	(1.46)
	D - Jacques J-550	3.21	(1.43)
17 June 1980	A - 448 Sigco	4.02	(1.79)
	B - USDA 894	3.68	(1.64)
	C - 454 Sigco	3.45	(1.54)
	D - Jacques J-550	3.1	(1.38)
26 May 1981	D - Jacques J-550	2.47	(1.11)
7 July 1981[a]	D - Jacques J-550	1.29	(0.58)

[a]Planting date comparable to planting date for double-crop system

On each plot treatment in 1980, three random preharvest and header-loss samples were taken with a 0.23-m^2 (2.5-ft^2) [0.3 m X 0.76 m (12 in. X 30 in.)] metal frame across a row at an arbitrary point. The number of sunflower seeds inside the frame were counted, recorded, and later converted into a weight. The threshing and separating losses were determined by catching the material from a 13.94-m^2 (150.0-ft^2) area [3.05 m X 4.57 m (10 ft X 15 ft)] behind the combine on a canvas apron. The sunflower seeds that remained attached to the heads were counted as a threshing loss. The loose seeds remaining with the crop residue were counted as separating losses, which included the seeds that fell from the rear of the straw walkers and the cleaning shoe. After the combine passed over the randomly selected 13.94-m^2 area, the header losses were measured. Each sample was obtained by placing the metal frame across the next-to-the-outer rows at an arbitrary point and counting and recording the number of seeds inside.

In 1980 the sunflowers were harvested with a John Deere 6600 combine with a six-row 853A ROW-CROP head*. The sunflowers were harvested at approximately 12% MCWB. The combine was operated at 4.8 km/h (3.0 mph).

In 1981 the sunflower harvesting experiment was designed as a randomized block, with four treatments and three replications. The treatments were the two dates of planting listed in Table 1 and two types of combine headers for harvesting sunflowers--John Deere Six-row 853A ROW-CROP header* and the HEADSNATCHER* (General Dryer Corp., Clarkfield, MN).

The 1981 test procedure was the same as the 1980 test procedure, except that only preharvest and header losses were measured; and, the header losses were measured only for the second date of planting.

The mean total sunflower-seed yields for the 1980 and 1981 harvesting seasons are listed in Table 1, and the harvesting losses for 1980 are listed in Table 2. In 1980 the data tend to indicate that the total harvest losses are influenced by variety, even though there were not significant differences, statistically. In 1981 the mean loss for the John Deere header (data not shown) was 2.17%, and the mean loss for the HEADSNATCHER was 2.06%. The statistical analysis of header losses in 1981 indicated that the losses with the two types of combine header were not significantly different.

Table 2. Sunflower Harvest Losses for 1980 at Ames, Iowa

Planting date	Variety	Preharvest[a] loss	Header[a] loss	Threshing[a] loss	Separating[a] loss
		(%)	(%)	(%)	(%)
9 June 1980	A - 448 Sigco	2.66	5.99	0.12	1.94
	B - USDA 894	2.35	5.44	0.04	2.19
	C - 454 Sigco	2.64	5.13	0.04	1.58
	D - Jacques J-550	1.19	4.87	0.09	1.32
17 June 1980	A - 448 Sigco	1.75	8.07	0.25	1.80
	B - USDA 894	2.81	5.77	0.06	2.71
	C - 454 Sigco	3.10	3.20	0.16	1.44
	D - Jacques J-550	1.26	3.79	0.05	1.12

[a]John Deere 660 combine with 853A ROW-CROP header

QUANTITIES OF VEGETABLE OIL REQUIRED FOR CULTURAL PRACTICES

To determine the quantities of vegetable oil needed as liquid fuel for the field cultural operations, it was assumed that--

1. The central Iowa cash grain farm has 180 ha (450 acre) of land available for crop production.
2. Yields for the various potential crops--corn, soybeans, wheat, and sunflowers are given in Table 3.

*Reference to a company or product name is for specific information only and does not imply approval or recommendation of the product by Iowa State University or the U.S. Dept. of Agriculture to the exclusion of others that may be suitable.

Table 3. Crop Yields for Various Crop Rotation Systems in Central Iowa

Crop rotation system	Corn[a] t/ha (bu/acre)	Soybeans[a] t/ha (bu/acre)	Winter wheat[b] t/ha (bu/acre)	Sunflowers[c] t/ha (ton/acre)
Conventional system:				
W	8.7(139)	2.7(40)	3.70(55)	1.50(0.67)
X	8.7(139)	2.7(40)	--	--
Y	8.7(139)	--	--	3.00(1.34)
Z	8.5(136)	--	--	--
No-till system:				
W	8.5(136)	2.7(40)	3.70(55)	1.50(0.67)
X	8.5(136)	2.7(40)	--	--
Y	8.5(136)	--	--	3.00(1.34)
Z	7.8(124)	--	--	--

W--Soybeans - winter wheat + sunflowers - corn (4 crops in 3 years)
X--Corn - soybeans (2 crops in 2 years)
Y--Corn - sunflowers (2 crops in 2 years)
Z--Corn (continuous)

[a]D. Erbach. 1982. Tillage for continuous corn and corn-soybean rotation. TRANS. of the ASAE (in press).

[b]R. E. Atkins et al. 1978. Wheat and barley variety tests--1975-78. Iowa Coop. Ext. Serv. Pahmphlet AG-6-8.

[c]Adapted from Table 1.

3. Corn is harvested at an average of 24% moisture content-wet basis (MCWB) (Van Ee and Kline 1979).
4. Soybeans are harvested at an average of 13% MCWB. If additional moisture removal from soybeans is needed, an unheated air-drying system is used.
5. All of the agricultural equipment involved in the cultural practices for the crop and in collecting corncobs is powered by engines that burn liquid fuels (diesel--vegetable oils).
6. The vegetable oil has a heat value of 36270 kJ/liter (130000 Btu/gal) whereas the heat value of diesel fuel is 38180 kJ/liter (137000 Btu/gal).
7. Off-farm supplies made from petroleum--or from other energy sources such as fertilizer, pesticides, or electricity--is not included in the energy analysis (Fluck and Baird 1980).
8. Vegetable oils are not used for drying high-moisture corn.

The energy requirements for the conventional and no-till cultural systems were determined for corn, soybeans, winter wheat, and sunflowers.

The conventional system includes the following field operations:

1. Broadcasting the potassium and phosphorous fertilizer on all land.
2. Using a moldboard plow for tillage of all land.
3. Applying anhydrous ammonia fertilizer for corn, winter wheat, and sunflowers.
4. Using tandem disk secondary tillage with herbicide application--two passes.
5. Planting crop and applying insecticides.
6. Cultivating corn, soybeans, and sunflowers with sweep-type cultivator --one pass.
7. Harvesting with a combine.
8. Hauling grain to bin site.

9. Miscellaneous operations, such as traveling to and from the fields. The quantity of energy used for these operations is assumed to be 10% of the total sum of the energy required for the above items.

The no-till system includes all items for the conventional systems except for items 2 and 4. The herbicide is applied during planting.

The energy requirements for the conventional and no-till cultural systems (Table 4) were determined in terms of diesel fuel quantities needed per unit of land area (Ayres 1976).

With the previously stated assumptions, the total energy requirements may be determined for corn, soybeans, winter wheat, and sunflowers for these two tillage systems. The quantity of vegetable oil needed for the farming operations may be determined from the total energy requirements and converted to quantities of sunflower seeds or soybeans needed for blends or pure vegetable oil. The energy requirements needed are presented in Table 5.

From Table 5, with a 30% extraction rate for sunflowers and a 10% extraction rate for soybeans, it was calculated that--

1. With the conventional cultural systems and full-season sunflower production, a maximum of 38 t (42 ton) of sunflower seed or approximately 15% of the farm area is required to replace the total farm diesel fuel needs.
2. With the no-till cultural systems and full-season sunflower production, a maximum of 16 t (18 ton) of sunflower seed or approximately 6% of the farm area is required to replace the total farm diesel fuel needs.
3. With the conventional cultural systems, a maximum of 114 t (126 ton) of soybeans (approximately 24% of the farm area) is required to replace the total farm diesel fuel needs.
4. With the no-till cultural systems, a maximum of 48 t (54 ton) of soybeans (approximately 10% of the farm area) is required to replace the total farm diesel fuel needs.

ECONOMIC COSTS FOR CROP AND VEGETABLE OIL PRODUCTION

All assumptions, data, and costs that were used in the earlier parts of this paper were used for this economic analysis.

For this central Iowa farm, the cost of producing corn, soybeans, sunflowers, and winter wheat, as well as additional assumptions in the analysis, are presented in Table 6. It appeared that the choice of tillage system had only a small effect on the cost of producing corn, soybeans, winter wheat, and sunflowers. For this economic analysis, sunflowers were considered both as a single-crop and as a double-crop with winter wheat.

Based on the prices for the various commodities that were either bid or estimated for central Iowa on April 1, 1982 (Table 7), the only price that was higher than the cost of production was that for sunflowers. The sunflower price is most subject to question because few sunflowers are currently grown in central Iowa. Therefore, if managers are interested in producing sunflowers, they should investigate prices, production costs, and risks in detail to ensure a sound basis for a decision.

Linear programming models were used to compare the effects of various price and limitation situations, but their use provided no insights on crop mix not gainable through a close scrutiny of Table 7. However, the linear program proved useful for the examination of breakeven points for the substitution of fuels.

Table 4. Diesel Fuel Requirements for the Conventional and No-till Cultural Systems[a]

Operation	Corn: Conventional system liter/ha (gal/acre)	Corn: No-till system liter/ha (gal/acre)	Soybeans: Conventional system liter/ha (gal/acre)	Soybeans: No-till system liter/ha (gal/acre)	Winter wheat: Conventional system liter/ha (gal/acre)	Winter wheat: No-till system liter/ha (gal/acre)	Sunflowers: Conventional system liter/ha (gal/acre)	Sunflowers: No-till system liter/ha (gal/acre)
Broadcast dry fertilizer	1.40 (0.15)	1.40 (0.15)	1.40 (0.15)	1.40 (0.15)	1.40 (0.15)	1.40 (0.15)	1.40 (0.15)	1.40 (0.15)
Apply anhydrous ammonia	5.60 (0.60)	5.60 (0.60)	--	--	5.60 (0.60)	5.60 (0.60)	5.60 (0.60)	5.60 (0.60)
Moldboard plow	17.75 (1.90)	--	17.75 (1.90)	--	17.75 (1.90)	--	17.75 (1.90)	--
Tandem disk	13.10[b] (1.40)	--	13.10[b] (1.40)	--	13.10[b] (1.40)	--	13.10[b] (1.40)	--
Plant	5.60[c] (0.60)	6.10[d] (0.65)	5.60[c] (0.60)	6.10[d] (0.65)	5.60[c] (0.60)	6.10[d] (0.65)	5.60[c] (0.60)	6.10[d] (0.65)
Cultivate	4.20[e] (0.45)	3.75[f] (0.40)	4.20[e] (0.45)	3.75[f] (0.40)	4.20[e] (0.45)	3.75[f] (0.40)	4.20[e] (0.45)	3.75[f] (0.40)
Combine harvest	14.95 (1.60)	14.95 (1.60)	10.30 (1.10)	10.30 (1.10)	10.30 (1.10)	10.30 (1.10)	10.30 (1.10)	10.30 (1.10)
Haul	1.90[g] (0.20)	1.90[g] (0.20)	0.75[g] (0.08)	0.75[g] (0.08)	0.75[g] (0.08)	0.75[g] (0.08)	0.75[g] (0.08)	0.75[g] (0.08)
Miscellaneous	6.55[h] (0.70)	3.25[h] (0.35)	5.30[h] (0.57)	2.25[h] (0.24)	5.85[h] (0.62)	2.80[h] (0.30)	5.85[h] (0.62)	2.80[h] (0.30)
Total	71.05 (7.60)	36.95 (3.95)	58.40 (6.25)	24.55 (2.62)	64.55 (6.90)	30.70 (3.28)	64.55 (6.90)	30.70 (3.28)

[a] G. Ayres. 1976. Fuel required for field operations.
[b] Two passes with a tandem disk and herbicide application during the second pass.
[c] Planter with fertilizer and pesticide attachments, tilled seedbed.
[d] No-till planter with fertilizer and pesticide attachments.
[e] Sweep-type cultivator.
[f] Rolling-type cultivator.
[g] Field plus 0.8 km (0.5 mi) on gravelled road.
[h] Travel to and from the fields, etc.

Table 5. Vegetable Oil Requirements for Conventional and Cultural Systems

Fuel	Corn		Soybeans		Winter wheat		Sunflowers	
	Conventional system liter/ha (gal/acre)	No-till system liter/ha (gal/acre)	Conventional system liter/ha (gal/acre)	No-till system liter/ha (gal/acre)	Conventional system liter/ha (gal/acre)	No-till system liter/ha (gal/acre)	Conventional system liter/ha (gal/acre)	No-till system liter/ha (gal/acre)
100% diesel	--	--	--	--	--	--	--	--
75% diesel 25% vegetable oil	18.65 (2.00)	9.70 (1.05)	15.35 (1.65)	6.45 (0.70)	16.95 (1.80)	8.05 (0.85)	16.95 (1.80)	8.05 (0.85)
50% diesel 50% vegetable oil	37.30 (4.00)	19.40 (2.10)	30.70 (3.30)	12.90 (1.40)	33.90 (3.60)	16.10 (1.70)	33.90 (3.60)	16.10 (1.70)
100% vegetable oil	74.60 (8.00)	38.80 (4.20)	61.40 (6.60)	25.80 (2.80)	67.80 (7.20)	32.20 (3.40)	67.80 (7.20)	32.20 (3.40)

Assumptions: (1) Diesel fuel--38180 kJ/liter (137000 Btu/gal)
(2) Vegetable oil--36270 kJ/liter (130000 Btu/gal)
(3) Vegetable oil weight--0.923 kg/liter (7.70 lb/gal)
(4) Oil extraction rate--30% (30 kg-oil/100 kg-seed) for sunflowers and 10% for soybeans

Table 6. Estimated Total Costs of Crop Production[a]

Item	Corn		Soybeans		Winter wheat[b]		Sunflowers[c]	
	Conventional system $/ha ($/acre)	No-till system $/ha ($/acre)	Conventional system $/ha ($/acre)	No-till system $/ha ($/acre)	Conventional system $/ha ($/acre)	No-till system $/ha ($/acre)	Conventional system $/ha ($/acre)	No-till system $/ha ($/acre)
Growing	86.25	48.20	72.90	34.85	86.75	47.45	86.25	48.20
	(34.90)	(19.50)	(29.50)	(14.10)	(35.10)	(19.20)	(34.90)	(19.50)
Harvesting	145.70	145.70	52.25	52.25	104.90[g]	104.90[g]	52.25	52.25
	(58.95)	(58.95)	(21.15)	(21.15)	(42.45)	(42.45)	(21.15)	(21.15)
Seed, chemicals, etc.	291.90	316.60	162.40	187.10	97.50	122.20	139.15	163.85
	(118.10)	(128.10)	(65.70)	(75.70)	(39.45)	(49.45)	(56.30)	(66.30)
Farm operating[d] overhead	59.80	51.05	36.20	34.85	36.35	34.85	35.20	33.85
and insurance	(24.20)	(23.65)	(14.65)	(14.10)	(14.70)	(14.10)	(14.25)	(13.70)
Labor[e]	42.40	31.75	29.65	19.05	52.75	40.40	29.65	19.05
	(17.15)	(12.85)	(12.00)	(7.70)	(21.35)	(16.35)	(12.00)	(7.70)
Subtotal of nonland	625.90	600.70	353.40	328.10	378.25	349.85	342.55	317.20
costs	(253.25)	(243.05)	(143.00)	(132.75)	(153.05)	(141.55)	(138.60)	(128.35)
Land charge	308.95	308.95	308.95	308.95	185.35	185.35	308.95[f]	308.95[f]
	(125.00)	(125.00)	(125.00)	(125.00)	(75.00)	(75.00)	(125.00)	(125.00)
Total cost	934.85	909.65	662.35	637.00	563.65	535.20	651.50	626.15
Unit-of-land basis	(378.25)	(368.05)	(268.00)	(257.75)	(228.05)	(216.55)	(263.60)	(253.35)
Unit-of-mass basis	10.75	10.70	24.55	23.60	15.25	14.45	21.70	20.90
$/100 kg ($/bu)	(2.70)	(2.70)	(6.70)	(6.45)	(4.15)	(3.95)	(9.85)[h]	(9.45)[h]

[a]W. Edwards and H. Thompson. 1982. Estimated costs of crop production in Iowa--1982.

[b]Assumed to be equivalent to oats production.

[c]Assumed to be equivalent to soybean production.

[d]Overhead costs are 10% of the growing and harvesting costs plus seed, chemicals, etc., and the insurance cost is $3/acre.

[e]Labor costs @ $6/hr.

[f]For the double crop system of winter wheat and sunflowers, the land charge is $123.55/ha ($50/acre). The total cost of production on a unit-of-land basis is $436.40/ha ($176.60/acre) for the conventional system and $406.10/ha ($164.35/acre) for the no-till system.

[g]Includes drying the grain.

[h]Sunflowers $/100 kg ($/100 lb)

Table 7. Profit for Crop Production Systems

System	Cost/unit	Price/unit	Yield	Profit
	($/100 kg)	($/100 kg)	(t/ha)	($/ha)
Corn--Conventional				
W	10.75	9.45[a]	8.7	-113.10
X	10.75	9.45[a]	8.7	-113.10
Y	10.75	9.45[a]	8.7	-113.10
Z	10.70	9.45[a]	8.5	-106.25
Corn--No-till				
W	10.70	9.45[a]	8.5	-106.25
X	10.70	9.45[a]	8.5	-106.25
Y	10.70	9.45[a]	8.5	-106.25
Z	10.70	9.45[a]	7.8	-97.50
Soybeans--Conventional				
W	24.55	22.05[b]	2.7	-67.50
X	24.55	22.05[b]	2.7	-67.50
Soybeans--No-till				
W	23.60	22.05[b]	2.7	-41.85
X	23.60	22.05[b]	2.7	-41.85
Wheat--Conventional				
W	15.25	12.75[c]	3.7	-92.50
Wheat--No-till				
W	14.45	12.75[c]	3.7	-62.90
Sunflowers--Conventional				
W	21.70	22.00	1.5	0.45
Y	21.70	22.00	3.0	0.90
Sunflowers--No-till				
W	20.90	22.00	1.5	0.45
Y	20.90	22.00	3.0	0.90

W--Soybeans - winter wheat + sunflowers - corn (4 crops in 3 years)
X--Corn - soybeans (2 crops in 2 years)
Y--Corn - sunflowers (2 crops in 2 years)
Z--Corn (continuous)

[a]If the normal price/unit for corn is $10.85/100 kg ($3/bu), the profit for the conventional and no-till systems is $8.7/ha and $12.75/ha, respectively.

[b]If the normal price/unit for soybeans is $25.70/100 kg ($7/bu), the profit for the conventional no-till systems is $31.05/ha and $56.70/ha, respectively.

[c]If the normal price/unit for winter wheat is $14.70/100 kg ($4/bu), the profit for the conventional and no-till systems is -$20.35/ha and $9.25/ha, respectively.

If the sunflower model is correct, the most profitable (least loss) alternative would be the corn-sunflower rotation (system Y) with no seedbed preparation before planting. When sunflowers were limited to 10% of the land area, the next most profitable system which was used on the remaining 90% of the land area was the corn-soybean rotation (system X). The continuous corn system (system Z) did not compete with systems Y or X because of lower yields even though a charge for extra nitrogen fertilizer (compared to system X) was not included. System W with four crops in three years, did not compete with systems X or Y because of the severe drop in yield for the double-crop sunflowers and the lack of profitability of the wheat even at the higher prices.

Table 8 presents the cost of producing on-farm-pressed oil. Smith et al. (1982) determined that it cost approximately $0.35/liter ($1.25/gal) to squeeze and filter sunflower oil (sunoil). The same cost was used for soybean oil (soyoil). When prices for the base commodities, soybeans, and sunflower seeds are depressed, the real cost of the oil may be somewhat lower because the raw material (seed crop) cost would be the selling price rather than the production cost. For example, if soybeans were selling for $22.05/100 kg ($6/bu), the cost of on-farm-pressed soyoil (reduction in revenue) to the farm would be the potential selling price of the soybeans required for one liter [$2.05] plus the press charge [$0.35/liter] minus any credit for soybean meal or $2.40/liter for soyoil with no credit for soybean meal, or $0.35/liter for soyoil with a $2.05/liter credit for soybean meal. The meal would be fed to livestock on the farm, to replace purchased protein, or sold on the market. The cost of diesel fuel was $0.26/liter ($1/gal) in early 1982.

The current market price for soyoil is roughly one-half that for sunflower oil; therefore, purchased oil would be soyoil rather than sunoil unless technical difficulties greatly increased the cost of using soyoil compared to sunoil.

Table 8. Input Costs for Farm-Produced Vegetable Oils

System	Unit	Cost/unit	Pressing costs	Cost/unit without meal credit	Cost/unit with meal credit
	(100 kg)	($/100 kg)	($/liter)	($/liter)	($/liter)
Soybeans--conventional	0.09	24.55	0.35	2.55	0.50
Soybeans--no-till	0.09	23.60	0.35	2.45	0.40
Sunflowers--conventional	0.03	21.70	0.35	1.00	0.65
Sunflowers--no-till	0.03	20.90	0.35	0.98	0.63

CONCLUSIONS

1. Based on two years of data, the average yield for full-season sunflowers was 3.3 t/ha (1.48 ton/acre).
2. Sunflower harvest losses can be held under 5% with currently available headers and attachments.
3. For the conventional crop production systems (four crops), an average of 64.55 liters/ha (6.9 gal/acre) of diesel fuel or 67.8 liters/ha (7.2 gal/acre) of vegetable oil was required for field operations.
4. For the no-till crop production systems (four crops) an average of 30.7 liters/ha (3.3 gal/acre) of diesel fuel or 32.2 liters/ha (3.4 gal/acre) was required for field operations.

5. At the depressed 1982 prices, full-season sunflowers would have been the most profitable cash crop in central Iowa.
6. From a cost of production standpoint, sunoil will have difficulty competing with soyoil as a home pressed fuel in central Iowa if there is a use for the soybean meal on the farm.

REFERENCES

1. Atkins, R.E., L.W. Moeller and M.C. Lenz. 1978. Wheat and barley variety tests--1975-78. Iowa Coop. Ext. Serv. Publ. AG-6-8.

2. Ayres, G.E. 1976. Fuel required for field operations. Iowa Coop. Ext. Serv. Pamphlet Pm-709.

3. Beard, B.H. 1981. The sunflower crop. Scientific American 244(5): 150-161.

4. Bruwer, J.J., B.V.D. Boshoff, F.J.C. Hugo, J. Fuls, C. Hawkins, A.N.v.d. Walt, A. Engelbrecht and L.M. du Plessis. 1980. Diesol--An alternative fuel for compression ignition engines. Agricultural Energy--Biomass Energy and Crop Production. ASAE National Energy Symposium 2: 385-390.

5. Edwards, W.E. and H.G. Thompson. 1982. Estimated costs of crop production in central Iowa--1982. Iowa Crop Ext. Serv. Publ. Fm-1712.

6. Erbach, D. 1982. Tillage for continuous corn and corn-soybean rotation. TRANS. of the ASAE (in press).

7. Fluck, R.C. and C.D. Baird. 1980. Agricultural Energetics. AVI Publishing Co., Inc.

8. Green, F.L. 1975. Energy potential from agricultural field residues. General Motors Corp. Manufacturing Development Paper No. MD75-021. Warren, MI.

9. Machine Design Staff. 1981. Food production: Going for a low-energy diet. Machine Design 53(9): 44-48.

10. Nelson, L.F., W.C. Burrows and F.C. Stickler. 1975. Recognizing productive energy-efficient agriculture in the complex U.S. food system. ASAE Paper No. 75-7505. St. Joseph, MI.

11. Quick, G.R. 1980. Farm fuel alternatives. Power Farming Magazine (2): 10-17.

12. Smith, D.R., W.G. Lovely and W.F. Buchele. 1982. Expelling and combustion of the sunflower oil. ASAE Paper No. MC82-146. St. Joseph, MI.

13. Van Ee, G.R. and G.L. Kline. 1979. Management strategies for corn production and low-temperature drying. ASAE Paper 79-3032. St. Joseph, MI.

BUFFALO GOURD: POTENTIAL AS A FUEL RESOURCE ON SEMI-ARID LANDS

P. Gilmer Young, Robert P. Morgan and
Eugene B. Shultz, Jr.

ABSTRACT

Buffalo gourd, (_Cucurbita foetidissima_), is a wild, hot-dry-land plant native to the semi-arid regions of North America. Its triglyceride oil and fermentable starch make it a potential biomass energy source. These products, along with the seed meal and foliage, also offer the potential for cultivation in semi-arid regions of the developing world as a food and feed source. Alternatively, the plant may help to maintain economic vitality in regions such as the Texas High Plains, where declining water supplies threaten present irrigation practices. Technical feasibility, impacts, commercialization requirements, and research needs are discussed.

CHARACTERISTICS OF THE BUFFALO GOURD

Plant Morphology

A member of the pumpkin and squash genus, _Cucurbita foetidissima_ produces a round, hard-shelled fruit, or pepo, approximately 5-7 cm. in diameter. Single, mature plants are capable of producing 200 or more pepos in a single growing season. The seed, which accounts for one-third of the weight of the pepo, contains 30-40% oil and 30-35% protein. Seed production ranges from 100-200 seeds per fruit, with an average weight of 4 g. per 100 seed, in recent tests at the University of Arizona (Bemis, et al., 1979).

The buffalo gourd is perennial due to its large, starch-containing storage roots. Four-year-old roots may weigh over 40 kg. (fresh root basis). Although the vines are killed by temperatures below 0°C, the hardier roots may survive temperatures as low as -25°C, particularly if there is an insulating snow cover (Bemis, 1979). Most reproduction is asexual, resulting from adventitious rooting from vine nodes. Large, homogeneous colonies may develop from single plantings. Vine growth is prodigious; one specimen produced 220 meters of vine, with an estimated 15,000 leaves, in a five-month growing season (Dittmer and Talley, 1964). Most plants are monoecious. Pollination in the wild is accomplished by the squash and gourd bees, _Peponapis_ and _Xenoglassa_ (Bemis, 1979). Large genetic variations exist among varieties from diverse stands, indicating great potential for selecting desired characteristics through hybridization (Scheerens, et al., 1978).

Virtually all of the seed oil is found in the embryo, which makes up approximately two-thirds of the dry weight of the seed. Extraction is possible with a mechanical press or solvent extraction, and with or without the

---

The authors are: P. Gilmer Young, Graduate Research Assistant, Robert P. Morgan, Professor and Chairman, and Eugene B. Shultz, Jr., Professor, Dept. of Technology and Human Affairs, School of Engineering, Washington University, St. Louis, Missouri, 63130.

removal of the seed coat. The fatty acid distribution of buffalo gourd oil (BGO) shows some similarities to those of soybean oil (SBO) and high-linoleic sunflower oil (SFO):

	Fatty acid, percent			
	Sat'd	18:1	18:2	18:3
Buffalo gourd oil (BGO)*	14	27	59	-
Soybean oil (SBO)**	14	23	51	7
Sunflower oil (SFO)**	10	20	66	-

(18:1 = oleic, 18:2 = linoleic, 18:3 = linolenic acids)

*Scheerens, 1978

**Pryde, April 1981.

Energy Potential

The Iodine Value of buffalo gourd oil has been measured as 129.9 (Vasconcellos, 1980) which is in the typical range for soybean oil and at the low end of the typical range for high-linoleic sunflower oil (Pryde, April 1981). The concentration of phosphatides, 0.8% (Vasconcellos, 1980), is similar to that of sunflower oil and substantially lower than that of soybean oil (Pryde, April 1981). These characteristics suggest the potential for utilization of buffalo gourd oil as a diesel fuel substitute or extender. Pryde (October, 1981) has suggested that Iodine Values should be in the range of 80 to 145. As yet, we are not aware of long-term tests of buffalo gourd oil in any type of diesel engine. When these tests are carried out, we would anticipate results similar to those with sunflower oil and degummed soybean oil.

Limited studies indicate that a yield of 2000 kg/ha of buffalo gourd seed per year is possible (Bemis, 1982, Curtis, 1974, and Shahani, 1951). Assuming an oil content of 33% (Scheerens, 1978), oil yield would be 660 kg/ha. This suggests a potential yield of energy of about 10 to 11 million Btu/acre-yr as vegetable oil. By contrast, U.S. sunflower oil yields per acre have averaged 525 pounds (Pryde, April 1981), about 9 to 10MBtu/acre-yr.

A second energy source is ethanol fermented from the starch of the buffalo gourd root. Starch makes up 64.6% of the dry root weight (Berry, et al., 1976). A fermentation and distillation test using buffalo gourd starch gave promising results (Gathman, 1979). Expected yields of starch are 6750 kg/ha-yr (Bemis, 1979). Assuming a typical yield of one liter of ethanol per 1.8 kg of starch (Fuel from Farms, 1980), this would result in 3750 liters of ethanol per hectare-year, representing a potential energy yield of about 34 million Btu/acre-yr in the form of ethanol.

Food and Feed Source

Buffalo gourd seed oil and root starch are both edible and have potential as food sources. The seed meal contains a high-quality protein similar to those in soybean, cottonseed, and sunflower meal, and may be used as a food or feed source. The high fiber content restricts the usage of the seed hull to ruminants. The fruit pulp and vines may have potential as forage (Bemis, 1979), but the vines may have to be returned to the soil to control wind erosion. Ruminants appear to tolerate cucurbitacins that are present.

Cultivation

Buffalo gourd is highly drought-tolerant and has very low water requirements compared to most commercial crops (Young, 1982). Preliminary studies indicate that a minimum of 250 mm annually is required to produce a crop. Plants can survive temperatures of 40°C or higher on as little as 150 mm, but would rarely flower and produce fruit under these conditions (Vasconcellos, 1980). Yields would increase with more than 250 mm of water, but extended periods of excessive moisture causes root rot. Plants in test plots were unresponsive to nitrogen fertilizer, possibly because the extensive root system was able to extract sufficient amounts from the soil depths. Long-term depletion of subsurface nutrients may require eventual usage of supplements (Bemis, 1982). Buffalo gourd is immune to cotton root fungus, and is highly resistant to cucumber beetles, which plague other cucurbits (Shahani, 1951).

Bemis (1978) and co-workers have proposed and are testing a novel cropping system to optimize root starch yields, and to prevent overcrowding of the plantation. Because of reproduction from the extensive and numerous runners, development of roots could severely crowd the plantation if roots are not harvested periodically. The concept calls for the digging of alternate one-meter-wide swaths annually, to remove all roots in the swath. The next year, the harvested swath is regenerated by asexual rooting of the vines that creep over from the unharvested swath. Although root harvesting in the fall will disturb the soil, raising the possibility of wind erosion of the soil during the winter fallow, we think it should be possible to use vine residues to cover the swaths that have been dug so that the effects of sustained high winds in such places as the Texas High Plains might be mitigated.

Comparison with *Euphorbia lathyris*

The vine growth of the buffalo gourd is extensive, providing a dense mat of vegetation to protect underlying soil. Harvest of the oil-bearing pepos and starchy roots does not require the removal of these vines. This gives the gourd a distinct advantage over another proposed energy crop, *Euphorbia lathyris* (gopher plant), whose oil is distributed throughout the whole plant. The entire *Euphorbia lathyris* plant must be removed from the fields and transported to a centralized processing plant: the residue, which makes up the bulk of the plant's mass, is unlikely to be transported back to the fields in order to restore the lost nutrients and provide erosion protection. By remaining in the fields, buffalo gourd vines can maintain soil fertility and tilth, and reduce erosion damage.

POSSIBLE APPLICATIONS IN THE THIRD WORLD

As a source of oil, starch, and seed meal, buffalo gourd may be a valuable crop for semi-arid regions of the developing world. Many developing countries are entirely dependent on foreign sources of liquid fuels. At the same time, firewood, their primary energy source, is being rapidly depleted. Buffalo gourd offers two potential energy sources, oil and alcohol, which are renewable and can be developed locally. As a nutrition source, buffalo gourd can supplement diets with highly needed oil and protein (Curtis, 1946). Field studies in Lebanon in the early 1970's gave promising results (Curtis, 1974), but we know of no other studies that have been carried out in developing nations.

THE POTENTIAL FOR BUFFALO GOURD IN THE TEXAS HIGH PLAINS

The Texas High Plains makes up the eastern two-thirds of the Llano Estacado, or Staked Plains, which stretches across the eastern part of New Mexico and much of the Texas Panhandle. Although its climate is semi-arid, the region is a major agricultural production center, thanks to its fertile soil and underground water resource, the Ogallala aquifer. A vestige of the wet Pleistocene era, the Ogallala is a vast underground reservoir, reaching from Texas to the South Dakota border. Recharge to the aquifer, however, is limited to the amount of rainfall which percolates down to the water table, estimated to be about two cm per year (Bell and Morrison, 1978). Throughout most of the Texas High Plains, water is being pumped out at a much faster rate than this. Studies indicate that the water table is receding at an average rate of 60 cm/yr (Sweazy, 1978). The uneven distribution of water within the aquifer gives some farmers more water than others, but most experts agree that much of the aquifer in the Texas Panhandle will be totally exhausted in the next 30-50 years. One study found that sixteen of forty-five High Plains counties would have no significant water reserves after 2031 (Sweazy, 1978).

Effects of Rising Energy Costs

The cost of natural gas, the primary fuel used to pump irrigation water, has risen from $14.13/1000m^3 in 1974 (Sweazy, 1978) to $155.03/1000m^3 in 1981 (Harmon, 1982). Similar increases have occurred in other energy and energy-related inputs, including diesel fuel, herbicides, and fertilizer. As a result, even farmers with large amounts of water under their land often cannot afford to pump it up to their crops.

Such problems can be devastating to the economy of a region heavily dependent on the farming industry for jobs. The Texas High Plains is the site of the world's largest cottonseed processing center (in Lubbock) and one of the world's greatest concentrations of cattle feedlots. In addition to these areas, jobs in meat packing, transportation, cotton ginning, and sales of farm equipment, seed, fertilizer, and herbicide are closely tied to the success of farming in the High Plains.

Currently, the major crops of the Texas High Plains are cotton, grain sorghum, corn, and wheat. All require irrigation. Lesser amounts of alfalfa, soybean, sunflower, and other crops are cultivated as well. Much of this agricultural production goes to feed cattle at local feedlots. Two and one-half million cattle were marketed in 1975, with a value of $1.1 billion (Weber and Watson, 1976).

The Rationale for Considering Buffalo Gourd

Cultivation of buffalo gourd appears to be suited to the climate, soil, and economic character of the Texas High Plains. Buffalo gourd products are compatible with present agricultural markets in that region. Other than cotton, the region's major products are vegetable oil, cattle feed, and starch, which are also the primary products from buffalo gourd. Therefore, buffalo gourd might potentially replace conventional sources without major disruptions in the existing market structure (Young, 1982).

Our preliminary study of rainfall history in the region suggests that buffalo gourd might produce economic yields without irrigation, relying only upon annual rainfall for needed moisture. Even if irrigation has to be utilized, the plant would require smaller amounts than conventional crops. This potential could significantly reduce the rate of depletion of the Ogallala aquifer and ultimately preserve the economic vitality of the region (Young, 1982).

As a source of oil and ruminant feed, buffalo gourd would help to sustain the important oil and feedlot industries in the area. Conservation of water would also help to preserve municipal water supplies and allow industrial water users, such as secondary oil recovery operations, to continue. Even those economic sectors which are not directly dependent on the Ogallala aquifer, such as service industries, would benefit from the continuation of a stable economic environment (Young, 1982).

REQUIREMENTS FOR COMMERCIALIZATION

Buffalo gourd was originally suggested as a potential food source by Dr. L. C. Curtis (1946). Much of the present research is being conducted by the University of Arizona Department of Plant Sciences and Department of Nutrition and Food Science. Research began there in the early 1960's (Bemis and Nelson, 1963). Current work includes hybrid development, nutritional research, seed production, and analysis of cultivation methods (Bemis, 1979, 1982). The Center for Development Technology at Washington University in St. Louis is analyzing buffalo gourd as one of many potential oilseed crops for multiple uses in the U.S. (Morgan and Shultz, 1981 and Shultz, et al., 1982).

Nevertheless, much more work is necessary before buffalo gourd can be cultivated and marketed. The first task is to test the oil as a substitute for diesel fuel, and to analyze potential processing methods. Studies are needed to develop fermentation methods and examine the economics of buffalo gourd root starch as a feedstock for the production of ethanol. Feeding studies on laboratory animals are necessary to determine the nutritional value of the components of the plant. Estimations of potential yields of oil, starch, and seed meal are required, along with examination of irrigation, fertilizer, and herbicide use.

Although some promising buffalo gourd field tests were carried out by Shahani (1951) some years ago on the nearby Rolling Red Plains (Chillicothe, Texas), testing in several locations directly within the Texas High Plains should begin, to confirm the anticipated yields from the University of Arizona cropping system. Specific local markets should be identified, and the costs of production and economic value of buffalo gourd products determined.

Additional investigations are needed to study the potential environmental impacts of buffalo gourd production. The primary impact is the extent to which the rate of depletion of the Ogallala aquifer can be reduced. The effect of buffalo gourd production on soil erosion should also be analyzed. Of particular importance is the impact of production on marginal lands, where the soil is most vulnerable to improper farming practices.

Government Incentives

The buffalo gourd is one of four arid-land plants that would receive substantial government support under the Arid Lands Renewable Agricultural Resources Corporation Act of 1981, sponsored in the U.S. House of Representatives by George E. Brown, Jr., (D-Calif.) and and Kika de la Garza (D-Tex.). The Senate version is sponsored by Senators DeConcini and Goldwater of Arizona (Arid Lands Act, 1981). If this proposed legislation becomes law, price guarantees, loan guarantees, direct loans and joint ventures would be made available to encourage the production of buffalo gourd along with guayule, jojoba, and *Euphorbia lathyris*. The object is to secure the domestic availability of these renewable resources of fuels and chemical feedstocks, in the interest of national security. Although the proposed legislation in its present form emphasizes non-fuel uses, it might be modified to provide substantial support for fuel uses, such as diesel fuel substitute or extender and root starch alcohol from buffalo gourd.

The evolution of a buffalo gourd market is inhibited by a problem which is common to all new crops. Farmers will not grow buffalo gourd until they are assured of a demand for its products, and no market will develop until there is a sufficiently large and stable supply of crops to purchase. The Arid Lands Act, which is patterned after the Synthetic Fuels Corporation Act of 1980, would hope to break this impasse by providing financial incentives to both farmers and processors to demonstrate the feasibility of buffalo gourd production, encourage farmers to begin cultivation, provide necessary funds for the construction of processing plants, create a supply of raw materials sufficiently large to establish a market for buffalo gourd products, and, it is hoped, "foster greater energy security and reduce the Nation's economic vulnerability to disruptions in imported energy." (Arid Lands Act, 1981). The Arid Lands Act is still under consideration in Congressional committees. Final action will probably not be taken until 1983.

CONCLUSIONS

Buffalo gourd is a potential source of triglyceride seed oil, root starch, and high-protein seed meal for semi-arid regions throughout the world. Cultivation in the Texas High Plains may help to reduce the rate of depletion of finite water supplies, and thus preserve the economic vitality of the region. The Iodine Value and fatty acid distribution of buffalo gourd are similar to those of soybean oil and sunflower oil, and the phosphatide content is low. These characteristics suggest that buffalo gourd is likely to exhibit diesel behavior similiar to soybean oil and sunflower oil. Oil yields per acre are expected to be higher than those from soybean and sunflower, in addition to a substantial yield of fermentable root starch. Although much is now known about the buffalo gourd, mainly due to the efforts of the University of Arizona research team headed by W. P. Bemis, much additional applied work is needed in the context of the Texas High Plains. There is a possibility that this work might receive Federal support if the Arid Lands Act or its successor becomes law.

REFERENCES

1. Arid Lands Renewable Resources Corporation Act of 1981. July 10, 1981. Congressional Record-Senate. S7482-7498.

2. Bell, A. E. and S. Morrison. June 1978. Analytical study of the Ogallala aquifer in Lubbock County, Texas. Report No. 216. Texas Department of Water Resources, Austin, TX. 63 p.

3. Bemis, W. P. and J. M. Nelson. 1963. Interspecific hybridization within the genus Cucurbita.I: Fruit set, seed, and embryo development. J. of the Arizona Academy of Science 2:104-107.

4. Bemis, W. P., J. W. Berry, C. W. Weber and T. W. Whitaker. 1978. The Buffalo Gourd: a new potential horticultural crop. HortScience 13(3):235-40.

5. Bemis, W. P., J. W. Berry and C. W. Weber. 1979. The Buffalo Gourd, a potential arid land crop. New Agricultural Crops, edited by G. A. Ritchie. Westview Press, Boulder, Colorado.

6. Bemis, W. P. 1982. University of Arizona, Dept. of Plant Sciences. Personal communication.

7. Berry, J. W., C. W. Weber, M. L. Dreher and W. P. Bemis. 1976. Chemical composition of Buffalo Gourd, a potential food source. J. of Food Sci. 41:465-66.

8. Curtis, L. C. 1974. An attempt to domesticate a wild, perennial, xerophytic gourd, Cucurbita foetidissima. Progress Reports I-IV. The Ford Foundation, New York, N.Y.

9. Curtis, L. C. 1946. The possibilities of using species of perennial

cucurbits as sources of vegetable fats and protein. Chem. Dig. 5:221, 223-224.

10. Dittmer, H. J. and B. D. Talley. 1964. Gross morphology of tap roots of desert cucurbits. Bot. Gaz. 125:121-126.

11. Fuel from Farms: A guide to small-scale ethanol production. February 1980. Solar Energy Research Institute, Report No. SERI/SP-451-519.

12. Gathman, A.,D. Grygiel, A. Kartman, N. MacDonald, D. Marburger and J. W. Berry. 1979. A solar energy fermentation process for fuel alcohol production from xerophytic starches. Report to the National Science Foundation. 28 p.

13. Harmon, W. March 15, 1982. Agricultural economist, Texas Agricultural Experiment Station, Lubbock, Texas. Telephone interview.

14. Morgan, Robert P. and Eugene B. Shultz, Jr. September 7, 1981. Fuels and chemicals from novel seed oils. Chem. and Engineering News 59: 69-77.

15. Pryde, E. H. April 1981. Vegetable oil vs. diesel fuel: chemistry and availability of vegetable oils. Presented at the Workshops on Alcohol and Vegetable Oil as Alternative Fuels, U.S. Dept. of Agriculture, Northern Regional Research Center, Peoria, Illinois. Pages 217-232 of the proceedings.

16. Pryde, E. H. October 1981. Vegetable oil fuel standards. Appendix 26 of proceedings, Vegetable Oil as Diesel Fuel Seminar II, U.S. Dept. of Agriculture, Northern Regional Research Center, Peoria, Illinois.

17. Scheerens, J. C., W. P. Bemis, M. L. Dreher and J. W. Berry. 1978. Phenotypic variation in fruit and seed characteristics of buffalo gourd. J. of the Am. Oil Chem. Soc. 55:523-525.

18. Shahani, H. S., F. G. Dollear, K. S. Markley and J. R. Quinby. 1951. The Buffalo Gourd, a potential oilseed crop of the southwestern drylands. J. of the Am. Oil Chem. Soc. 28:90-95.

19. Shultz, Eugene B., Jr., W. P. Darby, Harold M. Draper, III, and Robert P. Morgan. January, 1982. Widespread utilization of novel oilseeds on marginal lands: potential benefits and risks. Presented at the annual meeting of the American Association for the Advancement of Science, Washington, D.C.

20. Sweazy, R. M. 1978. Groundwater resources of the Texas High Plains: State-of-the-art and future research needs. Water Resources Center, Texas Tech University. Lubbock, Texas. 97 p.

21. Vasconcellos, J. A., W. P. Bemis, J. W. Berry and C. W. Weber. July 1980. The Buffalo Gourd, _Cucurbita foetidissima_ HBK, as a source of edible oil. Report No. 3343, Arizona Agricultural Experiment Station, Tucson, Arizona. 26 p.

22. Weber, E. G. and K. B.Watson. October 1976. Fact sheet. Water, Inc. Lubbock, Texas. 2 p.

23. Young, P. G. 1982. The potential for the development of Buffalo Gourd production in the High Plains of Texas. M.A. Thesis. Dept. of Technology and Human Affairs, Washington University, St. Louis, Missouri.

CHARACTERIZATION OF VEGETABLE OILS FOR USE AS FUELS IN DIESEL ENGINES

Thomas W. Ryan, III
Senior Research Engineer

Timothy J. Callahan
Research Engineer

Lee G. Dodge
Research Engineer

Southwest Research Institute
San Antonio, Texas

ABSTRACT

The current specifications for petroleum fuels have evolved over the history of the petroleum industry and the development of the internal combustion engine. Present day fuel specifications are based on a wealth of empirical data and practical experience. A similar data base is only now being developed for the specification of vegetable oil fuels for diesel engines. Four different types of vegetable oil (soy, sunflower, cottonseed and peanut) have been obtained, each in at least three different stages of processing. All of the oils (14) have been characterized with respect to their physical and chemical properties. The spray characteristics of five of the oils have been determined at a variety of fuel temperatures using a high-pressure, high-temperature injection bomb and high-speed motion picture camera. These same oils have been tested in a direct injection farm tractor engine. The engine data consists of the normal performance measurements as well as the determination of heat release rates from cylinder pressure data.

INTRODUCTION

The use of various vegetable oils as fuels for diesel engines is not a new concept. Over the years since the invention of the diesel engine, however, the development of the engine has been based on the availability of petroleum-derived diesel fuel which in turn has been tailored to meet the needs of the current engines. During this period, a wealth of empirical knowledge has been developed which serves as the data base for the current diesel fuel specifications. Periodically the vegetable-oil fuel concept has been reintroduced, usually during periods of petroleum shortages. In most cases the interest faltered due to renewed availability of more economical petroleum-derived fuels.

The 1973 oil embargo signaled the beginning of a new period of petroleum shortages. As a result, the international interest in the use of vegetable oil as diesel fuel has been once again renewed. Much of the current work is being performed in countries which have little or no internal petroleum resources, such as South Africa (Augo 1981) and Australia (Ward 1981 and Quick 1980). Work has been progressing in the United States at Ohio State University (Engelman et al. 1978), North Dakota State University (Kaufman 1981 and Ziejewski and Kaufman 1982), Southwest Research Institute (Fort et al. 1982), University of Idaho (Peterson et al. 1981), University of Alabama

(Adams 1982), and at several engine manufacturers including International Harvester (Baranescu and Lusco 1982), John Deere (Barsic and Humke 1981), Caterpillar (McCutchen 1981), and Perkins (Bacon et al. 1981).

The outstanding question regarding the use of vegetable oils as fuels for diesel engines relates to the observed durability problems. These problems appear to be related to the high viscosity of the vegetable oils. Similar problems occur with the use of high viscosity petroleum fuels in engine systems not designed to handle viscosities higher than that of DF-2. The durability problems encountered with the use of vegetable oils appear to be very strong functions of engine type, with direct-injection engines being more susceptible than the indirect-injection engines. Direct-injection engines are much more dependent upon the ability of the injection system to accomplish a high degree of atomization of the fuel than are the indirect-injection engines (Obert 1970).

The differences between the performance of the vegetable oils and diesel fuel can also be related to the difference in the chemical strucure of the vegtable oils. The chemical differences could affect not only the combustion of the oils but also the injection characteristics through preflame reactions.

OBJECTIVES

The specific objectives of the program, sponsored by the U.S. Department of Agriculture are, to:

- Characterize vegetable oils from different sources and with different degrees of processing.
- Determine the injection characteristics of the various test oils.
- Evaluate the effects of the test oils on engine combustion, performance, and degradation (deposits, lube oil contamination, etc.)

The results presented in this paper are preliminary in that the work has been completed on 5 of 14 test oils selected for examination in the overall program. The results for the other test oils and the development of the preliminary specification will be presented in the near future.

PHYSICAL AND CHEMICAL CHARACTERIZATION

Four different types of vegetable oils have been chosen for use in this program; they include soy, sunflower, cottonseed and peanut oil. The oils were obtained from commercial suppliers under varying degrees of processing. A total of 14 different oils were obtained for use on this program. Table 1 is a listing of the various test vegetable oils and the degree of processing of each of the samples.

Table 1. Vegetable Oils Selected for Testing and Evaluation

Degree of Processing	Soybean	Sunflower	Cottonseed	Peanut
Crude	X	X	X	X
Degummed	X	X		
Once Refined	X	X	X	X
Deodorized		X	X	
Fully Refined				X
Hydrotreated	X			

As indicated previously, all of the work has been completed on five of the oils, including the four soybean oils and deodorized sunflower oil.

The goal of physical and chemical characterization of the test oils was to document the properties of oils as well as to determine the applicability of the various ASTM fuel tests and the AOCS vegetable oil tests to the specification of vegetable oil fuels. Where necessary, the standard methods were modified in order to accommodate the peculiarities of the vegetable oils. Table 2 is a listing of the results of several of the determinations.

The standard methods will not be discussed in this paper. Those tests which required modification will, however, be discussed in some detail. It was not possible to perform the stability (ASTM-D2274) and steam jet gum (ASTM-D381) determinations on the vegetable oils. The stability test (ASTM-D2274) is designed to measure the stability of distillate fuels under accelerated oxidation conditions. When attempted with the vegetable oils the high viscosities of the oils made the initial filtration impossible and dilution with a solvent would have made the test meaningless. The steam jet gum determination (ASTM-D381) was designed to determine the existent gum in motor gasolines. When applied to the vegetable oils, it was found that the vegetable oils would not evaporate at the conditions specified in the procedure. It was felt that modification of the procedure to accomplish the vaporization would have increased the severity of the test and resulted in polymerization of oils, making the results very difficult to interpret.

Boiling point distributions (distillation) by ASTM-D2887 are normally performed on petroleum products with final boiling points below 538 C. The technique consists of a gas chromatographic technique in which a 183 cm column of silicone gum rubber on diatomaceous earth is operated through a

Table 2. Vegetable Oil Fuel Properties

Property	Fuel					
	Typical DF-2	Deodorized Sunflower	Crude Soybean	Degummed Soybean	Once Ref Soybean	Lt Hydro Soybean
Copper Corrosion	1A	1A	FP	FP	FP	FP
Heat of Comb (MJ/kg)	44.95	40.14	39.39	39.39	40.20	39.82
Ash(wt%)	0.02	0.01	0.08	0.04	0.01	0.01
Water(wt%)	0.01	0.02	0.036	0.068	0.066	0.030
Free Fatty Acid(wt%)	NA	0.056	0.081	-	0.022	0.027
Surface Tension (N/m)	0.025	0.033	0.032	0.032	0.032	-
Peroxide No.	NA	72.5	43.2	53.8	162.0	8.14
Particulates (mg/100ml)	0.5	1.6	101.0	14.0	1.0	2.4
Flash Point(°C)	56	323	312	315	320	325
Cloud Point(°C)	-13	-7	-6	-7	-1	31-37
Pour Point(°C)	-18	-9	-10	-10	9	2
Viscosity - See Table 6						

prescribed temperature program. The boiling range is determined by comparing the retention times of the fuel components to standards generated using known mixtures of hydrocarbons. In order to perform the test on the vegetables oils the column length was reduced to 30.5 cm and the samples were diluted 50 percent with carbon disulfide. The results of the boiling point determinations are somewhat questionable when applied to the vegetable oils due to the strong possibility of chemical changes in the samples. The distillation data presented in Table 3 do show that the range is much higher and the distribution somewhat different than the baseline DF-2.

Table 3. Distillation and JFTOT Analyses

	Percent Evaporated					JFTOT	
Soybean	IBP	10%	50%	90%	FBP	Visual	Time to Plugging(min)
Crude	359.6	556.8	605.1	626.5	679.5	4+	11
Degummed	360.7	556.8	602.3	623.7	679.1	4	15
Once Refined	360.7	563.3	605.6	626.5	681.4	0	139
Light Hydro	359.4	559.7	601.4	624.2	681.4	1	88
Sunflower							
Deodorized	360.0	558.2	603.2	626.0	681.9	3	71
Baseline DF-2	187.8	222.2	263.3	300.0	321.1	-	-

The Jet Fuel Thermal Oxidation Tester (JFTOT-ASTM-D3241-77) provides a measure of the tendencies of gas turbine fuels to deposit decomposition products on components of the fuel system. The apparatus consists of a temperature-controlled sump to contain the fuel, a pump to move the fuel through the system at a fixed volumetric flow rate, a precision stainless steel filter where fuel degradation products may become trapped, and an aluminum heater tube where thermal oxidation products may deposit. The essential information derived from normal operation of the JFTOT consists of a visual rating of the amount of deposits detected on the heater tube and the rate of plugging of the filter.

Visual inspection of the filters indicated that particulates not being filtered out prior to the tests were responsible for the plugging of the JFTOT filter. The JFTOT results presented in Table 3 indicate that the propensity of the oils to form thermal decomposition deposits decreases with increasing processing. A visual rating of 4 indicates severe deposits and a rating of 0 indicates no deposits. The time to filter plugging is an indication of the amount of suspended material contained in each sample.

Particulate contamination (ASTM-D2276) is generally performed on aviation turbine fuels. The procedure consists of filtering a known volume of fuel through a membrane filter. The contaminant is determined by comparing the weight of the test filter to that of a control filter located below the test filter. The vegetable oils would not flow through the membrane filter. Two modifications were made to make the test applicable to vegetable oils while still providing results which are comparable to petroleum fuels. The samples of vegetable were diluted up to 70 percent with either toluene or heptane before filtering using the standard two filter apparatus. Instead of filtering the fixed volume of sample, filtration proceeded until the filter plugged. Knowing the volume of sample which passed through the test filter it was possible to present the results in the normal units of mg/liter. As can be seen in the results presented in Table 2, the particlate matter decreases as the level of processing increased.

The fatty acid analysis was performed using the gas chromatographic technique described by McCreary et al. (1978). Table 4 is a listing of the results in terms of weight percent of the various constituent fatty acid. As was expected the processing had very little effect on the various soybean oils until the hydrotreating stage; at that point there was a marked shift from the two and three double bond structures to the single double bond structure. As is normally found, the sunflower oil contained more linoleic acid and less oleic acid than the soybean oil.

Table 4. Constituent Fatty Acid Analysis of the Various Test Fuels

	Constituent Fatty Acid (Weight Percent)						
	C_{14} Myristic	C_{16} Palmitic	C_{18} Stearic	$C_{18:1}$ Oleic	$C_{18:2}$ Linoleic	$C_{18:3}$ Linolenic	C_{20}-C_{24}
Soybean							
Crude	0.2	9.7	4.6	23.6	56.7	4.2	0.9
Degummed	0.5	10.2	4.4	24.6	55.5	3.9	1.0
Once Refined	0.4	10.7	4.8	23.8	55.2	4.4	0.7
Light Hydro	0.4	9.5	4.9	50.3	32.8	1.3	0.8
Sunflower							
Deodorized	0.5	6.0	4.8	18.9	67.9	0.3	1.7

The results of the elemental analysis are presented in Table 5. There were no unexpected results from these analyses; the crude and degummed soybean oils both contained some phosphorus compounds which were apparently removed upon further processing.

Table 5. Elemental Composition of the Test Vegetable Oils

	Carbon	Hydrogen	Oxygen	Nitrogen	Sulfur	Phosphorus	Chlorine
Soybean							
Crude	76.9	11.3	10.7	0.009	N.D.	0.027	N.D.
Degummed	77.3	11.5	10.7	0.004	N.D.	0.015	N.D.
Once Refined	77.4	11.5	10.5	0.001	N.D.	N.D.	N.D.
Light Hydro	77.4	11.7	10.7	N.D.	N.D.	N.D.	
Sunflower							
Deodorized	77.7	11.5	10.4	0.004	N.D.	N.D.	N.D.

Table 6 is a listing of the viscosities and densities of the various oils, both as functions of temperature. The viscosity data could be plotted as straight lines on the log-log viscosity-temperature coordinates.

Table 6. Viscosity and Density at Various Temperatures

	Viscosity (cS)			Density(gm/ml)		
	20 C	40 C	100 C	20 C	40 C	100 C
Soybean						
Crude	65.8	32.2	7.7	0.921	0.908	0.869
Degummed	66.5	31.3	7.6	0.922	0.909	0.871
Once Refined	64.3	30.6	7.5	0.919	0.908	0.868
Light Hydro	-	37.5	8.4	0.916	0.902	0.867
Sunflower						
Deodorized	67.1	31.7	7.7	0.921	0.907	0.868
Baseline DF-2	-	2.4	-	-	0.851	-

The standard cetane rating method and equipment were modified to allow variation in the fuel temperature. Normally the test is performed with the fuel temperature at the injection nozzle set at approximately 38 C. The cetane rating of degummed soybean oil at 38 C was found to be 36.6. Raising the fuel temperature to 145 C and using the same compression ratio as was used in the 38C test resulted in a cetane rating of 39.3. The reference fuels also demonstrated a high "cetane number" at the elevated fuel temperature. It appears that the effects of viscosity on spray characteristics and evaporation do affect the cetane rating; these effects apparently being demonstrated by the test fuels as well as the reference fuels.

INJECTION AND ATOMIZATION EXPERIMENT

A device has been designed and built at Southwest Research Institute which provides a means for examining the diesel injection event in an environment which is thermodynamically very similar to that encountered in an engine. The system consists of a bomb, a fuel injection system and a high-speed motion picture camera.

The purpose of the bomb is to allow observation of the characteristics of diesel type fuel injection sprays in a high-temperature and high-pressure environment. The design temperatures and pressures are 500 C and 4.14 MPa, respectively. The temperature is maintained using electrical resistance heaters. A quiescent, inert atmosphere of nitrogen is used to prevent autoignition of the test fuel. The bomb has a cylinderical geometry with quartz-glass end plates which allows direct visual observation through the bomb.

The injection nozzle (Bosch No. 438057) is installed in the center section in such a way that one of the four spray cones is perpendicular to the axis of the cylinder. The other three holes in the nozzle tip are plugged to prevent fuel impingement on the window. The nozzle holder includes a water cooling jacket to allow control of the fuel temperature in the injection nozzle. The possible range on the fuel temperature is from below atmospheric to 150 C.

The fuel injection system consists of the Bosch nozzle and a Bosch jerk pump plunger and barrel. The nozzle is equipped with a needle lift indicator and a line pressure transducer. The drive system consists of a cam shaft driven at constant speed by an electric motor. A latching mechanism on the cam follower makes it possible to vary and control the number of injections from two (2) to ten (10).This system accurately reproduces the injection event in an engine.

The high-speed movies are taken using a Hycam II, 16 mm high-speed motion picture camera. The camera is capable of running 11,000 full frames per second or, with the installation of a quarter framing control head, at 44,000 quarter frames per second. The camera is equipped with a Com-Nikkor lens (focal length of 88 mm f/2.0, fixed magnification of 0.2X). The camera was positioned such that the lens was focused on the center of the spray. Back lighting was provided through the other window of the bomb using a 1000w Tungsten-Halogen lamp. Four-hundred and fifty foot rolls of Kodak type 2498 RAR reversal, 16 mm film were used for all of the movies. The film speed, for all of the movies, was set at 25,000 quarter frames per second, or 40μs between frames. The camera is equipped with an internal switch which closes when the camera is up to speed. This switch was used to control the firing of the injection pump drive.

Movies of the sprays of the five oils have been obtained at two different fuel temperatures. The conditions in the bomb at the time of injection were always set at 4.14 MPa and 482 C. The fuel temperatures for the tests were 40 C and 145 C. The 40 C temperature was selected to provide a direct comparison between the vegetable oils and the baseline DF-2 which was also tested at 40C. At 145 C the viscosities of all of the vegetable oils are in the range of 4.0 cS, approximately the same viscosity as the baseline DF-2 at 40 C (2.4cS). Thus, comparisons are possible at the same fuel temperatures and also at equal fuel viscosities.

Injection and Atomization Results

All of the test vegetable oils displayed similar injection characteristics. Figure 1 shows the penetration distance versus time data for both the diesel fuel and the various test oils. In Figure 1a the vegetable oils at 40 C all showed somewhat slower penetration rates as compared to the DF-2 at 40 C. In all of the tests at 40 C the fuel traveled the entire width of the viewing area and impinged on the opposite wall of the bomb. Figure 1b shows the comparison between the DF-2 at 40 C and the vegetable oils at 145 C. At the elevated oil temperature the penetration rates for the vegetable oils approached that of the DF-2. The vaporization of the vegetable oils was greatly enhanced by raising the oil temperature from 40 C to 145 C; at 145 C none of the vegetable oils impinged on the opposite wall of the bomb.

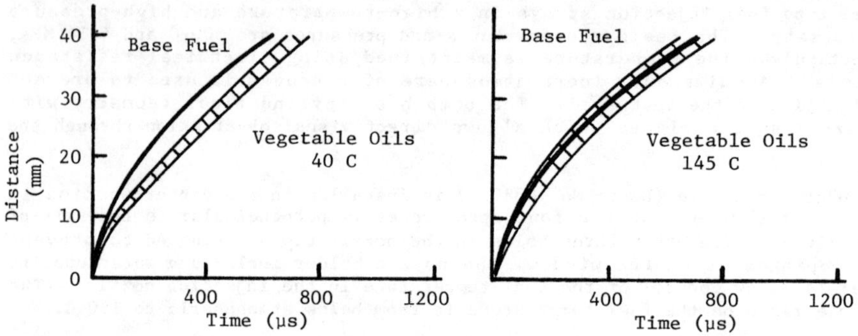

FIGURE 1. SPRAY PENETRATION DATA FOR THE VEGETABLE OILS AND THE BASELINE DF-2

The width of the spray cone, or the cone angle, decreased as the temperature of the oil increased (up to a 45 percent decrease with the crude soy oil). The light hydrotreated soy oil had the smallest cone angle while the crude soy oil displayed the largest cone angle.

ENGINE EXPERIMENT

A caterpillar, Model 3306 DI/TC, direct-injection engine was selected for use in this study. Table 7 is a listing of the specifications for the test engine.

Table 7. Engine Specification

Configuration	Inline 6 cylinder
Displacement (liters)	10.45
Combustion Chamber	Direct injection, "Mexican hat" piston
Bore x Stroke (cm)	12.1 x 15.2
Maximum power (kw)	154 @ 2000 rpm
Maximum torque (J)	866 @ 1500 rpm
Compression Ratio	14.6
Injection Nozzle	multi-orifice pencil

The injection system consists of inline jerk pumps with helical return ports. The rack (rack and pinion) acts directly on the pump plungers to rotate them, thus determining the point of opening of the helical return port (controlling the end of injection).

The test matrix used for engine evaluation of the test vegetable oils consisted of two speeds with four load conditions at each speed. The power settings were determined using the baseline DF-2; these same power settings were then used with each of the test fuels.

The vegetable oils were heated to 145 C for the engine experiments (viscosities of the test oils were approximately equal to that of base fuel at 40 C, the fuel temperature for the base fuel test). The fuel consumptions rates were determined gravimetrically using a progammable electronic weigh platform. The intake air flowrate was measured by a vortex-shedding flow meter. Gaseous exhaust emissions (NO/NO_x, CO_2, CO, O_2, and unburned hydrocarbons) were determined using Beckman exhaust emissions analyzers. The results from the analysis of the exhaust gases were used to calculate the air-fuel ratio. The calculated air-fuel ratios agreed favorably with those determined from the fuel and air flow measurements.

Number one cylinder of the test engine was equipped with a water-cooled pressure transducer for measuring the pressure in the combustion chamber. The pressure data was supplied to a digital high speed data system. The data system and the operating software were previously described by Ryan et al. (1980). Basically, the first law of thermodynamics is applied to contents of the combustion chamber in order to calculate an apparent rate of heat release ($\dot{Q}$). Diagrams of the heat-release rate versus crankangle are very useful in diagnosing the combustion event. The centroid of the heat release diagram($\bar{\dot{Q}}$,$\bar{\theta}$) provides a very good indication of late combustion and slow burning (Ryan et al. 1982).

The engine experiments were performed to determine the effects of the various oils on the combustion event and to determine the effects of heating the oils (reduced viscosity) on combustion and the deposit problems. Deposits were qualitatively examined for each oil by removing an injection nozzle for inspection and cleaning between each test.

Engine Results

Evaluation of the engine performance was based on the Indicated Specific Energy Consumption (ISEC), having units of MJ/kw-hr. Although the engine power setting was based on the brake measuements, all evaluations were performed using the indicated power, calculated from the cylinder pressure data. It was hoped that the combination of the atomization data and the engine combustion performance data would point out those fuels which would have durability problems due to either inadequate atomization and/or poor combustion.

Rather than presenting the engine data for all modes, one mode (rated torque setting at 1500 rpm) was selected as representative of the other modes. Figure 2 is a bar chart showing the ISEC, the apparent combustion efficiency, and the centroid of the heat release diagram for the base fuel and the five vegetable oils. Based on the specific energy consumption, the degummed soybean oil performed as well as the basefuel. The deodorized sunflower oil appears to be the worst of those tested to-date, with an energy consumption 10 percent higher than the base fuel. These results are further reflected in the apparent combustion efficiency data presented in Figure 2. The apparent combustion efficiency is computed as the ratio of the total apparent heat release (integration of the heat release diagram) to the energy content of the fuel injected during one cycle.

Based on thermodynamic consideration of the diesel cycle, the maximum work is theoretically obtained when the energy is instantaneously input to the working fluid when the piston is at top dead center (constant volume combustion). The centroid of the hypothetical heat release diagram would be $\bar{\dot{Q}}=\infty$ and $\bar{\theta}$=TDC. In reality, the centroid of the heat release diagram is always located after top center and within the range of 30-40 J/degree. The centroid data presented in

Figure 2 shows that the worst fuel (deodorized sunflower oil) displayed a centroid which occurred very late in the cycle and with very low Q.

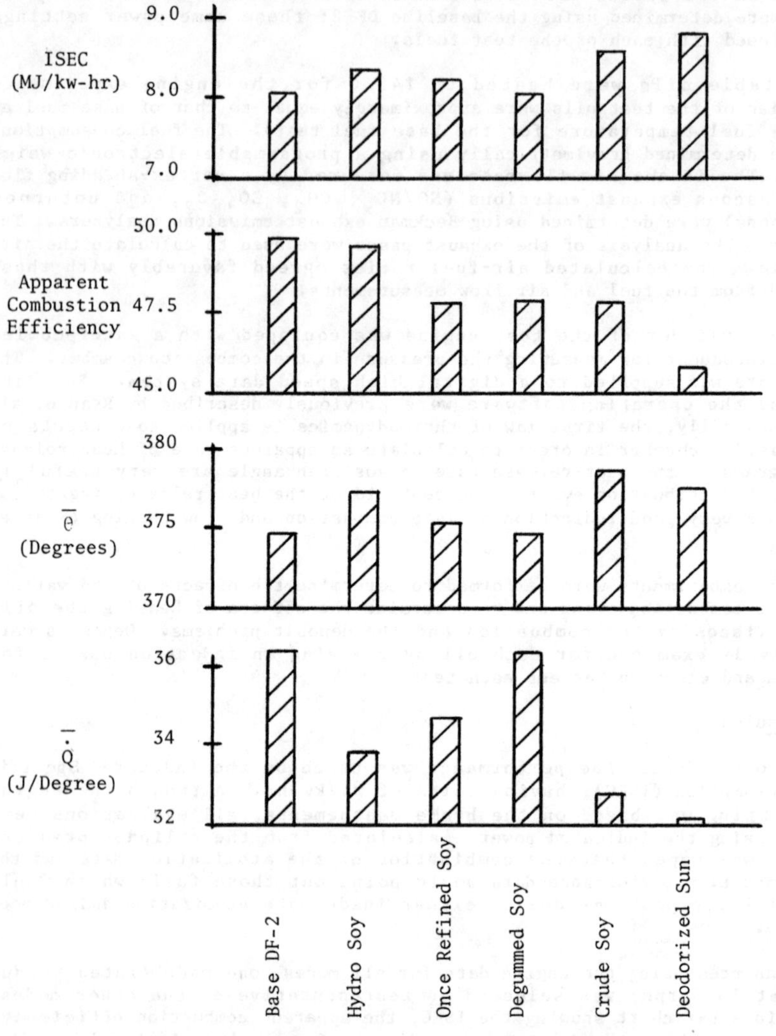

FIGURE 2. ENGINE COMBUSTION AND PERFORMANCE DATA FOR THE RATED TORQUE POWER SETTING

Figure 3 is a comparison of the heat release diagrams for the baseline DF-2 and the deodorized sunflower oil. The centroids are indicated for both fuels. As can be seen the initial spike in the diagram for the sunflower oil is much lower than that of the DF-2. This indicates that the vegetable oil has a much smaller premixed combustion stage. In addition, the diffusion stage of combustion appears to be flatter (slower) for the sunflower oil than for the DF-2. These trends indicate that the sunflower oil both vaporizes and burns at a slower rate than the DF-2. Similar trends were present in the data for the crude soybean oil. All of the other test oils displayed combustion characteristics more similar to those of DF-2.

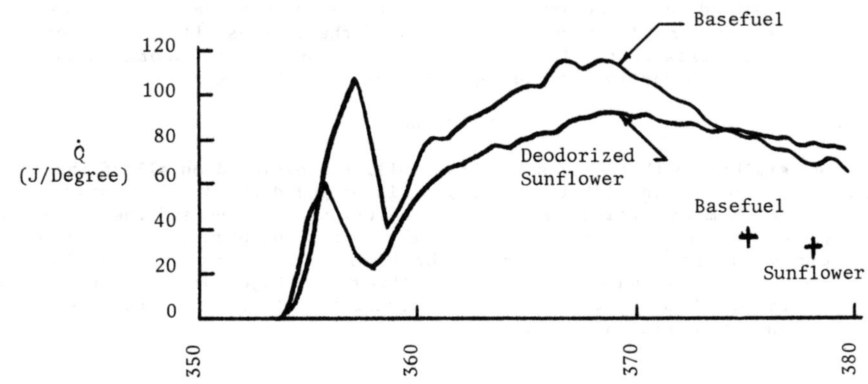

FIGURE 3. HEAT RELEASE RATE DIAGRAMS FOR THE BASELINE DF-2 AND DEODORIZED SUNFLOWER OIL

Inspection of the injection nozzles after each test indicated that fuels which had combustion characteristics similar to diesel fuel also displayed the same tendency to form deposits as the DF-2. The two fuel (crude soybean and deodorized sunflower) which had the worst combustion characteristics also displayed the strongest tendencies to form deposits on the injection nozzle.

DISCUSSIONS

The characteristics of the vegetable oils provided some insight into mechanisms leading to the durability problems encountered when used as fuels for diesel engines. The JFTOT analysis indicated that some oils are thermally unstable at temperatures normally encountered in a fuel system. This instability could obviously lead to carbon and varnish deposits in regions of high temperatures.

The injection studies pointed out that the vegetable oils behave very different from petroleum derived fuels. Normally as the viscosity of a fuel is decreased the penetration rate decreases and the spray cone angle increases. With the vegetable oils, however, the penetration rates increased and cone angles decreased as the viscosities were reduced by increasing the temperatures of the oils. This is also evident in the fact that the vegetable oils all had lower penetration rates and large cone angles at 40 C than the DF-2. There were very obvious differences in the spray characteristics of the vegetable oils as the oil temperature was increased from 45 C to 145 C. At the higher temperature, there was a pronounced increase in the vaporization rate, much more than would be expected with 100 C increase in fuel temperature (the atmosphere in the bomb was always held at a temperature 482 C).

The engine data indicates that heating the oil does reduce the carbon deposit problem and does offer the possibility of obtaining satisfactory performance in the engine. It does appear, however, that the deposit and other durability problems are related to viscosities of the oils as well as the differences in the chemical structure of the oils as compared to DF-2. The amount of deposits which have been observed in this study are much lower than the amounts reported by others in similar engines. The fact that some of the oils (deodorized sunflower) displayed a greater propensity to form deposits than others, when all were tested at essentially the same viscosity, indicates that some of the deposit problem is related to the chemical structures of the oils. The preliminary results indicate that the heat release rate calculation provides a method for screening the fuels for the performance characteristics as well as the durability characteristics of the various oils. The logic of this statement rests in the fact that all of the durability problems result from incomplete combustion of the vegetable oils in the engine.

FUTURE WORK

Testing, similar to that discussed above, will be completed on all of the oils listed in Table 1. In addition, a program is proposed which will examine the effects of chemical structure on the injection processes and the mechanism leading to carbonization and deposits in the combustion chamber. A hypothesis has been presented which says that the high temperatures and the highly elongational flow encountered at the exit from the injection nozzle causes changes in the chemical structure which affect the atomization and subsequent ignition and combustion of the oils.

REFERENCES

1. Adams, J. F. Peters, B. J. Schroer and M. C. Ziemke. 1982. Investigation of plant oil (soybean) as a diesel fuel extender. Presented at American Oil Chemists' Society Annual Meeting. Toronto, Canada.

2. Bacon, D.M., F. Brear, I.D. Moncrieff and K.L. Walker. 1981. The use of vegetable oils in straight and modified form as diesel engine fuels. Third International Conference on Energy Use Management. Berlin (West). Pergamon Press. 1525-1533.

3. Baranescu, R. A. and J. J. Lusco. 1982. Sunflower oil as a fuel extender in direct injection turbocharged diesel engines. SAE 820260.

4. Barsic, N.J. and A.L. Humke. 1981. Performance and emissions characteristics of a naturally aspirated diesel engine with vegetable oil fuels. SAE 810262.

5. Bruwer, J.J. and F.J.C. Hugo. World overview of plant oils for fuel: status of research and applications technology. Third International Conference on Energy Use Management. Berlin (West). Pergamon Press. 287-295.

6. El-Hamdy, A.H. and E.G. Perkins. 1981. High performance reversed phase chromatography of natural triglyceride mixtures: critical pair separation. JAOCS. 867-872.

7. Engelman, H.W., D.A. Guenther and T.W. Silvis. 1978. Vegetable oil as a diesel fuel. ASME 78-DGP-19.

8. Fort, E.F., P.N. Blumberg, H.E. Staph and J.J. Staudt. 1982. Evaluation of Cotton Seed Oils as Diesel Fuel. SAE 820317.

9. Goering, C.E., A.W. Schwab, M.J. Daugherty, E.H. Pryde and A.J. Heakin. 1981. Fuel properties of eleven vegetable oils. ASAE 81-3579.

10. Hawkins, C.S., J. Fuls and F.J.C. Hugo. 1981. Laboratory procedures for investigating some fuel properties of sunflower oil esters in diesel engines. Third International Conference on Energy Use Management. Berlin (West). Pergamon Press. 1611-1617.

11. Humke, A.L. and N.J. Barsic. 1981. Performance and emissions characteristics of a naturally aspirated diesel engine with vegetable oil fuels - Part 2. SAE 810955. Special Publications 495. Diesel Combustion and Emissions Part III.

12. Kaufman, K.R., M. Ziejewski, M. Morohl and A.E. Jones. 1981. Sunflower oil and methyl ester as fuels for diesel engines. Third International Conference on Energy Use Management. Berlin (West). Pergamon Press. 1635-1643.

13. McCreary, D.K., W.C. Kossa, S. Ramachandran and R.R. Kurtz. 1978. A novel and rapid method for the preparation of methyl esters for gas chromatography: application to the determination of the fatty acids of edible fats and oils. J of Chromatographic Science. 16: 329-331.

14. McCutchen, R. 1981. Vegetable oil as a diesel fuel - soybean oil. Third International Conference on Energy Use Management. Berlin (West). Pergamon Press. 1679-1686.

15. Obert, E.F. 1973. Internal combustion engine and air pollution. Intext Educational Publishers, New York, NY.

16. Peterson, C.L., G.L. Wagner and D.L. Auld. 1981. Performance of vegetable oil substitutes for diesel fuel. ASAE 81-3578.

17. Quick, G.R. 1980. An in-depth look at farm fuel alternatives. Power Farming Magazine.

18. Ryan, T.W., W.E. Likos, and C.A. Moses. 1980. The use of hybrid fuel in a single-cylinder diesel engine. SAE 801380

19. Ryan, T.W., T.J. Callahan, W.E. Likos, C.A. Moses, and L.G. Dodge. 1982. Alternative (hybrid) fuels for highway transportation. Final report on DOE Contract DEAC04-79-CS54240.

20. Van der Walt, A.N. and F.J.C. Hugo. 1981. Cyclic load testing of agricultural tractors in fuel research - instrumentation and equipment for a test facility. Third International Conference on Energy Use Management. Berlin (West). Pergamon Press. 1919-1925.

21. Van der Walt, A.N. and F.J.C. Hugo. 1981. Diesel engine tests with sunflower oil as an alternative fuel. Third International Conference on Energy Use Management. Berlin (West). Pergamon Press. 1927-1933.

22. Ward, J.F. 1981. Agriculturally based oils as engine fuels for the Australian and South Pacific region. Third International Conference on Energy Use Management. Berlin (West). Pergamon Press. 1935-1938.

23. Ziejewski, M. and K.R. Kaufman. 1982. Endurance test of a sunflower oil/diesel fuel blend. SAE 820257.

SUNFLOWER OIL AS A FUEL FOR COMPRESSION IGNITION ENGINES

A. R. Tahir
Assoc. Member ASAE

H. M. Lapp
Member ASAE

L. C. Buchanan
Assoc. Member ASAE

Many energy fuels are being investigated as potential substitutes for current high cost fuels obtained from diminishing conventional sources. Energy conservation and fuel extender programs, such as gasohol, are expedient and important short term measures. They are, however, inadequate to satisfy long range energy needs and to gain independence from petroleum-based fuels. It is therefore of great urgency that all potential fuel alternatives be recognized and examined. Vegetable oils may provide one such alternative and their potential has been examined in past years. Strayer (1980) and Quick (1980) referred to work done as early as 1928 by Gautier in France on the utilization of various vegetable oils for energy fuel. Sunflower oil has emerged to a central focus position among other vegetable oils being examined. Since the Province of Manitoba is the major producer of sunflower oil in Canada (approximately 96%) this study was initiated to investigate the fuel-related physical properties and performance characteristics in compression ignition engines.

Chemical Composition of Vegetable Oils

The molecule of a vegetable oil is usually composed of triglycerides (about 95%) and non-triglycerides (about 5%). Triglycerides are esters composed of one molecule of glycerol and three molecules of fatty acids (Formo et al. 1979).

The chemical structure of triglycerides is given below:

$$\begin{array}{llcll} CH_2 - OH & H - O - (CO) - R & & & CH_2 - O - (CO) - R \\ | & & & & | \\ CH - OH & + \; H - O - (CO) - R' & \longrightarrow & 3HOH \; + & CH - O - (CO) - R' \\ | & & & & | \\ CH_2 - OH & H - O - (CO) - R'' & & & CH_2 - O - (CO) - R'' \\ \text{Glycerol} & \text{Fatty Acids} & & \text{Water} & \text{Triglycerides} \end{array}$$

Fatty acids are composed of a carboxyl group and a hydrocarbon chain.

$$\begin{array}{cc} H - O - \overset{\overset{\displaystyle O}{\|}}{C} \text{------------} & (CH_2)_n - CH_3 \\ \text{Carboxyl end} & \text{Hydrocarbon chain} \end{array}$$

Fatty acids comprise from 94 to 96% of the total mass of a triglyceride molecule, therefore the proportions of these acids determine the physiochemical properties (e.g. viscosity, calorific value) of an oil.

The most common fatty acids composition (determined by the Department of Plant Science, University of Manitoba) in sunflower oil was found to be (1) palmitic acid, 6.0% (2) stearic acid, 4.0% (3) oleic acid, 18.5% and (4) linoleic acid, 71.0%. The free fatty acids in sunflower oil are usually 0.5% or more. The quantity of free fatty acids predominantly affect the flash point of a veget-

The authors are: A. R. Tahir, Former Graduate Student, H. M. Lapp, Professor and L. C. Buchanan, Professor, Agricultural Engineering Department, University of Manitoba, Winnipeg.

able oil. The proportions of non-triglycerides (hydrocarbons, waxes and sterols) influence the cloud point whereas the proportion of tochopherols, an antioxidant non-triglyceride, defines the oxidation stability of a vegetable oil (Formo et al. 1979).

Equipment and Test Procedures

Almost all fuel-related physical properties of crude-degummed sunflower oil were determined by using ASTM standard test procedures specified for diesel fuel.

Kinematic Viscosity: The laboratory was not equipped with an apparatus which could directly measure the kinematic viscosity and a two-step viscosity determination was therefore followed. For temperatures higher than 20°C the saybolt viscosity was determined by using a Saybolt Universal Viscometer. The efflux time in seconds for a 60 ml sample to flow through a calibrated orifice was measured under carefully controlled conditions. For temperatures lower than 20°C the dynamic viscosity was determined by using a Synchro-Lectric Viscometer. The resistance offered by sunflower oil to the disc spindle rotation was measured in mPa.s. The saybolt viscosity and the dynamic viscosity were converted into the kinematic viscosity.

Cloud and Pour Points: A sample of crude-degummed sunflower oil was filtered into a glass jar which was then placed in a cooling bath after placing a thermometer in the jar. At each temperature decrease of 1.0°C the sample was inspected for haze formation. In the case of pour point the sample of sunflower oil was heated to 46°C and then it was cooled to 35°C in air. The test jar with thermometer was placed in the cooling bath and inspected after every temperature decrease of 3.0°C for the ceasing of fuel flow.

Flash Point: A sample of sunflower oil was poured into the flash cup, its lid was locked in position and it was then placed on the heater. The sample heating rate with the stirrer in operation was controlled at the rate of 5 to 6°C per minute. A test flame of about 6 mm in diameter was exposed to the vapors after every one degree celsius increase in the temperature while the stirrer was stationery to determine the vapor ignition temperature.

Distillation Curve: A 100 ml sample of sunflower oil was carefully poured into the distilling flask. After setting the thermometer in the flask it was placed over a heater. The condenser bath was filled with chopped ice and water to condense the vapors of sunflower oil. Heating was regulated according to the ASTM prescribed standards. The temperature readings corresponding to the percent recovery were taken and used to plot a graph between temperatures and percent recovery.

Heat of Combustion (Calorific Value): The gross heat of combustion of sunflower oil was determined by igniting a small sample (about one gram) in an Oxygen Bomb Calorimeter. A thermometer accurate to two decimal places was used to sense the temperature rise due to heat released by the combustion process. The gross heat was calculated from temperature, fuel and bomb data using formulae provided with the calorimeter. The gross heat of combustion is the heat released by the combustion of a unit mass of fuel in a constant volume bomb with substantially all of the water condensed to the liquid state. The net heat of combustion is the gross heat minus the latent heat of condensation of water vapor.

API Gravity and Density: The gravity of a liquid fuel is usually expressed in degrees API according to a scale designed by the American Petroleum Institute. The following relationship was used to convert API gravity into density (kg/L):

$$\text{API gravity, degrees} = \frac{141.5}{\text{Sp. gr. } (15.56°\text{C})} - 131.5 \qquad (1)$$

Both the density and API gravity were determined at 15.56°C, a specified temperature for the liquid fuels. The API gravity of sunflower oil and diesel fuel was determined by the hydrometer method while the density of the sunflower oil's methyl ester was determined by the gravimetric method.

<u>Other Fuel-Related Properties</u>: The fuel-related properties of sunflower oil which could not be determined in the department laboratory because of the lack of equipment were determined by the Alberta Research Council (ARC). Some of the properties determined by the ARC are discussed in the results and discussion section.

<u>Transesterification of Sunflower Oil*</u>: The methyl ester of sunflower oil was prepared by esterification of sunflower oil in the Department of Chemistry, University of Manitoba. The reaction can be summarized by the following:

$$\begin{array}{lclcl}
CH_2 - O - CO - R & & & & \\
CH_2 - O - CO - R' & + \; 3\,NAOCH_3 & \xrightarrow{\text{heat}} & 3CH_3 - O - CO - R & + \; \text{Biproducts} \\
CH_2 - O - CO - R'' & & & & \\
\text{Crude-degummed} & \text{Sodium} & & \text{Methyl Ester} & \\
\text{Sunflower Oil} & \text{Methoxide} & & & \\
(0.456 \text{ mole}) & (1.2 \text{ mole}) & & (82\% \text{ yield from oil}) &
\end{array}$$

Some physical properties of methyl ester, determined by the same procedures as those used for sunflower oil, are discussed in the results and discussion section.

<u>The Affects of Sunflower Oil on Diesel Fuel Systems</u>: A nozzle testing apparatus, donated by Powell Equipment Limited, Winnipeg, consisting of an injection nozzle, fuel filter, vacuum gage, pressure gage, plastic tubing and fuel buckets was modified to investigate the affects of sunflower oil on a diesel fuel system. The parameters measured were the mass of fuel discharged over a specified time period, nozzle valve opening pressure and the fuel line vacuum which may develop due to fuel filter blockage. At first, the apparatus was operated on No. 2 diesel fuel for 34 hours to obtain the control data for the system. The same apparatus was then switched to sunflower oil to obtain the comparable data over a period of 210 hours.

<u>Sunflower Oil Performance as a Fuel in a Diesel Engine</u>: Power output, specific fuel consumption and thermal efficiency of crude-degummed sunflower oil and No. 2 diesel fuel were determined by operating a Buda diesel engine. Since the engine was very old and a complete complement of equipment was not available in the department to test a small engine, a procedure was developed to make a comparison of certain engine and fuel-related parameters. A 4 kW generator was modified to operate as a dynamometer. The engine was a one cylinder Buda with a bore and stroke of 8.73 cm, precombustion chamber design and a maximum power of 7.5 kW. It was operated on diesel fuel first and then on sunflower oil.

Results and Discussion

<u>Kinematic Viscosity</u>: The kinematic viscosity of crude-degummed sunflower oil and its methyl ester was determined over a temperature range of -15.5 to 100°C and 1.0 to 70°C respectively. The viscosity results at selected temperatures are compared with No. 2 diesel fuel in Fig. 1. The kinematic viscosity values at temperatures below 25.7°C were determined by the Industrial Technology Centre, Winnipeg, with an "RVF" Brookfield viscometer. At 37.78°C, sunflower oil (33-45 mm^2/s) was found to be about 14 times more viscous than No. 2 diesel fuel (2.40 mm^2/s). The viscosity (5.7 mm^2/s) of the methyl ester was found to be about twice that of diesel fuel. At 0°C diesel fuel has a viscosity of 6.42 mm^2/s compared to 157.28 mm^2/s for sunflower oil: greater by a

*Method followed by M. Z. Khan, a Ph.D. student in the Department of Chemistry to prepare methyl ester.

factor of 24. In cold temperatures this higher viscosity of sunflower oil will be a serious problem affecting satisfactory engine fuel system operation. The kinematic viscosity values determined by the authors agrees well with the results obtained by other researchers.

Cloud and Pour Points: The cloud and pour points, -7.77 and -15.0°C respectively, observed for crude-degummed sunflower oil are a little higher than for No. 2 diesel fuel at -9.0 and -18.5°C respectively. However, the cloud and pour points at -2.8 and -4.0°C respectively of the methyl ester vary to a greater degree when compared with diesel fuel. The cloud point indicates the start of clogging in both fuel lines and filters because of the wax formation while the pour point indicates the cessation of fuel flow. The cloud point determined for sunflower oil agrees with other researchers whereas pour point differs with the results of Ramdeen et al. (1981) who reported -8.72°C as being the pour point of alkali-refined sunflower oil. The difference in the results may be due to the difference in the degree of refinement of the oil.

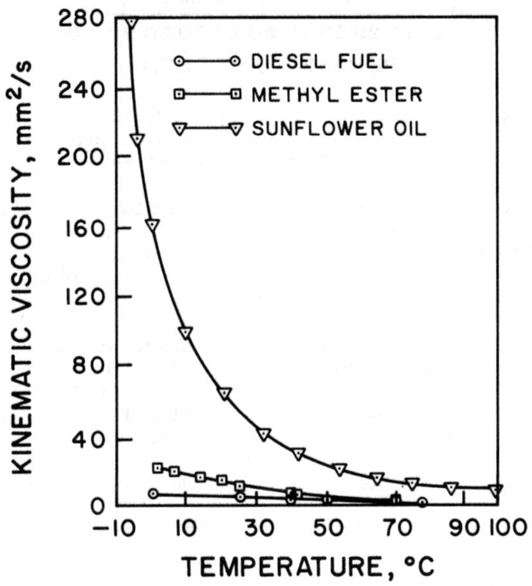

Fig. 1. Kinematic Viscosity of No. 2 Diesel Fuel, Crude-Degummed Sunflower Oil and its Methyl Ester.

Flash Point: The observed flash points for sunflower oil and No. 2 diesel fuel were 214°C and 68°C respectively. This great difference can be attributed to the longer carbon chain (C57) and the degree of unsaturation (3 to 6 double bonds) in the case of sunflower oil. The flash point was determined by using a Pensky-Martens closed tester. The high flash point of sunflower oil suggests reduced fire hazards will be present in handling and storage.

Distillation Curve: The distillation curves of crude-degummed sunflower oil and its methyl ester were obtained by the National Testing Laboratories Limited, Winnipeg. The distillation curves of No. 2 diesel fuel, methyl ester and the first two points of the curve of sunflower oil are shown in Fig. 2. At 214°C the initial boiling point of sunflower oil was observed and at about 316°C only 10 ml of darkened distillate passed over along with heavy dense white fumes and a very irritating odor. This heavy smoking and discoloration of the sample was an indication of the chemical decomposition of the fatty acids. At this stage, the distillation was discontinued which was considered to be at a premature stage according to ASTM regulations. The distillation of the methyl ester was satisfactory up to 85% recovery of the sample after which thermal cracking took place.

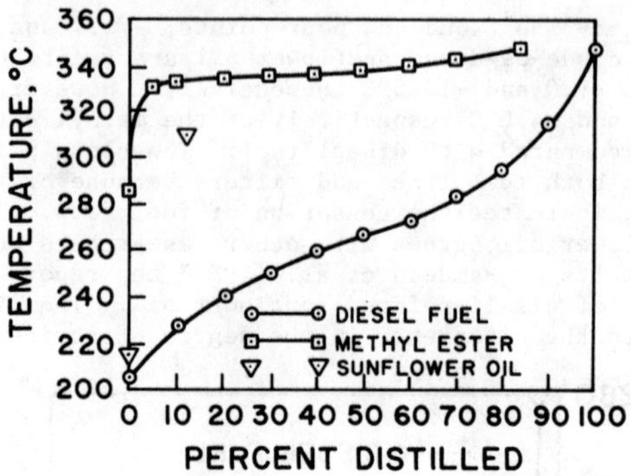

Fig. 2. Distillation Curves of No. 2 Diesel Fuel, Crude-Degummed Sunflower Oil and its Methyl Ester.

The distillation curve of sunflower oil shows some similarities to the curve determined by Ramdeen et al. (1981) but it does not coincide with the results of Bruwer et al. (1980) and Quick (1980). Variation in the results can be attributed to the ratio of linoleic and oleic fatty acids which is higher for the results obtained by Ramdeen et al. (1981).

Heat of Combustion: Heat of combustion or the calorific value of a fuel is an important measure since it is the heat energy produced by the fuel within the engine cylinder which enables the engine to do work. The heat of combustion of No. 2 diesel fuel, crude-degummed sunflower oil and its methyl ester are shown in Table 1. The gross calorific value of sunflower oil is noted to be

Table 1. Gross Calorific Value of Diesel Fuel, Sunflower Oil and its Methyl Ester.

Fuel Type	MJ/kg	MJ/L
No. 2 diesel	45.676	38.779
Crude-degummed sunflower oil	39.646	36.633
Methyl ester	40.0	35.17

about 87% of that of diesel fuel on a mass basis and about 95% on a volume basis. The methyl ester has about 12 and 9% lower heat content than diesel fuel on a mass and volume basis respectively. Lower heat content in the cases of sunflower oil and its methyl ester will result in a higher fuel consumption than for diesel fuel to maintain the same power output. The gross calorific value obtained agrees very closely with other researchers.

API Gravity and Density: Both the API gravity and density are referenced to a temperature of 15.56°C. Since the laboratory was not equipped with a constant temperature bath, a graphical procedure was adopted to determine the API gravity of the fuels at the reference temperature. The best fit lines of hydrometer readings at different temperatures, with a correlation factor of 0.999, are represented in Fig. 3. The results of API gravity and density are contained in Table 2. With respect to density, crude-degummed sunflower oil and its methyl ester are respectively 8.3% and 3.6% heavier than No. 2 diesel fuel.

The higher density of sunflower oil will mean increased costs for handling and transportation. The values of density and API gravity are almost the same as determined by other researchers.

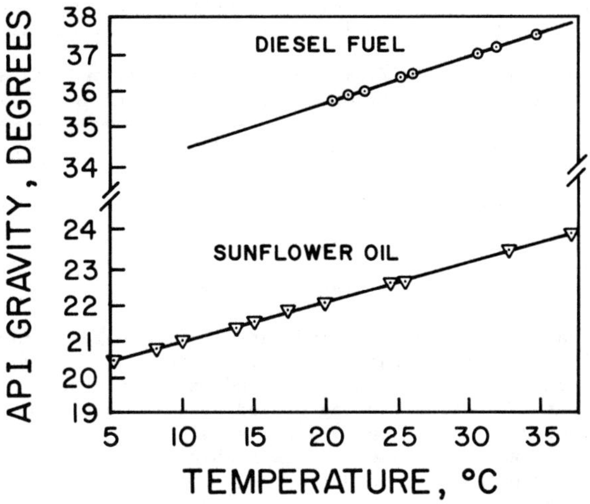

Fig. 3. API Gravity of No. 2 Diesel Fuel and Crude-Degummed Sunflower Oil by Hydrometer Method.

Table 2. API Gravity and Density of Diesel Fuel, Sunflower and its Methyl Ester at 15.56°C.

Fuel Type	API gravity (degrees)	Density kg/L
No. 2 diesel	35.13	0.849
Crude-degummed sunflower oil	21.65	0.924
Methyl ester	30.58	0.873

Cetane Number: The cetane number of crude-degummed sunflower oil was determined by the Alberta Research Council (ARC), Edmonton using the CFR test method. The cetane number of sunflower oil is 34 compared with 48 for No. 2 diesel fuel. A notable difference in the cetane numbers of the two fuels may be due to their different chemical compositions and the degree of unsaturation and refining. The long chain fatty acids with 3 to 5 double bonds, in the case of sunflower oil, have more binding force in the atoms of the molecules than diesel fuel which need high temperatures to vaporize. The cetane number determined by the ARC does not differ very much with the results of other researchers. The difference in the cetane ratings may be due to different growing climates which have an appreciable effect on the chemical composition of sunflower oil.

Oxidation Stability: The oxidation stability is an important parameter relating to fuel characteristics of vegetable oils. Oxidation increases the viscosity and deteriorates the flavor of vegetable oils with the passage of time. Many vegetable oils are poorer in oxidation stability than sunflower oil (Cobia and Zimmer 1978). Over the long run, sunflower oil due to oxidation tends to leave heavy gum and wax deposits on stationary metal components and fuel tanks. The oxidation stability value determined by the ARC, using an accelerated method, is 78.3 mg/100 ml which is lower than rapeseed oil (86.8 mg/100 ml) but higher than canola oil (43.6 mg/100 ml).

The Affects of Sunflower Oil on Diesel Fuel Systems: The affects of sunflower oil on a typical diesel fuel system were examined by operating a closed circuit system (test bench). The average mass of fuel discharged and the nozzle valve opening pressure measured at six hour intervals were recorded in Table 3. It is noted that the mass of sunflower oil discharged in one hour is considerably higher than that of diesel fuel. In the case of No. 2 diesel fuel almost

Table 3. Average Data Obtained from a Bench Test of 34 and 210 Hours Duration for Diesel Fuel and Sunflower Oil Respectively in a Fuel System.

Statistical Analysis	Sunflower Oil Discharge kg/h	Sunflower Oil Pressure kPa	No. 2 Diesel Discharge kg/h	No. 2 Diesel Pressure kPa
Average	0.4885	10.223	0.410	10.025
Standard Deviation	0.002	62	0.0013	34

no pressure and vacuum build-up were indicated by the gages through 34 hours of operation. In the case of sunflower oil, however, a pressure build-up of about 200 kPa in the line between the pump and injector nozzle remained almost the same throughout the entire testing period of 210 hours. This higher pressure can be attributed to the higher viscosity value of sunflower oil. After 210 hours the vacuum gage indicated an observable deflection but it could not be recorded due to the coarse graduations of the gage scale. This vacuum may be due to fuel filter blockage.

Performance Characteristics of Sunflower Oil: Some engine and fuel-related performance characteristics of crude-degummed sunflower oil and No. 2 diesel fuel were determined by operating the Buda diesel engine. The parameters obtained from the fuel performance evaluation tests were power output, specific fuel consumption and thermal efficiency which are shown graphically in Fig. 4 and Fig. 5. Figure 4 indicates that the difference in the power output

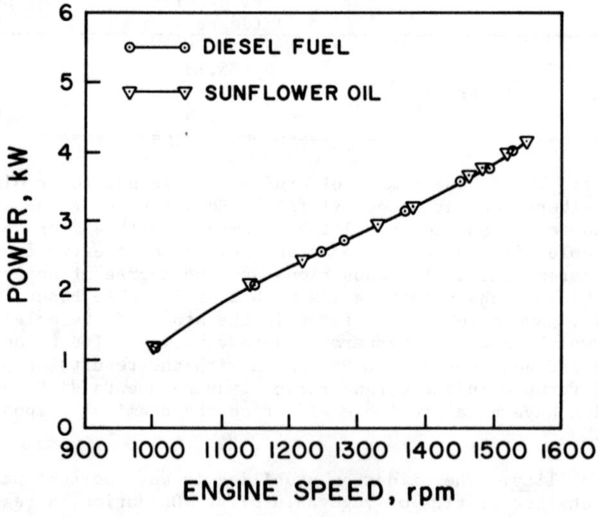

Fig. 4. Power Performance Curve of No. 2 Diesel Fuel and Crude-Degummed Sunflower Oil.

with diesel fuel and sunflower oil over a one hour interval at various engine speeds is negligible. Figure 5 shows that at about 60% of maximum power the specific fuel consumption and thermal efficiency with sunflower oil are 6% and

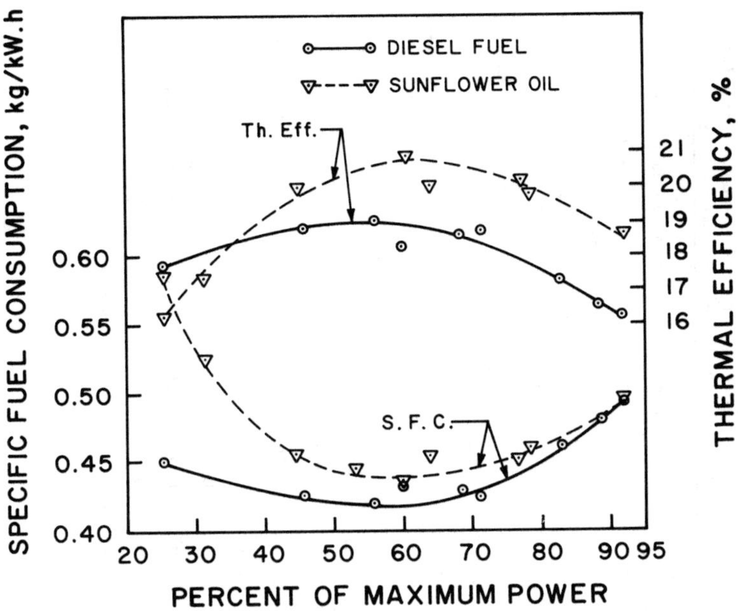

Fig. 5. Specific Fuel Consumption and Thermal Efficiency of No. 2 Diesel Fuel and Crude-Degummed Sunflower Oil.

2% respectively higher than No. 2 diesel fuel. The higher specific fuel consumption in the case of sunflower oil is due to its lower energy content; about 13% less than No. 2 diesel fuel. It was observed during the fuel performance tests that sunflower oil was not able to flow through the secondary filter even at room temperature without the addition of a fuel transfer pump in the fuel system. Engine starting problems and heavy knocking were also encountered.

Summary of the Results

All the results, determined by the Alberta Research Council and the authors, for sunflower oil are summarized and compared with No. 2 diesel fuel in Table 4. The following conclusions can be drawn from the results:

(a) The most detrimental parameter in the use of sunflower oil is its higher viscosity which is about 14 times higher than diesel fuel at 37.78°C. The higher viscosity will lead to blockage of fuel lines and filters, higher nozzle valve opening pressures and poor atomization. The poor atomization will lead to injector coking and contamination of the lubricating oil.

(b) The problem of higher viscosity can be solved by transesterification of sunflower oil to its methyl ester. Although the viscosity curve of the methyl ester up to 0°C is quite close to No. 2 diesel fuel, its pour point (-4.0°C) is higher which indicates the cessation of fuel flow just a few degrees celsius below this temperature. It will be satisfactory in tropical climates provided it is economical to use.

(c) The cetane number of sunflower oil is a little less than the minimum value of 40 for No. 2 diesel fuel. Knocking and starting problems were encountered when burnt in the compression ignition engines due to its low cetane rating and its high boiling point.

(d) Specific fuel consumption was higher due to its lower energy value whereas thermal efficiency was satisfactory when compared with No. 2 diesel fuel.

Table 4. Crude-Degummed Sunflower Oil Properties Determined by the Alberta Research Council (ARC) and Authors.

Fuel Property	Sunflower Oil		No. 2 Diesel		ASTM No.	Probable source of variation
	ARC	Author	Determined	Published[a]	ARC/AUTHOR	in sunflower oil results
K. Viscosity, mm^2/s (40/40.56°C)	30.6	31.17	-	$2.0^{b}-4.3^{c}$	D524/D2983	Different methods used
Cloud Point, °C	-6.0	-7.77	-9.0	-	D2500/D2500	Precision of temperature measuring device
Pour Point, °C	-12	-15.0	-18.5	-	ISO3016[d]/D97	Oxidation of oil and different methods
Flash Point, °C	318	214.44	67.5	-	D92/D93	Different methods used
Calorific value, MJ/kg	-	39.646	45.676	-	-/D240	
API gravity, degrees (15.56°C)	21.3	21.65	35.13	-	D287/D287	Oxidation of oil and precision in taking readings
Density, kg/L (15.56°C)	0.926	0.924	0.849	-	D287	Negligible difference
Cetane No.	34	-	-	$40^{b} - 48^{c}$	D613	
Oxidation Stability, mg/100 ml	78.3	-	-	-	D2274	
Carbon Residue, %	0.39	-	-	0.35^{c}	D526	
Ash Content, %	0.40	-	-	0.01^{c}	D482	
Acidity, mg_{KOH}/g	1.1	-	-	-	D976	
Sulfur, %	0.2	undetectable	-	0.50^{d}	D1552/D129	Different methods used
Corrosion	1a	-	-	No. 3	D130	

[a] From Tractors and Their Power Units by Liljedahl et al. (1979)
[b] Minimum
[c] Maximum
[d] Not an ASTM method and also unknown to the author

(e) Oxidation of sunflower oil left heavy gum and wax deposits on the stationary engine parts and the test bench equipment.
(f) Lower levels of corrosion can be expected on metal parts due to the lower sulfur content in the sunflower oil.
(g) Fire hazards associated with fuel handling will be reduced because of the higher flash point of sunflower oil.

REFERENCES

1. Annual Book of ASTM Standards. 1980. Petroleum products and lubricants. Part 23. Amer. Soc. Testing and Materials.

2. Bruwer, J.J., B.V.D. Boshoff, F.J.C. Hugo, L.M.D. Plessis, J. Fulls, C. Hawkins, A.N.V. Watt and A. Englbrecht. 1980. Sunflower seed oil as an extender for diesel fuel in Agricultural Tractors. Div. Agr. Eng. Dept. of Agriculture and Fisheries. Private Bag X515, Silverston, South Africa. 7 p.

3. Chowbury, D.H. 1942. Indian vegetable fuel oils for diesel engines. Gas and oil power 37(5), Whitehall Press, England. pp. 80-85.

4. Cobia, D.W. and D.E. Zimmer. 1978. Sunflower production and marketing. Ext. Bul. 25. North Dakota State Univ., Fargo, N.D. 73 p.

5. Formo, M.W., E. Jungerman, F.A. Norris and N.O.V. Sanntage. 1979. Bailey's industrial oil and fat products. Vol. 1 (4th Ed.), John Wiley and Sons, N.Y., Toronto. pp. 177-227.

6. Quick, G.R. 1980. Developments in use of vegetable oil as fuel for diesel engines. Amer. Soc. Agr. Eng., St-Joseph, Mich. Paper 80-1525. 15 p.

7. Ramdeen, P., L.F. Backer, K.R. Kaufman, H.L. Kucera and C.W. Moilanen. 1981. Some physio-chemical tests of sunflower oil and No. 2 diesel oil as fuels. Amer. Soc. Agr. Eng., St-Joseph, Mich. Paper NCR 81-009. 21 p.

8. Strayer, R.C., E. Coxworth and R.G. Chopiuk. 1980. Canola oil as a fuel for diesel engines. Can. Soc. Agr. Eng. Annual Meeting. Edmonton, Alberta. Paper 80-302. 7 p.

PRODUCTION AND FUEL CHARACTERISTICS OF VEGETABLE OIL FROM OILSEED CROPS IN THE PACIFIC NORTHWEST[1]

D. L. Auld, B. L. Bettis, and C. L. Peterson[2]

ABSTRACT

The purpose of this research was to evaluate the potential yield and fuel quality of various oilseed crops adapted to the Pacific Northwest as a source of liquid fuel for diesel engines. The seed yield and oil production of three cultivars of winter rape (*Brassica napus* L.), two cultivars of safflower (*Carthamus tinctorius* L.) and two cultivars of sunflower (*Helianthus annuus* L.) were evaluated in replicated plots at Moscow. Additional trials were conducted at several locations in Idaho, Oregon and Washington. Sunflower, oleic and linoleic safflower, and low and high erucic acid rapeseed were evaluated for fatty acid composition, energy content, viscosity and engine performance in short term tests. During 20 minute engine tests power output, fuel economy and thermal efficiency were compared to diesel fuel.

Winter rape produced over twice as much farm extractable oil as either safflower or sunflower. The winter rape cultivars, Norde and Jet Neuf had oil yields which averaged 1740 and 1540 L/ha, respectively. Vegetable oils contained 94-95% of the KJ/L of diesel fuel, but were 11.1-17.6 times more viscous. Viscosity of the vegetable oils was closely related to fatty acid chain length and number of unsaturated bonds (R^2=.99). During short term engine tests all vegetable oils produced power outputs equivalent to diesel, and had thermal efficiencies 1.8-2.8% higher than diesel. Based on these results it appears that species and cultivars of oilseed crops to be utilized as a source of fuel should be selected on the basis of oil yield.

---

[1]Contribution of Deps. Pl. and Soil Sci. and Agric. Eng., Univ. of Idaho, Moscow, ID 83843.

[2]Associate Professor of Plant Breeding and Genetics, Research Associate and Professor of Agricultural Engineering, Univ. of Idaho, Moscow.

INTRODUCTION

Agricultural production on the 5.1 million cultivated hectares in the Pacific Northwest currently requires 840 million liters of liquid fuel annually (6). In recent years, the agricultural industry of this region has been threatened by the increased costs and limited supplies of conventional petroleum fuels. The production of liquid fuels from agricultural products could insure a continuous supply of fuel for this industry.

Sunflower (*Helianthus annuus* L.), safflower (*Carthamus tinctorius* L.) and winter rape (*Brassica napus* L.) produce the highest yields of vegetable oil in the Pacific Northwest (6). The objective of this study was to evaluate the potential oil production of these crops and to determine the effect of their differential fatty acid composition on viscosity, heat content, and fuel performance during short-term diesel engine tests.

MATERIALS AND METHODS

The three species of oilseed crops were evaluated under dryland conditions at Moscow, Idaho from 1978 to 1981. Several cultivars of each species were grown in separate replicated nurseries. Individual plots were 3.5 m^2. Cultural practices which insured maximum seed yield were standardized in all nurseries. At maturity the seed of each plot was harvested and oil content was measured with a Newport Analyzer, MKIIIA (Newport of North America, Inc, Villanova, PA). Total oil yield was adjusted by multiplying by 0.75 for safflower, 0.70 for sunflowers, and 0.85 for winter rape. These values reflect actual yields obtained from a small scale extraction procedure at the University of Idaho.

Nurseries of the three species of oilseed crops were also planted at several locations in Idaho, Washington and Oregon during 1979 and 1980 as part of a project supported by the Pacific Northwest Regional Commission. Nurseries were maintained using optimum management practices including irrigation at those sites where cereal production required irrigation. Data acquisition was similar to the procedure described above.

Five vegetable oils were obtained from commercial processors. High erucic acid rapeseed oil was obtained from Coast Trading Company, Great Falls, Montana; low erucic acid rapeseed oil from C.S.P. Foods, Ltd., Saskatoon, Saskatchewan, Canada; sunflower oil from Cargill, Inc., Minneapolis, Minnesota; and oleic and linoleic safflower oils from Pacific Vegetable Oils International, Richmond, California. All oils had been filtered and alkali refined by the respective processors. Each oil was evaluated to determine 1) fatty acid composition; 2) the physical characteristics of specific gravity, viscosity, heat of combustion; and 3) the fuel performance characteristics of power, fuel consumption, specific energy and thermal efficiency.

In order to determine fatty acid contents, each oil was converted to its respective fatty acid esters by the transesterification technique of Mason and Waller (5). Additional water washing and drying over anhydrous sodium sulfate was conducted to prepare samples pure enough for gas liquid chromatography (GLC) analysis. Chromatographic analyses were performed on a Packard-Becker Model 419 gas chromatograph with a flame ionization detector. Glass columns 1829 x 6 mm O. D. were packed with 10% SP-2330 on 100-120 mesh Chromasorb WAW (Supelco, Inc., Belefonts, PA). The injection port, column, and detector were maintained at 220, 195 and 235 C respectively. Nitrogen at 20 ml/min., was used as a carrier gas.

Physical characteristics of the five vegetable oils and No. 2 diesel fuel were evaluated in accordance with the American Society for Testing and Materials (ASTM) guidelines for petroleum products. Specific gravity was determined using hydrometers as per ASTM D28-67 (1). The kinematic viscosity of the oils was measured at 37.78 C using Cannon-Fenske viscometers as per ASTM D445-79. Heat of combustion was determined in a Parr Model 1241 bomb calorimeter with an adiabatic jacket. For this purpose samples were enclosed in gelatin capsules to prevent volatilization during handling. ASTM D240-76 was used as a procedural guide.

Each oil was evaluated as a fuel in short-term engine tests on a Ford, 4-cylinder, 2818 cc, diesel engine equipped with separate fuel filters for diesel and vegetable oils. This engine has hole-type, direct fuel injection nozzles. Each vegetable oil was used to power the engine for 20 minutes during which (i) power and (ii) fuel consumption were measured at 5 minute intervals and (iii) the temperature of the incoming air, oil, engine coolant, fuel and exhaust gases were monitored with thermocouples and automatically recorded at 1 minute intervals. Diesel fuel was used to power the engine for 20 minutes before and after each vegetable oil test and the same data was recorded to detect any changes in engine performance during the testing procedure. Specific energy, a measure of power output per volume of fuel, and thermal efficiency were derived from these data. After each oil test, the vegetable oil fuel filter housing was cleaned and the spin-on type element was replaced. Filters were located immediately ahead of the fuel pump to minimize mixing of the oil types between tests. Each oil was tested three times and the data were analyzed using a completely randomized design.

RESULTS

Crop Production

Under dryland conditions at Moscow, Idaho the three winter rape cultivars produced an average of 1790 L/ha of farm extractable vegetable oil (Table 1). The safflower and sunflower cultivars produced an average of 725 and 640 L/ha of farm extractable vegetable oil, respectively. Winter rape produced the highest oil yield of the three species evaluated. The three cultivars of winter rape ranged from 1530 to 2020 L/ha of extractable oil. Selection of high yielding cultivars such as Norde and Jet Neuf would greatly improve oil yields. Norde produces high concentrations of erucic acid and Jet Neuf produces high levels of oleic acid.

In the regional trials winter rape produced an average of 1560 L/ha of farm extractable oil compared to 660 and 615 for safflower and sunflower, respectively (Table 2). Winter rape produced over twice as much oil as the other two species due to its high seed yield, relatively high oil content and high efficiency of extraction (0.85). The winter rape cultivars Norde and Jet Neuf also produced the highest yields in the regional trials.

Physical Properties

Vegetable oils consist primarily of triglycerides with fatty acid chains 16 to 22 carbons in length (8). Classification of the vegetable oils by their predominant fatty acid resulted in three groupings (Table 3). Both sunflower and linoleic safflower oil had high concentrations of linoleic acid. Oleic safflower and oleic spring rape oils were high in oleic acid and erucic winter rape oil was high in erucic acid. The erucic winter rape oil contained significant amounts of fatty acids longer than 18 carbons, with 64% of the fatty acids being either erucic acid or its biosynthetic precursor eicosenoic acid. Only sunflower and oleic spring rape oils contained linolenic acid. All oils contained significant amounts of oleic and linoleic acids.

The vegetable oils did not differ in specific gravity, but all were 7 - 9% heavier than diesel fuel (Table 4). The heat of combustion did not differ between vegetable oils, but vegetable oil contained only 94 - 95% as many KJ per liter as diesel. Viscosity of the five vegetable oils was 11 to 17 times higher than that of diesel fuel. Multiple regression analysis indicated that 99% of the variation in viscosity of the vegetable oils was due to the fatty acid chain length, the number of unsaturated bonds and the interaction between these two components (Fig. 1).

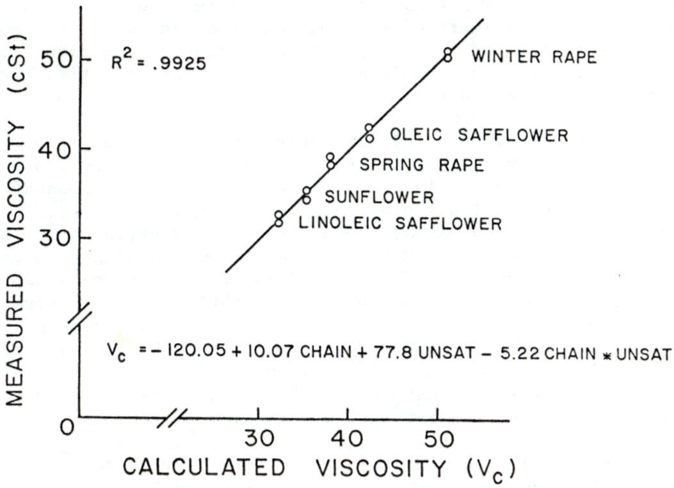

Fig. 1: Comparison of estimated viscosity with actual viscosity of five vegetable oils. HEAR and LEAR refer to winter rapeseed oil with high and low erucic acid concentrations, respectively.

Eq. (1).

$$\text{Chain} = \frac{(\text{Chain Length}) \quad (\%\ \text{Fatty Acid})}{100}$$

$$\text{Unsat} = \frac{(\text{Unsaturated bonds}) \quad (\%\ \text{Fatty Acid})}{100}$$

$$\text{Viscosity} = -120.05 + 10.07\ \text{Chain} + 77.8\ \text{Unsat} - 5.22\ \text{Chain} * \text{Unsat}.$$

Short Term Engine Performance

During short term engine tests, the average power output of the engine did not vary significantly for different fuel types (Table 5). Fuel consumption did not differ between fuels, but No. 2 diesel had the lowest rate of fuel comsumption by weight and the highest rate of consumption by volume (Table 5). The lower rate of consumption of diesel fuel by weight was expected due to its lower specific gravity but it is not clear why a greater volume of diesel fuel was consumed since diesel fuel contains more energy per unit volume than do vegetable oils. The vegetable oils did not differ in their power produced per unit of fuel (specific energy). All vegetable oils except linoleic safflower oil produced more power per liter than did diesel fuel. Thermal efficiency of all vegetable oils was significantly higher than that of diesel, indicating that the chemical energy of vegetable oil was more efficiently converted to mechanical power than that of diesel fuel. This could be why both vegetable oil and diesel produced equivalent power outputs even though the vegetable oils contained 5% less energy per volume than diesel fuel. Oleic safflower oil had a higher thermal efficiency than linoleic safflower oil even though they were produced by varieties from the same genus and species.

DISCUSSION

Under both dryland conditions at Moscow and at regional trials located throughout the Pacific Northwest winter rape produced the highest oil yields. Successful production of vegetable oil as a fuel is dependent on selecting the highest yielding cultivars of the best adapted species. The use of winter rape cultivars such as Norde and Jet Neuf would require the diversion of less land to produce farm fuel and could make vegetable oil competitive with diesel at a much lower price than other oilseed species.

The energy content of vegetable oils is very similar to diesel, but vegetable oils are 11-17 times more viscous. The high viscosity of vegetable oils may be overcome in several ways. Increased concentrations of monounsaturated and shorter chain fatty acids would reduce viscosity. Fatty acid composition of safflower and rapeseed oils have been altered by genetic manipulation (3, 4). Sunflowers grown at higher latitudes produce a higher concentration of polyunsaturated fatty acids (7). The minimum viscosity obtainable through genetic manipulation would still result in vegetable oils that were several times more viscous than diesel fuel. This viscosity would probably cause fuel flow problems under cold operating conditions. Chemical modification of the vegetable oils to form methyl or ethyl esters of the fatty acids could decrease the viscosity still further (2), but this process may be too energy intensive and complex to be practical in farm size operations.

Although small statistical differences were found between vegetable oils and diesel fuel for fuel consumption, specific energy and thermal efficiency, for practical purposes vegetable oils appear to be equivalent to diesel fuel in short term engine operation. However, long periods of using vegetable oil fuel may cause the formation of gum and carbon deposits which may cause mechanical problems in the engine. Further testing is necessary to determine the true reduction in engine life that occurs as a result of using vegetable oils as a fuel. The high viscosity or level of polyunsaturation of vegetable oils may be a major factor in carbon deposits in the combustion chamber and exhaust ports. A study in South Africa indicated that higher viscosities resulted in incomplete atomization of the fuel (2). Incomplete atomization of high viscosity fuels in turn prevents complete combustion of the larger droplets and results in carbon deposits. The utilization of additives that minimize viscosity and/or inhibit the formation and deposition of gum in the injectors and combustion chambers may be necessary to allow long term engine operation on vegetable oils.

Despite these problems there is still potential for the use of vegetable oils as an alternative to diesel fuel. Compared to alcohol the production and processing of vegetable oils is far more energy efficient. Vegetable oils can be utilized in unmodified diesel engines and a small percentage of a farm's acreage could produce enough fuel to supply farm needs. Further research should concentrate on developing fuel additives or pre-treatments to reduce engine deposits; developing extraction and processing techniques that would allow vegetable oil extraction and utilization at a local level; and identification of crop species and specific cultivars that produce optimum yields of fuel quality vegetable oil.

REFERENCES

1. Annual Book of ASTM Standards. 1979. Petroleum products and lubricants (1). Amer. Soc. for Testing and Materials. Philadelphia Part 23.

2. Bruwer, J. J., B. Van D. Boshoff, F. J. C. Hugo, L. M. du Plessis, J. Fuls, C. Hawkins, A. N. Van der Walt and A. Engelbrecht. 1980. Sunflower seed oil as an extender for diesel fuel in agricultural tractors. Div. of Agric. Eng. Dep. of Agric. and Fish. Rep. of South Africa.

3. Calhoun, W. 1977. The lipid crops. In Calhoun, W. Industrial Crops. Oregon State Univ. Bookstore, Corvallis.

4. Downey, R. K. and B. L. Harvey, 1963. Methods of breeding for oil quality in rape. Can J. Plant Sci. 43:271-275.

5. Mason, M. E. and G. R. Waller. 1964. Dimethoxypropane induced transesterification of fats and oils in preparation of methyl esters for gas chromatographic analysis. Anal. Chem. 36:583-586.

6. Peterson, C. L., D. L. Auld, V. M. Thomas, R. V. Wither, S. M. Smith and B. L. Bettis. 1980. Vegetable oil as an agricultural fuel for the Pacific Northwest. Idaho Agric. Exp. St. Bull. No. 598.

7. Robertson, J. A., W. H. Morrison, III and R. L. Wilson. 1979. Effects of planting location and temperature on the oil content and fatty acid composition of sunflower seeds. USDA, SEA, Ag. Res. Results ARR-S-3. New Orleans.

8. Zadernowski, R. and F. Sosulski. 1978. Composition of total lipids in rapeseed. Oil Chem. Soc. 55:870-872.

Table 1. Farm extractable vegetable oil production of selected cultivars of safflower, sunflower, and winter rape in experimental plots at Moscow, Idaho

Species Cultivar	Test Year 1978	1979	1980	1981	Average
	L/ha[1]				
Safflower					
S208	730	930	--	440	700
UC-1	760	990	--	500	750
Sunflower					
Hybrid 894	850	790	460	511	650
Nk 254	780	850	420	458	630
Winter Rape					
Norde	--	2720	1640	1700	2020
Dwarf Essex	--	1970	1420	1190	1530
Jet Neuf	--	2050	1730	1690	1820

[1]Determined by multiplying total oil production measured by nuclear magnetic resonance by 0.75 for safflower, 0.70 for sunflowers and 0.85 for winter rape.

Table 2. Farm extractable vegetable oil production of selected cultivars of safflower, sunflower, and winter rape in experimental plots at several locations in the Pacific Northwest during 1979 and 1980.

Species Cultivar	Seed Yield	Oil Content	Oil Yield[1]
	Kg/ha	-%-	L/ha
Safflower (14 locations)			
S208	2180	39.0	690
UC-1	1880	40.0	630
Sunflower (7 locations)			
Hybrid 894	1800	47.5	580
Nk 254	2020	46.4	650
Winter Rape (9 locations)			
Norde	3730	45.4	1740
Dwarf Essex	3200	45.7	1390
Jet Neuf	3750	42.9	1540

[1]Determined by multiplying total seed oil production measured by nuclear magnetic resonance by 0.75 for safflower, 0.70 for sunflowers and 0.85 for winter rape.

Table 4. Specific gravity, viscosity, and heat of combustion of five vegetable oils and No. 2 Diesel.

Oil Type	Specific Gravity		Kinematic Viscosity		Heat of Combustion	
	-g/ml-	-ratio-[1]	mm^2/s	-ratio-[1]	-KJ-	-ratio-[1]
Sunflower	0.92	1.08	34.9d[2]	12.0	2081b[2]	0.95
Linoleic Safflower	0.93	1.09	32.3e	11.1	2082b	0.95
Oleic Safflower	0.92	1.08	42.1b	14.5	2064b	0.94
Oleic Spring Rape	0.92	1.08	39.0c	13.4	2096b	0.95
Erucic Winter Rape	0.91	1.07	51.0a	17.6	2086b	0.95
No. 2 Diesel	0.85	1.00	2.9f	1.00	2196a	1.00

[1]Relative to No. 2 diesel.
[2]Means within a column not followed by the same letter differ at the 0.05 level of probability according to Duncan's new multiple range test.

Table 5. Power output, fuel consumption, specific energy and thermal efficiency of five vegetable oils and No. 2 diesel during short term engine tests.

Oil Type	Power	Fuel Consumption		Specific Energy	Thermal Efficiency
	-Kw-	-Kg/hr-	-L/hr-	-Kw-hr/L-	%
Oleic Safflower	30.0 a[1]	10.2 a[1]	11.0 a[1]	27.2 a[1]	27.0 a[1]
Linoleic Safflower	29.5 a	10.4 a	11.2 a	26.4 ab	26.0 b
Erucic Winter Rape	29.4 a	10.0 a	10.9 a	26.9 a	26.6 ab
Oleic Spring Rape	29.7 a	10.2 a	11.1 a	26.9 a	26.5 ab
Sunflower	29.5 a	10.1 a	10.9 a	27.0 a	26.7 ab
No. 2 Diesel	29.5 a	9.7 a	11.4 a	25.9 b	24.2 c

[1]Means within a column not followed by the same letter differ at the 0.05 level of probability according to a protected Duncan's multiple range test.

Table 3. Fatty acid composition of five vegetable oils as determined by gas chromatography.

Oil Type	Palmitic 16:0[1]	Stearic 18:0	Oleic 18:1	Linoleic 18:2	Linolenic 18:3	Eicosenoic 20:1	Erucic 22:1
	------------	------------	% by weight	methyl esters[2]	------------	------------	------------
Sunflower	6.4	4.2	23.9	61.4	3.0	--	--
Linoleic Safflower	7.1	2.5	13.3	76.6	--	--	--
Oleic Safflower	5.4	2.1	75.3	16.1	--	--	--
Oleic Spring Rape	4.4	1.8	59.8	20.6	11.1	--	0.9
Erucic Winter Rape	3.2	1.1	14.8	12.7	--	17.7	46.7

[1]Number of carbons and unsaturated bonds, respectively, for each fatty acid.

[2]May not sum to 100% due to the presence of other minor fatty acids.

VEGETABLE OIL FUEL STANDARDS*

E. H. Pryde
Executive Affiliate ASAE

The concept that engine design is all-important in the use of vegetable oils as a diesel fuel has been confirmed by J. J. Bruwer in the Republic of South Africa. One hundred percent sunflower oil cannot be used safely in a direct injection diesel engine but can be used in an indirect injection, naturally aspirated and air-cooled engine for long periods of time. Questions still remain unanswered on what the American farmer can do to fuel his direct injection farm tractor during a diesel oil shortage. Blends containing about 20% vegetable oil in the diesel oil might be used safely, but such blends would not be a solution to the problem during extended shortages.

Short-term engine tests indicate good potential for whole vegetable oil fuels. Long-term endurance tests show that there are serious problems in injector coking, ring sticking, gum formation, and lubricating oil thickening and gelation. These problems are related to the high viscosity and non-volatility of the vegetable oils, which result in inadequate fuel atomization and incomplete combustion. Fines from seed hulls and meal cause similar problems and must be removed by filtration through extra-fine filters.

Methods being investigated for reducing vegetable viscosities include: dilution with diesel oil, other petroleum products, alcohols, and various additives; heating the fuel lines; using simple esters in place of triglyceride oils; and forming microemulsions with aqueous ethanol. The last two approaches are being investigated at NRRC.

Once the major problem of viscosity has been overcome, a number of secondary problems will require attention such as the effects of free-fatty acids, phosphatides, and degree of unsaturation. Also needed will be studies on additives to improve cyclinder lubrication, fuel atomization, combustion, viscosity, and oxidative stability of the vegetable oil.

The American farmer continues to seek energy self-sufficiency. However, on-farm production of vegetable oil fuel raises a number of questions including: economic feasibility, quality control, engine warranties, storage and palatability of feed products from expeller presses, and recovery and marketing of byproducts such as glycerol.

The author is: E. H. PRYDE, Research Chemist, Northern Regional Research Center, Agricultural Research Service, U.S. Department of Agriculture, Peoria, Illinois 61604.

*The mention of firm names or trade products does not imply that they are endorsed or recommended by the U.S. Department of Agriculture over other firms or similar products not mentioned.

Questions on use of standard vegetable oil fuels and on fuel specifications need attention. Vegetable oil refiners can supply oils having various degrees of refining and a high order of reproducibility; such oils can be considered standard. Vegetable oil specifications for fuel use can be developed and proposed, but many final specifications can be developed only after completion of exploratory engine tests. An important specification that can be suggested now is the measurement of suspended matter by turbidimetric or nephelometric means. All vegetable oil fuels should be clear and bright and free of turbidity.

DISCUSSION

Suggested standards for vegetable oils and ester fuels, as well as ASTM specifications for No. 2 diesel oil, are listed in Table 1. Some comments seem desirable for some of the properties listed in the table and are given in the following paragraphs.

Physical Properties

Cetane number: The cetane number of several vegetable oils as determined by an indirect method has been reported to be in the range of 35-40. The direct method reportedly gives a cetane number that is about 10 units lower, and the value of cetane number ratings as applied to vegetable oil fuels has been questioned. A more reliable criterion for successful combustion in a diesel engine may be how well the fuel is atomized upon injections from a standard injector tip.

Cloud point: Many vegetable oils have some triglycerides with saturated fatty acids, e.g., soybean oil, and such triglycerides have higher melting points than the triglycerides with mostly unsaturated fatty acids. Consequently, many vegetable oils have relatively higher cloud points than diesel oil. If the high cloud point is undesirable, the vegetable oil can be winterized at 4-5°C and then filtered to remove most of the saturated fat, or about 10% of the starting oil in the case of soybean oil. The suggested specification of 20°C would exclude such fats as tallow and babassu, coconut, and palm oils, all of which are solids at 20°C. Of course, these oils have been used as diesel fuel in the past, but the fuel tank and lines had to be heated.

Distillation temperatures: The ASTM D-975 specifications require that the 90% point for the distillation fall in the 282-338°C temperature range. Distillation of vegetable oils and esters will not take place at these temperatures at atmospheric pressure. Instead, decomposition (pyrolysis) will gradually begin at pot temperatures of 300°C and the "distillate" will consist of decomposition products; the 90% point will never be reached. In "distilling" a vegetable oil at atmospheric pressure, the first drop of distillate may occur at about 150°C, the first 10% at 150-300°C, and by the 80% point a vapor temperature of more than 350°C will have been reached. Nevertheless, a specification of a 100-250°C distillation range at 2 mm of mercury with a distillation residue maximum of 10% may be a useful guide for characterizing the identity of a vegetable ester.

Flash point: Vegetable oils and esters can be distilled only under reduced pressure. At atmospheric pressure, decomposition rather than volatilization will start at about 300°C, so the flash points for vegetable oils and esters will be considerably higher than that for diesel oil. The presence of low-boiling solvents or additives will lower the flash points. Free fatty acids in the vegetable will also lower the flash point because of the greater volatility of the free fatty acids compared to the oil.

Pour point: The same comments for cloud point apply to pour point.

Table 1. Suggested Standards for Vegetable Oil and Ester Fuels vs. Diesel Oil Specifications

Property	Test method[a]		Suggested standards		ASTM specifications for no. 2 diesel oil (D975)
	AOCS	ASTM	Oil	Ester	
Physical					
Cetane number (min.)	-	D613	(30-40)	(30-40)	40
Cloud point (max.)	Cc6	D2500	20°C	20°C	6°C above 1/10 percentile minimum ambient temperature[b]
Distillation temperatures, 90% point	-	D86			
min.			(>300°C)	(>300°C)	282°C
max.					338°C
Flash point (min.)	Cc9a	D93	(∿300°C)	(∿300°C)	52°C
Pour point (max.)	-	D97	-5°C	-5°C	-
Specific gravity, 15/15°C	Cc10a	D1217	0.910-0.930	(0.8-0.9)	-
Turbidity	-	-	-	-	-
Viscosity, cSt @ 100°F	-	D445	30-50	(3-6)	1.9-4.1
Compositional					
Ash (max.)	Ca11	D482	0.05%	0.05%	0.01%
Carbon residue on 10% residuum (max.)	-	D524	-	-	0.35%
Copper strip corrosion (max.)	-	D130	-	-	No. 3
Free fatty acid (max.)	Ca5a	-	0.2%	0.2%	-
Insolubles (max.)	Ca3	D128	0.001%	0.001%	-
Iodine value	Cd1	-	80-145	80-145	-
Moisture (Karl Fischer) (max.)	Ca2e	D1744	0.2%	0.2%	-
Phosphorus (max.)	Ca12	D1091	0.02%	0.02%	-
Sulfur (max.)	-	D129	-	-	0.50%
Volatile matter (max.)	Ca 2b, c, or d	-	0.3%	0.3%	-
Water and sediment (max.)	-	D1796	-	-	0.05%
Wax (max.)	-	D3117	0.02%	0.02%	-

[a] AOCS = American Oil Chemists' Society; ASTM = American Society for Testing and Materials. Values in parentheses are actual values or ranges and not necessarily a suggested specification.

[b] For example, the 10th percentile minimum temperature for Northern Illinois is 16°F for November and 12°F for March, bracketing the winter period when there would be minimum tractor activity in the field.

Turbidity: Turbidity of a vegetable oil, whether caused by vegetable waxes or water or insoluble matter, is a good sign that trouble may be ahead in its use as a fuel. The same comment applies to diesel oil. Specifications should be set up for both vegetable oil and diesel oil fuels.

Viscosity: This is perhaps the most important property of all with respect to fuel properties; the high viscosity of vegetable oils compared to diesel oil is their most distinguishing feature. A suggested specification of 30-50 cSt at 100°F would include most vegetable oils except castor oil. Castor oil has been used as a fuel, but only when heated to reduce viscosity. Esters have viscosities (4-7 cSt at 100°F) almost within the specification range for No. 2 diesel oil. There is evidence that fuels with high viscosities, up to perhaps 8-10 cSt at 100°C, may be acceptable for some diesel engines.

Free fatty acids: Crude vegetable oils ordinarily have less than 1% free fatty acid, except for crude palm oil which may have as much as 10-15%. Free fatty acids have as much fuel value as the oil, but their effect on corrosion of some engine parts is unknown. At elevated temperatures, fatty acids will react with many metals, and fatty acid metal salts could be introduced into the engine cylinder could increase wear. Free fatty acids can be removed by physical (steam) or alkali refining procedures, or they can be neutralized by ammonia or amines to form ammonium soaps.

Iodine value: The higher the iodine value, the greater the unsaturation (number of double bonds) in the vegetable oil. Unsaturation is a desirable property to maintain liquidity at low temperatures but undesirable with respect to oxidative stability. The suggested specification range of 80-145 would exclude the high-melting, saturated fats and also the varnish-forming linseed oil (IV $\geqq$155).

Phosphorous: Vegetable oils contain phospholipids, such as phosphatidyl choline, that can pick up moisture from the atmosphere to form insoluble gums that form deposits in storage tanks and lines. The phosphatides can be removed easily by treatment with small amounts of water followed by centrifugation or by treatment with aqueous alkali, which removes both the phosphatides and free fatty acid, followed by centrifugation. The effects of the phosphorus compounds on combustion or on the engine parts are unknown. The amounts of phosphatides contained in the vegetable oils varies from oil to oil. Sunflower oil contains relatively small amounts (up to 0.5%) of phosphatides, but soybean oil contains relatively large amounts (up to 2.0%). To prevent problems in fuel tanks, lines, and filters, the phosphatides should be removed from soybean oil. Whether they need to be removed from other oils having lesser amounts remains to be determined.

Wax: Vegetable waxes consist of the esters of long-chain fatty acids and alcohols and are distinct from the long-chain paraffin waxes that appear in diesel oil. The waxes in crude sunflower oil amount to about 0.1%, more than in most of the other commercial vegetable oils. At this concentration, the waxes give a turbid appearance to the oil, but they gradually settle out at ambient temperatures. The waxes will dissolve when the oil is heated. They can be removed by winterization, by cold refining, or by cold washing with detergent. The presence of phosphatides may interfere with the wax removal. Waxes, if not dissolved in the oil, will interfere with fuel filtration, but their effect on engine operation is unknown. Waxes can be determined by turbidimetric methods, and a specification for a maximum allowable concentration seems desirable.

A wax appearance point, i.e., the temperature at which a wax will crystallize out from the oil when cooled, determined by the ASTM procedure D3117 will give good indication of the wax content in a qualitative but not quantitative sense.

CONCLUSIONS

It is apparent that vegetable oils and their esters cannot meet ASTM specifications D975 for No. 2 diesel oil for use in the diesel engine. Vegetable oil modification or engine design modification may make it possible eventually for vegetable oils to become suitable alternative fuels. However, until the modifications have been tested thoroughly and generally accepted, vegetable oils must be recognized as experimental fuels. Therefore, the suggested specifications for vegetable oils and esters in Table 1 merely serve as a means for determining the quality or even identity of the experimental fuels and do not serve as specifications for fuel use, in contrast to the specifications for diesel fuel which are aimed specifically toward fuel use.

The suggested standards in Table 1 are offered only as a starting point for further discussion, and by no means are they to be taken as definitive.

VEGETABLE OILS - A NEW ALTERNATIVE

Saad Romano*

This paper relate:

1. The use and production of M.E.V.O. as fuel in diesel engines and the effect of the catalyst proportion, alcohol and V.O on the transesterification process.

2. Simple control methods during industrial preparation and the bahavior of V.O. and M.E.V.O on accelerated oxidation test to determine the maximum contration of V.O. in M.E.V.O. that do not cause problems on the injectors.

3. The behavior of M.E.V.O. and V.O. on parafinic and naphtenic lubricants, with high T.B.N. and without organo-metallic compounds, using antioxidants as B.H.T to reduce the oxidation effect.

** V.O. = Vegetable Oil
M.E.V.O. = Methylic ester of vegetable oil.

The use of vegetable oils, animal fats, etc. as source of energy, has already been known by men for a long time. The utilized process was the direct combustion and the combustion itself occurred in the presence of air and at normal pressure conditions. The gases of combustion were: CO_2, CO, H_2O. Because the combustion conditions occurred at an atmospheric temperature and pressure, and the formed gases were dissipated or dilluted in the atmosphere: few were the inconveniences found or observed.

The use of vegetable oils "in nature" or refined as an alternative to the partial or total substitution of Diesel-Oil brings some inconveniences, due to the nature of the oil, its molecular weight, its boiling point and also to the combustion conditions that are very different from the ones in the open air. We have to take into account that a motor is formed by metals, which - during combustion - may serve as catalysts of some by-products, polymer, etc., or still be corroded by the direct reaction with a fatty acid.

Besides all this, we have to consider that the combustion occurres under pressure and the formed gases will have a certain residence time inside the motor before they are expelled.

Another important detail is the oxidation that a vegetable oil may suffer under these combustion conditions, due to the high temperatures and to the existence of a reasonable amount of Oxygen in the presence of metals, which causes a higher vicosity in the oils. The animal or vegetable oils are found as Glycerine esters and have high boiling and flash points. They may suffer a thermal decomposition with the liberation of the Glycerine, and fatty acid. The fatty acid may suffer oxidation or burn but the glycerine ,

* Director of Chemical Spcialties - Brazil

which, due to the combustion in the motor, may produce large amounts of instable Acrolein that can perfectly polymerize with the formation of Disacryl, which is a plastic solid.

It is clear that the thermal decomposition of an oil, which is a Glycerine ester, in the absence of Oxygen leads to different products than the ones formed in its presence.
The report is a study on the use of Methyl and Ethyl Esters as fuels for Diesel engines.
The idea of transforming a V.O. in its methyl ester, has the following goals:

1. Total elimination of the Glycerine
2. Lowering the boiling point of the oil
3. Lowering the flash point of the oil
4. Lowering its viscosity
5. The use of the Glycerine in the Glycerol-Chemistry
6. The use of the by-products by the soap and detergent industry.

The esters, in general, change their boiling and flash points, according to the molecular weight of the utilized alcohol.

Preparation of Methyl Esters of Vegetable Oils:

The V.O. are a mixture of Glycerine esters (mono, di und tri Glycerides)and have a calorofic value that varies, according to the number of carbon and hydrogen atoms in the carbon chain.

Basically, the preparation of methyl esters of V.O. is a transesterification reaction. This reaction occurs in the presence of adequate catalysts.

We may represent the transesterification of oils as following : (Fig. 1)

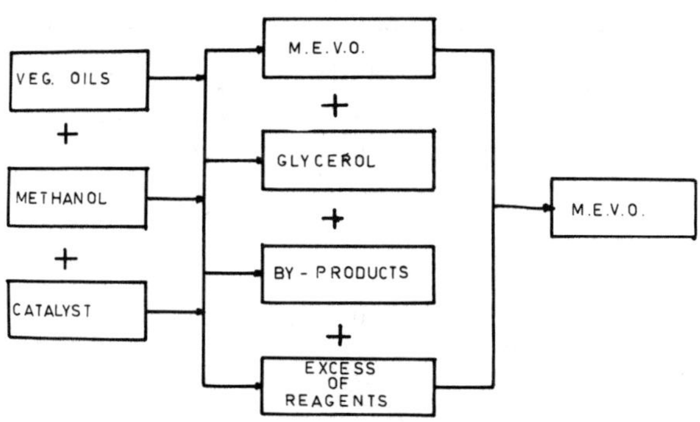

As we note, for the transesterification to occur, there is a need for an excess amount of methyl alcohol besides the catalyst. The excess alcohol is necessary for a higher yield of ester, i.e., it is responsible for a greater percentage of broken Glycerine/fatty acids linkage. But the exageration in the excess of alcohol causes some troubles, such as higher consumption of energy in its recovery and also greater accidental loses. The time of reaction varies according to the excess of alcohols, to the amount of catalysts utilized and also to the temperature in which the reaction occurs.

On table 1 we may observe the amount of catalysts consumed after 20 minutes of reaction and also the yield expressed in percentage of the broken Glycerine/Fatty Acids Linkages as a function of the temperature and of the concentration of Alcohol and Catalyst.

Table 1. Amount of NaOH destroyed after ethanolysis for 20 minutes.

Original Catalyst concentration (%) (a)	Alcohol to Fat ratio (b)	Temperature ºC	Catalyst consumed (%)	Glycerol Linkages Broken(%)
0,1	1:1	50	90,4	49,6
0,2	1:1	50	.85,2	57,7
0,4	1:1	50	89,9	62,7
0,2	2:1	50	59,0	80,9
0,4	2:1	50	63,2	86,8
0,8	2:1	50	60,2	96,7
0,2	2:1	70	98,4	76,3
0,2	2:1	30	52,4	69,0

(a) Percentage of catalyst, calculated on the oil
(b) Expressed in alcohol equivalent by Fatty Acid equivalent

The yield of the esters is affected by the presence of water or humidity, both in the oils and alcohols. On table 2 we present the effects of water in the transesterification of 100 grs. of coconut oil with ethyl alcohol.

Table 2. Effect of water on Ester Interchange reactions

Water in Oil (g) (%)	Ethyl Alcohol 99,5% by weight (g)	Water in Alcohol	Amount of Added water (g)	Total amount of water (g)	NaOH used (g)	% Glycerine yield
0,21	30	0,15	0,00	0,36	0,5	55,4
0,21	30	0,15	0,15	0,51	0,5	40,3
0,21	30	0,15	0,30	0,66	0,5	33,9
0,21	30	0,15	0,60	0,96	0,5	17,6
0,00	30	0,15	0,00	0,15	1,0	88,2
0,21	30	0,15	0,00	0,36	1,0	81,9
0,21	30	0,15	0,30	0,66	1,0	54,1
0,00	40	0,20	0,00	0,20	1,0	94,0
0,21	40	0,20	0,00	0,41	1,0	91,9
0,21	40	0,20	0,40	0,81	1,0	83,9
0,00	30	0,15	0,00	0,15	2,0	95,4
0,21	30	0,15	0,00	0,36	2,0	92,8
0,21	30	0,15	0,30	0,66	2,0	73,2

The best conditions of work with Methyl Ester are that the temperature of reaction is around 70ºC, and the heating achieved by reflux of approximately IH30 (agitation accelerates the rate of reaction).

As products of the reaction we obtain the methyl ester, the Glycerine and by-products, which changes, according to the utilized catalyst. For example, if we use metallic sodium, we will obtain Sodium Salts of fatty acids that are no more than soaps.

As a second phase of the process, we have the purification, which must be proceeded always by the recovery of the excess of methyl alcohol. This methyl alcohol is isolated from the mixture by distillation, either at reduced pressure or at atmospheric pressure. Diagram is on fig. 2.

Fig. 2

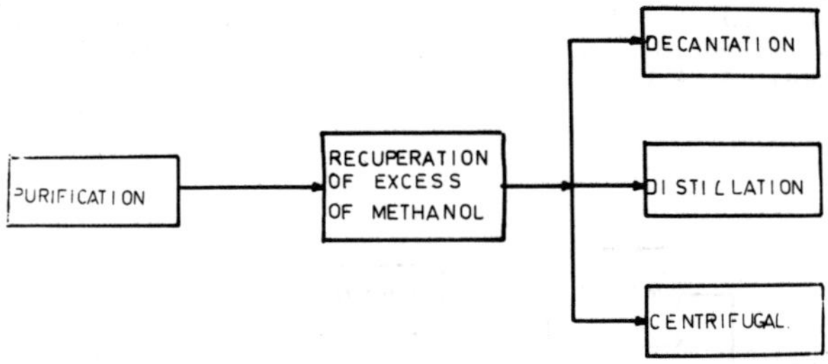

Usually, due to the difference in density, the Glycerine and the by-products tend to decant as we stop the stirring. Their separation from the ester is done by the inferior part of the reactor, but not all of the catalyst, the Glycerine and the by-products can be separated from the ester by this way. So a purification of the ester is necessary, and we represent it in fig.3 as follows:

Fig. 3

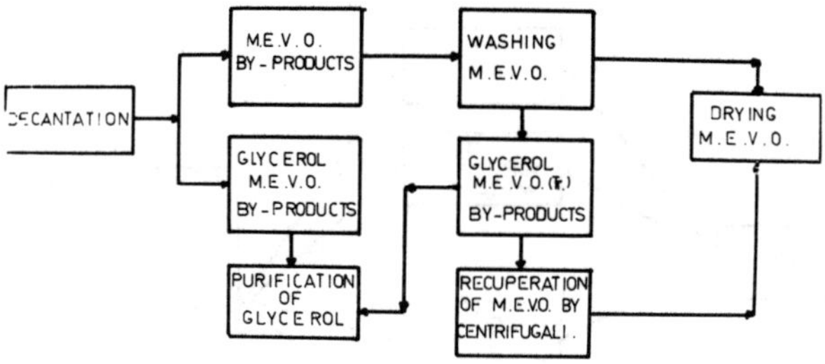

The withdraw of the Glycerine traces and of the catalyst and by-products is done by washing the methyl ester with water, because this ester is insoluble in water, whereas the other products are not.

The separation of the introduced water and also of the by-products is done by decantation or centrifugation: the last system is faster and most efficient. By this process we take off 99% of the introduced water, the rest has to be eliminated by heating at reduced pressure, so as to avoid exporing the methyl esters to high temperatures in the presence of air, otherwise they could be oxydated.

The separation of the methyl ester by centrifugalization seems to be interesting at first sight, because we remove most of the by-products and the Glycerine. But the residual catalyst is not eliminated in this method. This residue of catalyst may cause serious problems, both in the motors and in the storage, because in a basic media there is an attack to the metals, such as aluminum and zinc, with the formation of Aluminates and Zincates. In prolongued storage, in presence of air humidity, we have a move on the equilibrius towards the formation of soap.

A a third option we may separate the ester by vacuum destillation, having the fractionation of the ester as an advantage, that is, the separation of the heavier esters from the lighter ones. In the distillation we will have a small amount of Glycerine carried, that must be eliminated by simple decantation. Diagram is not fig. 4

Fig. 4

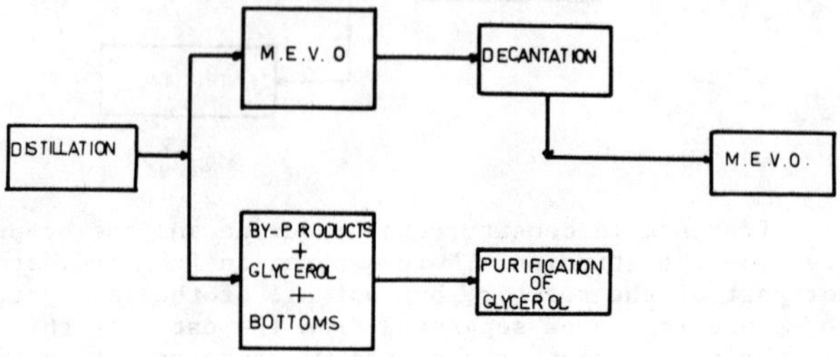

As for the catalysts, we may divide them in four main groups (fig. 5)

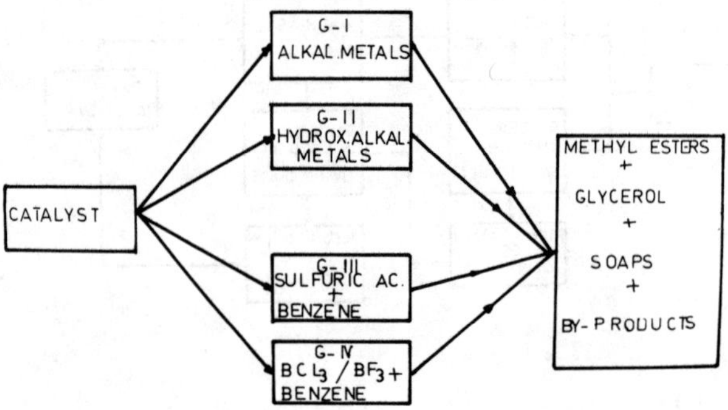

The choise of the catalyst if made according to its costs, availability, desired final by-products, facility of handling and to main restrictions, due to the types of oils used. Restrictions fig. 6)

Fig. 6

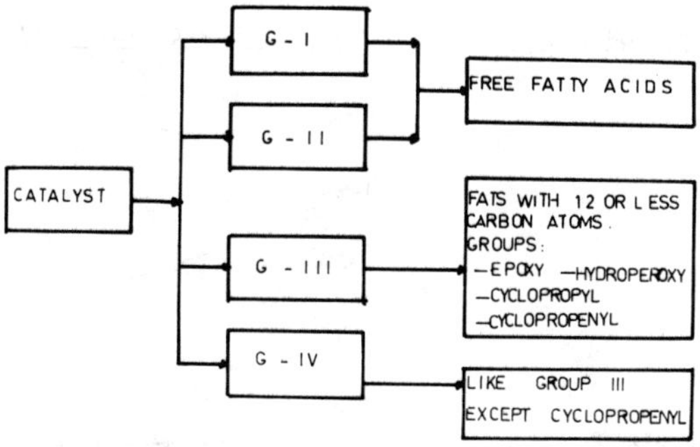

So, that we would follow the transesterification process with relatively simple methods, controlling the esterification grade, we tried to evaluate some variations of the physical characteristics. For that, we prepared mixtures of methyl-ester of soybean oil, reesterified and purified in the laboratory with soybean oil and we obtained the following results:

Table 3. Change of physico-chemical characteristics of M.E.S.O.

V.O. %	M.E.S.O. %	Density (g/cm3	Viscosity Ford cup 4	Refractive Index n_D^{25}
100	0	0,9206	19"	1,4753
90	10	0,9156	18"	1,4738
80	20	0,9146	16"	1,4720
70	30	0,9116	15"	1,4705
50	50	0,9045	13"	1,4670
40	60	0,8995	12,8"	1,4657
30	70	0,8975	12,5"	1,4640
20	80	0,8935	12,0"	1,4622
10	90	0,8894	11,7"	1,4607
5	95	0,8884	11,4"	1,4595
2,5	97,5	0,8874	11,4"	1,4582
1,0	99,0	0,8864	11,4"	1,4590
0,0	100,0	0,8864	11,4"	1,4590

The samples must be without free methyl alcohol, if not, the obtained results would be quite different, because the refractive index of methyl alcohol is 1,3390, its density 0,7915 g/cm3and its viscosity Ford Cup 4 at 25ºC is 10".

The viscosities and refraction indexes were determined at 25ºC (Fig. 7 and Fig. 8).

Fig. 7

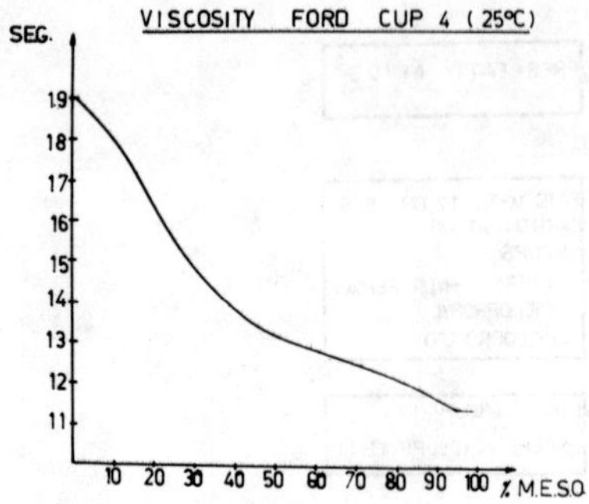

Fig. 8

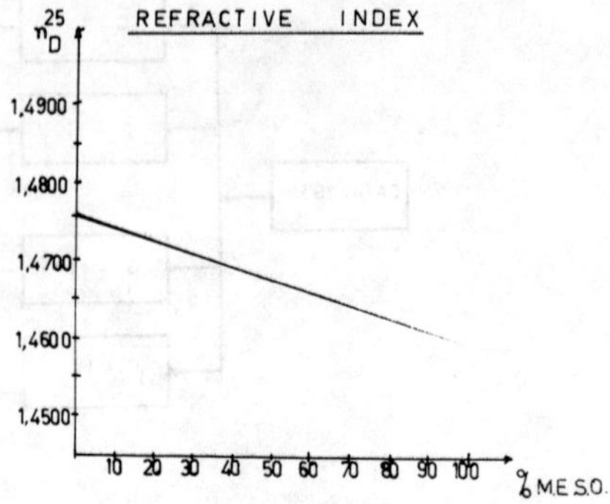

The densities were determined with the use of Pycnometer and analytical balance at 20ºC (fig. 9)

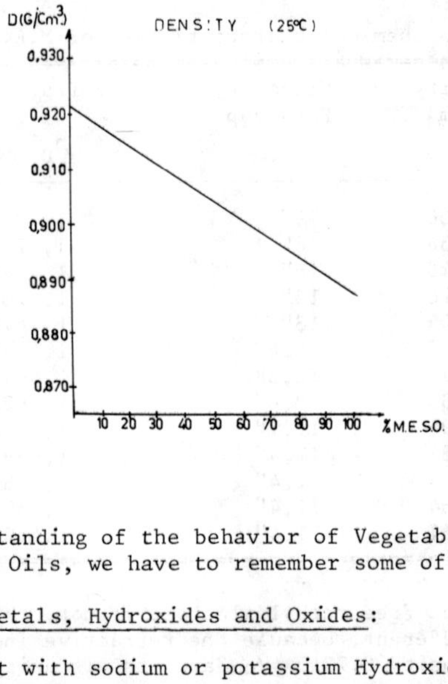

For a better understanding of the behavior of Vegetable Oils and Methylic Esters of Vegetable Oils, we have to remember some of their characteristics:

1. Reactions with metals, Hydroxides and Oxides:

   Fatty acids react with sodium or potassium Hydroxides, oxides and carbonates, to produce soaps.
   Reactions of the fatty acids with the corresponding compound of such metals as zinc, lead, manganese, cobalt, or tin are more difficult, but are redily accomplished at elevated temperatures.
   Commercially, these reactions are carried out on large scale, particularly in the manufacture of lubricating grease and metallic dryers for paints and other protective coatings, and also in the preparations of thickening lubricating oils. Remember, that the engine is composed by many of these metals and the reaction of combustion of the fuel occurs at elevated temperatures, that can promote the reactions outlined above.

2. Pyrolysis of Methyl and Ethyl Esters:

   Methyl Esters are more thermally stable than Ethyl Esters or other simple

esters that contain a β-Hydrogen in the alcoholic portion of the carbalkoxy group. Ethyl Esters are capable of relatively efficient pyrolysis at about 400°C to ethylene and the corresponding fatty acid.
Methyl Esters can not pyrolyze to olefinic derivates. On heating Methyl Oleate in the absence of air at 280°C, Sen Gupta isolated Methyl decanoate, 9 octadecene, dymethyl 9 octadecene 1,18 dioate, 9,17 hexacosadiane, methyl 9,17-hexacosadienoate and dimethyl 9,17-hexacosadiene-1,26 dioate as the principal decompostion products.

3. Atmospheric Oxidation:

The oxidation of vegetable oils is due to:

a) Bacterial action,
b) Enzyme-catalized hydrolysis and oxidation,
c) Direct chemical attack by oxygen of the air - autoxidation

Autoxidation can not be checked easily since it is a chemical reaction with a very low activation energy, estimated at 4-5-Kcal/mole and 6-14 Kcal/mole for the two reactions. Polyunsaturated components of fats are oxidized more rapidly than monounsaturated and saturated components. (table 4)

Table 4. Relative rates of oxidation of fatty acids

Fatty Acid	Common Designation	Temp: 20°C	37°C	40°C	100°C
Stearic	C : 18:0	-	-	-	0,6
Oleic	C : 18:1	4	-	-	6
Linoleic	C : 18:2	48	42	-	6,4
Linoleic	C : 18:3	100	100	100	100
Arachidonic	C : 20:4	-	199	-	-
α-Eleostearic	C : 18:3	-	-	515	-

The rates of autoxidation of methyl oleate, methyl linoleate, methyl linolenate are 1:12:25, so vegetable oils rich in polyunsaturated fatty acids are more exposed to autoxidation than vegetable oils rich in saturated or unsaturated fatty acids.

When the autoxidation of fat is followed experimentally, either by measuring the amount of oxygen absorbed or by determining the peroxide value of the fat, it is found that the course of oxidation exhibits two distinct phases. During the inicial phase, oxidation proceeds at a relatively slow and more or less uniform rate. Then, a certain critical amount of oxidation has occured, the reaction enters a second phase, characterized by a repidly accelerating rate of oxidation and by an eventual rate many times greater than that observed in the inicial phase. The inicial period of relatively slow oxidation of a fat is called the induction period.
The amount of oxygen that must be absorved to produce rancidity varies considerably according to the composition of the oil, to the presence of natural or added antioxidants, to the metal content (especially copper, iron and heavy metals) and to the temperature and conditions under which oxidation is carried out. Fats high in Oleic acid and low in linoleic, linolenic or other polyunsaturated acids become rancid after the absorption of less oxygen than fats in which the ratio of these acids is reversed
For determining of the rates of auto-oxidation between soyabean oil and the respective methyl ester, we have affectuated some accelerated oxidation tests by aerating the samples (30 l/h) at 140°C and by determining the acid value, iodine value and viscosity Ford cup 4 and the results are expressed by Fig. 10, 11, 12.

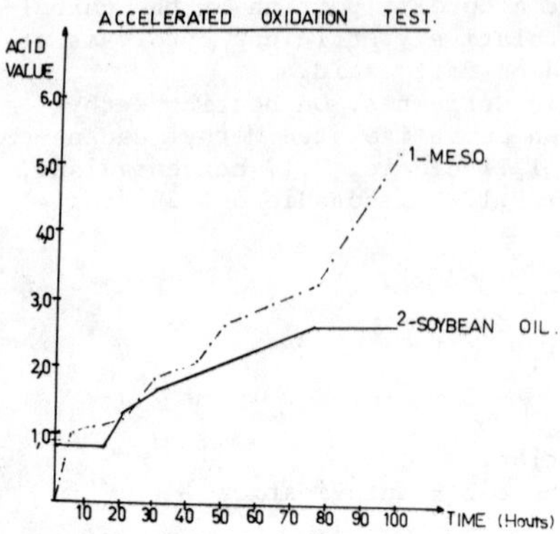

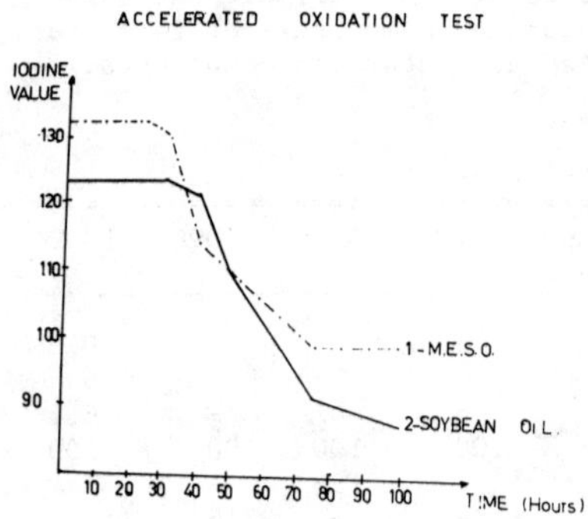

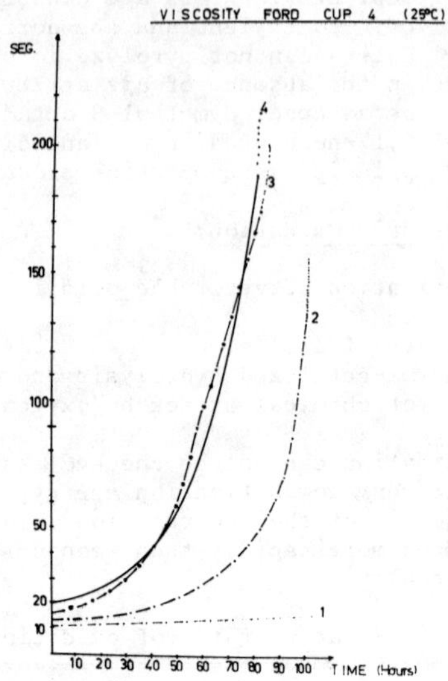

1) M.E.S.O 100%
2) M.E.S.O/Soyabean oil (50/50%)
3) M.E.S.O/Soyabeanoil (30/70%)
4) Soyabean Oil 100%

We can see that the soybean oil has an induction period of about 50 hours, where the viscosity increases 154% and enters to a second phase of the reaction, where viscosity reaches an increase of about 2.820%, at the end 100 hours. On the other side M.E.S.O. on the same period his viscosity increases only 46%.

Behavior of lube oil in Diesel engines using M.E.V.O and V.O. as a fuel

By using methylic esters of vegetable oils as a fuel in Diesel engines we note that after 200 or 250 hours the lube oil lost its characteristics of lubricating, by the formation of a gelatinous deposit and that causes the metals to wear out. We studied the problem by trying to reproduce it in the laboratory. We executed an accelerated oxidation test of the pure lube oil to have a standard to compare with. In all tests we aereted the samples (flow air 30 l/h)at 140ºC, simulating conditions of the engine. By contaminating the same lube-oil with 3% of M.E.S.O. (V/V) and taking off samples to determine viscosity, TBN, TAN, Insolubles in heptane and benzene, in this concentration of M.E.S.O. in lube oil, there is no significant differences between the two tests than by increasing the concentration of M.E.S.O. and repeating an other test, using a mixture of 50% of M.E.S.O. and 50% of lube oil, the results are reported in table 5.

Table 5: Lube Oil (1:1) M.E.S.O.

Time Hours	Viscosity at 40ºC	100ºC	T.A.N.	T.B.N.	Insoluble in Heptane	Benzene
00H00	15,23	3,82	1,36	4,52	-	-
05H00	15,60	3,97	1,49	4,63	-	-
25H00	17,42	4,26	2,13	4,17	-	-
50H00	21,12	4,80	3,09	0,63	-	-
75H00	-	-	-	-	-	-
100H00	30,30	6,02	4,14	0,30	-	-
125H00	36,10	6,55	5,02	0,81	5,50	0,02
150H00	41,84	7,17	5,84	0,61	7,19	0,01
175H00*	47,15	7,71	6,08	0,51	8,18	0,09
200H00*	49,64	8,06	6,26	0,19	12,22	0,05
225H00*	50,63	8,16	6,26	0,18	14,75	0,06
250H00*	52,65	8,33	8,70	0,33	15,66	0,01

* samples with sediments

These results induced that the organo-metallic compounds are responsible for the problems with lube oil. Then a new accelerated tests was run, using a mixture of 50% of M.E.S.O and 50% of basic lube oil without any additive; the results demonstrated no significative difference. So we conclude that the presence or absence of organo-metallic compounds did not modify the results, then we began a new serie of tests by using antioxidants like B.H.T and the results are reported on table 6.

Table 6. Lube Oil (1:1) M.E.S.O. + 25ppm B.H.T.

Time (Hours)	Viscosity at 40ºC	100ºC	T.A.N.	T.B.N.	Insoluble in Heptane	Benzene
00H00	14,58	3,77	0,91	5,35	-	-
05H00	15,29	3,88	0,85	4,82	-	-
25H00	16,99	4,18	1,52	2,96	0,01	-
50H00	19,33	4,55	2,02	2,39	0,13	-
75H00	22,31	4,96	2,38	2,03	0,77	0,02
100H00	24,02	5,13	2,15	1,89	1,43	0,01
125H00	32,00	6,23	2,99	1,42	2,58	0,01
150H00	38,73	6,96	3,98	1,04	5,04	0,01
175H00*	46,88	7,88	5,08	0,77	7,26	0,01
200H00*	60,27	9,30	6,07	0,43	10,16	0,02
225H00*	60,04	9,17	6,89	0,20	12,96	0,03
250H00*	57,66	8,95	7,74	0,10	15,79	0,03
275H00*	57,89	8,97	8,77	0,00	19,86	0,04
300H00*	60,04	9,08	8,31	0,00	21,80	0,04

* samples with sediments

These results induced that the B.H.T. is not suitable as an antioxidant in this case of lubricating oil. To have a standard to compare between the behavior of the lubricant in the presence of M.E.S.O. and Soybean oil, we executed a new test using soybean oil, mixed with the lube oil (1:1) The result are reported on table 7:

Table 7. Lube Oil 50% + Soybean Oil 50%

Time (Hours)	Viscosity at 40ºC	Viscosity at 100ºC	T.A.N.	T.B.N.	Insoluble in Heptane	Insoluble in Benzene
00H00	50,39	8,97	1,29	5,58	-	-
05H00	51,80	9,11	1,51	4,71	-	-
25H00	59,14	9,86	1,70	2,05	-	-
50H00	67,44	10,71	1,83	1,74	-	-
75H00	78,15	11,87	2,01	1,63	-	-
100H00	88,65	12,70	2,36	1,45	-	-
125H00	106,93	14,60	2,58	1,20	-	-
150H00	133,13	17,12	2,85	0,97	-	-
175H00	199,13	22,62	4,09	0,75	-	-
200H00	326,43	32,04	4,84	0,54	-	-
225H00	995,02	78,23	6,18	0,19	-	-
250H00	4.921,33	266,45	6,98	0,00	-	-
275H00	-	-	-	-	-	-
300H00	-	-	-	-		-

- samples without sediments

The results induced that the configuration of lube oil is important, because we have the formation of compounds that are insoluble in the mixture of lube oil and M.E.S.O. after oxidation, then we executed a new test, using a lube oil rich in naftenic compounds and the results demonstrated no significative difference.

Many other test have to be run to study and understand the reaction of oxidation of M.E.V.O. in presence of lube oils to try to solve this problems.

References:

1. Snell, F.D., Ettre, L.S. ENCYCLOPEDIA OF INDUSTRIAL CHEMICAL ANALYSIS, Volume 12 : pages 472-493

2. Formo, M.W. et al, BAILEY'S INDUSTRIAL OIL AND FAT PRODUCTS, pages 105-155.

3. Wright, H.J. et al, OIL & SOAP, May 1944, pages 145-148.

4. Fenge, R.O., THE JOURNAL OF THE AMERICAN OIL CHEMIST'S SOCIETY, March 1949 pages: 97-102.

5. THE MECK INDEX - Eighth edition - 1968.

6. EXPERIMENTAL RESULTS:

   - Industria Agro-Química Braido
   - Shell Brasil S.A. - (Petróleo)
   - Chemical Specialties Indústria e Comércio Ltda.

FATTY ESTERS FROM VEGETABLE OILS FOR USE AS A DIESEL FUEL*

B. Freedman

E. H. Pryde
Executive Affiliate ASAE

ABSTRACT

Transesterification of sunflower and soybean oils to fatty esters has been carried out to study reaction variables such as: (1) molar ratio of alcohol to vegetable oil, (2) alcohol type (methanol, ethanol, and 1-butanol), (3) catalyst type (alkaline and acidic), and (4) reaction temperature (60°, 45°, and 32°C).

These studies showed that ester formation was 90 to 98% complete at the respective molar ratios of methanol/sunflower oil of 4:1 and 6:1. All three alcohols produced high yields of esters. Alkaline catalysts were much more effective than acid catalysts. At both 45° and 60°, 97% of methyl esters were produced in 1 hr.

INTRODUCTION

Because vegetable oils have much greater viscosities and are much less volatile than No. 2 diesel oil, problems have arisen when vegetable oils have been used as a substitute for diesel fuel. One promising solution to this problem, as shown by research in South Africa (Fuls and Hugo 1981) and elsewhere (Bacon et al. 1981), is to use fatty esters that can be obtained from the vegetable oils by transesterification. The physical properties of these fatty esters are much closer to those of No. 2 diesel oil, and they should make suitable fuels.

The transesterification reaction itself has been studied extensively. In the 1940's more than a dozen U.S. patents were granted to Dupont, Colgate, and others dealing with the conversion of vegetable oils to esters. Typical of these patents is the alkali-catalyzed alcoholysis of triglycerides to yield the corresponding alkyl esters (Bradshaw and Meuly 1944). Several review papers summarize both older (Formo 1954) and more recent findings (Sridharan and Mathai 1974). Our objective was to redefine the critical elements of the reaction with the aid of modern instrumentation. By use of an Iatroscan TH-10, an instrument which combines thin-layer chromatography (TLC) with a flame ionization detector (FID), product composition could be

---

The authors are: B. FREEDMAN, Research Chemist, and E. H. PRYDE, Research Leader, Oilseed Crops Laboratory, Northern Regional Research Center, Agricultural Research Service, U.S. Department of Agriculture, Peoria, IL 61604.

*The mention of firm names or trade products does not imply that they are endorsed or recommended by the U.S. Department of Agriculture over other firms or similar products not mentioned.

quantitated. By this technique we were able to pinpoint those reaction variables that had an important bearing on fatty ester production.

EXPERIMENTAL

Materials

Soybean and sunflower oils were refined, edible-grade oils. Alcohols were either spectrograde or were distilled prior to use. Alcohols were stored over molecular sieves (Linde 4A) to remove traces of water. Catalysts were the purest grades commercially available.

Methods

Transesterification reactions were conducted in glass, three-necked flasks equipped with mechanical stirrer, thermometer, and reflux condenser. The flasks were immersed in silicone oil baths, the temperatures of which were controlled to ca. ± 0.1°C by an I^2R Therm-O-Watch. Reaction temperatures were generally adjusted to 4°C below that of the boiling point of the alcohol. When alkaline catalysts were employed, 0.5 ml samples were withdrawn from the reaction after 1, 2, 4, 30 and 60 min. The reaction was quenched by mixing the samples with cold water (Gauglitz and Lehman 1963).

TLC on silica-coated glass plates was utilized where semi-quantitative analyses sufficed. The solvent system used was petroleum ether (bp 35-60°C)-ether-acetic acid 90:10:1. For quantitative analyses we used the Iatroscan TH-10 (Lachat Chemicals, Inc., Mequon, Wisconsin). Standard curves were first developed for pure standards (NuChek Prep, Inc., Elysian, Minnesota) with stearyl alcohol as an internal standard (Van Tornout et al. 1979). Product composition was calculated automatically by a computer-assisted program. Three to four replicates were used for each sample that yielded ester compositions with a relative standard deviation of ca. 1-4%.

RESULTS AND DISCUSSION

Effect of Stoichiometry on Product Composition

One of the most important variables that affects ester formation is the molar ratio of alcohol to vegetable oil. The stoichiometry of this reaction requires 3 moles of alcohol per mole vegetable oil to yield 3 moles of fatty ester and 1 mole of glycerol. We have previously reported the effect of the molar ratio of methanol/soybean oil on methyl soybean ester formation (Freedman and Pryde 1981). This study has now been extended to the reaction of methanol with sunflower oil. The results are shown in Fig. 1. Because of their compositional similarity, sunflower oil and soybean oil gave a series of curves which were strikingly similar. To obtain 90 or 97% ester conversions, respective molar ratios of 4:1 (33% excess) or 6:1 (100% excess methanol) are required. At the theoretical ratio of 3:1 and below the percentage of ester decreased rapidly with a corresponding increase in tri-, di- and monoglycerides.

Comparison of Methyl, Ethyl and Butyl Alcohols

One characteristic of alkali-catalyzed transesterifications of vegetable oils is the speed with which the reaction occurs. Gauglitz and Lehman (1963) demonstrated that methanolysis of a fish oil triglyceride (menhaden oil) with sodium methoxide was essentially complete in 2-4 min. Our kinetic studies showed (Fig. 2) that butyl, methyl and ethyl esters of sunflower oil were formed to the extent of 88, 80, and 76%, respectively, after 1 min.

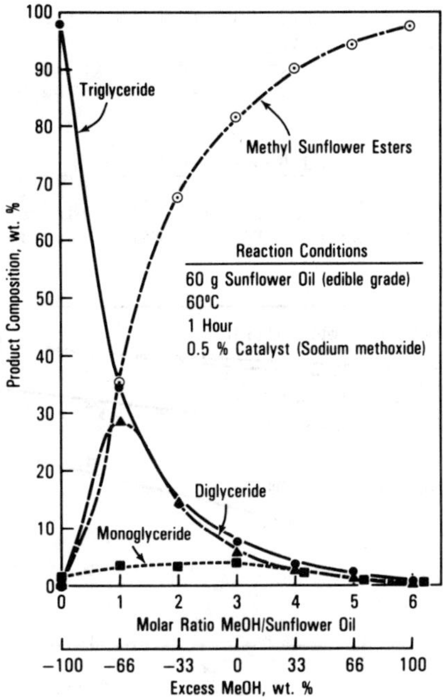

Fig. 1. Product Composition in Transesterified Sunflower Oil--Effect of Stoichiometry.

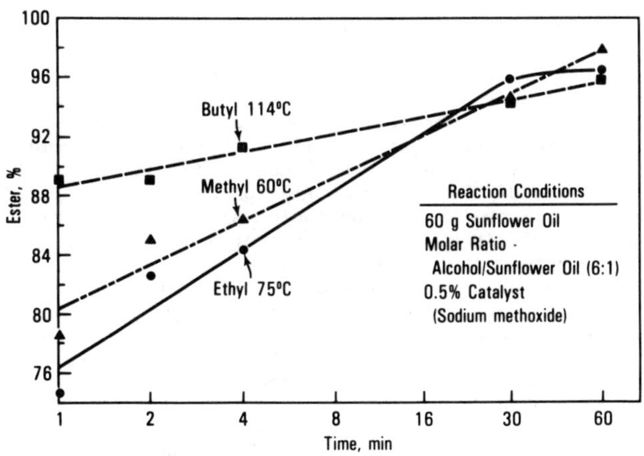

Fig. 2. Kinetics of the Transesterification of Sunflower Oil to Methyl, Ethyl and Butyl Esters.

After 1 hr, the percentage of methyl, ethyl and butyl esters were 98, 96 and 96, respectively. Thus, although there were some differences in ester formation after 1 min, these differences were insignificant after 1 hr. The eventual selection of one of these three esters will be based on cost and performance considerations.

Comparison of Sodium Hydroxide and Sodium Methoxide as Catalysts

These two catalysts were compared at 6:1 and 3:1 molar ratios of methanol/vegetable oil. As shown in Fig. 3, relatively little difference

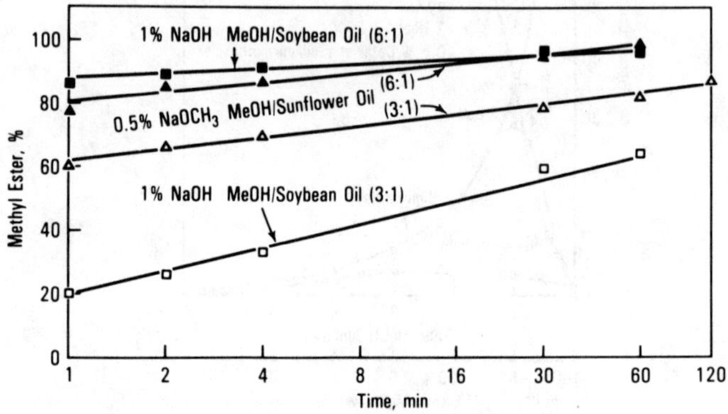

Fig. 3. Comparison of Transesterification Catalysts: Sodium Methoxide vs. Sodium Hydroxide.

was seen in their ability to promote methanolysis at the 6:1 ratio. At the 3:1 ratio, however, sodium methoxide (0.5%) was superior to sodium hydroxide (1%) during the entire 60-min reaction period. The lower cost of sodium hydroxide would probably make it the preferred catalyst in large-scale operations.

Comparison of Alkaline and Acidic Catalysts

One advantage is apparent for using an acidic catalyst over an alkaline catalyst: if the vegetable oil had a free fatty acid content of >1%, the alkaline catalyst would be destroyed, whereas the acidic catalyst would still be effective. For alkali-refined oils with free fatty acid contents of <1%, alkaline catalysts are preferred because the transesterification is completed in a fraction of the time required for acid catalysis. Reid (1911) concluded that alcoholysis using sodium alkoxides took place 4,000 times as fast as alcoholysis using an equivalent amount of HCl. In Fig. 4 a comparison was made between two alkaline catalysts vs. concentrated sulfuric acid. For the alkaline catalysts where a 6:1 molar ratio of methanol/vegetable oil was employed, transesterification was essentially complete in 1 hr. In this same period of time, only a trace of ester was formed with the sulfuric acid catalyst. A 30:1 molar ratio of methanol/soybean oil and 44 hr of heating were required to obtain a good conversion with the acid catalyst. Butyl alcohol afforded 98% ester in 4 hr with 1% sulfuric acid, presumably because of the higher reaction temperature (114°C) compared to that used with methanol (60°C). For large-scale production, acid-resistant metal tanks would increase the cost of acid-catalyzed transesterifications.

Effect of Temperature on Ester Formation

Feuge and Gros (1949) investigated the effect of three temperatures, 70°,

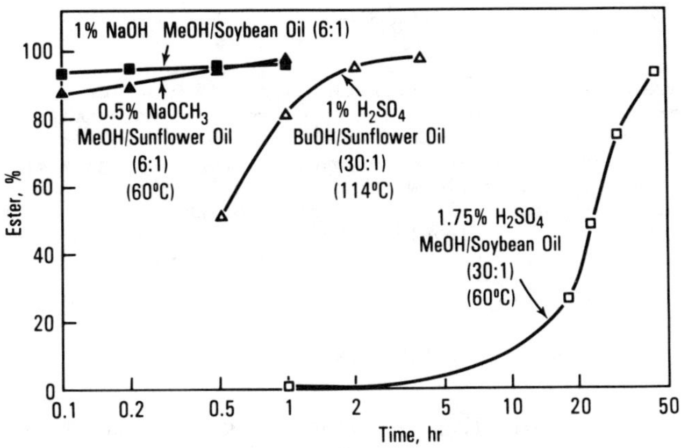

Fig. 4. Comparison of Alkaline and Acidic Transesterification Catalysts.

50° and 30°C, on the yield of glycerol produced during the ethanolysis of peanut oil (6:1 molar ratio) with 0.2% sodium hydroxide. They concluded the maximum glycerol yield (91%) was obtained at 50°; during this same reaction time of 16 hr, 86% glycerol was obtained at 30°C. This experiment, as well as the works of others, suggested that transesterification can occur at moderate temperatures. We therefore investigated how the methanolysis of soybean oil (6:1 ratio, 1% sodium hydroxide) would be influenced by temperatures of 60°, 45° and 32°C. The results of these experiments (Fig. 5)

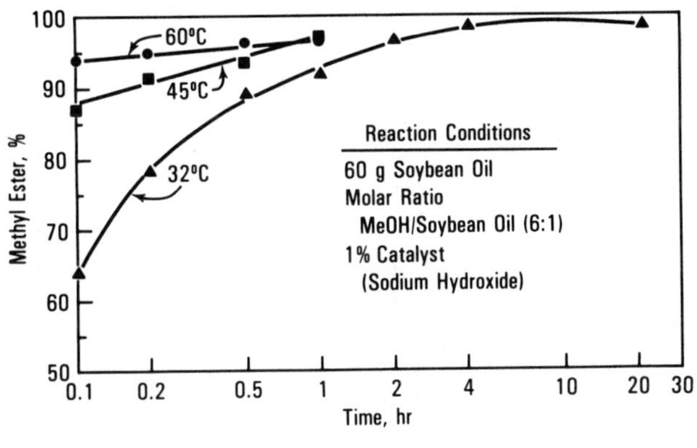

Fig. 5. Effect of Temperature on Methyl Ester Formation.

showed that after 1 hr, the % ester formed at 45° was identical to that at 60° (97%), whereas the % ester at 32° was only slightly lower (92%). By continuation of the 32°C reaction for 4 or more hr, the yield of ester was raised to 99%. These studies indicate that transesterification can proceed satisfactorily at 32°C in several hr.

CONCLUSIONS

Our studies have shown that transesterification of soybean or sunflower oil to methyl ester requires at least a 4:1 molar ratio of methanol/vegetable oil for adequate conversions. Methyl, ethyl and butyl alcohols all transformed sunflower oil to the corresponding esters in high yields. Alkaline catalysts were generally much more effective than acid catalysts. Sodium hydroxide would probably be the catalyst of choice because of its low cost and effectiveness at low levels. The transesterification of vegetable oils to fatty esters can be successfully performed at moderate temperatures (32°C) in about 4 hr.

REFERENCES

1. Bacon, D. M., F. Brear, I. D. Moncrieff and K. L. Walker. 1981. The use of vegetable oils in straight and modified form as diesel engine fuels. Reference 3, pp. 1525-1533.

2. Bradshaw, G. B. and W. C. Meuly. 1944. Preparation of detergents. U.S. Pat. 2,360,844.

3. Fazzolare, R. A. and C. R. Smith, (Eds.). 1981. Beyond the Energy Crisis--Opportunity and Challenge Volume III. Third International Conference On Energy Use Management. Berlin (West). October 26-30. Pergamon Press, Oxford.

4. Feuge, R. O. and A. T. Gros. 1949. Modification of vegetable oils. VII. Alkali catalyzed interesterification of peanut oil with ethanol. J. Am. Oil Chem. Soc. 26: 97-102.

5. Formo, M. W. 1954. Ester reactions of fatty materials. J. Am. Oil Chem. Soc. 31: 548-559.

6. Freedman, B. and E. H. Pryde. 1981. Fatty esters from soybean oil for use as a diesel fuel. Vegetable Oil as Diesel Fuel Seminar II, Northern Agricultural Energy Center, Peoria, Illinois. October 21-22.

7. Fuls, J. and F. J. C. Hugo. 1981. On-farm preparation of sunflower oil esters for fuel. Reference 3, pp. 1595-1602.

8. Gauglitz, E. J., Jr. and L. W. Lehman. 1963. The preparation of alkyl esters from highly unsaturated triglycerides. J. Am. Oil Chem. Soc. 40: 197-198.

9. Reid, E. E. 1911. Studies in esterification. IV. The interdependence of limits as exemplified in the transformation of esters. Am. Chem. J. 45: 479-516.

10. Sridharan, R. and I. M. Mathai. 1974. Transesterification reactions. J. Sci. Ind. Res. 33: 178-187.

11. Van Tornout, P., R. Vercaemst, H. Caster, M. J. Lievens, W. D. Keersgieter, F. Soetewey and M. Rossenev. 1979. Use of 1-octadecanol as an internal standard for plasma lipid quantitation on chromarods. J. Chromatogr. 164: 222-277.

SUNFLOWER OIL METHYL ESTER AS DIESEL FUEL

D.J. Hassett R.A. Hasan

ABSTRACT

Methyl ester formation represents one approach to overcome the problems associated with the relatively high viscosity of sunflower oil when used as a diesel fuel replacement. Sunflower oil methyl ester is being prepared at the University of North Dakota Engineering Experiment Station. Physical and chemical properties of this material at varying levels of refinement and purity will be used to define fuel properties. Engine testing is being carried out to determine if the fouling characteristics of methyl ester are significantly less than those of sunflower oil.

INTRODUCTION

The concept of using vegetable oils as diesel fuels is an old one going back to at least 1928 (Walton, 1938). In all the years since that time very little has been done using vegetable oils as diesel fuels, primarily because of the cheap seemingly endless supply of petroleum based fuels. Like many ideas which have not undergone the changes and improvements brought about by years of use, the problems evident in 1928 are with us today. Some of the advantages and disadvantages of using sunflower oil and sunflower oil methyl ester as diesel fuels are shown in Table 1.

The major problem to be dealt with when considering vegetable oils as diesel fuels is the reactivity of unburned fuel both in the cylinder and in the crankcase oil where raw and partly burned fuel can show up in a number of ways. Unburned or partly burned fuel in the cylinder is the cause of varnish and injector coking. One way to reduce this cylinder wall varnish buildup and injector fouling is to reduce fuel viscosity, and thus improve injector performance. A finely atomized fuel completely burned will not gum up or foul any engine parts. Reducing fuel viscosity by methyl ester formation to improve injector performance will be a partial, if not total solution to some of the engine fouling problems so evident when vegetable oils are used as a replacement diesel fuel or fuel extender.

Methyl ester formation can introduce another beneficial property to sunflower oil which is the particular oil we are studying. Methyl esters have a reduced tendency to polymerize three-dimensionally like triglycerides and show better long term storage with a reduced tendency toward oxygen polymerization. This change in the oil may help reduce crankcase oil thickening if fuel contamination should occur.

The authors are: D.J. HASSETT, Research Chemist, Engineering Experiment Station; and R.A. HASAN, Assistant Professor, Chemical Engineering Department, University of North Dakota.

This study is being conducted under a grant from the United States Department of Agriculture, ARS Northern Agricultural Energy Center at Peoria, Illinois

EXPERIMENTAL PLAN

Synthesis of Methyl Esters

Sunflower oil methyl ester to be used as a fuel for our research is being synthesized in nine liter batches in the following manner.

Nine liters of sunflower oil are mixed with 1456 grams of methyl alchol to which 8.4 grams of sodium has been added. The mixture is heated to 62-65°C and held at this temperature with stirring for two hours. Heating and stirring are stopped and at this point glycerol, the primary reaction by-product, begins to separate as a layer at the bottom of the reaction vessel. The crude methyl ester is drawn off of the top of this reaction mixture and washed with warm water to remove suspended glycerol, unreacted methanol and catalyst. This product, after drying with anhydrous sodium sulfate, is the material which we refer to as "fuel grade" methyl ester.

This particular ratio of sunflower oil to methanol (1 mole of sunflower oil to 4.8 moles of methanol) was chosen because the 1.6 times excess methanol over the ideal 3 moles gives a good yield of methyl ester (87 percent to 90 percent) while allowing the glycerol to separate for easy removal, (Price, 1942).

We are snythesizing our own fuel in order to have control over synthesis parameters. It is a part of our experimental plan to test at least two grades of methyl ester. Fuel grade methyl ester prepared as described above is being engine tested and compared with distilled, essentially 100 percent pure methyl ester. This was prepared by vacuum distillation of our fuel grade methyl ester. A chemical analysis of the starting product (unrefined sunflower oil) along with some analytical data on our methyl ester is given in Table 2. Temperature viscosity curves for sunflower oil, sunflower oil methyl ester and number 2 diesel fuel are shown in Fig. 1. These indicate that methyl ester is about twice as viscous as diesel fuel and about one-tenth as viscous as sunflower oil. The curve for methyl ester stops at about 40°F because it is at about this temperature that crystals begin to form in the fuel mixture. Like sunflower oil, methyl ester would not be a good cold weather fuel. Heated tanks and fuel lines would be required for use below 40°.

Engine Testing

Fuel grade sunflower oil methyl ester is being engine tested in a single cylinder direct injection type diesel engine. A Lister type LT-1 engine rate at 5.25 BHP is being used. Testing on this fuel is being done to rate performance with respect to injector fouling and varnish buildup on cylinder walls. Testing is being carried out at moderate loading and moderate engine speed so an evaluation can be made on a relatively small amount of fuel. The fuel grade methyl ester will be rated in comparison to diesel fuel and refined sunflower oil. We expect through this testing to be able to determine the degree of refinement needed to utilize methyl ester as a diesel fuel. We will also be evaluating the effect of residual catalyst on injector and engine parts.

SUMMARY

The ongoing program of engine testing is expected to be completed sometime around July 1. At that time data collected should indicate the usefulness of viscosity reduction as a means to overcome engine fouling problems associated with vegetable oil diesel fuels. Knowing the degree of refinement necessary to produce a fuel grade methyl ester will allow an economic evaluation of production costs for this fuel. This we feel is important for although economics may not be the driving force in pursuing a means of energy independence for farmers, it will be an important factor.

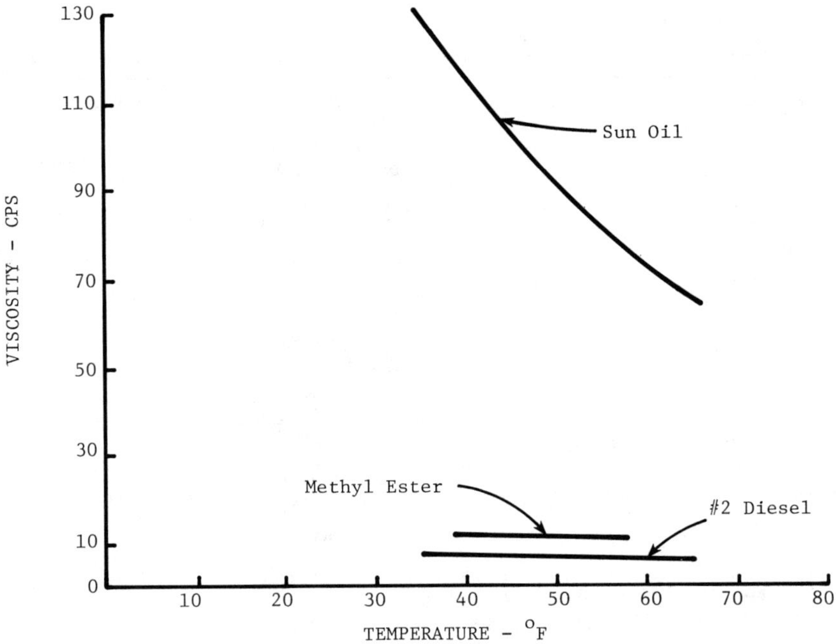
VISCOSITY - CPS
130
110
90
70
50
30
10
0
Sun Oil
Methyl Ester
#2 Diesel
10
20
30
40
50
60
70
80
TEMPERATURE - °F

Fig. 1

Table 1.

Sun Oil		Methyl Ester	
Advantages	Disadvantages	Advantages	Disadvantages
Low Engine Knock	High Injector Fouling Under Low Load	Distillation and Viscosity Curves Approximate Diesel Fuel	Polymerized Oil?
Only 3% Lower Combustion Efficiency	Polymerized Oil	Low Smoke	Reported Injector Sticking Problems
Easy Starting	High Engine Deposits	Better Thermal Efficiency	
		Low Injector Fouling?	
		Easy Starting	

Table 2. Chemical and Physical Data

Chemical Analysis of Crude Sunflower Oil Used to Synthesize Methyl Ester

	Wt. %
Palmitic Acid	7.0
Stearic Acid	4.7
Oleic Acid	19.8
Linoleic Acid	68.5
Arachidic Acid	<1

Free Fatty Acid Determination

	% FFA As Oleic Acid
Sunflower Oil	1.46
Sunflower Oil Methyl Ester (fuel grade)	.20
Sunflower Oil Methyl Ester (distilled)	.19

Iodine Number (calculated)	142

REFERENCES

1. Price, Donald. 1942. U.S. Patent 2543 421 Transesterification Process.

2. Walton, John. 1938. The Fuel Possibilities of Vegetable Oils. Gas and Oil Power 33: p. 167-168.

TRANSESTERIFICATION OF VEGETABLE OILS FOR FUELS

Paul F. Kusy
Manager, Organic Materials Research
Deere & Company
Moline, Illinois

In the United States both soybean and sunflower oils, because of their relatively high yield per hectare and widespread production, are being studied as potential diesel fuels by many researchers. Oils of other crops such as peanuts and cotton are also being evaluated; however, their yields are lower at present.

High viscosity appears to be a fundamental property at the root of many problems associated with direct use of these oils as fuels. High viscosity has the effect of increasing droplet size on injection into the cylinder and results in poor combustion. This, in turn, results in the formation of deposits in the combustion chamber and introduction of unburned fuel into the lubricating oil, or fuel dilution. Because of their unsaturation, fuel dilution by vegetable oils results in excessive thickening of the lube oil and the attendant problem of assuring adequate engine lubrication over a wide temperature range.

Also limiting the use of these vegetable oils as fuels is their tendency to solidify at normal winter operating temperatures. This problem may be partially overcome by some type of heating or dilution with diesel fuel. Soybean oil becomes solid between −10°C and −16°C and sunflower oil solidifies at −18°C.

One suggested approach to reduce viscosity and its associated problems is to transesterify the vegetable oils with either methanol or ethanol.[1]* The objective of the work reported herein was to make a preliminary evaluation of the feasibility of producing a suitable diesel fuel by transesterifying soybean and sunflower oils using ethanol. Both the oils and the ethanol are obtainable from farm crops. In addition, we were interested in the preliminary performance of such ester fuels in a direct injection diesel engine.

TRANSESTERIFICATION

The conversion of an organic acid ester into another ester of that same acid is defined as transesterification.[2] In the case of soybean or sunflower oils (triglycerides) being transesterified with ethanol, the reaction is as follows with scaled-up proportions from those used in our work. (The actual amounts used in our work are shown in Table 3.)

$$\underset{\text{250 liters}}{\text{Triglyceride}} + \underset{\text{135 liters}}{\text{ethanol}} \underset{}{\overset{\text{catalyst}}{\rightleftharpoons}} \underset{\text{250–275 liters}}{\text{ethyl ester fuel}} + \underset{\text{19 liters}}{\text{glycerol}} + \underset{\text{89 liters}}{\text{ethanol}}$$

The approximate percent of acids typically found in each oil used as well as their saponification number and iodine number are shown in Table 1.[3]

The transesterification methods used in this work are those Bradshaw and Meuly of DuPont Company[4] and Trent of Colgate-Palmolive Peet Company[5] patented in the 1940s.

---

*Numbers in brackets refer to references at the end of this paper.

TABLE 1. TYPICAL PROPERTIES AND ACID COMPOSITION OF SOYBEAN AND SUNFLOWER OILS

	SOYBEAN OIL, %	SUNFLOWER OIL, %
FATTY ACID		
Myristic	1	
Palmitic	8	6.4
Stearic	3	1.3
Oleic	26	21.3
Linoleic	49	66.2
Linolenic	11	<0.1
Arachidic		4.0
Behenic		0.8
SAPONIFICATION NUMBER	189–195	188–194
IODINE NUMBER	127–138	125–136

PROCESS USED

Where Bradshaw and Meuly utilized methanol as the alcohol for transesterification, ethanol was used in this experiment. In our work based on Bradshaw and Meuly's patents, 1.75 equivalents of ethanol and approximately 0.5 percent (based on the weight of oil) of sodium methoxide in methanol were used with both purified and unpurified soybean and sunflower oils. To catalyze the reaction, a 25 percent solution of sodium methoxide in methanol was used.

Trent's patents suggested using three to five times the stoichiometric amount of alcohol as methanol and, again, ethanol was substituted. The same catalyst, 0.5 percent of sodium methoxide in methanol, was used and the soybean and sunflower oils used were both purified and unpurified.

Following the transesterification reaction, both Bradshaw and Meuly, and Trent suggest the use of an organic or inorganic acid to render the catalyst inert and allow recovery of unreacted alcohol. This practice was also used in our work.

PRETREATMENT OF OIL

The first trials of transesterification were carried out using "purified" oils since it had been suggested that ester yields would be reduced should any gums, free acids, water or phospholipids be in the vegetable oils.[6]

The oil was purified by using an initial phosphoric acid wash to remove any proteinaceous material and phospholipids, specifically with a two percent solution of phosphoric acid, heated to approximately 70°C, with stirring, and then allowed to cool. Free acid content of the raw oils may vary up to several percent by volume. The supernate was separated and treated with anhydrous calcium oxide (CaO) to remove trace amounts of water and free fatty acids or phosphoric acid which remained. The vegetable oil was then filtered or the supernate decanted to separate the solid calcium oxide and any calcium salts of acids. The transesterification reaction was carried out with and without this pretreatment on the vegetable oils.

REACTANTS

The crude soybean and sunflower oil reactants used were both purified as above, and in later trials the unpurified oils were used. The ethanol was absolute grade Specially Denatured Alcohol (SDA)-12A (95.2 percent ethanol, 4.8 percent benzene). Both sodium ethoxide powder and sodium methoxide (25 percent in methanol) were used as catalysts. Since water and free acid in the reactants slow the reaction, we determined the amounts in the reactants. Table 2 shows the water content of the reactants and the acid number of the crude oils and soybean oil purified as described above.

The acid number of 2.12 for crude sunflower oil corresponds to an approximate acid content of 1.1 percent by weight based on oleic acid (MW 282.45).

TABLE 2. WATER CONTENT AND ACID NUMBER OF REACTANT MATERIALS

MATERIAL	PERCENT WATER[a]	ACID NUMBER[b]
Crude sunflower oil	0.068	2.12
Crude soybean oil	0.082	0.56
Purified soybean oil	—	~0.06
"Absolute" ethanol SDA 12A	0.146	—

a. Karl Fischer
b. ASTM D1386

REACTION CONDITIONS

Following our review of the work of Bradshaw and Meuly, and Trent, and after some preliminary work in the laboratory, a scheme was developed to carry out the transesterification on a continuous basis. A diagram of the procedure is shown in Figure 1. The procedure was followed in a stepwise manner except in cases where less alcohol was used, as in the work of Bradshaw and Meuly, and the centrifuging steps were not repeated. These separations, for the most part, were carried out in a separatory funnel after going through the full preparatory procedure.

REACTION PROCEDURE

Reactions were carried out in a 1000 ml reaction vessel equipped with a condenser and thermometer. The solid catalyst, sodium ethoxide, was dissolved in the alcohol and added with stirring to the oil in the flask. Sodium methoxide was also used with equally good results. The amount of catalyst used was 0.3 to 0.5 percent based on the weight of the oil. This was suggested by Dr. E. H. Pryde[7] and also cited in the patent literature.[4,5] Reaction parameters varied in the transesterification reaction were temperature, time, and amount of alcohol. An excess of alcohol, from three to five times the stoichiometric amount, was used in most cases. In the experiments to evaluate Bradshaw and Meuly's method, approximately 1.75 times the stoichiometric amount of oil was used.

Upon completion of the reaction the catalyst was deactivated by adjusting the solution pH to five to six using 1:1 sulfuric acid. Excess alcohol was then distilled off and the mixture transferred to a separatory funnel to separate the glycerol from the ester. Acidification prevents reversal of the transesterification reaction back to the triglycerides (oils) by neutralizing the catalyst, any soaps formed and facilitates separation of the glycerol.

FINAL TREATMENT OF ETHYL ESTERS

Residual glycerine and salts from sulfuric acid and catalyst in the ester layer were removed by washing the ester layer with water until it tested neutral. There is a tendency for emulsions to form upon excessive washing. A single trial using a few grams of sodium chloride was added to the emulsion and it immediately broke to form a clear water layer and clear ester. After the washings any residual water and alcohol were removed by heating the esters to approximately 100°C under a nitrogen purge or it was dried by adding anhydrous calcium oxide plus calcium chloride and then filtered to separate the drying agents.

FIGURE 1. SCHEME FOR CONTINUOUS TRANSESTERIFICATION OF VEGETABLE OILS

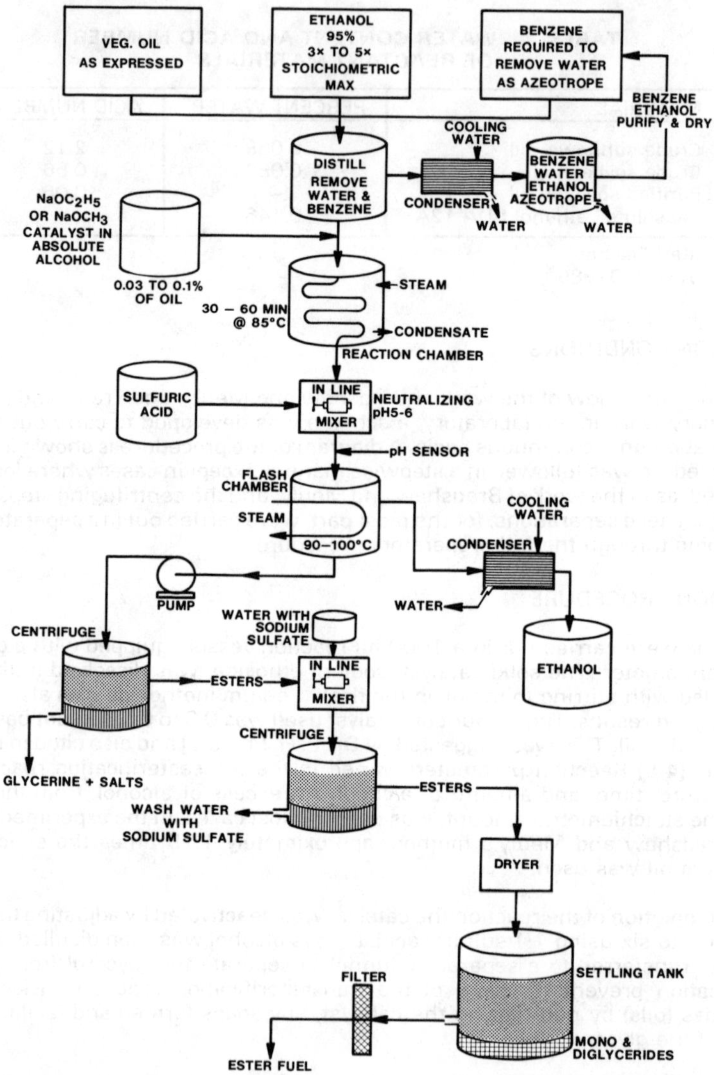

DISCUSSION OF RESULTS

Nine batches of ethyl soyate and four batches of ethyl sunflowerate were prepared, six using pretreated oils to remove undesirable components and seven crude as received. Table 3 shows the variables used in preparing each batch and the estimated conversion to esters, glycerol and unreacted alcohol recovered, based on the assumed molecular weight for each oil. Increasing the amount of catalyst, and even reaction temperature did not significantly alter yields. Measuring the amount of free glycerol

TABLE 3. REACTION CONDITIONS USING ANHYDROUS ETHYL ALCOHOL SDA 12A

Batch No.	Vegetable Oil	Vegetable Oil ml	Alcohol[c] ml	% of Theory[d] for Alcohol	Reaction Temp. °C	Reaction Time Minutes
1	Soybean[a]	250	225	500	50	30
2	Soybean[b]	250	135	300	50	30
3	Soybean[b]	250	135	300	85	30
4	Soybean[b]	250	200	435	83	60
5	Soybean[b]	250	250	555	78	60
6	Sunflower[b]	250	200	435	83	30
7	Sunflower[b]	250	225	500	83	60
8	Soybean[a]	250	77.6	175	83	60
9	Soybean[a]	250	77.6	175	83	60
10	Soybean[a]	250	50.7	110	23	60
11	Sunflower[a]	250	77.6	175	83	60
12	Sunflower[a]	250	135	300	83	60
13	Soybean[a]	250	73.9	160	83	60

Batch No.	Vegetable Oil	Catalyst[e]	Final Ester Yield ml	Free Glycerol[f] Yield ml	Alcohol Recovered after Reaction ml	Estimated Conversion %
1	Soybean[a]	2.5 g	275	15	178	78
2	Soybean[b]	1.8 g	275	15	88	78
3	Soybean[b]	1.8 g	275	15	86	78
4	Soybean[b]	2.5 g	250	15	154	78
5	Soybean[b]	2.5 g	250	15	198	78
6	Sunflower[b]	2.5 g	250	13	154	68
7	Sunflower[b]	2.5 g	250	16	168	83.5
8	Soybean[a]	5 ml	275	18.5	30.5	96
9	Soybean[a]	5 ml	h	8	55	41.7
10	Soybean[a]	5 ml	j	3	—	—
11	Sunflower[a]	5 ml	275	18.2	30.5	95.5
12	Sunflower[a]	5 ml	275	19.2	90	99
13	Soybean[a]	k	225	9	—	47

a. Crude
b. Pretreated
c. Alcohol 95.2% ethanol, 4.8% benzene
d. Based on assumed molecular weight of soybean and sunflower oil as 871 g/mole
e. Gram (g) amounts sodium ethoxide, ml amounts 25% sodium methoxide in methanol
f. Estimated at 19.2 ml for 100% reaction. Also contained some salts (sodium sulfate).
h. No final product separated. Did not add acid to deactivate catalyst – reaction reversed during distillation.
j. Ingredients mixed together, then centrifuged for 60 minutes reaction time – product not worked up.
k. Catalyzed with 1.15 g of anhydrous sodium hydroxide

isolated after removing the alcohol is the easiest way to approximate the degree of transesterification of the oils. For all the reactions we ran we estimated a yield of 19.2 ml of glycerol would give a 100 percent conversion of oil to transesterified ethyl esters based on a molecular weight of 871 for both soybean and sunflower oils (Table 3). It is interesting to note that good conversions were obtained using untreated soybean and sunflower oil at low reaction temperature (50° and 83°C) and 30 and 60 minutes contact time. Varying the amount of ethanol from three to five times the stoichiometric ratio to one of oil did not appear to have a significant effect on yield of glycerol in the soybean oil reaction. Extending the reaction time and increasing the ratio of ethanol over the stoichiometric amount did, however, markedly increase the glycerol yield (21.5 percent) with the pretreated sunflower oil (Batches 6 and 7).

Utilizing sodium methoxide catalyst with both crude soybean and sunflower oil gave excellent yields compared to the sodium ethoxide catalyzed preparations when the full procedure was followed. Altering the procedure by not deactivating the catalyst (Batch 9) or not having excess alcohol or heating (Batch 10) gave greatly reduced yields. In the one experiment (Batch 9, Table 3), using crude soybean oil and ethanol, where excess alcohol was distilled from the reaction mixture without the addition of a deactivating acid, the amount of glycerol obtained was approximately 41.7 percent of that expected. The product evidently contained many mono- and diglycerides which resulted in an inseparable emulsion of the product with the water used to wash it. Addition of an equivalent volume of petroleum ether to the oil used aided in separating part of the product, but a significant emulsion still remained which would not separate even when centrifuged at 1,000 rpm for 30 minutes. Using anhydrous sodium hydroxide to catalyze the reaction, only a 47 percent yield was obtained with crude soybean oil (Batch 13).

A gelatinous precipitate in some ethyl soyate and some ethyl sunflowerate preparations, which developed on standing, indicates that its use as a diesel fuel could cause serious fuel line clogging problems unless it is further purified by cooling and filtration. Centrifugation of the precipitated material indicated solids to be about four percent of the total by volume. Infrared analysis of the solid showed it to be a mixture of mono- and diglycerides. The single batch of ethyl sunflowerate which gave an estimated 99 percent yield developed a very faint haze after standing for about one week. Other batches of 95 percent plus yields of both ethyl soyate and ethyl sunflowerate developed a slight precipitate after standing about one week. Those batches which had low yields (less than 50 percent) developed considerable precipitates after just a few hours standing.

During our early preparation work we were able to obtain a sample of pure ethyl soyate from Emery Industries for comparison purposes. Table 4 shows the composition, by weight percent, of some of the ethyl esters in the products we made by transesterification compared to the Emery product made by direct esterification of the fatty

TABLE 4. COMPARISON OF TRANSESTERIFIED SOYBEAN ESTERS WITH A DIRECT ESTERIFIED ESTER

ACIDS	TRANSESTERIFIED			DIRECT (EMERY INDUSTRIES)
	BATCH 1	BATCH 2	BATCH 4	
Saturated				
Palmitic C16	10.0	10.0	9.8	10.0
Stearic C18	5.0	5.5	5.1	5.7
Unsaturated				
Oleic C18:1	22.1	21.6	20.7	22.2
Linoleic C18:2	51.6	50.3	49.4	51.2
Linolenic C18:3	8.0	8.3	9.2	6.5

acids, a more positive procedure for producing a pure ester. Data developed using gas chromatography shows the distribution of acids to be quite similar for the two procedures. Variations likely indicate difference in the starting oil products used.

Other data developed on the pure ethyl soyate are shown in Table 5. These include density, viscosity and cloud, flash and fire points, carbon residue, cetane number, heat of combustion and distillation data. Notice that the cloud point of pure ethyl soyate is 8°C which restricts its usefulness as a fuel. These data provide a point of reference because direct esterification of fatty acids produces a product with a minimum of contamination by unreacted material or by-products.

TABLE 5. PROPERTIES OF A PURE ETHYL SOYATE

TEMPERATURE	DENSITY, g/ml
20°C	.875
24.5°C	.872
40°C	.820
100°C	.802
Viscosity, cST	
@ 40°C	4.39
@ 100°C	1.82
Cloud Point, °C	8
Flash Point, °C	182.2
Fire Point, °C	198.9
Carbon Residue, % Wt.	0.01
Cetane Number	50.0
Heat of Combustion (Gross, Btu/lb)	17,260

Distillation Data (ASTM D-86) °C	
328	IBP
343	10
346	20
347	30
348	40
350	50
352	60
353	70
353	80
362	90
382	95
386	FBP
Recovery, %	97

TABLE 6. FUEL DATA FOR TRANSESTERIFIED PRODUCTS

	ETHYL SOYATE			ETHYL SUNFLOWERATE			
	BATCH 1 °C	BATCH 3 °C	BATCH 8 °C	BATCH 6 °C	BATCH 7 °C	BATCH 11 °C	BATCH 12 °C
Percent Distilled							
IBP	325	320	336	235	295	340	328
10	344	345	340	345	344	345	333
20	349	348	346	345	344	345	343
30	349	350	346	347	346	345	345
40	349	350	346	347	347	348	345
50	353	352	346	347	347	350	347
60	354	353	347	347	347	350	347
70	356	353	347	347	347	351	347
80	357	358	349	349	350	357	347
90	375	368	362	352	360	358	358
FBP	376	372	363	358	361	358	363
Recovery	96%	95%	96%	96%	94%	96%	94%
Flash Point °C				187.5	182.5		
Fire Point °C				198.8	196.2		
Cloud Point °C			+8		14	+8	+8

Distillation, flash point, fire point and cloud point data on several of the products prepared by transesterification are shown in Table 6. Ethyl soyate products had boiling ranges similar to those of the pure product (Table 5). The final boiling point for our products was not as high as the pure ester, and some of the ethyl sunflowerate products had somewhat lower initial boiling points. The lower initial boiling points of these products are likely due to a small amount of some unidentified lower boiling ester in the product compared to the ethyl soyate. Since sunflower oil may contain both arachidic and behenic acids, one would expect a higher final boiling point than observed. Gas chromatograph analysis, however, did not show the presence of any other esters either with less than 16 or more than 18 carbon atoms.

The distillate from these tests had an odor of alkenes which would indicate some decomposition. Figures 2 and 3 show, for soybean and sunflower oils, the distillation curves respectively for the oil, the ethyl ester and ASTM No. 2 diesel fuel. Although

FIGURE 2. DISTILLATION CURVE FOR SOYBEAN OIL

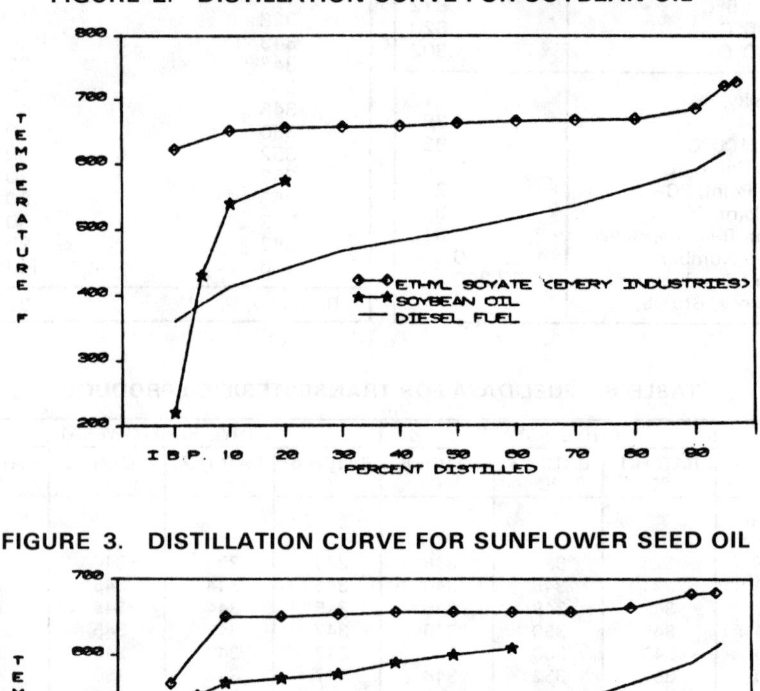

FIGURE 3. DISTILLATION CURVE FOR SUNFLOWER SEED OIL

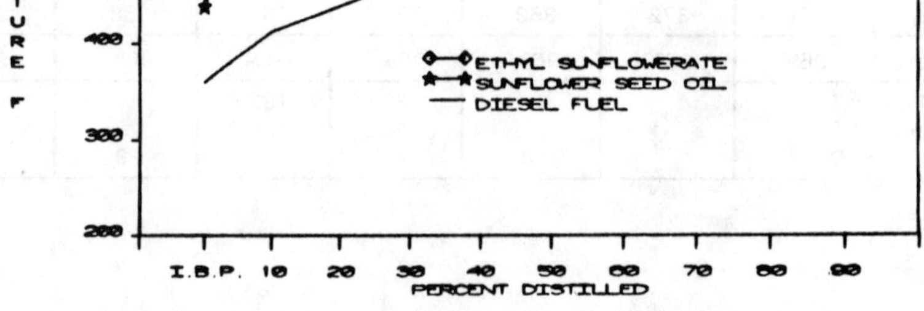

the ethyl esters are significantly lower in viscosity than the oils themselves, the esters are higher boiling. This is somewhat contradictory on the basis of molecular weight differences; however, the boiling curves for the oils suggest decomposition and should not be considered the true boiling point of the oils themselves.

LONG-TERM STORAGE OF FINAL PRODUCTS

Some of the ethyl soyate and sunflowerate esters were left to stand for eight months without the addition of any oxidation inhibitors. A slight rancidity was noted but no additional precipitate had formed. For maximum protection against oxidation of the esters, bubbling dry nitrogen or carbon dioxide through the product, then storing it under an oxygen-free atmosphere would be required. Nitrogen is preferred. The addition of butylated hydroxy toluene (BHT) (2,6-di-tert-butyl-para-cresol), at between 200 to 400 PPM is a practical alternative where storage in an inert atmosphere is not possible.

LOW TEMPERATURE PROPERTIES

Samples of the laboratory and Emery products were cooled to −30°C (−22°F) overnight. All samples became solid at that temperature and returned to liquid state upon reaching room temperature again. Unreacted or partially reacted vegetable oil in the laboratory soybean ethyl esters, which had not settled out previously, did so upon cooling.

As is shown by the cloud points of the Emery ethyl soyate (Table 5), some of our ethyl soyate and ethyl sunflowerate products (Table 6) are probably not entirely free of mono-, di-, or triglycerides since their cloud points are higher than Emery's pure product. Such minor impurities are not expected to have a significant effect on combustion characteristics.

COMMENTS ON CONTINUOUS TRANSESTERIFICATION PROCESS

One problem we encountered in the lab was to purify and dry the alcohol benzene mixture for reuse, but that can be performed by standard procedures. However, in a continuous process such as suggested in Figure 1, this problem could likely be handled in the same manner as that used currently in the manufacture of absolute alcohol. The removal of the excess alcohol after neutralization proceeds smoothly and rapidly. After the reaction is completed, the glycerol/salt portion settles rapidly in a separatory funnel which, with the great difference in density (.87 vs. 1.26), will separate very fast with centrifuging. Using a water wash with some salt, such as sodium sulfate or sodium chloride, greatly facilitates emulsion breaking and allows for a clean separation leaving behind the full mixture of esters, unreacted oil and partially reacted mono- and diglycerides.

When we suspected water in our sunflower oil and alcohol, we added excess alcohol and benzene to the flask and performed the azeotropic distillation using a one-half meter fractionating column. When the distillate temperature rose to 78.5°C, we discontinued the distillation, added catalyst and carried out the reaction. The procedure worked well. The temperature in the flask reached a maximum of about 140°C during the azeotropic distillation.

FUEL ASPECTS OF PRODUCTS MADE IN THE LABORATORY

Much concern has been raised over the effect of fuel dilution and blow-by on the engine lubricating oils. To simulate the effect of fuel dilution on oil thickening, a comparative 100 hour laboratory oxidation stability test was run using a premium-quality, SAE 30W/API CD engine oil to which was added one percent pure ethyl soyate and one percent crude soybean oil.

The data in Table 7 shows that the soybean oil increases the viscosity of the engine oil more than the ethyl soyate ester. We would expect blow-by products from incomplete combustion to have a greater effect on thickening since they would likely contain significant amounts of dienes (double unsaturation). These products would tend to oxidize as well as polymerize causing severe thickening of the engine oil. Since the ethyl esters have a somewhat lower viscosity than the vegetable oils, we would expect them to combust more completely and thus produce less blow-by.

TABLE 7. OXIDATION OF 1% DILUTION OF 30W PREMIUM CD OIL WITH CRUDE SOYBEAN OIL AND ETHYL SOYATE

MATERIAL	VISCOSITY, CENTISTOKES			
	INITIAL		AFTER 100 H @149°C	
	40°C	110°C	40°C	110°C
Premium 30W-CD oil	105.35		106.70	
Premium 30W-CD oil		11.38		11.55
Ethyl Soyate	4.39	1.82		
Crude soybean oil	30.63	7.39		
1% ethyl soyate in 30W-CD oil	99		103.55	
1% ethyl soyate in 30W-CD oil		11.06		11.38
1% soybean oil in 30W-CD oil	99.76		105.05	
1% soybean oil in 30W-CD oil		11.28		11.52

ENGINE TESTS

The Emery ethyl soyate was used to fuel a direct injection engine in a John Deere 4640 tractor in a brief test. We measured PTO power over a range of engine speeds and loads from high idle to 1500 rpm. The principal finding was that engine torque was five percent lower with the ethyl soyate than with No. 2 diesel fuel, reducing the PTO power accordingly. The fuel temperature during the test was 5°C higher for the ethyl soyate, accounting for a portion of the torque reduction. The engine started readily on the ethyl soyate at the ambient temperature of 20–25°C. The exhaust odor was similar to deep fat frying, not objectionable but not pleasant. Visible smoke was similar to operation on diesel fuel.

RECOMMENDATIONS AND CONCLUSIONS

1. A process for transesterification was tested in several steps and it appears highly likely that the process can be made continuous to produce fuel quality grade ethyl esters.

   The ethyl esters of transesterified soybean and sunflower oils are significantly lower in viscosity than the oils themselves and thus should be more suitable diesel fuels.

   The esters should also prove to be satisfactory fuel extenders when mixed with No. 1 or No. 2 diesel fuel.

   An analysis should be made to determine the cost of producing esters and their relative economic value as fuel.

2. The neat ethyl esters will not be suitable for use during low temperature months. Below their cloud points they will begin to congeal, causing filter clogging until warmed up.

   The addition of low temperature fuel flow improvers may increase the low temperature operating ranges of ester/diesel fuel blends. This should be evaluated.

3. A current production engine started and ran well on ethyl soyate, but no long-term tests were made.

   Because of the much higher boiling range of both the ethyl soyate and sunflowerate and higher degree of unsaturation than No. 2 diesel fuel, long-term engine tests should be conducted to determine any effect of blow-by or fuel dilution on viscosity change of engine oils.

SUMMARY

A continuous procedure was developed and tested, in a stepwise manner, for the transesterification of soybean and sunflower oils using ethanol. Good yields of ethyl soyate and sunflowerate were achieved, and the products made agreed very closely with those made by a direct esterification of the acids of vegetable oils and ethanol. The viscosity of the esters was considerably less than that of the oils and more nearly like that of diesel fuel.

Because the ethyl soyate and sunflowerate have many components which solidify at relatively high temperatures, cloud points of the fuels are about 8 to 12°C, which indicates they would not be readily usable at or below that temperature without dilution with No. 1 or No. 2 diesel fuel and/or the addition of additives.

ACKNOWLEDGEMENTS

The author thanks Mary Russell, Phil Carter and John Vrombaut for their assistance in performing many of the preparations and the associated analytical procedures reported.

REFERENCES

1. L. J. Bruwer, *Sunflower Seed Oil as Our Extender for Diesel Fuel in Agricultural Tractors;* Division of Agricultural Engineering, Republic of South Africa.

2. McGraw-Hill Dictionary of Scientific and Technical Terms. 1978: Editor-in-Chief D. N. Lapedes. McGraw-Hill Book Company: P. 1651.

3. The Merck Index. 9th Edition. 1976. Merck and Company: P. 8502 and P. 8788.

4. Preparation of Detergents. G. C. Brudschaw and W. C. Meuly. U.S. Patent No. 2,360,844 (October 24, 1944).

5. Process of Treating Fatty Glycerides. W. R. Trent. U.S. Patent Nos. 2,383,632 and 2,383,633 (August 28, 1948).

6. K. T. Zilch, Emery Industries, Inc., Cincinnati, Ohio. Private Communication.

7. E. H. Pryde, USDA, Peoria, Illinois. *Preparation of Methyl Esters by Transesterification with Methanol.* Private Communication.

NATIONAL ECONOMIC IMPLICATIONS OF SUBSTITUTING PLANT OILS FOR DIESEL FUEL*

Glenn S. Collins, Ronald C. Griffin and Ronald D. Lacewell**

ABSTRACT

A regional field crop and national livestock econometric model (TECHSIM) was used to examine the impacts of diverting plant oils (cottonseed and soybeans) to use as a diesel fuel replacement. Two scenarios which represented a five and ten percent replacement of agriculture's diesel fuel use by plant oils were simulated. Producers shift into cotton and soybean production and out of corn, small grains and grain sorghum. Significant price shifts were estimated for cottonseed and soybean meal and oil, fed beef, pork and sheep. The annual reduction in social well being was estimated at about $.5 billion and over $1 billion for replacement of 5 and 10 percent, respectively of agricultural's diesel fuel use by plant oils.

I. INTRODUCTION

National interest in developing renewable biomass fuels continues to be strong given uncertainties involving the future availability of petroleum-based fuels. Although extensive interest has primarily centered around ethanol and methanol production, there is growing interest in using plant oils as diesel fuel extenders or replacements. Unlike ethanol and methanol, plant oils offer the potential for direct use as a diesel fuel substitute without requiring extensive engine adjustments or modifications. However, some technical aspects of using plant oils for diesel fuel require further study. For example, preliminary evaluations at Texas A&M of sunflower oil/diesel fuel blends in a single-cylinder engine indicate that as sunflower oil content is increased, torque and power output decrease and fuel consumption and engine knock increase. However, it appears that improved engine performance on sunflower oil fuel is expected with throttle and other engine adjustments (Lepori 1982).

Assuming that technical problems can be overcome, economic issues need to be addressed. Currently, plant oils cost two to three times as much as diesel fuel on a BTU basis (Figure 1). This relationship between plant oil prices and diesel prices has existed for many years, and it is apparent that plant oils are not inexpensive substitutes for diesel fuel. However, the influence of international oil cartels on the price and availability of crude oil may change this relationship. Should crude oil supplies be temporarily dis-

* Technical article _____ of the Texas Agricultural Experiment Station. The work upon which this report is based was supported in part by the Texas Agricultural Experiment Station and the Center for Energy and Mineral Resources, Texas A&M University and the Texas Energy and Natural Resources Advisory Council, Austin, Texas.

** Glenn S. Collins, Assistant Professor; Ronald C. Griffin Assistant Professor; and Ronald D. Lacewell, Professor, Department of Agricultural Economics, Texas A&M University.

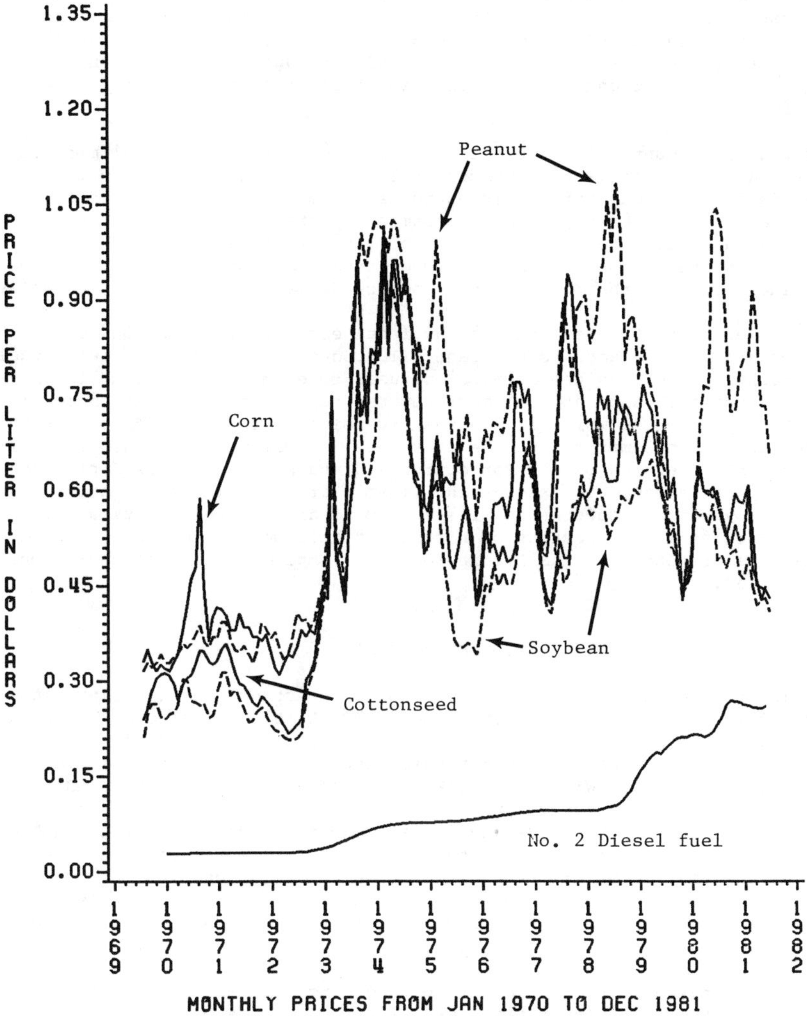

Fig. 1. Diesel Equivalent prices[1] for major plant oils.

[1]All plant oil prices are for crude oils and are adjusted to represent the energy equivalent of one liter of diesel fuel. Diesel prices from 1970 through 1977 are average annual prices. All prices are wholesale for bulk quantities and exclude taxes.

rupted, or perhaps if the price of crude oil gets high enough, the use of plant oils may be economically feasible. Of course, the mere availability of such substitutes can have a beneficial influence by holding crude oil prices below a certain level. Thus, with the possibility of short term crunches, and because of the exhaustibility of oil and gas supplies, it is prudent to evaluate the implications and impacts associated with using plant oils as a diesel fuel.

If substantial amounts of plant oils are diverted from traditional markets for usage as fuel, the market impacts will be large and wide ranging. Hence, efforts designed to quantitatively estimate the multi-market impacts of utilizing plant oils as energy sources becomes important. Many producer and consumer groups would be affected, some positively, others negatively. The importance of these impacts makes it imperative that some empirical results be obtained prior to the resolution of policy decisions involving plant oils.

This study examines the regional and national economic implications of diverting alternative amounts of plant oil production to replace diesel fuel. In the following section the methodology used to examine this issue is presented. A regional and national agricultural econometric simulation model (TECHSIM) is presented which allows the diversion of soybean and cottonseed oil from traditional markets to use as a fuel. In Section III, a hypothetical plant oil diversion policy for cottonseed and soybean oil is formulated. This policy is designed to examine the impacts of diverting alternative quantities of soybean and cottonseed oils to use as diesel fuel substitutes. Then, in Section IV the results of simulating this policy for two different scenarios is presented. Finally, in Section V implications and conclusions are discussed.

II. METHODOLOGY

To examine the regional and national economic impacts of substituting plant oils for diesel fuels, an econometric model (TECHSIM) developed at Texas A&M was used (Collins 1980). TECHSIM is a regional field crop and national livestock model which was designed to use extraneous information to evaluate technological changes. Within TECHSIM, extraneous information is used as a basis to shift relevant supply and/or demand functions to reflect a change in technology. Then, the relative aggregate economic impacts of the technological induced shift can be estimated by simulating future values of endogenous variables that may occur with and without the technological change. This approach is less subjective than a partial budgeting approach, and less expensive and time consuming than regional or national linear programming approaches.

TECHSIM is represented by 13 field crop producing regions shown in Figure 2 and four national livestock categories. The field crops included in the model are: corn, small grains (wheat, barley and oats), grain sorghum, cotton lint, cottonseed and soybeans. The model also contains the forward meal and oil products of cottonseed and soybeans. The national livestock products in the model include fed beef, non-fed beef, pork and sheep.

TECHSIM is a recursive model which centers around three types of estimated equations: planted hectares, yield and demand. The planted hectares and demand equations are primarily represented by a Generalized Leontief structure. For example, the Leontief form for the regional hectares equations is derived from the following regional indirect profit or rent[1/] function:

$$\tilde{R}(R_1,\ldots,R_n) = \sum_i R_i * \tilde{A}_i(R_1,\ldots,R_n) = \sum_i \sum_j B_{ij} R_i^{\frac{1}{2}} R_j^{\frac{1}{2}} \qquad \text{Eq. (1)}$$

1/ The use of the terms rent, profit or net-returns have the same meaning in this study. All of these terms are assumed to depict the short-run returns to fixed production factors.

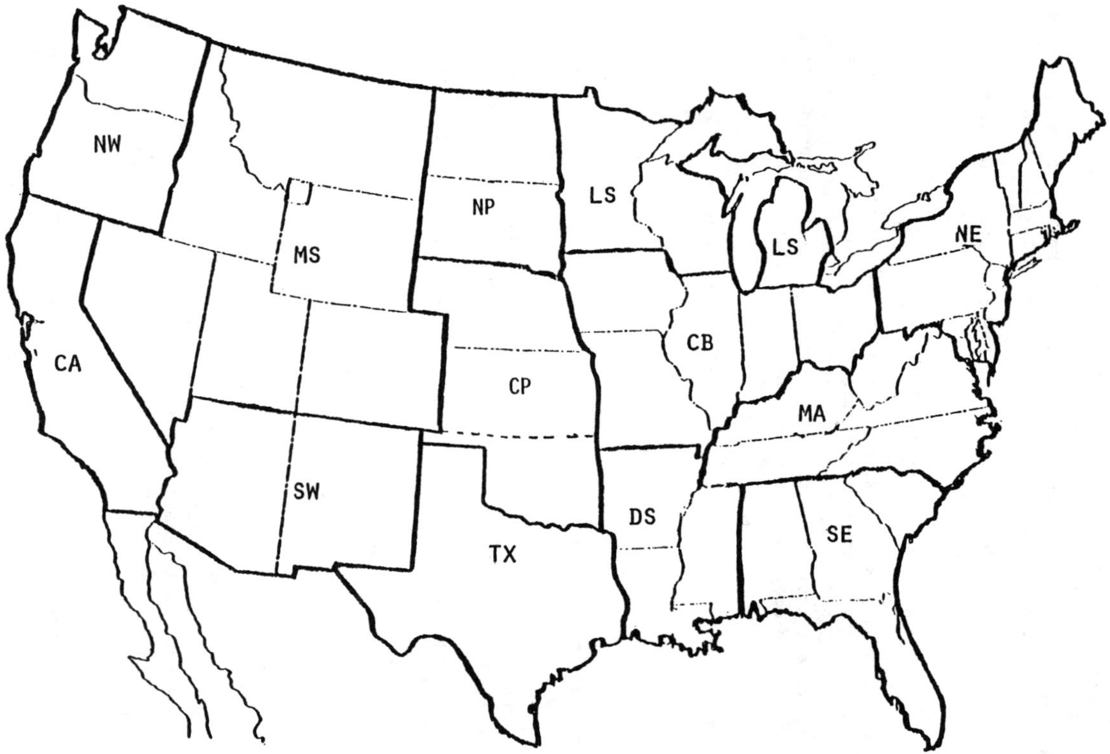

Fig. 1. TECHSIM production regions.

where:

$\tilde{R}$ = total regional expected crop net-returns;
R_i = P *Y - VC = expected net-returns per hectare of crop i;
R_j = expected net-returns per hectare of crop j;
$\tilde{A}_i$ = planted hectares for crop i;
P_i = expected product price for crop i;
Y_i = expected yield for crop i;
VC_i = expected variable production costs per hectare of crop i; and
B_{ij} = the estimated indirect profit coefficients.

Using the envelope theorem (Silberberg 1974) the behavioral choice equations for planted hectares are given by:

$$\frac{\partial \tilde{R}}{\partial R_i} = \tilde{A}_i(R_1,\ldots,R_n) = \sum_j B_{ij} R_j^{\frac{1}{2}} R_i^{-\frac{1}{2}} \qquad \text{Eq. (2)}$$

The model assumes that producers within a region make planting decisions based upon the relative expected net-returns of alternative field crops within the region. The recursiveness of TECHSIM is based upon the expected net-returns which are equivalent to the previous years net-returns. This is a special case of the adaptive expectations hypothesis proposed by Hicks (1938).

Along with the planted hectares equations, yield equations are represented by the form:

$$Y_i = \tilde{Y}_i(T) \qquad \text{Eq. (3)}$$

where:

Y_i = yield of crop i; and
T = year.

These yield trends are based on the productivity changes that have occurred during the model's estimation period, 1961-1977.

The demand equations like the hectares equations are represented by the Generalized Leontief form:

$$\tilde{D}_i(P_1.,\ldots,P_n) = \sum_j C_{ij} P_j^{\frac{1}{2}} P_i^{-\frac{1}{2}} \qquad \text{Eq. (4)}$$

where:

$\tilde{D}_i$ = demand for commodity i;
P_i = price of commodity i;
P_j = price of commodity j; and
C_{ij} = the estimated demand coefficients.

All of the demand equations with the exception of final consumer demands of the livestock sector are intermediate demands. This implies that additional processing of these commodities are required before ultimately reaching the final consumer.

In this study, shifts resulting from diversion of soybean and cottonseed oils to substitute for diesel are introduced into the model through direct purchases. Once these oils are purchased, the model traces through the impacts and provides estimates of regional and national planted hectares, yields, production and supplies of each field crop. The model also gives estimates of national demands (domestic, export, private and government stocks) of all commodities in the model. To access the welfare implications of diverting soybean and cottonseed oils, TECHSIM estimates the short-run rents or profits of all industries in the model and the agricultural sector of the U.S. economy. These results are obtained by simultaneously solving all markets for prices such that each commodity's total supplies are equal to total demands.

III. PLANT OILS PROGRAM

In light of the current prices of plant oils there is little evidence that plant oils will become a competitive economic substitute for diesel fuels in the near future. However, if there is a temporary shortage of diesel fuels then government, industry or agriculture may alleviate these shortages by purchasing plant oils in the market place. These purchases could be processed to provide a substitute for diesel fuel. To minimize costs of such a program, the objective would strive to purchase a target level of plant oils subject to their domestic market prices. In this situation, the program would minimize:

$$L = P_c * Q_c + P_s * Q_s + \lambda(X - a*Q_c - b*Q_s) \quad \text{Eq. (5)}$$

where:

- P_c = price of cottonseed oil, cent/kg;
- P_s = price of soybean oil, cent/kg;
- Q_c = quantity of cottonseed oil purchased, m kgs;
- Q_s = quantity of soybean oil purchased, m kgs;
- X = target BTU content of plant oils to be purchased, m. BTU.;
- a = BTU content of cottonseed oil, BTU/kg; and
- b = BTU content of soybean oil, BTU/kg; and
- λ = a Lagrangian multiplier.

Minimizing L yields the following first order conditions:

$$\frac{\partial L}{\partial Q_c} = P_c - \lambda * a = 0 \quad \text{Eq. (6)}$$

$$\frac{\partial L}{\partial Q_s} = P_s - \lambda * b = 0 \quad \text{Eq. (7)}$$

$$\frac{\partial L}{\partial \lambda} = X - a*Q_c - b*Q_s = 0 \quad \text{Eq. (8)}$$

The least-cost expansion path for purchases of plant oils is obtained by solving equations (6) and (7):

$$P_c * b - P_s * a = 0 \quad \text{Eq. (9)}$$

The BTU content of cottonseed and soybean oils is shown in Table 1.

Table 1. BTU Content of Selected Plant Oils

Plant Oil	BTU Per Liter	Liter Per Kilogram	BTU Per Kilogram
Cottonseed Oil	29,409	1.26	37,095
Soybean Oil	29,574	1.26	37,263

To simulate a program of purchasing plant oils to substitute for diesel, be it government or the private sector, equations (8) and (9) were added to the price solution algorithm in TECHSIM. These two additional equations were satisfied in addition to the orginal condition of each commodities total supply equaling total demand.

Since X is exogenous to the model, different assumptions and conditions can be simulated by changing the level of total BTU that would be purchased as well as the frequency of purchases. For example, the government or industry may intervene only on spot occasions when a shortage arises, or they may make purchases on a yearly basis to meet a targeted percentage of total agriculture diesel fuel usage. The latter condition appears more reasonable than the former given the initial cost and time of setting up such a program.

Assuming private industry or the government decides to make purchases on a yearly basis, an important question is to determine the size of the target purchases. For this study, two scenarios were simulated based upon the total BTU usage of diesel by U.S. agriculture. According to Kolmer (1979), the total distillate usage of diesel fuel in 1979 was approximately 277 trillion BTU. To prevent large shocks to domestic plant oil markets, 5 and 10 percent of the total distillate usage was chosen to be the target purchase policies.

RESULTS

TECHSIM was simulated for the years 1982-1990 with a base simulation and each policy simulation. The base simulation assumes that no plant oils will be diverted to manufacture diesel fuel, whereas each policy simulation reflects a program position of purchasing plant oils to meet targeted diesel fuel shortages. After simulating the base and policy simulations, TECHSIM then computes the differences between the base simulation and each of the two policy simulations. These results are depicted below for 1982 and 1990.

Planted Hectare Shifts

Regional planting shifts due to simulating the two policy alternatives are shown in Table 2 for the years 1982 and 1990. All results corresponding to 1982 reflect the initial impact of diverting plant oils, whereas 1990 depicts the final year impact. As expected, most regions planted more cotton and soybeans as a result of purchases of cottonseed and soybean oils. The largest single shift of planted hectares for both diversion policies was in the Corn Belt (CB) region. In this region, producers shifted from corn production and primarily into soybeans. Nationally, producers planted more hectares for all field crops except corn and grain sorghum. However, grain sorghum planted hectares decreased only in 1982 under the ten percent diversion policy. Summing the changes in national planted hectares shows that approximately 381,000 and 311,000 more hectares were planted in 1982 and 1990 with a five percent diversion policy. Similarily, approximately 666,000 and 556,000 more hectares were planted in 1982 and 1990 when a ten percent diversion policy is adopted. Although these appear to be large hectare shifts they represent only an additional 3 to 6 percent increase in planted hectares.

Net-Return Shifts

The total regional and national net-returns or rents to fixed production factors are shown in Table 2. For each region these net-returns represent the sum of the changes in net-returns for all field crops in a region. The national net-return figures represent the summation of all regions net-returns.

All regions experienced an increase in net-returns for both diversion policies. The largest dollar increase was in the Corn Belt (CB) region, where producers gained 584 and 504 million dollars in 1982 and 1990 with a five percent diversion policy, and 1181 and 1072 million dollars in 1982 and 1990 under a ten percent diversion policy. This region accounted for approximately one-third of the total increase in national net-returns for both policies. In general, the Corn Belt (CB), Delta States (DS), Central Plains (CP) and Texas (TX) producers gain the most from diverting cottonseed and soybean oils to use as a diesel fuel substitute.

Price Shifts

The price shifts for each commodity in TECHSIM are depicted in Table 3. Due to the purchasing activities for plant oils, the prices of cottonseed and soybean oils increased for both policy scenarios. This resulted in price increases for cottonseed and soybeans. These price increases provided incentives to increase planted hectares of plant oil crops. However, since meal

Table 2. Regional Shifts in Planted Hectares and Net Returns: 1982 and 1990

Policy	Year	US	NW	CA	MS	SW	CP	NP	TX	LS	CB	DS	SE	MA	NE
<u>Five Percent Diversion</u>															
Planted Hectares (1000)															
Corn	1982	-309	0	-1	-4	0	-31	-3	-8	-19	-242	0	0	-1	0
	1990	-421	0	-1	-8	0	-80	-8	-10	-37	-267	0	-6	-2	-2
Small Grains	1982	73	11	2	30	-2	-63	0	-17	17	62	6	7	10	10
	1990	47	10	2	31	-2	-104	0	-13	29	64	6	6	9	9
Grain Sorghum	1982	6	0	0	0	-10	10	0	-13	0	11	4	2	2	0
	1990	28	0	0	0	-9	15	0	-9	0	19	8	2	2	0
Cotton	1982	62	0	9	0	10	5	0	18	0	0	10	8	2	0
	1990	51	0	8	0	8	5	0	21	0	0	4	5	0	0
Soybeans	1982	549	0	0	0	0	12	7	13	103	379	-2	11	22	4
	1990	606	0	0	0	0	10	11	10	91	434	-12	21	35	6
Net Returns (m. $)	1982	1702	31	78	59	39	159	95	135	117	584	242	71	82	10
	1990	1492	27	75	51	38	144	86	134	101	504	196	61	70	11
<u>Ten Percent Diversion</u>															
Planted Hectares (1000)															
Corn	1982	-576	0	-2	-6	0	-57	-7	-16	-24	-464	0	0	-1	1
	1990	-722	0	-2	-11	0	-135	-13	-19	-57	-477	0	-3	-3	-2
Small Grains	1982	84	18	2	47	-6	-119	-1	-32	24	101	9	9	16	16
	1990	33	17	3	51	-5	-212	-1	-26	44	111	11	10	15	15
Grain Sorghum	1982	-6	0	0	0	-22	16	-1	-34	0	20	9	2	4	0
	1990	28	0	-1	0	-21	23	-1	-31	0	36	17	2	4	0
Cotton	1982	141	0	21	0	23	11	0	46	0	0	19	17	4	0
	1990	136	0	19	0	22	12	0	50	0	0	16	14	3	0
Soybeans	1982	1023	0	0	0	0	28	18	24	203	673	-25	32	59	11
	1990	1081	0	0	0	0	21	21	20	180	765	-41	38	65	12
Net Returns (m. $)	1982	3488	52	185	101	96	309	165	314	256	1181	480	149	171	29
	1990	3179	48	179	91	93	281	154	307	225	1072	419	133	151	26

is a joint product with oil, the prices of cottonseed meal and soybean meal decreased. Similarily, since cottonseed and cotton lint are joint products, an increase in production of cotton to produce more cottonseed depressed the price of cotton lint.

The prices of other field crops such as corn, small grains and grain sorghum increased as a result of producers shifting from these crops and into cotton and soybeans. However, the price shifts for these field crops are relatively small under both program scenarios.

The price shifts for the livestock commodities result from depressed prices for cottonseed and soybean meal feed-stuffs. This encourages livestock producers to increase the number of animals fed which leads to higher livestock production levels and lower prices. Only non-fed beef increased in price depicting the models shift from non-fed livestock production to fed livestock production.

Table 3. Price Shifts of U.S. Farm Field Crops and Livestock (cents/kg)

Commodity	Five Percent Diversion		Ten Percent Diversion	
	1982	1990	1982	1990
Corn	.18	.11	.20	.12
Small Grains	.62	.55	1.03	.96
Grain Sorghum	.28	.28	.51	.51
Cotton Lint	-2.75	-3.15	-7.58	-8.07
Cottonseed	8.28	8.55	21.08	21.11
Cottonseed Meal	-5.60	-5.53	-8.28	-8.37
Cottonseed Oil	19.95	22.06	53.11	54.56
Soybeans	2.50	1.87	4.41	3.61
Soybean Meal	-7.01	-7.28	-10.42	-10.99
Soybean Oil	44.88	41.00	78.22	73.68
Fed Beef	-8.82	-2.20	-11.02	-2.21
Non-Fed Beef	2.21	8.81	4.41	11.02
Pork	-74.95	-61.72	-105.82	-90.39
Sheep	-61.73	-50.71	-85.98	-74.95

Welfare Shifts

In Table 4 the change in total well being or welfare of each diversion policy is illustrated. The first measure depicts the change in total crop producer net-returns or rents. These are the rents resulting from summing the change in total regional field crop rents. Total crop forward industry rents are the rents to intermediate industries and the final consumer surpluses, or measures of consumers' well being of those industries which utilize field crops as inputs which are not accounted for in other rent measures in Table 4. The total meal and oil industry rent depicts the rent to meal and oil processors plus forward industry rents and final consumer surpluses which utilize oil and/or meal as an input. The other rent measures in Table 4 reflect the change in rents to other agricultural industries in TECHSIM. The total cost of purchasing plant oils reflects the purchase price times the quantity purchased for both cottonseed and soybean oils. The total welfare measure is the sum of all rent measures in Table 4, less the cost of purchases for each policy. For a detailed explanation of how these rents were computed see Chavas and Collins (1982).

As shown in Table 4, all producer/consumer classifications lose with the exception of field crop producers and final consumers of livestock products. The crop forward industries loss results from increases in more crop prices than decreases. For example, only cotton lint out of the field crops shows a price decrease. Furthermore, since field crops are basically intermediate products, higher crop prices imply higher input costs to those industries which process these products. Thus, in the absence of higher forward product prices for those industries which use field crops as inputs, rents to these industries will fall.

Table 4. Shifts in Total Agricultural Industry Rents and Plant Oils Program Cost

Rent or Cost Measure	Five Percent Diversion 1982	Five Percent Diversion 1990	Ten Percent Diversion 1982	Ten Percent Diversion 1990
Total Crop Producer Rents (m. $)	1702	1498	3487	3179
Total Crop Forward Industry Rents (m. $)	-919	-777	-1654	-1460
Total Meal and Oil Industry Rents (m. $)	-939	-788	-1938	-1730
Total Livestock Producer Rents (m. $)	-74	-50	-102	-73
Total Livestock Wholesale and Retail Producer Rents (m. $)	-88	-57	-121	-84
Total Livestock Final Consumer Surpluses (m. $)	167	109	230	160
Total Program Cost of Purchasing Plant Oils (m. $)	435	444	1118	1130
Total Welfare = Sum of All Rents Less Program Purchase Cost (m. $)	-586	-509	-1216	-1138

Although not shown, the meal and oil processors porportion of the total meal and oil industry rent is positive. However, the increased prices of cottonseed and soybean oils also result in higher input costs to those industries which utilize these oils as inputs. Consequently, the total meal and oil industry rent is negative, even though it was offset somewhat by positive rent increases in the cottonseed and soybean meal markets.

Total livestock producer rent is negative although some feed-stuffs decreased in price and livestock production increased. Thus, the porportional decrease in livestock prices was greater than the lower feed input costs resulting in rent losses. In addition to crop producers, final consumers of livestock products showed a positive increase in welfare due to lower livestock prices.

The program cost of purchasing plant oils ranged from approximately half billion dollars per year with a five percent diversion, to over a billion dollars per year under a ten percent diversion policy. Although this study does not take into account the program revenue which would result from selling plant oils as diesel fuel, even if the program broke even (value of sales equalled cost of procurement and processing of plant oils for use as a fuel), the total economic impact would be negative for both policy alternatives.

V. IMPLICATIONS AND CONCLUSIONS

The recent national interest in developing plant oils as a substitute for diesel fuel will not occurr in the private sector given the current disparity between plant oil and diesel fuel prices. The price of diesel fuel would have to double or triple before plant oils would become competitive on an economic basis. Even if a non-temporary shortage arises, the private sector of the economy may not be able to allocate resources to maintain reasonable supplies for diesel fuel. Thus, the government may use its resources to impose a program to meet targeted diesel shortages by substituting plant oils for diesel. However, the results of this study indicate that the impacts on domestic markets will be large. Furthermore, the initial cost of setting up a program, plus the operational costs will be high to U.S. taxpayers. For example, just to meet ten percent of Agriculture's diesel usage would require an approximate yearly outlay of over a billion dollars. Furthermore, impacts of such a program will result in large price and profit shifts throughout the agricultural sector. For example, field crop producers and oilseed processors stand to gain but most other agricultural industry profits will be reduced. Thus, it appears that a government sponsored plant oil diversion policy would be both costly and possibly inequitable.

REFERENCES

1. Chavas, J. P. and G. S. Collins. 1982. Welfare measures from technological distortions in general equilibrium. Southern Economics Journal. 48:745-753.

2. Collins, G. S. 1980. An econometric simulation model for evaluating aggregate economic impacts of technological change on major field crops. Unpublished Dissertation. Texas A&M University.

3. Hicks, J. R. 1938. Value and capital: an enquiry into some fundamental principles of economic theory. Oxford Press.

4. Kolmer W. 1979. The energy requirements of U.S. agriculture. Monthly Energy Review. Department of Energy.

5. Lepori, W. 1982. Personal communication. Associate Professor. Department of Agricultural Engineering. Texas A&M University.

6. Silberberg, E. 1974. A revision of comparative statics methodology in economics or how to do comparative statics on the back of an envelope. Journal Economic Theory. 7:159-172.

ECONOMIC IMPLICATIONS FOR THE POTENTIAL DEVELOPMENT OF A VEGETABLE OIL FUEL INDUSTRY

John R. Dunn Kenneth C. Schneeberger

If we know anything for certain in the energy area, it is that things are uncertain. Who would have thought three years ago when President Carter implemented Special Rule 9 to move diesel fuel to shortage areas that we would be experiencing the recent surpluses and declining prices. This demonstrates the risks inherent in making intermediate and long-term projections about fuels and many other commodities.

It is highly important, however, to understand the forces that may be expected to influence future development of a vegetable oil fuel market whether or not we are able to project specific economic outcomes. The purposes in this paper are, then, to (1) summarize the domestic and international oilseed situation with emphasis on trends which will affect the long-run supply and demand for oilseeds; (2) describe the existing oilseeds processing sector so as to focus on the existing linkage between food and potential fuel markets for vegetable oils; and (3) present a basic framework for analyzing the supply, demand, and price effects of significant use of vegetable oil as a fuel.

Historically, the wholesale price of vegetable oils has exceeded that of diesel fuel by a substantial margin. Although the gap has closed, vegetable oils are not competitively priced with diesel fuel today. If vegetable oils are eventually to play a significant role in the U.S. and world diesel fuel supply, economic changes which permanently close this difference in relative prices must occur. The permanency of this change is of great importance. Unless those economic agents that might accept a market role of bringing vegetable oils into the fuels stream are convinced of the long run feasibility of that role, they are not likely to invest in the facilities and equipment required to do the job.

Thus, an economic analysis of the potential of vegetable oils to extend diesel supplies must focus on how vegetable oil price, determined by the supply of oilseeds and the demand for vegetable oils, will move relative to the price of diesel fuel, which is determined by a different set of supply and demand conditions.

It is useful to make several assumptions in assessing how vegetable oils might be incorporated into the fuels system.

First, we assume domestically produced vegetable oils are interchangeable in food and fuel uses. Some of the vegetable oils are unlikely to be used for fuel given current technology, however, we assume that research can solve this problem.

Second, we assume that profit is the primary motivating factor for farmers and oilseed processors. Non-monetary goals, such as energy independence may

---

The authors are: John R. Dunn, Agricultural Economist, Agricultural Cooperative Service, U.S. Department of Agriculture, and Kenneth C. Schneeberger, Professor, Agricultural Economics, University of Missouri.

exist for some individuals. Over the long haul, however, the ability to generate profits will be critical.

Third, the market for fuel use of vegetable oils and the markets for food and other uses cannot be effectively segmented. Supply and demand situations within one market will directly impact what happens in the other markets. The economic linkage between the food, fuels, and other vegetable oil markets is price. The agents of this linkage, that is, the persons or firms who determine how the physical product will move in response to price, are the processors.

Finally, we assume that prices for raw oilseeds and substitute crops and for processed oilseed products are determined in effectively competitive markets. That is to say that no firm, or group of firms acting in concert, are in a position to administer price. Price will be set and move in response to changes in supply and demand conditions.

We make no assumption about where the price of diesel fuel will go, it could hold steady, rise dramatically, or even fall farther over the long-term. It is the relative prices of vegetable oils and diesel which is of primary concern when the two commodities are considered substitutes.

CHARACTERISTICS OF OILSEED SUPPLY AND DEMAND

The U.S. oilseeds complex is a set of horizontally and vertically related industries encompassing a wide variety of activities, but tied by a common foundation, oil-bearing farm commodities. The importance of the oilseeds complex to the U.S. farm and domestic economy has grown considerably over the past two decades. Oilseed commodities represented 10.4 percent of total farm cash receipts in 1980. Over half a million farms grow soybeans, more than for any other cash commodity except corn.

Oilseed commodity and product exports were worth $8.9 billion in 1979, representing 25.6 percent of total U.S. agricultural exports. World export trade of oilseed commodities and products contributed a significant $7.9 billion to the U.S. balance of payments.

The moving forces in the U.S. oilseeds complex are the domestic and world demands for edible fats and oils and for protein. Food related demand is supplemented by a much smaller, though significant industrial demand for oils and resins. Demand for oilseeds and oilseed products at every vertical level in the oilseeds complex is derived from final retail demand for oilseed products. These products cover many uses including margarine, salad oil, shortening, animal feed, meat substitutes, infant formula, candies, medicines, paints, and soaps.

Domestically produced oilseeds include soybeans, cottonseed, sunflowerseed, flaxseed, and peanuts. These commodities must compete for markets among themselves, and to varying degrees, with other sources of fats and protein. Substitute fat and protein sources include both vegetable and animal fats such as palm, copra, lard, coconut, and fish oil and meal. Many substitutes are produced throughout the world, reflecting the global nature of the markets for fats, oils, and proteins.

The U.S. is the largest oilseed producing nation. We produced 35 percent (55.9 of 159.3 million metric tons) of world oilseeds in 1980-81. U.S. production was distributed among the oilseeds as follows:

Oilseed	Production	Share of U.S. Production
	Metric tons	Percent
Soybeans	48,772,000	87.2
Sunflower	1,748,000	3.1
Cottonseed	4,056,000	7.3
Flaxseed/Safflower	306,000	.5
Peanut	1,047,000	1.9

U.S. oilseed production is dominated by soybeans. Production doubled in the 1960's and doubled again in the 1970's. This was due largely to increased acreage (from 24.4 million acres in 1950 to 70.1 million in 1980), but was accompanied by an annual yield trend increase of 1/3 bushels per acre. Soybeans replaced small grains and some corn in the Midwest (see map). In the Delta and Southeast, soybeans replaced cotton, corn, and oats and are now planted on 60 percent of the cropland acres.

Sunflower was the only other "growth" oilseed in the past decade. Cottonseed production, a function of the demand for cotton, held relatively constant. Flaxseed production declined, being largely replaced by sunflower. Peanut production fluctuated, but was an insignificant factor in the vegetable oil market because of the demand for the edible nut. Less than 20 percent of peanuts are processed into oil.

Many economists believe that the boom period of growth for oilseeds in the U.S. is over. Growth in production is expected to continue over the next two decades, but at a slower and steadier pace. Future growth in oilseed production will depend on the profitability to the farmer of producing oilseeds relative to the profitability of planting substitute crops. The primary competition will be between soybeans and sunflower and various food and feed grains such as corn and barley in the North and cotton and wheat in the South. Most of this intercrop competition will be for existing farm acreage--new acreage will be greatly limited. Genetic breakthroughs which significantly raise output per acre would change the picture by allowing large increases in total oilseed production without a decrease in the output of other crops.

Worldwide, the major vegetable oils in terms of volume produced in 1979/80 were soybean, sunflower, palm, rapeseed, peanut, cottonseed, coconut, and olive. Total world production of vegetable oils was 39.3 million metric tons in 1979/80, up 72 percent from 1969/70. The U.S. produced about a third of the total world production of vegetable oils, but accounted for over half of the exports.

Roughly 40 percent of world oil production moves in export markets each year. Soybean oil accounts for over 40 percent; sunflower, palm, and coconut account for 10-15 percent each; and rapeseed and peanut oils 5-10 percent each. Cottonseed oil accounts for less than 5 percent of world vegetable oil trade.

The projections are for increasing international competition between oils in the 1980's. This is due to an expected leveling off in demand for both meal and oil coupled with continuing growth in supply of palm oil (Malaysia), sunflower oil (Russia), and soybean oil (U.S., Brazil, and Argentina).

The major determinants of demand worldwide for vegetable oils are price, (relative to non-vegetable sources of fats and oils), incomes, and population. The dominant determinant in any particular market is related to the level of development of a country. Industrialized countries generally have high fat diets, high income, and low population growths. These factors combine to create a rather stable demand for fats and oils which is somewhat insensitive to swings in vegetable oil prices.

Developing and underdeveloped countries typically have higher population growth rates, low incomes, and low levels of fats consumption in their diets. Demand for fats and oils in these countries tends to be most unstable, more subject to swings in prices, exchange rates, and shortages of hard currency. Potential growth in demand for fats and oils is the greatest in these areas.

Demand for U.S. produced vegetable oils will be affected by all of these factors. The value of the U.S. dollar relative to other currencies must be considered of great importance in analyzing demand for exports of U.S. vegetable oils. Under a situation of a highly valued U.S. dollar, the price of U.S. export oils to the rest of the world is high and the price of substitute oil imports is low, reducing total demand for U.S. vegetable oils.

The impact of the use of vegetable oils as fuels must be viewed the same way as all other potential uses for vegetable oils from a demand analysis perspective. The amount of vegetable oil used as a fuel will be a decreasing function of vegetable oil price. The fuel demand component must be added to the food and other technical uses components to give total demand for vegetable oils. The major difference between fuel demand for vegetable oil and the other components of demand is likely in the sensitivity of demand to changes in price. Since vegetable oils can provide only a small portion of total fuel requirements, increases in the price of vegetable oil relative to the price of diesel would cause large decreases in the quantity of vegetable oil demanded for fuel use. On the other hand, vegetable oils have a dominant role in supplying total requirements for edible fats and oils. This makes substitution away from vegetable oils to other sources of edible fats and oils less possible. Food demand for vegetable oils is less sensitive to changes in price as a result.

Pitcher, of the Foreign Agriculture Service, USDA, recently reviewed the world oilseeds situation and made projections for soybean oil during the 1980's. The key points of his analysis were:

1. The outlook for soybean oil use is dependent on

    a. The supply of competitive vegetable oils

    b. The ability of major developing countries to import, i.e., world economic conditions

    c. The demand for vegetable oils for industrial purposes, including use as a fuel.

2. Most developed countries have excess oil supplies. Import demand must come from about 15 developing countries.

3. Increasing international competition between oils is expected in the 1980's. This is due to an expected leveling off in demand for both meal and oil coupled with continuing growth in supply of palm oil (Malaysia), sunflower oil (Russia), and soybean oil (U.S., Brazil, and Argentina).

4. The biggest unknown is fuel use of vegetable oils. Brazil has a target of 16 percent of diesel use replaced by vegetable oil by 1985. That is the equivalent of 3 million tons (780 million gallons) of soybean oil taken off world markets. The Philippines has initiated a coco-diesel blend program that is projected to use 60,000 metric tons of coconut oil annually.

THE PROCESSING INFRASTRUCTURE

The linkage between the farm and the domestic and foreign food sectors is through an industry composed of firms which process raw oilseed commodities. In many cases in the U.S., firms involved in oilseed processing are also involved in raw commodity assembly and transportation to and through final export markets.

The U.S. oilseed processing industry had a daily crushing capacity of approximately 128,000 metric tons in 1980. Soybeans dominate crush capacity with 82 percent. Cottonseed has 14 percent, followed by sunflower and flaxseed with a combined 3 percent.

Oilseed Production Areas, 1979

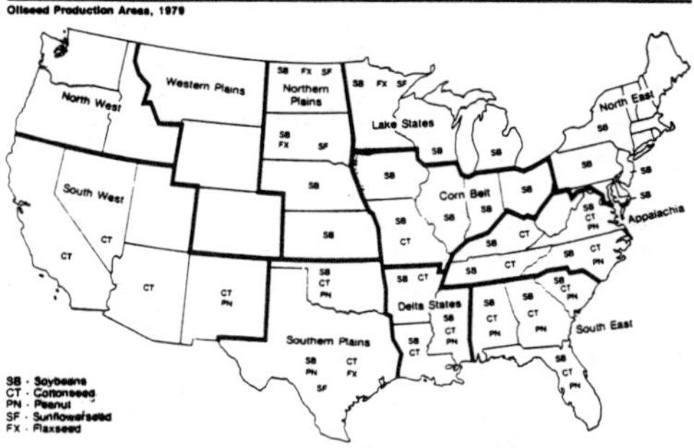

The distribution by region of capacity closely tracks the distribution of production (table 1). The combined Cornbelt, Lakes States and Northern Plains regions have 58 percent of U.S. capacity and produce 64 percent of the total U.S. oilseed output. The combined Southeast, Appalachia, and Delta States regions have 33 percent capacity and produce 28 percent of total output.

Table 1. Share of U.S. Oilseed Production and Crushing Capacity, 1979

Region	Percent share of production	Percent share of processing capacity
Cornbelt	46.8	44.5
Delta States	14.2	12.8
Lake States	8.9	6.0
Northern Plains	8.7	7.1
Appalachia	7.3	11.6
Southeast	6.6	8.1
Southern Plains	3.5	5.9
Southwest	2.2	3.4
Other	1.8	0.6

U.S. soybean crushing capacity has increased steadily from 900 million bu. per year in 1971 to an estimated 1,425 million in 1980 (table 2), closely paralleling soybean production. The soybean crushing industry had moderate excess capacity throughout the decade. Much of the tendency of the industry to expand in spite of low levels of capacity utilization may be attributable to expectations of continued growth in export segments of demand.

Growth in capacity has been accompanied by large increases in plant size and decreases in plant numbers. Average plant size nearly doubled in the 1970's to 14.4 million bu. per year in 1979. This was due to exit by many small mechanical plants and exoansion at the larger solvent extraction plants. Few new plants were constructed.

The cottonseed crushing industry also changed significantly over the past decade. The number of plants declined from 115 in 1972 to 74 in 1980. Total cottonseed crush was fairly constant. The regional shift in cottonseed crushing continued away from traditional areas of the Southeast and Delta States toward the Southern Plains and Southwest. The average cottonseed mill in the U.S. was 243 tons per day in 1979. Plants in the Southwest and

Southern Plains averaged 388 and 286 tons, which are considerably larger than the 185 ton average in the older cotton areas of the South.

Table 2. U.S. Soybean Crushing Industry Data, 1971-1980

Marketing year	U.S. annual crush capacity	U.S. soybean crush	Capacity utilization	Plants	Average plant size
	Million bushels		Percent		Mil. bu/day
1971	900	700	80.0	123	7.3
1974	1,050	701	66.8	108	9.7
1976	1,200	790	65.8	103	11.6
1979	1,350	1,125	83.2	94	14.4
1980 1/	1,425	1,020	71.6	-	-

1/ Preliminary

Sunflower crushing has grown considerably in the last five years, with total capacity doubling since 1976 to approximately 4,700 tons per day. Some cottonseed plants crushed sunflower earlier in the 1970's, however, as production increased and concentrated in Minn., N.D., and S.D., these plants' roles declined.

The U.S. processing industry is a mix of multinational firms, farmer cooperatives, vertically integrated food and feed manufacturers, and commodity marketing firms. The ten firms which operate more than one soybean plant own a total 83 percent of U.S. capacity. The largest 20 processors operate 75 plants with about 96 percent of U.S. soybean processing capacity.

Eight firms operate more than one cottonseed plant, two of which are also among the top 20 soybean processors. Three of the largest four sunflower processors also operate soybean processing plants. Several single-plant processing firms operate in each of the commodity areas. These plants tend to be smaller and older, with a few notable exceptions.

U.S. processors must compete, directly and indirectly, with processors located in non-U.S. countries including both net-importing and net-exporting regions. Some of these processors are multinationals with capacity in the U.S. as well, but many others are operated by foreign food corporations, cooperatives, trading companies, and a variety of State-owned entities.

Processing plants are located in both producing and consuming countries. The 1979 world soybean processing capacity of about 280,000 tons per day was 58 percent in soybean exporting countries and 42 percent in importing countries. The U.S. had the largest share of capacity with 38 percent, followed by Brazil with 16 percent, and Argentina with 5 percent.

A wide variety of close substitutes, many potential origins, reasonably public pricing systems, and individual buyers and sellers with divergent and competitive interests mean that prices for vegetable oils in the U.S. are directly affected by world supply and demand. U.S. firms will not, therefore, be able to rationalize a valuation system for their vegetable oil which is independent of world price levels.

Oilseed Processing Margins

The market value of oilseed processing services is determined at any given location and time as the difference between the market value of products per unit of raw oilseed and the market value of the unprocessed commodity. Known as the processing margins, this value may exceed equal, or be less than the actual cost of processing. A plant will be operating at a loss unless both the fixed and variable plant costs are met. In the short run, a plant may operate at a loss as long as variable costs are covered. However, total costs must be met over the long run or the plant will not be operated.

Over the long haul, competition between plants will cause processing margins to converge toward the cost level achieved by the most efficient plants. Plants with consistently higher costs will be weeded out of the industry. Processing margins for soybeans have varied considerably over the past few years from over $1.00 to less than 10 cents. For the past year or so, processing margins have been very poor. Near term prospects for improvement don't look favorable.

Four year trends in soybean processing costs are shown in table 3 for major cost categories. The data are simple averages from 13 solvent extraction plants with an average daily capacity of 1300-1500 tons. The category "other" includes repairs, taxes, insurance, sewage, water, depreciation, and miscellaneous costs. The cost figures represent the narrowest definition of processing costs. Other costs of operation include storage, finance charges, transportation, administration, and corporate overhead. These additional costs must be paid out of the processing margin as well.

Table 3. Soybean Processing Costs, 13 Plant Average, 1977-1981

Manufacturing cost category	Crop year			
	1977/78	1978/79	1979/80	1980/81
	Cents per bushel			
Labor	7.68	8.43	9.02	9.68
Fuel and power	10.93	11.23	11.78	13.72
Solvent	1.46	1.81	2.82	3.66
Other	12.92	11.74	14.53	13.52
Total	32.99	33.21	38.15	40.58

Processing margins are also of great concern for sunflower, flaxseed, and peanut processors since market prices exist for those commodities in their unprocessed state and for their processed products. Cottonseed, as a by-product of cotton production, does not have a meaningful market price for the unprocessed seed, thus the concept of a processing margin is much less clear.

Oilseed processors will base their procurement, operating, and marketing decisions on their processing margins. They will seek to maximize processing margins by adjusting utilization of capacity and by directing their output of processed products toward the most lucrative markets. If the demand for one of the outputs is altered significantly, that change will be reflected in the processing margins of individual plants. By responding to such changes, the oilseed processing industry will direct the flow of vegetable oils between alternative uses in accordance with relative price.

THEORETICAL ANALYSIS OF A VEGETABLE OIL FUEL MARKET

The U.S. oilseeds market can be graphically analyzed using a demand component model developed by Houck for soybeans (figure 1). Total supply and demand of raw oilseeds is shown in the upper right. In the remaining graphs of the upper tier, total demand for raw oilseeds is segmented into three components: domestic crush, raw oilseed exports, and raw oilseed stocks. Demand for domestic crush is segmented into meal and oil components in the middle and lower tiers. Finally, meal and oil demands are broken into their export, stocks, and domestic use components. This model shows the U.S. oilseed market as a system of interrelated markets. Changes in demand in one market can ripple through the entire system of markets.

The addition of a fuel use market for U.S. vegetables is shown graphically by an outward shift in the domestic oil demand curve (dashed). The curve may become less steep as well, reflecting the higher sensitivity of fuel-use oil demand to price changes. The one-year ripple effect of this increase in oil demand may be seen through the following sequence: (1) total oil demand shifts outward, leading to an outward shift in demand for domestic crush; (2) increased demand for domestic crush will increase processing margins, inducing an increase in total demand for oilseeds (dashed line) in the upper right graph; (3) given a fixed supply of oilseeds in a single year, the increase in demand will lead to a price increase for raw oilseeds; and (4) the effect of the increased price may be traced back into the meal and oil markets. The larger crush will create an excess supply of meal, lowering its price. Part of the increase in domestic demand for oil will be fulfilled by the larger crush volumes. The rest will be offset by the higher oil prices.

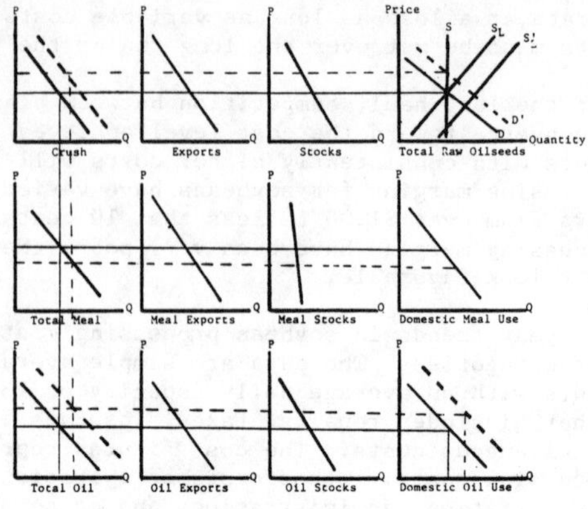

Figure 1. Demand Component Graphical Oilseed Model

The one-year impact of a vegetable oil fuels market will be to increase the price of vegetable oils. Over several years, the supply of oilseeds can't be considered fixed, rather is an increasing function of price, as shown by S_L. The higher oilseed prices caused by the fuel demand for vegetable oils will attract a larger volume of oilseed production. This will moderate the original price increase.

These implications of a vegetable oil fuels market are straightforward if there is little or no increase in oilseed production productivity. It is possible, however, that genetic or other improvements could increase productivity--that is, increase output for a given amount of land and other inputs. A productivity increase may be shown by an outward shift of the supply curve to S_L'. If the productivity gain is large, the full impact of the increased fuel use demand on price for vegetable oil could be completely offset. In that case, prices for beans, meal, and oil could all fall.

It is possible that fuel use demand for vegetable oils could continue to grow as opposed to the one-time increase previously assumed. This situation could result from an expanding market for diesel fuels due to increased popularity of diesel powered automobiles. This situation may be shown graphically by shifting the domestic oil use demand curve further outward, along with appropriate outward shifts of total oil demand, crush demand, and total raw oilseeds demand. The continuing increase in demand results in a steady upward pressure on vegetable oil price which may or may not be offset by any increased production and productivity.

Estimates from the Theoretical Model

The graphical presentation above indicated the direction of change in oilseed market prices and quantities from the creation of a vegetable oil fuels market. Levin and Meyers developed some estimates of the magnitude of these changes over an eight-year period. They limited their analysis to soybeans and assumed U.S. demand for soy oil increased by 227,000 metric tons, remaining at that level for eight years. That is equivalent to converting about 10 percent of U.S. agricultural diesel use to a 25-75 vegetable oil-diesel mix.

The short run impact of increased vegetable oil demand was increased domestic crushing, reduced export use, and reduced stocks. In year 1, oil prices increased 5 cents per pound (38.5 cents per gallon) and meal prices fell. Soybean prices increased by a net $0.30 per bushel.

In year 2 production increased due to larger planted acreage in response to the higher year 1 price. Some of the price increase of year 1 was diluted as a result.

In the long run (7-8 years) prices were pushed back further while acreage in soybeans increased by 1 million acres. Oil prices remained $0.03 per pound above what they would be without the fuel market. Net income from soybean production, however, is not much larger than without a VOF market due to the lower prices for meal.

The Levin-Meyers formulation represents a first approximation of the long run impact of using vegetable oils as fuels. It is limited by the small fuels market that is assumed and it ignores the possibility of increasing productivity.

CONCLUDING NOTES

We reach no firm conclusions on how vegetable oils might fit into the diesel fuel market. That will rest on how the relative prices of vegetable oils and diesel move--which are subject to great uncertainty.

The likely impact of a developing vegetable oils fuel market would be to increase vegetable oil prices. The size of the increase will depend on how large the fuel market demand ultimately becomes, and thus on the price of diesel fuel. It will also depend on how well oilseed production can be adapted, technologically, and in acreage, to meet the needs of a large fuels market while maintaining its critical role in the foods sector.

Other factors could influence the outcome. Government programs of taxes, quotas, or subsidies could affect vegetable oil supply and/or demand. International trade practices could change, altering the flow of oils between markets. New fuels or technologies could be developed rendering the vegetable oil fuels obsolete.

There are many uncertainties in assessing the economic picture for vegetable oil use as a diesel fuel substitute. However, we need to understand how the various factors will come together to determine the outcome. If we have this understanding we can better judge what will evolve.

REFERENCES

1. Dunn, J.R., et. al. 1982. Cooperative Involvement and Opportunities in Oilseeds. ACS, USDA. ACS Research Report No. 13.

2. Hazera, J. and E. Fryar. 1981. Regional Soybean Production Since 1960 and the Outlook for the 1980's. Fats and Oils Situation. ERS, USDA. FOS-305.

3. Houck, J.P., et. al. 1972. Soybeans and Their Products. Univ. of Minnesota Press, Minneapolis, Minn.

4. Levins, R.A. and W. H. Meyers. 1981. Market Implications of Soybean Oil Use as a Diesel Fuel Substitute. AAEA Summer Meeting. Clemson, S.C.

5. Meyers, W.H. and D. D. Hacklander. 1979. An Econometric Approach to the Analysis of Soybeans and Soybean Product Markets, NED, ESCS, USDA Staff Report.

6. Pitcher, S. 1981. Soybeans in the 1980's: Expanding Competition and Changing Trade Patterns. Foreign Agriculture. FAS, USDA, Vol. XIX, No. 7: 7-10.

OILSEED AND OTHER BIOMASS FUEL PENETRATION IN THE AGRICULTURAL SECTOR UNDER ALTERNATIVE PRICE ASSUMPTIONS FOR COMPETING FUELS

David J. Bjornstad, Edward L. Hillsman, and Richard C. Tepel*

Although the use of oilseeds and other biomass fuels as energy sources for farm operations has an attractive intuitive appeal, successful commercialization of these energy sources requires that they offer cost advantages over competing fuels. This paper examines the rate at which biomass fuels become cost-competitive as fuel prices for diesel oil, fuel oil, gasoline, natural gas, LPG and electricity (central fuels) increase. While oil from oilseeds substitutes primarily for diesel fuel in tractors and trucks, it may also be used in place of fuel oil for certain space heating applications and could be used to power electrical generators. To the degree possible, it is desirable to consider all possible substitutions.

In this analysis, fuel demands are projected for ten farm operations in the year 1990 and are matched with biomass fuel substitutes from ten feedstocks and nine processes. In all, 61 feedstock/process combinations are available to substitute for the central fuels. This information is processed using a linear programming model that seeks to meet fuel demands at minimum cost. The model is solved for a series of central fuel real price increases holding costs of biomass fuel substitutes constant to determine the relative increases in central fuels necessary to bring about biomass fuel penetration. Two alternative assumptions are made about rates of biomass fuel penetration. The first assumption allows penetration to be limited by projected feedstock availability and construction leadtimes, while the second omits these limits to estimate maximum penetration rates. Results reported include the percent of total central fuel and individual central fuels for which biomass substitutes become cost-competitive under the range of central fuel prices, and detailed information about oilseed penetration.

BIOMASS FUEL ALTERNATIVES

Figure 1 shows the feedstock/process combinations selected for use in this study. Because emphasis is placed on using biomass fuels through a

*The authors are staff members of the Energy Division, Oak Ridge National Laboratory. Oak Ridge National Laboratory is operated by Union Carbide Corporation for the Department of Energy. They wish to express their gratitude to Ms. Lillian Cochran for typing this paper and to Ms. Jeanne Rainey for assistance in preparing the data.

†The work reported herein draws heavily upon a larger study of on-farm biomass fuel use conducted at the Oak Ridge National Laboratory. For greater detail regarding this analysis, the reader is referred to D. J. Bjornstad, E. L. Hillsman, and R. C. Tepel, et al. <u>The On-Farm Use of Biomass Fuels: Market Penetration During Normal and Fuel Emergency Conditions,</u> forthcoming, Oak Ridge National Laboratory, Oak Ridge, Tennessee.

Feedstock	Process-Product								
	Vegetable Oil (Press)	Vegetable Oil (Solvent)	Butanol	Ethanol	Gasoline (MTG Process)	Low BTU Gas (Gasifier)	Methane (anaerobic digestion)	Direct Combustion	Electricity
Corn	X	X	X	X	X	X		X	X
Sugar beets			X	X	X	X		X	X
Sugar cane			X	X	X	X		X	X
Sunflower Seeds	X	X							X
Soybeans	X	X							X
Peanuts	X	X							X
Wood			X			X		X	X
Corn residue			X	X	X	X		X	X
All other crop residue			X	X	X	X		X	X
Animal residue						X	X	X	X

Fig. 1. Biomass Feedstock - Process Combinations

decentralized distribution system, two types of facilities are considered. The first is on-farm production units and includes oil presses, low-Btu gasifiers, biogas digestors and direct combustion units. The second type of facility is a farmer's co-operative scale production facility that can capture some economies of scale, but is often more costly than a large centralized facility such as those now used to produce vegetable oils for food or chemical industries. The primary data describing these technologies is cost per unit output, where costs are calculated as first-year capital charges, plus all allocable operating expenses, less any by-products of value. All costs are based on commercial operations. Homemade or makeshift applications are not considered.

OVERVIEW OF MODEL AND ASSUMPTIONS

This study assumes that farming behavior can be represented nationally through an aggregate farm sector that adjusts fuels used in response to energy price changes, but does not adjust total quantities of energy demanded. Farmers are assumed to make these choices based solely on costs, and to choose least cost fuel combinations unless constrained to do otherwise. Farms are assumed to be strictly limited as to fuel substitution choices. Vegetable oil and butanol may be substituted for diesel and fuel oil; ethanol and MTG-processed ethanol (gasoline) may be substituted for gasoline; and, low-Btu gas and biogas may be substituted for natural gas and LPG. All fuels except gasoline and ethanol may be used to generate electricity. For space heating and grain drying, biomass products may be burned directly. No other substitutions are assumed. Except for central fuel prices, all other prices are held constant. For example, where relevant, it is assumed that farmers can dispose of biomass fuel by-products (such as sunflower seed presscake) in unlimited amounts at constant prices.

These assumptions are operationalized through a linear program that consists of an objective function describing the costs of supplying energy for end-use activities and a set of constraints describing the amounts of energy demanded and the limits on supplying them. The objective function may be written as

$$\text{minimize } Z = \sum_{i=1}^{n} \sum_{j=1}^{m} C_{ij} X_{ij} \qquad \text{Eq. (1)}$$

where i denotes a process for obtaining energy, and j denotes an end-use demand for energy. For example, i might denote oil pressed from sunflower seeds, or ethanol fermented from sugar beets, and j might denote diesel fuel used in tractor operations or LPG required for brooding. C_{ij} is the cost of supplying a unit of energy to meet a unit of demand, and X_{ij} is the number of units of energy supplied by i to j. For the centrally-delivered fuels, the costs used are the projected market-clearing price levels while for the biomass fuels, the costs are the estimated unit production costs.

The model contains three types of constraint equations. The first type specifies the amount of each kind of centrally-delivered energy demanded in each end-use activity, the biomass technologies that can produce energy to substitute for this energy, and the efficiency with which the biomass energy substitutes for the centrally-delivered energy in that application. These constraints may be written as

$$\sum_{i=1}^{n} E_{ij} X_{ij} \geq D_j \qquad \text{for all } j \qquad \text{Eq. (2)}$$

where i again is an energy source, D_j is the amount of energy needed in activity j, X_{ij} is the number of energy units obtained from source i in meeting demand j, and E_{ij} is the efficiency of substitution.

The second type of constraint specifies the availability of biomass feedstocks that can be converted to energy. These constraints may be written as

$$\sum_{i \in I_k} \sum_{j=1}^{m} T_{ij} X_{ij} \leq A_k \qquad \text{for all k} \qquad \text{Eq. (3)}$$

In this case, k denotes a physical resource, such as fuel oil, wood, or sunflower seeds, and A_k is the number of physical units of resource k available. There is one constraint of this type for each biomass feedstock, and one for each centrally-delivered fuel. The I_k indicates the set of processes that draw upon the resource k to produce energy, while for a commercial fuel this would simply be purchases of the fuel. The subscript j allows the fuel to substitute in all of the demands that can use it. X_{ij} is, again, the number of units of energy obtained from process i for demand j, and T_{ij} is the number of physical units of the resource required to produce a unit of energy.

The third type of constraint limits the rate of penetration of biomass energy for reasons other than feedstock availability. These constraints include the ability of the existing farm machinery to use ethanol in high concentrations and delays in constructing the biomass processing equipment. Ethanol is allowed to replace no more than 10 percent of the volume of gasoline, or 5 percent of its energy content, in any farm activity.

The ethanol constraint may be further modified, however, to account for the lag in developing the process equipment at the farm or co-op level. This delay is likely to be at least nine months, even during the most urgent of circumstances. The equipment thus would operate for only the last three months of the first year the process becomes cost-competitive. Over a decade of slow adjustment to slowly increasing real energy prices, this might mean that only 25 percent of the on-farm demand for ethanol would be met in 1990, increasing in later years. The ethanol constraint is written to reflect this as follows

$$\sum_{i \in I_e} X_{ij} \leq (.05)\ (.25) D_j \qquad \text{Eq. (4)}$$

where I_e is the set of processes that yield ethanol and the remaining variables are defined as before.

Other penetration constraints are similarly formed assuming ethanol, butanol, synthetic gasoline and vegetable oil from solvent extraction could be available for 25 percent of demand, biogas for one-half of demand, vegetable oil from on-farm presses for 75 percent of demand, and low-Btu gas for 83 percent of demand. Direct combustion is cost-competitive now for some applications and may penetrate fully by 1990.

The model is solved under two sets of assumptions. The first assumption is that feedstocks and penetration rates are each binding. This is a conservative assumption that provides a likely lower bound to penetration. The second assumption removes these constraints, though it keeps such technical constraints as that governing ethanol/gasoline mixing. This provides an upper bound to penetration.

PRICE RATE INCREASES

Of the components that govern biomass fuel penetration, prices of competing fuel types are the most uncertain. This occurs in large part because of the influence of world oil prices on domestic energy markets. Because of the uncertainty and the fact that different biomass feedstock/process combinations become cost-competitive for different central fuels at different central fuel prices, we examined nine different average annual real rates of increase in central fuel prices (0%,3%,6%....24%). The price increases were applied uniformly to each of the six central fuels. Because the rates are assumed to be average annual increases persisting until 1990, small averages can lead to large real price increases by 1990. A six percent averge annual rate of increase in real energy prices would make real energy prices 80 percent higher in 1990 than they would be if they merely increased at the rate of inflation. A 12 percent real annual increase would double real energy prices by 1990, an 18 percent rate would quadruple them, and the extreme case of 24 percent would increase real energy prices eightfold.

RESULTS

Figures 2 and 3 show the empirical results from operating the model at each rate of price increase alternatively assuming that feedstock and penetration constraints do and do not apply. Figure 3 indicates the percent of centrally-delivered fuel that on-farm biomass fuels could replace were each average annual rate of price increase to persist until 1990. The results indicate that at current prices about 10 percent of a central fuel could be replaced and that with a fifteen percent average annual increase all could be replaced, though constraints would likely limit the rate at which replacement would occur.

Breaks in Fig. 2 appear because individual fuels become cost-effective at different prices. These are shown in Fig. 3 where the constrained rate of penetration for each rate of price increase is indicated by the solid line and the unconstrained rate by the broken line. Currently, fuel oil, natural gas and LPG in grain drying and space heating applications could be replaced by biomass fuels at cost-savings. This replacement would entail burning wood or crop residues in place of the central fuel. Oilseeds become cost-effective replacements for diesel oil at a six percent average annual rate of price increase, i.e., less than a doubling of real prices over the decade. This is a significant finding since by 1990 diesel oil is projected to make up 35 percent of total agricultural energy use, the largest single energy source for the farm sector. It should also be noted that gasoline, which makes up 33 percent, will be replaced by ethanol only if a 15 percent average annual price increase is sustained.

Oilseed penetration first occurs when the average annual real price increase is six percent. At this point soybean oil, produced by the co-op scale mechanical process supplemented by solvent extraction, is the most cost-competitive substitute for diesel oil. Sunflower seeds and corn processed in this manner would also be cost-competitive, but the other feedstock/processes considered in this study would require a higher rate of increase in energy prices to penetrate. If soybeans were used this extensively, the impact on product markets would be substantial. Meeting all agricultural diesel demands in 1990 would require the entirety of the soybean crop projected for that year, plus an additional 14 percent. If this shortfall were met with sunflower seeds it would require 77 percent of the projected 1990 crop. However, were the lower penetration limit to apply, roughly 25 percent of diesel demand would be met, requiring 28 percent of the soybean crop. The on-farm mechanical press process of

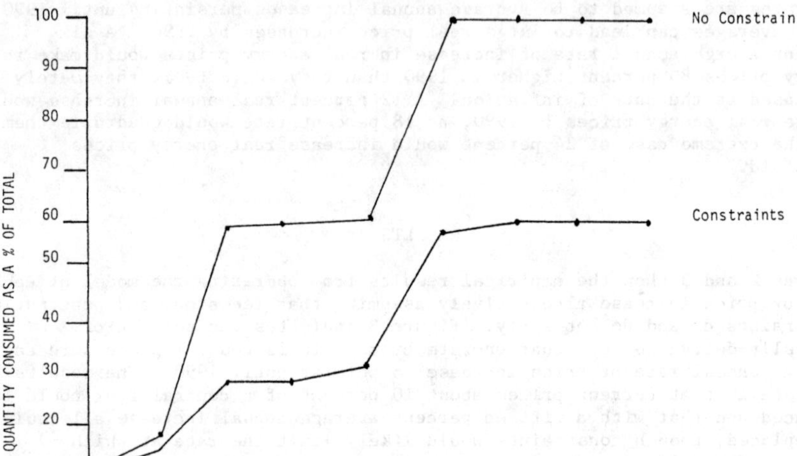

Fig. 2. Biomass Fuels as a Percent of Total Energy Consumed On-Farms in 1990 Under Alternative Price Growth Rates

Note: Constraints case is limited by feedstock availabilities and penetration limits.

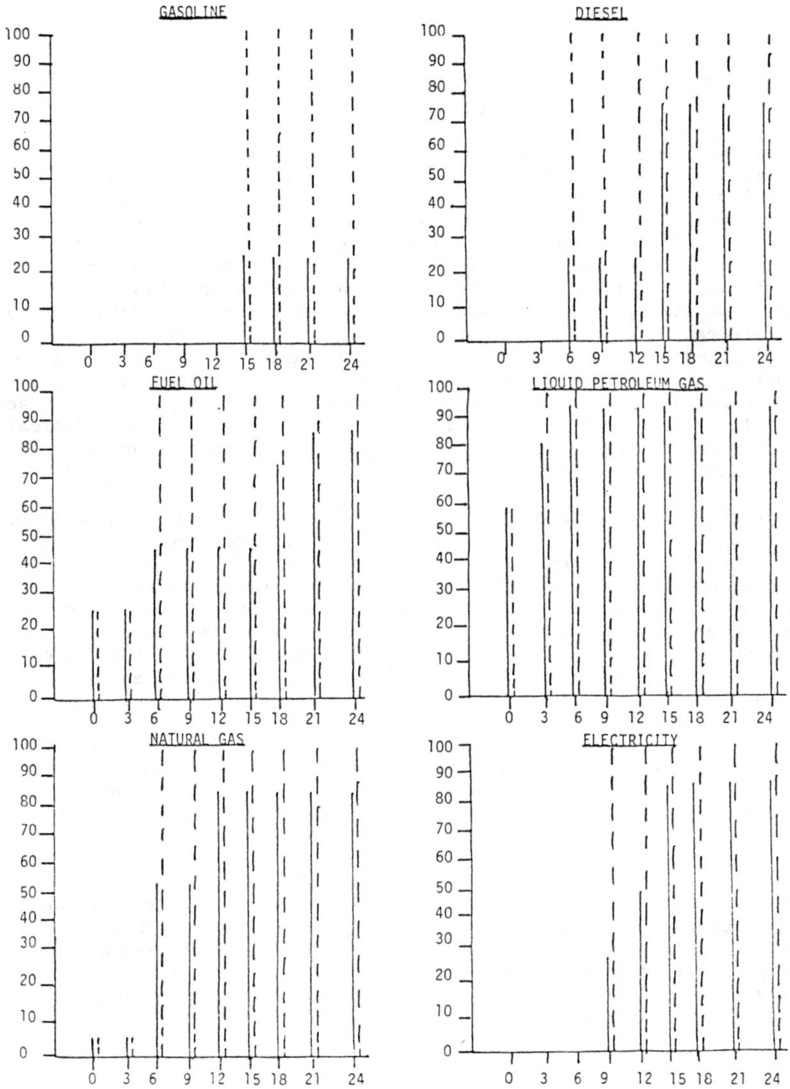

Fig. 3. Percent of Central Fuels Replaced by Biomass Fuels On-Farms in 1990 Under Alternative Price Growth Rates

Note: Solid lines show constrained case, broken lines show unconstrained case. Vertical axis shows quantity of fuel replaced, horizontal axis shows % average annual year price increase.

producing vegetable oil without solvent extraction would yield a cost-competitive energy product only if real energy price increases averaged 15 percent or more per year. Pressing of sunflower seed would yield the least expensive oil of the on-farm processes, but this would still be nearly three times as expensive as the soybean oil produced by the solvent process.

CONCLUSIONS

The data developed for this study indicate the potential importance of biomass fuels in the agricultural sector. At present, many biomass products could be burned to dry grain and heat structures, replacing LPG, fuel oil, and natural gas. The penetration of oilseeds will require substantial relative price increase for central fuels, averaging six percent over the next decade. Such an adjustment could occur slowly allowing markets to adjust gradually or it could occur rapidly, as in 1974 and 1979. If adjustment occurs slowly, penetration will likely occur as described in our upper limit case. Adjustments made in response to rapid price increases will be limited by feedstock availabilities and leadtimes for planning and constructing as described in our lower case. Price increases in the six percent range are not unreasonable but they are probably higher than those expected in the next decade if conditions in energy markets remain "normal." Under non-normal circumstances, such as import oil market disruptions, they are well within the feasible range.

If penetration of oilseeds occurs, our data indicate that soybeans will be preferred on a cost basis, though for practical purposes sunflower seeds and corn, of the major crops, are roughly equivalent. This conclusion is based on assumptions concerning feedstock and by-product prices that are somewhat speculative. One could imagine price assumptions that would reverse the order of penetration. It is clear that on-farm processes will not compete with central solvent extraction processes. Our co-op scale solvent processing assumption is conservative; larger facilities could yield economies of scale, but would incur higher transportation costs.

We have only reported national results from the model. In different regions of the nation, where soybeans are insufficient to meet on-farm demands for fuel, corn, sunflower, or other oilseeds would be used instead, and the national use of soybeans for fuel would be lower than that reported for the nation. We have verified this for selected regions of the country. Other factors could also affect our results. We have not, for example, considered many varieties of oilseeds which could offer advantages in individual circumstances, nor have we considered the "free labor" which some agrue exists on farms. Finally, there is an "independence" factor that might be considered. If oilseeds offer farms the opportunity to cut ties with far-off energy suppliers, be they Arab producers or domestic refiners, a premium price might be paid for vegetable oil. Such considerations would all increase vegetable oil penetration.

ECONOMICS OF ON-FARM SUNFLOWER PROCESSING

Delmer L. Helgeson and LeRoy W. Schaffner*

ABSTRACT

Sunflower oil is being researched as an extender or substitute for diesel fuel. Sunflower seed has a high oil content and an acre will produce about 60 gallons. Each Btu used to produce the seed and process sunflower oil will return about 5.78 Btu's. The price relationship per Btu of diesel to sunflower oil was 1:4.00 in 1979. This ratio declined to 1:1.80 in 1981. The on-farm processing cost for 4,800 gallons varied from \$2.82 to \$4.33 per gallon for three press sizes analyzed. Operating these presses 300 days annually the cost per gallon varied from \$1.74 to \$2.99.

Concern over increasing liquid fuel oil prices and scarcity of supply in the last decade has promoted interest in the development of alternative sources of liquid fuels. This interest lies in developing a renewable source of supply. Vegetable oils, particularly sunflower oil, has been one of the main alternative sources of diesel fuel for study.

Recent interest and attention has centered on the cost/price relationship of crude sunflower oil and #2 diesel price. By using comparative eight month average spot prices for crude sunlower oil and #2 diesel fuel (bulk price) for 1979 and annual average prices for both commodities for 1980 and 1981, a cost price ratio was calculated. Over time, this cost/price ratio has been moving steadily downward. For example, using eight month statistics for 1979, the cost/price ratio per Btu was 4.00, meaning the crude sunflower oil price was four times higher on a per Btu basis compared to #2 bulk diesel fuel. The cost price ratio shifted downward to 2.15 for 1980 and 1.80 for 1981. Questions are raised by farmers and others as to the cost of processing sunflower oil on the farm and the sunflower acreage required to supply oil for fuel. These areas will be discussed in this paper.

The tri-state area of North Dakota, Minnesota, and South Dakota produced about 99 percent of the oil type sunflower in the United States during the 1979 to 1981 period. During this same period North Dakota produced 63 percent of the oil type sunflower (Figure 1).

The four principal oil crops grown in the United States include soybeans, sunflower, cottonseed, and peanuts. The 1981 acreage and potential production of vegetable oil exceeded 85 million acres (Table 1). The data in this table assumes all of the production could be crushed for oil. These four oilseed crops utilized about 19 percent of the cropland acreage in the United States. Soybeans is the major oilseed crop with about 78 percent of the acreage and 84 percent of the oil production for these four crops.

If vegetable oils were to be used as an extender or substitute for diesel fuel, what is the consumption of diesel fuel in agricultural production? The

*Agricultural Economists, Agricultural Experiment Station, North Dakota State University, Fargo.

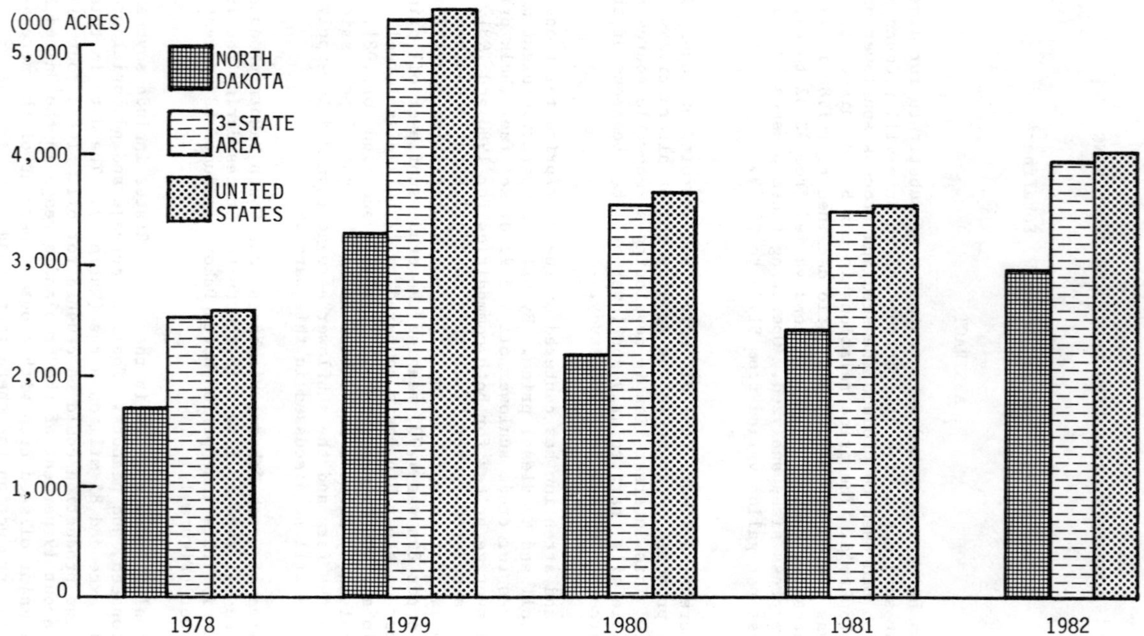

Fig. 1. Planted Acres of Oil Sunflower, North Dakota, Tri-State, and United States

Table 1. Harvested Acreage, Percent Oil in the Seed, and Potential Gallons of Oil, Oilseed Crops in United States, 1981

Crop	Harvested acres	Percent oil	Gallons oil	Percent of U.S. cropland acres
	(000)		(000)	
Soybeans	66,688	18	2,847,907	14.7
Sunflower (oil)	3,496	40	213,952	0.8
Cottonseed	13,820	17	166,731	3.1
Peanuts	1,488	31	158,985	0.3
Total	85,807		3,406,604	18.9

latest data available is a 1978 study by the United States Department of Agriculture (USDA 1980). The estimated use of diesel fuel for the United States was 3.3 billion gallons, 378 million in the tri-state area, and 126 million in North Dakota (Table 2). If vegetable oil substituted for 100 percent of diesel fuel it would take 3.6 billion gallons to supply United States agriculture. This is arrived at by adjusting for the difference in Btu's per gallon of diesel and vegetable oil (diesel fuel 140,000 Btu's/gallon and vegetable oil 130,000 Btu's/gallon).

The acreage of sunflower required to supply oil for agricultural production at varying rates of substitution for diesel fuel is illustrated in Table 3. The United States would require 60.0 million acres of sunflower to supply oil to substitute 100 percent for diesel fuel. This would require about 13 percent of the total cropland in the United States. The data in Table 3 are based on the average yield of sunflower in each area and an assumption of 37.6 percent oil yield from the seed. To provide enough oil in the tri-state area would require 6.9 million acres of sunflower or 10 percent of the total cropland acres in the area. North Dakota needs 2.3 million acres of sunflower to supply sunflower oil for 100 percent diesel fuel substitution or 8 percent of the cropland in North Dakota.

Data in Table 4 illustrate the competitive position of sunflower with other crops in east central North Dakota. The yield and price used were the 1977-81 average with production costs representing the average for the last three years. Sunflower ranked fourth in return to risk of the eight crops compared.

Table 2. Estimated Consumption of Diesel Fuel in Agricultural Production, North Dakota, Tri-State Area, and United States, 1978

Area	Diesel consumption[a]	Vegetable oil converted to diesel Btu's[b]
	(000 gals)	(000 gals)
North Dakota	125,940	135,625
Tri-State[c]	378,293	407,384
United States	3,307,747	3,562,113

[a]U.S. Department of Agriculture, Economics, Statistics, and Cooperative Services. 1980. Energy and U.S. Agriculture: 1974 and 1978. Statistical Bul. 632. Washington, D.C. 115 p.

[b]A gallon of diesel fuel had 140,000 Btu's compared with 130,000 for sunflower oil. A factor of 1.0769 was used to adjust sunflower oil to an equivalent diesel fuel Btu.

[c]The tri-state area includes North Dakota, Minnesota, and South Dakota.

Table 3. Sunflower Acreage Required to Produce Sunflower Oil to Substitute for Diesel Fuel at Varying Rates

Area	Rate of Substitution 10%	25%	50%	100%
	Acreage			
North Dakota	231,048	577,619	1,155,239	2,310,477
Tri-State	692,505	1,731,263	3,462,525	6,925,049
United States	6,078,691	15,196,728	30,393,456	60,786,911
Percent of U.S. Cropland	1.3	3.4	6.7	13.4

Table 4. Estimated Crop Costs and Returns per Acre, East Central North Dakota

Cost and return item	HRS wheat	Barley	Oats	Flax	Soybeans	Corn grain	Sun-flower	Dry beans
	bu.						cwt.	
Yield per Acre	35.8	58.3	69.2	14.8	25.0	70.0	14.4	15.00
Price per Unit[a]	3.21	2.16	1.28	6.06	6.31	2.42	10.15	19.30
Gross Income	114.92	125.93	88.58	89.69	157.75	169.40	146.16	289.50
Direct Cost[b]	59.62	59.00	49.69	49.63	65.39	84.69	75.02	83.56
Return Over Direct Cost	55.30	66.93	38.89	40.06	92.36	84.71	71.14	205.94
Indirect Cost[b,c]	68.23	65.74	64.71	60.52	79.04	77.45	67.77	80.86
Return to Risk	(12.93)	1.19	(25.82)	(20.46)	13.32	7.26	3.37	125.08
Rank	6	5	8	7	2	3	4	1

[a]1977-81 average price received by North Dakota Farmers.
[b]1980-82 average estimated production cost for east central North Dakota.
[c]Includes costs for machinery ownership, labor, land charge, and management.

It ranked better than the small grain crops and flax, and it is these crops that sunflower competes with primarily for acreage. Sunflower would rank near the top in west central and western North Dakota where soybeans, corn grain, and dry beans are not grown as extensively.

ENERGY CONVERSION

Another important relationship when evaluating an alternative energy source is to consider the energy inputs required to produce and process an alternative energy material. The conversion for sunflower oil, using current production and processing estimates results in an input/output ratio of 5.78 to 1 using an intermediate sized on-farm press as the base processing unit (Table 5). This states that for every Btu used to produce and process sunflower oil, 5.78 Btu's will be returned assuming a 1,450 pound sunflower yield per acre. If the energy derived from only the oil is used, 2.88 Btu's are returned for every Btu used in production and processing. These ratios have the potential for increasing with the prospects for genetic improvements and advances in cultural practices.

On-Farm Processing Costs

One of the potential sources of supply of sunflower oil for farmers is to process whole sunflower seeds on the farm. An average North Dakota farm is estimated to use approximately 4,800 gallons of diesel fuel annually, assuming diesel engines were used for all field, harvest, and livestock operations. An average size North Dakota farm, based on 1978 Census of Agriculture data, consists of 1,038 acres, composed of 737 acres in cropland or 71 percent, 245 acres (24 percent) in pasture, and 56 acres (5 percent) taken up by the farmstead, roads, and miscellaneous land usage.

Three sizes of screw-type presses were analyzed for on-farm usage ranging from 0.35 ton of whole sunflower seed per day, 1.67 tons per day, and 5.0 tons per day. The three presses were analyzed for processing 4,800 gallons of sunflower oil and for operating full time (300 days at 8 hours per day). The operation costs for the three presses include elevating seed to a hopper bin, cleaning the seed, elevating cleaned seed to another hopper bin, running seed through the screw-press to an oil settling tank, running the oil from settling tank through the filter, and moving the filtered oil to a storage tank. The meal would move to a storage area as it comes from the screw-press.

The equipment investment, including building, was $28,935 for the small press, $31,650 for the intermediate press, and $41,505 for the large press (Table 6). Building size was scaled to accommodate storage of high oil residual meal to a depth not to exceed five feet. Three auger design sets were used to elevate sunflower seed into the cleaner bin, to the press hopper bin, and to move the meal from the press to the meal storage area. Buildings, in each instance, were uninsulated steel structures. If heat was desired for the press area for cold weather operation, an additional investment allowance would need to be made for insulation and a heat source.

Table 5. Energy Inputs and Outputs for an Acre of Sunflower Yielding 1,450 Pounds per Acre

Input/output	Btu/acre
Energy input	
Production - 20.1 Gpa (gallons/ac) at 140,000 Btu/gal	2,814,000
Oil extraction[a] - 65.3 Gpa requiring 1,720 Btu/gal	112,316
Total	2,926,316
Energy output	
Oil - 65.3 gallons/ac @ 130,000 Btu/gal	8,489,000
Meal - 947 lbs/ac @ 8,900 Btu/lb	8,428,300
Total	16,917,300

$$\text{Ratio of } \frac{\text{Output}}{\text{Input}} = \frac{16{,}917{,}300}{2{,}926{,}316} = \sout{5.78{:}1}\ \ 3.65{:}1$$

[a]Small on-farm plant was assumed.

Table 6. Investment in Equipment and Building for Three Sizes of On-Farm Screw-Presses for Sunflower Seed Oil Extraction, North Dakota 1981

Item	Size of Press		
	0.35 Ton/Day	1.67 Tons/Day	5.0 Tons/Day
	dollars		
Equipment			
Seed Cleaner	1,030	1,030	1,170
Hopper Bins	1,760	1,760	2,600
Screw-press	7,165	7,955	14,745
Oil Settling Tank	230	270	370
Oil Filter	3,735	5,620	5,755
Oil Storage Tank[a]	2,600	2,600	2,600
Augers and Motors	1,165	1,165	1,165
Total Equipment Investment	17,685	20,400	28,405
Building	11,250[b]	11,250[b]	13,100[c]
Total Investment in Equipment and Building	28,935	31,650	41,505

[a]5,000 gallon capacity, a one year supply for the average farm.

[b]Steel building about 1,200 square feet, not insulated. The meal storage requires more space given the high oil content, meal should not exceed five feet in depth while in storage.

[c]Steel building about 1,400 square feet, not insulated.

Processing Cost of 4,800 Gallons

The processing of 4,800 gallons of sunflower oil would not require additional labor beyond the individual farm operator. The assumption used was the presses would operate eight hours per day plus an additional hour for starting and shutting the press down. The time required to process 4,800 gallons would be approximately 160 days for the small press, 31 days for the intermediate size and 10 days for the 5.0 tons per day press. Each press is capable of processing a minimum of 4,800 gallons of sunflower oil during the winter months.

The small press had an oil extraction efficiency of 82 percent while the medium and large presses had an extraction efficiency of 89 percent. These presses extract oil from whole sunflower seed without any cooking or heating process which makes oil extraction efficiency lower than commercial presses, resulting in meal with a higher oil content.

The estimated processing costs for 4,800 gallons of crude sunflower oil for each of the three sizes of presses amendable to on-farm processing is provided in Table 7. Total estimated processing cost per gallon of crude sunflower oil came to $4.33 for the 0.35 ton per day press, $2.82 for the 1.67 tons per day press, and $2.87 for the 5.0 tons per day press. These costs include a charge for all resources used in producing sunflower oil. The cost per gallon for the 5.0 tons per day press was higher than for the medium size press. This result occurs because the larger press was operated only 10 days to obtain 4,800 gallons which is not a sufficient volume to lower the fixed costs.

Depreciation and interest on the equipment was amortized over a 15-year period at 10 percent interest paid quarterly. It is recognized that the machines are used only a very limited time to produce 4,800 gallons of sunflower oil, 53 percent for the small press, 10 percent for the medium, and only 3.5 percent for the large press. If depreciation was prorated only for the time used, depreciation would be charged far beyond its likely technologically efficient

Table 7. Estimated Costs for Processing 4,800 Gallons of Sunoil on the Farm

Cost Item	Size of Press		
	0.35 Ton/Day	1.67 Tons/Day	5.0 Tons/Day
Variable Costs			
Equipment Repair[a]	$ 1,585	$ 360	$ 193
Building Repair[b]	225	225	262
Electricity @ $0.04/kilowatt hour[c]	189	85	85
Sunflower Seed @ $10.42/cwt.[d]	11,743	10,819	10,819
Total Variable Cost	13,742	11,489	11,359
Fixed Costs			
Equipment Amortized for 15 years[e]	2,289	2,640	3,676
Equipment Amortized for 25 years[f]	1,229	1,229	1,431
Insurance on Building and Equipment[g]	174	190	249
Management, Owner Operator Labor @ $5.00/Hour	7,200	1,400	470
Total Fixed Cost	10,892	5,459	5,826
Total Processing Cost	$24,634	$16,948	$17,185
Tons of Meal Produced	37.87	33.43	33.43
Credit for Meal @ $101.70/Ton[h]	$ 3,851	$ 3,400	$ 3,400
Total Cost to Process 4,800 Gallons of Sunoil	$20,783	$13,548	$13,785
Cost Per Gallon of Sunoil	$ 4.33	$ 2.82	$ 2.87
Cost Per Gallon Not Including Labor and Building Costs	$ 2.57	$ 2.27	$ 2.48
Processing Efficiency[i]	82%	89%	89%

[a]The annual repair cost was prorated by the time the machine was used, 53% of the small press, 10% of the medium press, and 3.5% for the large press. The annual equipment repair costs were charged on a percentage basis of newcost as follows: press 27%, filter 21%, other equipment 4%.
[b]Steel building repair charged at 2% of new cost.
[c]1981 commercial utility rate.
[d]1976-80 average price received by farmers.
[e]Depreciation and interest on equipment were amortized over 15 years at 10% interest paid quarterly. The 10% interest was the average paid for Baa industrial bonds for 1975-79.
[f]Depreciation and interest on the building were amortized over 25 years at 10% interest paid quarterly.
[g]$6.00/$1,000 charged on equipment and building.
[h]January-March 1981 average spot market price, Minneapolis, Minnesota.
[i]Percent of oil removed.

life. No salvage value was taken since after 15 years new technology may leave the equipment with very little value to recover.

As a means of providing for a wide range of alternatives in analyzing the cost structure, it was assumed that some farmers may already have a building suitable to house the press and a zero opportunity cost was considered for the operator's labor (Table 7). By deleting the fixed cost for a building and the labor charge, the resulting cost per gallon of crude sunflower oil is $2.57 for the smallest press, $2.27 for the medium press, and $2.48 for the large press.

Cost for Full Time Operation

When the on-farm presses are operated 300 nine -hour days per year, (eight hours for the press and one hour to start and stop the press at the beginning and end of the day), the volume of crude oil processed totals 9,000 gallons for the small press, 46,272 gallons for the medium press, and 138,696 gallons for the largest press (Table 8). This annual volume translates into a crude sunflower oil supply that would support the requirements of 1.87 average size North Dakota farms for the small press, 9.64 average size farms for the medium press, and 28.90 average size farms for the largest press.

The processing cost per gallon of crude sunoil decreases when the presses can be operated full time (300 days). The reduction in costs from part time to full time operation of the press is $0.40 per gallon for the small press, $0.79 per gallon for the medium press, and $1.13 for the large press. These cost reductions are for the situation where all costs are included.

The estimated costs for the three presses operating 300 days per year capture some economies of size. Economies of size, as used here, means reductions in total cost per unit of production resulting from changes in the size and quantity of resources employed. The estimated cost per gallon of crude sunflower oil is reduced from $3.93 per gallon for the small press to $1.74 per gallon for the large press. The assumption used for the presses, when operating 300 days, is that the farmer would collect his supply of crude sunflower oil when processed, so no additional storage tanks beyond 5,000 gallons were included in the investment costs.

Credit was given for the meal produced at $101.70 per ton in the cost analysis. This was the January-March 1981 average spot price at Minneapolis for 28 percent protein sunflower meal. Meal produced by the small on-farm presses is not the same quality as commercial meal because of the higher oil content. There is about 18 percent oil in the meal from the small press and 11 percent oil in the meal from the medium and large presses.

The presses processing 4,800 gallons of oil produced from 33 to 38 tons of meal depending on the size and efficiency of the press. The presses operating full time (300 days) produced from 71 to 966 tons of meal. The small press (0.35 ton per day) processing 4,800 gallons of sunflower oil could support from 122 to 367 animal units depending on whether the ration included one, two, or three pounds of meal per day. The large press (5.0 tons per day) could support 141 to 422 animal units producing 4,800 gallons or 3,578 to 10,733 animal units if operated 300 days a year. This suggests, that in most cases, the meal produced will have to find a market beyond the farm where it is produced.

The per gallon on-farm processing costs are still higher than the bulk price of #2 diesel fuel, a 1:1.80 ratio in 1981. The prospects for the cost/price ratio to equalize and actually reverse is very difficult to predict with any degree of accuracy. Such factors as the level of U.S. crude oil import dependence, conservation practices, increased domestic oil production, increased agricultural production and processing costs are all linked into such a forecast. The economic incentives for developing alternatives to diesel fuel are determined by the relationship of diesel prices to the cost of producing vegetable oils. Substitutes for diesel fuel will be produced when diesel price rises to a level equal to or above the cost of producing the substitutes.

Table 8. Estimated Costs for Processing 300 Nine-Hour Days by Three Sizes of Presses

Cost Item	Size of Press		
	0.35 Ton/Day	1.67 Tons/Day	5.0 Tons/Day
Variable Costs			
Equipment Repair[a]	$ 2,990	$ 3,601	$ 5,507
Building Repair[b]	225	225	262
Electricity @ $0.04/kilowatt hour[c]	165	822	2,446
Hired Labor @ $5.00/hour	6,300	12,100	13,030
Sunflower Seed @ $10.42/cwt.[d]	22,007	104,283	312,600
Total Variable Cost	31,687	121,031	333,845
Fixed Costs			
Equipment Amortized for 15 years[e]	2,289	2,640	3,676
Equipment Amortized for 25 years[f]	1,229	1,229	1,431
Insurance on Building and Equipment[g]	174	190	249
Management, Owner Operator Labor @ $5.00/Hour	7,200	1,400	470
Total Fixed Cost	10,892	5,459	5,826
Total Processing Cost	$42,579	$126,490	$339,671
Credit for Meal @ $101.70/Ton[h]	$ 7,217	$ 32,774	$ 98,242
Total Processing Cost	$35,362	$ 93,716	$241,429
Total Gallons of Sunoil Processed	9,000	46,272	138,696
Total Tons of Meal	70.96	322.26	966.00
Cost Per Gallon of Sunoil	$ 3.93	$ 2.03	$ 1.74
Cost Per Gallon Not Including Labor and Building Costs	$ 2.99	$ 1.97	$ 1.73
Processing Efficiency[i]	82%	89%	89%

[a]The annual equipment repair costs were charged on a percentage basis of new cost as follows: press 27%, filter 21%, other equipment 4%.
[b]Steel building repair charged at 2% of new cost.
[c]1981 commercial utility rate.
[d]1976-80 average price received by farmers.
[e]Depreciation and interest on equipment were amortized over 15 years at 10% interest paid quarterly. The 10% interest was the average paid for Baa industrial bonds for 1975-79.
[f]Depreciation and interest on the building were amortized over 25 years at 10% interest paid quarterly.
[g]$6.00/$1,000 charged on equipment and building.
[h]January-March 1981 average spot market price, Minneapolis, Minnesota.
[i]Percent of oil removed.

REFERENCES

1. Helgeson, D. L. and L. W. Schaffner. 1982. Economics of on-farm processing of sunflower oil. ND Farm Res. 34:4. Fargo, ND. 3-7 p.

2. Reff, T. 1979. 1980 crop production costs, east central North Dakota. Coop. Ext. Service, North Dakota State University, Fargo. 4 p.

3. Reff, T. and L. Schaffner. 1981. 1981 crop production costs, east central North Dakota. Coop. Ext. Service, North Dakota State University, Fargo. 4 p.

4. Reff, T. and L. Schaffner. 1982. 1982 crop production costs, east central North Dakota. Coop. Ext. Service, North Dakota State University, Fargo. 4 p.

5. U.S. Department of Agriculture, Economic, Statistics, and Cooperative Services. 1980. Energy and U.S. agriculture. Stat. Bul. 632. Washington, D.C. 115 p.

6. U.S. Department of Agriculture, Statistical Reporting Service. 1982. Crop production, 1981 annual summary. CrPr 2-1 (82). Washington, D.C. 42 p.

THE ECONOMICS OF ON-FARM PRODUCTION AND USE OF VEGETABLE OILS FOR FUEL

C. S. McIntosh, R. V. Withers, and S. M. Smith

ABSTRACT

The technology of oilseed processing, on a small scale, is much simpler than that for ethanol production. This, coupled with the fact that most energy intensive farm operations use diesel powered equipment, has created substantial interest in vegetable oils as an alternative source of liquid fuel for agriculture.

The purpose of this study was to estimate the impact on gross margins resulting from vegetable oil production and utilization in two case study areas, Latah and Power Counties, in Idaho. The results indicate that winter rape oil became a feasible alternative to diesel when the price of diesel reached \$0.84 per liter in the Latah County model. A diesel price of \$0.85 per liter was required in the Power County model before it became feasible to produce sunflower oil for fuel.

Recent concerns over the price and availability of liquid fuels have created substantial interest in alternative sources of fuel and in on-farm methods of production. Much of the interest and research to date has focused on alcohol production. Alcohol, however cannot be readily substituted for diesel fuel, and since most energy intensive operations in field crop production are accomplished using diesel engines, it is necessary to develop a more suitable alternative for diesel oil.

Vegetable oils may provide that alternative. Short run engine tests indicate that vegetable oils are suitable substitutes for, and can be mixed successfully with, diesel fuel without greatly affecting power output or fuel consumption (Peterson, et al.). These preliminary results, coupled with the fact that small scale oilseed processing technology is much simpler and cheaper than alcohol distillation, have created increasing interest in developing oilseed crops as alternative sources of liquid fuel.

Beyond technological considerations, however, it is essential that the production and extraction of the oil is economically feasible. The purpose of this paper is to examine the economic aspects of on-farm production and processing of three oilseed crops in Idaho. The specific objectives are to (1) develop crop enterprise budgets for oilseeds, (2) determine fixed capital investment requirements for an on-farm size oilseed processing facility, and (3) construct linear programming models to estimate the feasibility of on-farm production and processing of oilseed crops for fuel.

This study is part of an integrated research project at the University of Idaho, involving plant sciences, animal sciences, agricultural engineering, chemical engineering, and agricultural economics. The economic evaluation draws on physical data generated by all disciplines working on the project.

The authors are: C. S. MCINTOSH, Research Associate, R. V. WITHERS, Professor, and S. M. SMITH, Associate Professor in the Department of Agricultural Economics and Applied Statistics, University of Idaho.

Three oilseed crops have shown potential for production in Idaho. These are: 1) sunflower, 2) safflower, and 3) winter rape. Sunflower and safflower are relatively new crops in Idaho and are produced mostly on an experimental basis. Some areas of the state have, however, adapted these crops to their rotational planning. Winter rape has been produced in northern Idaho for several decades, but this crop is not well adapted to other areas of the state.

OILSEED PROCESSING TECHNOLOGY

There are three basic methods of extracting oils from oil bearing seeds. These are mechanical or full-press extraction, direct solvent extraction, and pre-press solvent extraction. The latter two methods employ solvent as a means of separating the oil from the seed. The solvent must then be distilled to recover the oil, which requires extensive capital investment in distillation equipment, and handling of highly volatile substances. The solvent extraction systems were considered to be inappropriate technology for small scale operations.

Mechanical or full-press extraction requires the least fixed capital investment of the three methods, and the process is easily understood. It is a continuous process using a screw press, consisting of a worm shaft rotating within a pressing cylinder or cage, which literally squeezes the oil from the seed. The oilseeds are usually conditioned by heating prior to extraction to facilitate the removal of oil from the seed fibers. The processing facilities with which this paper is concerned are of the mechanical, or full-press type.

THE COST DATA

Production cost data for the three oilseeds were gathered by personal interviews with growers. These data were used to develop enterprise budgets and representative farms for the two case study areas, Power County in southeast Idaho, and Latah County in the Palouse wheat producing area of north Idaho. The oilseed budgets are compared with budgets for other crops in these counties. (Tables 1 and 2).

Data concerning the fixed capital investment associated with constructing small scale vegetable oil extraction facilities were gathered from several equipment manufacturers and building cost guides (Boeckh, Dunning, Godfrey, Kagawa, Peters and Timmerhouse). Three sizes of facilities were studied with processing capacities of 40 kilograms per hour, 190 kilograms per hour, and 570 kilograms per hour. The first of these is typical of the size which could be maintained on an individual farm, while the latter two would be suited to a small group or cooperative investment. The initial investment cost of the small plant was estimated at $26,606, which includes processing equipment and oil storage tanks, but assumes that seed and meal storage facilities and a building for the plant already exist on the farm. (Table 3).

The University of Idaho Department of Agricultural Engineering has been operating a 40 kilogram per hour press periodically over the past year. Data concerning the operating characteristics of the press were collected during this time. Average extraction efficiencies of 70 percent for sunflower, 71 percent for safflower and 85 percent for winter rape were determined from these data, based on the percent of oil in the whole seed that was extracted. The linear programming models, explained more fully in the next section of this paper, are assumed to use a press of this type. The variable cost coefficients (Table 4) for oilseed extraction which appear in those models were derived, in part, from these data.

A high protein meal is produced as a by product of vegetable oil extraction. The value of the meal greatly influences the overall profitability of on-farm

Table 1. Estimated Per Hectare Costs for Selected Dryland Crops in Latah County, Idaho, 1981

Production Item	Sunflower	Safflower	Winter Rape
Direct Costs			
Seed	$ 20.76	$ 30.89	$ 4.15
Fertilizer	57.33	63.21	88.86
Herbicides	18.06	18.63	
Pesticides	24.34		15.76
Machinery	25.97	36.72	26.05
Tractors	33.06	23.13	30.17
Interest on Operating Capital	8.06	7.73	10.06
Gasoline	15.49	17.89	14.68
Total Direct Costs[a]	$ 203.07	$ 198.20	$ 189.72
Labor	$ 23.94	$ 20.58	$ 21.32
Total Variable Costs[a]	$ 227.01	$ 218.78	$ 211.04
Diesel Used (gallons)	16.33	14.06	14.73

[a]Total costs excluding diesel fuel.

Table 2. Estimated Per Hectare Costs for Selected Dryland Crops in Power County, Idaho, 1981

Production Item	Sunflower	Safflower
Direct Costs		
Seed	$ 20.76	$ 30.89
Fertilizer	61.03	40.28
Herbicides	18.63	18.63
Machinery	32.64	26.46
Tractors	34.08	32.47
Interest on Operating Capital	7.86	6.70
Gasoline	16.68	15.79
Total Direct Costs[a]	$ 191.68	$ 171.22
Labor	$ 23.72	$ 21.79
Total Variable Costs[a]	$ 215.40	$ 193.01
Diesel Used (gallons)	16.90	16.41

[a]Total costs excluding diesel fuel.

Table 3. Estimated Fixed Capital Investment for Full-press Extraction of Vegetable Oils[a] 40 kg./hr.

Item	Cost
Feed Hopper and Auger	$ 2,980
Seed Conditioner	5,110
Screw Press	4,120
Filter Press	3,940
Spare Parts[b]	1,556
Oil Storage Tanks	8,900
Total Fixed Capital Investment[a]	$ 26,606

[a]These values assume that the farmer has adequate storage facilities for seed and meal onhand, and that he assembles the plant in an existing building.

[b]Estimated spare parts for one year of operation.

oil extraction. The value of the meal was estimated for each study area based on protein content and the market price of soybean meal at 44 percent protein. Winter rape meal, however contains glucosinolates which make it unsuitable for feeding to monogastrics such as poultry and swine (Thomas). Winter rape meal was arbitrarily valued at 75 percent of its protein equivalent basis for this reason (Table 5).

THE MODELS

Linear programming models were constructed for Power and Latah Counties in Idaho. Both of these areas produce field crops under dryland conditions and have shown potential for producing oilseed crops. There are, however, large differences between the climatic characteristics of each region. The average annual precipitation for the Power County area is approximately 254 millimeters, while the Latah County area receives approximately 559 millimeters. Thus, the farming practices used in each area differ substantially. The low annual rainfall in Power County makes summer fallowing a common practice for moisture conservation. This practice is not prevalent in Latah County.

The representative farm developed for the Latah County model consists of 446 dryland hectares. The power County representative farm consists of 809 dryland hectares. In both models all land was assumed to be suitable for crop production, and of equal productivity. The yields in each model reflect those achieved by average growers in the region.

Five crop rotations were included in the Latah County model. These were: 1) winter wheat-peas, 2) winter wheat-spring barley-peas, 3) winter wheat-fallow-winter rape, 4) winter wheat-safflower, and 5) winter wheat-sunflower.

The Power County model also included five crop rotations. These were: 1) winter wheat-fallow, 2) winter wheat-spring barley-fallow, 3) winter wheat -spring wheat-fallow, 4) winter wheat-sunflower-fallow, and 5) winter wheat-safflower-fallow.

The price of diesel was allowed to vary upward in the models. This eventually forced the model farm to supply all of its diesel requirements by

Table 4. Estimated per Liter Costs of On-farm Vegetable Oil Extraction (excluding cost of oilseeds).

	Winter Rape	Latah County Sunflower	Safflower
Variable Costs			
Electricity	$ .0036	$ .0046	$ .0054
Labor	.0156	.0236	.0202
Repairs	.0529	.0523	.0534
Total Variable Costs	$ .0721	$ .0805	$ .0790
Fixed Costs			
Depreciation	$ .0882	$ .0871	$ .0890
Insurance	.0129	.0128	.0131
Taxes	.0085	.0084	.0086
Interest	.1132	.1119	.1142
Total Fixed Costs	$ .2228	$ .2202	$ .2249
Total Costs	$ .2949	$ .3007	$ .3039

	Power County Sunflower	Safflower
Variable Costs		
Electricity	$ .0104	$ .0126
Labor	.0273	.0202
Repairs	.0420	.0420
Total Variable Costs	$ .0797	$ .0748
Fixed Costs		
Depreciation	$ .0700	$ .0700
Insurance	.0102	.0102
Taxes	.0067	.0067
Interest	.0899	.0899
Total Fixed Costs	$ .1768	$ .1768
Total Costs	$ .2565	$ .2516

growing and processing oilseed crops. When the oilseeds were processed for use as a fuel the models assumed that all of the meal byproduct was sold. As the price of diesel varied, prices of all other petroleum based products were held constant.

The models employed in this analysis were designed to maximize gross margins. Gross margin is defined as income in excess of variable costs. This analysis is therefore short run in nature, and the solutions achieved

Table 5. Estimated Value of Oilseed Meals.

Latah County Model		Estimated Value[a] $/ton
Sunflower	28% Protein	196.25
Safflower	25% Protein	176.40
Winter Rape[b]	34% Protein	177.50
Power County Model		**Estimated Value[a] $/cwt.**
Sunflower	28% Protein	182.35
Safflower	25% Protein	163.17

[a]Estimated value is based on the protein content as a percentage of 44% soybean meal, valued at $308.70/ton in Latah County and $286.65/ton in Power County (eg. sunflower $\frac{308.70}{44} = \frac{x}{28}$).

[b]Winter rape meal has a limited value due to its glucosinolate content. It is arbitrarily valued at 75% of its protein equivalent basis.

indicate the profit maximizing combination of activities for each situation in the short run.

RESULTS

The Latah County model determined a profit maximizing solution which consisted of 223 hectares of winter wheat and 223 hectares of peas. Producing these crops required 25,130 liters of diesel and resulted in a gross margin of $101,431.00. When the model was forced to produce vegetable oils for fuel it chose winter rape as the most economical alternative. Producing winter rape oil reduced net income by $0.62 per liter produced. When the model was forced to supply 100 percent of its fuel requirements from vegetable oils the solution was 199 hectares of winter wheat, 199 hectares of peas, 34 hectares of winter rape, and 31 hectares of fallow. The fallow acreage appeared in this solution as a necessary condition of winter rape production. All of the rape produced as processed to provide 23,481 liters of fuel. The gross margin achieved by this combination was $86,966.38, a 15 percent reduction from the original solution.

When the price of diesel was allowed to vary the solution did not change until that price reached $0.84 per liter. At this price the same gross margin was achieved when all diesel was purchased as when all fuel was produced by growing and processing winter rape.

The Power County model determined the profit maximizing solution to consist of 269 hectares of winter wheat, 269 hectares of spring barley and 269 hectares of fallow. This combination resulted in a gross margin of $70,866.67 and consumed 30,862 liters of diesel fuel. When this model was forced to produce vegetable oils for fuel it chose sunflower which reduced net income by $0.57 per liter produced. When the model was forced to supply 100 percent of its fuel requirements from vegetable oils the profit maximizing solution became 269 hectares winter wheat, 137 hectares of barley, 133 hectares of sunflower, and 269 hectares of fallow. All of the sunflower produced was processed to provide 31,646 liters of fuel. The gross margin achieved by this combination was $52,712.47 a 26 percent reduction from the original solution.

When the price of diesel was allowed to vary the original solution remained valid until the price of diesel reached $0.85 per liter. At this price the same gross margin was achieved when all diesel was purchased, as when all fuel was produced by growing and processing sunflower.

Several key assumptions underlie the models. One is that the models utilize crop rotations. Therefore a solution implies that the rotations will be adhered to for the following seasons. Second, the models assumed vegetable oils to be perfect substitutes for diesel. A third major assumption was the ability of the farmer to process oilseed crops on a continuous, year round basis. The Latah County model required that the press be operated approximately 87 days for 12 hours per day, assuming a capacity of 41 kilograms per hour. The Power County model required operating the press 177 days for 12 hours per day assuming the same capacity. The systems can, however, be automated to such a degree that hiring extra labor to operate the press should not be necessary.

SUMMARY

The technology of oilseed processing on a small scale is much simpler than that for alcohol production, and vegetable oil substitutes more readily for diesel. This, coupled with the fact that most energy intensive farm operations use diesel fuel, has created substantial interest in vegetable oils as an alternative source of liquid fuel for agriculture.

The purpose of this study was to estimate the impact on gross margins of producing vegetable oil fuels for two case study farms. The results indicate that winter rape oil became a feasible alternative to diesel when the price of diesel reached $0.84 per liter in the Latah County model. A diesel price of $0.85 per liter was required before the Power County model farm began to produce fuel from sunflower. The income penalty from producing these oils was determined to be $0.62 per liter and $0.57 per liter respectively, for each area.

REFERENCES

Boeckh, 1980. "Boeckh Building Cost Guide", E.H. Boeckh Co., Milwaukee, Wisc.

Dunning, T.W., 1981. Anderson International Corp., personal correspondence, Strongsville, Ohio.

Godfrey, R.S., editor, 1981. "Building Construction Cost Data, 1981", Robert Snow Means Co., Inc., Kingston, Mass.

Kagawa, Y., 1981. Ce Co Co, Chuo Boeki Goshi Kaisha, personal correspondence, Osaka, Japan, Aug.

Peters, M.S., and K.D. Timmerhaus, 1980. "Plant Design and Economics for Chemical Engineers", McGraw-Hill Book Co., New York, NY, p. 147-225.

Peterson, C.L., et al., 1980. "Vegetable Oil as an Agricultural Fuel for the Pacific Northwest", University of Idaho, AES, Bulletin No. 598.

Powell, T.A., et al., 1980-a. "North Idaho Crop Enterprise Budgets", University of Idaho, CES, AES, M.S. No. 62, August.

Powell, T.A., et al., 1980-b. "Southeast Idaho Crop Enterprise Budgets", University of Idaho, CES, AES, M.S. No. 64, December.

Thomas, V.M., 1981. Department of Animal Sciences, University of Idaho, personal correspondence.

COMPARATIVE COMBUSTION STUDIES ON VARIOUS PLANT OIL ESTERS AND THE LONG TERM EFFECTS OF AN ETHYL ESTER ON A COMPRESSION IGNITION ENGINE

*C S Hawkins and *J Fuls

ABSTRACT

Combustion studies on both ethyl and methyl esters of various plant oils were carried out using the same engine for all the tests so that comparative studies could be achieved. Twelve esters were tested and the pertinent data was recorded. Some of the more important results are published in this paper to serve as a comparative guide to the study of plant oil esters as fuel. Bruwer et al (1980) suggested the use of plant oil esters to prevent injector coking in modern compression ignition engines. Very little information is available on the long term effects of such ester use. Cyclic endurance tests have been carried out on Perkins engines running on ethyl esters of sunflower oil. The exciting results of this work are reported, with specific reference to one tractor engine which has recorded more than 1 300 trouble free hours.

COMPARATIVE COMBUSTION STUDIES

Although esters of various plant oils have been tested as fuel in compression ignition engines these tests have been done at various times and in various places on fuels which may have been prepared quite differently from one another. Additionally, the engines used in the tests have varied.

To assist the Division of Agricultural Engineering in its programme of plant oil fuel research it was decided to prepare both ethyl and methyl esters of various available plant oils by a standard method so as to arrive at as uniform a product as possible allowing for variations in the reactions of the different plant oils.

The oils used were sunflower oil (degummed); maize oil (food grade); cotton oil (food grade); peanut oil (food grade); soya oil (crude) and castor oil (crude). It may be argued that the oils should have been of the same quality, but the ones used were the only ones obtainable within a reasonable time. The esters were prepared using sodium hydroxide as a catalyst after which they were neutralized with phosphoric acid. They were allowed to stand so that the salts could settle out and then the ester was drawn off and washed once with water.

Table 1 gives the measured properties of the various esters and of the diesel fuel used as a comparison. The prefix E is for

*Principal Engineers, Division of Agricultural Engineering, Department of Agriculture and Fisheries, Silverton, Republic of South Africa.

ethyl and M for methyl while the oils are denoted by SN for sunflower, COT for cotton, M for maize, GB for peanut, SY for soya and CAS for castor. D100 denotes diesel fuel.

Table 1. Properties of Diesel and the Various Plant Oil Esters

Fuel	Lower Heat Value MJ/kg	Viscosity at 40°C cSt	Coeff. of Expansion $X10^{-4}/°C$	Specific volume at 20°C mℓ/g	Ash Content % m/m	Catalyst Residue % m/m	TAN mgKOH/g
Diesel	43.57	4.1	8.317	1.174	0.01	-	<0.05
ESN	38.53	4.9	8.366	1.144	0.006	<0.0001	<5.0
MSN	37.69	4.4	8.237	1.130	0.001	<0.0001	<0.1
EM	38.05	5.2	8.562	1.156	0.003	0.00012	4.3
MM	38.03	4.5	8.632	1.160	0.005	0.00027	1.4
ECOT	37.95	5.0	8.734	1.143	0.02	0.001	4.0
MCOT	38.83	4.4	8.365	1.146	0.007	0.00027	2.0
EGB	38.55	5.1	8.546	1.161	0.003	0.00085	4.6
MGB	38.72	4.4	8.152	1.150	0.005	0.00024	1.0
ESY	37.75	4.7	8.823	1.176	0.003	0.0003	5.6
MSY	38.01	4.2	8.461	1.146	0.003	0.00017	2.9
ECAS	35.27	14.1	8.366	1.104	0.01	0.0015	4.0
MCAS	38.55	17.1	8.537	1.101	0.01	0.00029	6.6

As can be seen from this table the esters were all very similar to one another except for the esters of castor oil which were very much more viscous. For some unknown reason the heat value of the ethyl ester of castor oil was considerably lower than the fairly consistent value for the other esters.

The esterification of each fuel was checked by the thin layer chromotography method as described by Fuls (1981). In all cases the final product was considered to be of a high standard.

Test Procedures

The fuels were tested in a Perkins 4.236 engine fitted with a CAV DPA pump with fixed timing. Fuel consumption was measured volumetrically and repetitive consecutive readings of speed and torque were automatically taken during the period of fuel consumption measurement; then averaged to provide the speed, torque and power output figures. Injector needle lift and cylinder pressure traces were displayed on a storage oscilloscope and photographed in both single and multi-trace form for later analysis of maximum rate of cylinder pressure rise, dynamic injection timing and ignition delay. Temperatures were recorded during each test. Details of the equipment and test procedures have previously been published (Hawkins 1981).

COMBUSTION TEST RESULTS AND CONCLUSIONS

The results of the series of tests are contained in tables 2 and 3, but certain details (such as power output and torque) which were considered to be irrelevant or not of value have been excluded.

Power Output: Power output has not been included in the tables since maximum power only varied from a low of 44.4 kW to a high of 45.5 kW with diesel delivering 45.1 kW.

Maximum Torque: This varied quite a bit more; the highest being 225.9 Nm for diesel and the lowest 218.6 Nm for the ethyl ester of castor oil.

Smoke Values: As has been pointed out on previous occasions the smoke values in Hartridge Smoke Units are consistently lower for the plant oil esters than for diesel fuel at maximum power and maximum torque. What was perhaps surprising were the low figures recorded for the castor oil esters.

Specific Fuel Consumption: For the esters the SFC was consistently higher than that for diesel in all the tests, but the variation between esters was not large except in the case of the castor oil ester where the high specific fuel consumption can be attributed in part to the high viscosity of these esters.

Brake Thermal Efficiency: It was always higher than that of diesel and the smallest increase was that of methyl ester of castor oil which was only from 30.9 to 31.9%. The largest increase was for ethyl ester of soya oil where the increase was from 30.9 to a figure of 33.9%.

Exhaust Temperature: This was distinctly lower than that for diesel in all cases and was consistent throughout the range of outputs tested, as was indeed the case for most of the parameters recorded.

Dynamic Timing: For the different esters this was always a little faster than for diesel and can once again be attributed to viscosity differences. It must, however, be stressed that the differences were small and even the much higher viscosity of the castor oil esters did not make a great difference in dynamic timing.

Ignition Delay: The ignition delay of the various fuels varied by 2.0 degrees at maximum power being more or less equally above and below that of diesel, with similar results at maximum torque. The exceptions were again the two castor oil esters where the ignition delay did not seem to correlate with any physical properties of the fuel and could well be due to chemical differences.

Maximum Rate of Cylinder Pressure Rise: Excluding the castor oil esters this was consistently lower than that of diesel at maximum power, consistently higher at maximum torque and variable around the diesel figure for the intermediate power outputs. The castor oil esters had excessive rates of cylinder pressure rise and would certainly not be acceptable as fuels for compression ignition engines without the addition of cetane improvers. Whereas it is normally expected that increased ignition delay will lead to higher maximum rates of cylinder pressure

Table 2. Engine Performance Factors on Diesel and Various Plant Oil Esters

STATE	MAXIMUM POWER (Pmax)				MAXIMUM TORQUE (Tmax)				75% Pmax				50 Pmax				HIGH SPEED/ LOW LOAD			
FUEL	SFC g/kWh	Brake Thermal Eff. %	Exhaust Temp. °C	Smoke HSU	SFC g/kWh	Brake Thermal Eff. %	Exhaust Temp. °C	Smoke HSU	SFC g/kWh	Brake Thermal Eff. %	Exhaust Temp. °C	Smoke HSU	SFC g/kWh	Brake Thermal Eff. %	Exhaust Temp. °C	Smoke HSU	SFC g/kWh	Brake Thermal Eff. %	Exhaust Temp. °C	Smoke HSU
D100	267	30.9	712	59	281	29.4	666	86	260	31.8	506	6	298	27.7	403	4	418	19.7	306	2
ESN	285	32.8	682	26	292	32.0	650	56	286	32.6	491	3	327	28.6	387	1	458	20.4	296	1
MSN	287	33.3	689	29	292	32.7	651	53	284	33.6	492	3	324	29,5	388	3	453	21.1	300	2
EM	288	32.9	690	28	293	32.3	643	54	288	32.9	492	3	329	28.7	385	2	459	20.6	299	2
MM	287	33.0	682	28	292	32.4	642	54	287	33.0	489	3	327	29.0	383	2	464	20.4	297	2
ECOT	283	33.6	680	23	288	33.0	644	46	286	33.2	481	2	325	29.2	377	1	480	19.8	284	1
MCOT	289	32.0	680	24	295	31.4	640	50	289	32.1	490	3	332	28.0	375	2	466	19.9	291	1
EGB	287	32.5	681	25	293	31.9	639	48	288	32.4	486	3	329	28.4	379	2	467	20.0	292	1
MGB	283	32.8	683	23	290	32.1	642	49	284	32.7	484	3	324	28.7	380	2	453	20.5	296	1
ESY	282	33.9	682	22	286	33.3	641	47	282	33.8	486	2	323	29.6	379	1	457	20.9	295	1
MSY	282	33.6	677	19	287	33.0	639	44	285	33.3	484	2	324	29.3	378	1	464	20.4	293	1
ECAS	313	32.6	650	10	315	32.4	615	30	320	31.9	479	3	368	27.8	368	2	559	18.3	302	9
MCAS	310	31.9	672	16	311	31.8	636	36	313	31.6	483	3	360	27.5	376	3	526	18.8	298	4

Values have been corrected to 20°C inlet manifold temperature

Altitude : 1 330 m

Table 3. Dynamic Timing, Ignition Delay and Maximum Rate of Cylinder Pressure Rise

STATE	MAX. POWER (Pmax)			MAX. TORQUE (Tmax)			75% Pmax			50% Pmax			HIGH SPEED/ LOW LOAD		
FUEL	Dynamic Timing °BTDC	Ignition Delay °CA	Max ΔP kPa/°CA	Dynamic Timing °BTDC	Ignition Delay °CA	Max ΔP kPa/°CA	Dynamic Timing °BTDC	Ignition Delay °CA	Max ΔP kPa/°CA	Dynamic Timing °BTDC	Ignition Delay °CA	Max ΔP kPa/°CA	Dynamic Timing °BTDC	Ignition Delay °CA	Max ΔP kPa/°CA
D100	20.0	12.0	1453	20.5	11.2	1194	15.4	12.6	1115	12.8	12.3	1089	10.3	12.9	496
ESN	20.5	13.3	1056	20.6	11.1	1310	16.6	12.0	779	13.7	11.8	804	11.2	13.0	486
MSN	20.4	12.1	984	20.5	10.0	1246	16.3	10.4	860	13.5	10.9	750	11.5	10.5	451
EM	20.3	11.3	1075	20.9	11.4	1370	16.8	11.8	1036	13.4	11.2	1083	11.6	12.4	542
MM	20.5	12.2	1095	20.9	10.9	1304	16.4	11.6	1119	13.7	11.3	643	10.7	12.1	420
ECOT	20.4	12.0	1019	20.7	11.4	1386	16.9	12.7	979	13.7	12.2	781	10.9	13.6	474
MCOT	20.4	12.6	1226	20.8	11.1	1371	17.2	11.6	1035	13.8	12.8	911	11.0	13.0	466
EGB	20.5	12.0	1057	21.0	10.6	1354	16.6	11.9	885	14.1	11.5	840	11.0	12.3	637
MGB	20.5	11.7	1057	20.7	10.4	1200	16.3	10.7	945	13.0	12.2	910	11.5	11.9	450
ESY	20.6	13.3	1210	21.0	11.5	1413	16.3	12.6	930	14.0	12.7	905	11.1	13.0	468
MSY	20.5	12.0	1173	21.0	12.1	1516	16.5	12.4	1140	14.1	13.4	878	11.7	13.5	489
ECAS	20.6	15.7	1937	20.5	14.4	2390	17.0	16.3	1358	14.8	16.8	757	12.7	19.5	268
MCAS	20.8	14.4	1995	21.4	13.8	2130	17.4	15.3	1233	14.1	16.0	700	11.0	18.8	270
CAS+ 25%M	20.4	15.7	1491	21.3	14.6	1110							13.0	18.4	146

Max ΔP = Maximum rate of cylinder pressure rise

Values have been corrected to 20°C inlet manifold temperature

Altitude : 1 330 m

rise, this was not always the case in the tests performed on these esters and varying combustion characteristics of the different fuels must account for the differences.

Since castor oil and methanol mix so well it was decided to test a blend of the two. Having tested sunflower oil with success it was decided that sufficient methanol would be added to reduce the viscosity of the castor oil to approximately that of sunflower oil. It was necessary to add 25% methanol which was higher than desired. Nevertheless this blend was tested at maximum power, maximum torque and high speed/low load conditions.

The results are included in table 3 and are rather perculiar in that while ignition delay is very similar to that of the castor oil ester, the maximum rates of cylinder pressure rise are very much lower than for the castor oil esters and are in fact very similar to that of diesel. Despite the very reasonable rate of cylinder pressure rise, severe detonation was evident in the maximum power curve.

It is concluded from these tests that plant oil esters in general perform well as compression ignition fuels seen from the point of view of physical combustion characteristics. However it need not necessarily apply to the long term effects of the different esters and may well not have a bearing on the slight nozzle coking tendency of these fuels.

Experience with the use of sunflower oil esters (the ESN fuel is the standard one utilised in the long term tests reported later in this paper) indicates however, that from the combustion point of view, the various plant oil esters will perform fairly equally. However, physical characteristics such as cloud point, pour point, etc. may well dictate a preference for particular plant oil esters for particular climatic conditions.

It is certain that further research into the use of plant oil esters as compression ignition fuels is warranted and that future liquid fuel deficiency problems can be alleviated by the use of these fuels. It is also certain that the more the research that is carried out in the field of liquid fuels from plant oils, the greater the success that will be achieved.

LONG TERM EFFECTS OF AN ETHYL ESTER ON A C.I. ENGINE

When running an engine on an experimental fuel for an extended period various problems should be expected such as malfunctioning of the injection equipment, fuel dilution of the lubricating oil, burnt valves, corrosion and unacceptable engine wear. In the course of developing an acceptable ester fuel of sunflower oil all these problem areas were closely observed and the fuel specifications adapted accordingly:

Malfunctioning of Injection Equipment

Although the coking of injector tips normally associated with the use of neat plant oils can be overcome by using esters (Bruwer et al 1980) it should be emphasised that a high esterification yield in the order of 90% must be obtained. Tests conducted with an ester of 70% yield revealed unacceptable coking within 50 hours under part load conditions.

Sticking Injector Needles: After the preliminary success reported by Bruwer et al (1980) it was found that despite normal operation while the engine was running the injector needles stuck after a shut down time of about 48 hours.

The fuel was prepared by transesterification of raw degummed sunflower oil with ethanol using sodium hydroxide as catalyst. The catalyst was neutralised with nitric acid and centrifuged immediately prior to use. The reason for the needles sticking was found to be the fuel-soluble salts of sodium crystalising onto the bearing surfaces. (Fuls and Hugo, 1981) The needle sticking problem was subsequently overcome by using citric acid for pH adjustment, producing an insoluble salt which could fairly easily be stripped from the fuel.

Exhaust Dribble and Excessive Crank Case Oil Dilution

When exhaust dribble is noted within 30 minutes under low idling conditions it is a clear indication that excessive crank case oil dilution can be expected, resulting from poor combustion. The cause for this is either poor esterification yield or excessive free alcohol or both. (Hawkins et al, 1981)

Burnt Exhaust Valves

When running a Perkins 4.236 engine on a cyclic load using an ester prepared with sodium hydroxide as catalyst and citric acid for pH adjustment an apparent exhaust valve burn was observed within 36 hours. Careful inspection under a three-dimensional micro-scope revealed that this was caused by a white to yellowish deposit with a porous crystalline structure over the entire valve surface, thus preventing proper seating. These deposits were identified as sodium carbonates (NA_2CO_3) originating from the catalyst and the carbonate group in the citric acid. (Decker and Du Plessis, 1981) This problem was overcome by using phosphoric acid instead of citric acid. (Pryde, 1981).

Corrosion of Control Linkages in the Injection Pump

Serious corrosion of the protective coating on the control linkages of the injection pump was suspected when a gel-like substance was noted on various surfaces after submersion in the fuel for several weeks. This phenomenon still requires positive identification.

Corrosion Research: Detailed studies conducted by the University of Pretoria on the behaviour of various metals, rubbers and plastics revealed that there should be no marked corrosion in the pump, provided the pH of the fuel is maintained close to neutral and there is no aqua phase separation. Mass loss studies reported by Skinner (1982) indicated corrosion rates of 1 micrometer and less per annum on mild steel, cast iron, aluminium, brass, copper, galvanised steel, stainless steel and leaded steel. Skinner also investigated some pump control linkages, but could not find any evidence of corrosion nor was the gel deposit of a corrosive nature.

EXTENDED CYCLIC ENDURANCE TESTS

Very inspiring results were obtained with a Perkins 4.236 engine in a Massey Ferguson tractor and running on an ethyl ester of sunflower oil. The engine has completed 1 300 true hours (1 660 tachometer hours) at the time of compilation of this paper without any problems and is still performing normally.

The only modification to the engine was to hard-chrome the governor mechanism and control linkages because of initial concern about possible corrosion problems. Close tolerance parts were not chromed however, nor were the springs because of the likelyhood of altering the spring constants. The work since done by Skinner has proved this to have been unnecessary. Inspection of the uncoated pump parts by merely opening the pump top cover confirmed Skinner's findings.

The Fuel Used

The ester fuel used in the endurance tests is prepared on the basis suggested by Pryde (1981) with some modifications based on previous experience.

Fuel Preparation: The reactor used for ester preparation consists of a standard 2 200 litre farm fuel storage tank, a circulation pump and piping, heater bank and an automotive radiator on the top to serve as watercooled condenser.

Bulk fuel is prepared by dissolving 7,5 kg anhydrous sodium hydroxide in 500 litres of aerosol grade ethanol and then adding 1 000 litres of degummed sunflower oil. After overnight agitation at room temperature (25°C) the catalyst is neutralised with 12 litres of ortho-phosphoric acid. (A low pH is required to prevent soap formation during initial water washing). After another two hours of agitation the fuel is pumped into settling containers. At least six hours is allowed for the settling of glycerol and salts before returning the top phase into the reactor. The bulk of the alcohol is then stripped by refluxing and recovered for cleaning purposes. Residual free alcohol and salts are then washed out at 80°C with 35% water till a pH of above 6,5 is achieved in the wash water. After cooling the fuel is centrifuged prior to use.

Chemical and Physical Properties:

Table 1 lists the typical properties of the fuel used in cyclic endurance tests as compared to diesel fuel specifications.

Test Equipment

The power take off output of the tractor is absorbed by a M&W field dynamometer. The dynamometer is slightly modified to enable remote control by a micro processor in a central control room also serving four other identical test cells (Van der Walt and Hugo, 1981).

Test Procedure

The engine is run on a cyclic load comprising alternating loaded and unloaded cycling with the load varying from 30% through 75% to 100% of maximum power at rated speed. The cycle is repeated every 24 hours with 30 minute service shut down between cycles.

Extended shut down times of several days were also incorporated due to some public holidays, non-availability of bulk fuel and occasional equipment break-downs. Down-times like these are also essential to verify the overcoming of previous problems. It is also an improved resemblance of practical operating conditions.

In addition to daily performance monitoring, regular oil samples were forwarded to a specialised lubricating oil analysis laboratory (Wearcheck®) as well as the participating oil company (Castrol, SA) for evaluation of wear rates, contaminants and general oil condition.

Monitoring Lubricating Oil Condition: Because of the particular interest in the lubricating oil and the awareness of the detrimental effect of plant oil fuel dilution on crank case oil during extended operation, special watch was kept over the oil condition. Correlation with lube oil analyses confirmed that the simple blotting paper test for dispersancy sufficed as an early warning, field test procedure; When a drop of oil, deposited onto blotting paper, separates into a light ring surrounding a dark blob, the dispersancy has deteriorated and the oil should be drained immediately. (Stivi, 1981). Compare the break-down in dispersancy with the fall in Total Base Number (TBN) in table 2. Note also that the engine manufacturer's recommended oil change intervals were achieved.

In a further attempt to extend the serviceable life of the engine oil a few patented engine oil filtering units were installed on this engine but without positively identifiable benefit.

Table 1. Properties of Ester Fuel and Specifications of Diesel

		Diesel Specifications	Ester Properties
Density at 20°C	g/mℓ	0,82 min	0,874
Flash Point	°C	55 min	>100
Sulphur Content	% m/m	0,55 max	<0,02
Viscosity, Kinematic at 40°C	cSt	1,8-5,8	4,9
Carbon Residue	% m/m	0,2 max	0,05
Total Acid Number	mg KOH/g	0,5 max	8,3
Ash Content	% m/m	0,01 max	0,006
Water Content	% m/m	0,05 max	<0,1[a]
Cetane Number		48	> 48[b]
Gross Heat Value	MJ/kg	44,85 min	41,15

[a] Laboratory could not detect lower water content

[b] Combustion studies indicate a higher cetane number than diesel

Engine Performance

The engine performs surprisingly well despite the high acid number and doubt about the actual water content of the fuel.

Table 2. Abbreviated Tractor Log Sheet

True[a] Running Hours	Max Power kW	Fuel Cons. ℓ/h	Exh. Smoke HSU	Cyl. Blow-by ℓ/min	Lubricating Oil[b]			
					True Service Hours	Visc. 40°C cSt	Total Base Number	Dis-per-sancy
23	36,6	12,4	20	24	23	103	4	Good
152	35	12,8	20	22	25	93	9,6	Good
305	36,1	12,8	20	21	181	107	2,4	Poor[c]
459	36,4	12,8	26	20	155	93	6,4	Good
600	35,9	12,9	34	20	102	95	4	Good
718	37	12,8	22	20	93	85	4	Good
811	36,7	12,9	27	30	180	134	4	Good[d]
974	36,9	12,9	22	20	140	76	7,4	Good
1052	36,2	12,9	27	38[e]	217	112	5,3	Poor[f]
1244	35,7	12,8	16	25	142	No results yet		
1269	37	13	20	44[e]	167	No results yet		
0	38,6	12,8	62	Reference Performance on Diesel Fuel				

[a] For engine tachometer hours, multiply by 1,28

[b] Oil analyses by Wearcheck, South Africa

[c] Oil changed immediately

[d] Routine oil change interval: 230 tachometer hours

[e] Instrument contamination suspected

[f] Extended lube oil life resulting from increased sump capacity due to additional patented oil filter fitted

Inspection at 1100 Operating Hours (1400 Tachometer Hours)

Despite recommendations that fuel injectors be serviced at 600 operating hours, they were deliberately left untouched. At 1200 hours an inspection was undertaken out of sheer curiosity. The carbon deposit on the injector tips was comparable to that experienced with diesel fuel with all the holes still unobstructed. The cracking pressures were still as specified.

Inspection of the engine by means of an endoscope revealed the following:

Honing marks on the cylinder walls were still clearly visible. Valve seating surfaces appeared clean and in sound condition. Carbon deposits on the cylinder walls above the first compression ring travel were evident, but not in any way severe.

Deposits on the cylinder head and piston crown had a mere sooty dark grey appearance as though they had recently been cleaned.

An interesting dark straw to reddish brown deposit was found on the head and shoulders of the exhaust valves, resembling a ferritic oxide. Outlet ports had a black to dark grey sooty deposit and the inlet ports were absolutely free of deposits in contrast with neat sunflower oil operation in DI engines, indicating matched combustion and aspiration.

Deposits were found on the exhaust valve stem portion adjacent to the valve guide recess when in the closed position. It was interesting to note, at a chipped portion of the deposit, that it was nearly white in colour with a black surface, probably resulting from the lubricating oil.

Investigating the Effect of Catalyst Residues

In order to determine the upper limit of acceptable catalyst residue an engine was run on a 'high sodium residue' ester. Because the sodium-residue, or metal residue in general, is directly related to the ash content of the fuel, the maximum allowable ash content specified for diesel fuel was accepted as the criterion from the standpoint of permissible wear rates.

It is believed that any detrimental effects caused by combustion residues originating from the catalyst used in the fuel preparation will first be revealed in the combusion chambers and exhaust passages as was experienced with the apparent valve burn reported earlier in this paper.

After 81 hours of running on an ester with an ash content of 0,012% no traces of harm or foreign deposits could be found when inspected by means of an endoscope. The general appearance was very much the same as in the 1100 hour inspection of the long term engine.

These findings still have to be confirmed in extended duty cycle operation of several hundred hours.

CONCLUSIONS

There can be little if any doubt that an ester prepared from sunflower oil is a good substitute fuel for direct injection engines. There is also reason to believe that equivalent results can be obtained with esters of some other vegetable oils.

Lubricating oil

The only deleterious effect on the lubricating oil was found to be a loss of dispersancy with the result that if this was not regularly checked (the blotting paper test mentioned earlier in the paper was found to be sufficient) the oil galleries could easily become clogged.

Fuel preparation

When preparing an ester for fueling compression ignition engines the following factors should receive special attention:

Transesterification Yield: Poor yields result in injector coking and poor combustion with subsequent exhaust dribble, crank case oil dilution and subsequent oil deterioration. Poor yields are also reflected in the carbon residue. Yields of at least 90% are suggested.

Total Acid Number and pH: It is quite obvious that high acidity will result in serious corrosion in the presence of free water. A high base number or pH will lead to serious corrosion of aluminium parts in an aqua phase. It is evident that the fuel be kept as neutral as possible, pH values between 6,5 and 8 seeming to be acceptable. (pH measured in 50% wash water)

Free Water: The water saturation level of the fuel is temperature dependent. In order to prevent free water forming it is essential that the fuel be centrifuged at a temperature below expected ambient operating temperatures. Winterising is suggested as the ideal solution.

Catalyst Residues: Provided a neutralising agent is used which will form an insoluble salt, free from carbonate groups, it seems feasible to expect that an ash content of 0,01% could serve as criterion.

Free Alcohol: Apart from the obvious detrimental effect of excessive free alcohol on the combustion properties of the fuel, safety aspects in the storage and handling of the fuel necessitate a very low free alcohol content.

It was found that a free alcohol content as low as 5% lowers the flash point to room temperature. It is therefore suggested that alcohol be stripped completely from the fuel.

Other general fuel specifications do not seem to offer much cause for concern as many aspects are taken care of when observing the abovementioned guidelines.

Specifications for an Ester Fuel

By complying with the free alcohol specification the flash point will automatically be well above the specified minimum. The esterification yield specified also takes care of the gross heat value (not listed), viscosity and carbon residue. Sulphur content should not offer any problem.

Water saturation temperature is specified instead of water content because it was found that corrosion would only occur when an aqua phase separates from the fuel due to temperature drop.

The specified TAN (total acid number) may offer serious problems, depending on the reagents used in preparation but the specified water saturation temperature seems to take care of expected problems.

Table 3. Specifications for an Ester Fuel

Esterifications Yield	%	90 min
Free Alcohol		Nil
Density at 20°C	g/ml	0,8 min
Flash Point	°C	55 min
Sulphur Content	% m/m	0,55 max
Viscosity, Kinematic at 40°C	cSt	5 approx
Carbon Residue	% m/m	0,2 max
Total Acid Number	mg KOH/g	0,5 max
Ash Content	% m/m	0,01 max
Water Saturation Temperature	Below operating ambient	

Other specifications would also apply as for diesel fuel.

ACKNOWLEDGEMENTS

The valuable advice and assistance of Dr L.M. du Plessis and his personnel at the National Food Research Institute of the C.S.I.R. is gratefully acknowledged.

REFERENCES

1. Bruwer, J.J., B. van D. Boshoff, F.J.C. Hugo, L.M. du Plessis, J. Fuls, C.S. Hawkins, A.N. van der Walt, A. Engelbrecht. 1980. The utilisation of sunflower seed oil as a renewable fuel for diesel engines.

2. Castrol South Africa (Pty) Limited. 1982. Personal communication and reporting on oil analyses.

3. Decker, J. 1981. Fuel Research Institute of South Africa. Personal communication.

4. Du Plessis, L.M. 1981. National Food Research Institute, C.S.I.R., South Africa. Personal communication.

5. Fuls, J., F.J.C. Hugo. 1981. On-farm preparation of sunflower oil esters for fuel. Proceedings of the Third International Conference on Energy Use Management. Beyond the Energy Crisis. Pergamon Press. Volume III. 1595-1602.

6. Hawkins, C.S., J. Fuls, F.J.C. Hugo. 1981. Laboratory procedures for investigating some fuel properties of sunflower oil esters in diesel engines. Proceedings of the Third International Conference on Energy Use Management. Beyond the Energy Crisis. Pergamon Press. Volume III. 1611-1617.

7. Pryde, E.H. 1981. Northern Agricultural Energy Centre, Peoria, Illinois, U.S.A. Personal communication.

8. Skinner, W. 1982. Department of Chemical Engineering, University of Pretoria. Report on corrosion resistance of various materials to sunflower oil and esters.

9. Stivi, J. 1981. Shell, Brazil S.A. Personal communication.

METHYLESTERS OF PLANT OILS AS DIESEL FUELS, EITHER STRAIGHT OR IN BLENDS

Georg H. Pischinger
Rolf W. Siekmann
Alexandre M. Falcon
Flávio R. Fernandes

ABSTRACT

Engine and vehicle tests were carried out with three alternative Diesel fuels: straight methylester of soybean oil (MESO), 75-25 gasoil-MESO blend, and 68-23-9 gasoil-MESO-ethanol (anhydrous) blend. Fuel-relevant characteristics of the three Diesel alternatives are given, together with the phase diagram of the ternary blend. Power, torque and volumetric brake specific fuel consumption in an unmodified IDI Diesel engine reflect mainly the net volumetric heating values. Smoke decreases with the presence of oxygenate compounds as does the emission of CO, as measured on the chassis dynamometer. A rigorous durability bench test on straight MESO shows results entirely within VW specifications. Analyses have indicated that, for IDI engines, no lube-oil problems are anticipated. Investigation of compatibility of MESO with fuel system material reveals considerable similarity to gasoil, but some items may require adaptation or even substitution.

INTRODUCTION

In Brazil,in addition to its gasohol and straight ethanol vehicles, Volkswagen produces Diesel light commercials and trucks up to 13 tons GVW. This is why VW Brazil is carrying out an intensive research program into alternative Diesel fuels, with the accent on Brazil's very considerable biomass possibilities. What we are seeking is an alternative fuel which can be used either straight or blended into gasoil. Such a fuel should also be compatible with the existing fleet and the existing gasoil distribution network.

In this respect, plant oils, particularly in the form of their methyl- and ethylesters, show the highest potential among today's known biomass-based alternative Diesel fuels, and it was for this reason that we concentrated our work mainly on them.

Ethanol could be used in Diesel engines developed specifically for this, but certainly straight ethanol is incompatible with the Diesel engines on the road today. Nor is it possible to blend ethanol with gasoil, except with either chemical or mecha-

The authors are: G.H.Pischinger, Dr.techn.Dipl.-Ing., Manager of Research Division, R.W.Siekmann, Dr.-Ing. Dipl.-Chem., Manager Research Dept. of Energy Technology and Alternative Fuels, A.M. Falcon, Mech. Eng., Manager of Applied Research Dept., F.R. Fernandes, Mech. Eng., Applied Research Dept., Volkswagen do Brasil S.A.

nical help. However, since plant oils and their methyl- and ethylesters are miscible with both ethanol and gasoil, we also investigated the possibility of using them, both as a Diesel fuel extender and as a bridging agent between gasoil and ethanol.

TEST ENGINE & VEHICLES

Because the objective was to investigate the behavior of the alternative fuels in existing gasoil engines, the vehicles and engines used in testing are regular production models.

Tests are being carried out on Volkswagen 1.6-liter indirect injection (IDI) Diesel engines, either on dynamometers or installed either in VW Passat passenger cars or in a VW Type II light-duty delivery van:

Number of cylinders:	4 in line
Combustion process:	swirl chamber
Displacement:	1588 cm^3
Bore/Stroke:	76.5 mm/86.4 mm
Compression ratio:	23
Injection pump:	rotary type

MONOESTERS OF PLANT OILS

Initial tests with soybean and peanut oils (edible plant oils), which have already been published (/1/, /2/), have shown a great potential for plant oils in Diesel engines. Nevertheless, there are indications that some difficulties may be encountered in the use of unprocessed oils as Diesel fuel.

The property which is most obviously not compatible with today's existing equipment (engines and filling-station pump) is the high viscosity of plant oils. Also, it seems that the glycerol in the oils contributes to a certain extent to the formation of deposits in the combustion chamber. To counter these, one of the most promising approaches seems to be the use of monoesters obtained by the transesterification of the vegetable oils, a reaction in which the glycerol of the glycerides is substituted by a mono-alcohol, such as methanol or ethanol. Thus, the viscosities are decreased and the glycerol eliminated.

The production of methylesters presents advantages, when compared with that of ethylesters, because the latter demand more processing steps, which raises costs. It is for this reason that, for the time being, we have opted for the methylesters.

Some properties of the various methylesters, and of the respective basic plant oils from which they originate, are to be found in Table 1, together with those of gasoil. It can be seen that:
- The net volumetric calorific values of the plant oils and their esters are only a little lower than that of gasoil.
- The kinematic viscosity is reduced drastically to within the specifications for gasoil. The viscosities of the various esters show a narrower spread than do those of the respective plant oils.
- The cetane number of the esters is greater than the value currently specified for Brazil.
- The cloud points of the esters are not critical within the climatic conditions of Brazil and are, with the exception of the methylester of palm oil, below the current specification

for gasoil in Brazil.

- As for flash-point, the methylesters show values lower than those for the unprocessed oil, but they are much higher than the estimate for current gasoil in Brazil, which is 48ºC (because naphtha is blended into it as an extender).

Table 1. Properties of various plant oils and their methylesters, in comparison with the Brazilian gasoil specification

	Flash-Point (ASTM D93) ºC	Density 20/4ºC g/mℓ (ASTM D1298)	Viscosity at 37.8ºC mm^2/s (ASTM D445)	Cetane Number (ASTM D613)	Cloud Point ºC (ASTM D97)	Net Heating Value MJ/ℓ
Specs of Brazilian gasoil	–[b]	–[b]	1.6-6	min.45	9-19	~35.3
Peanut oil	237	0.921	41.2	41.5[a]	+ 3	34.1[a]
Methylester Peanut oil	176	0.883	4.9	54	+ 5	33.6
Soybean oil	219	0.923	36.8	38.5[a]	- 4	34.0[a]
Methylester Soybean oil	178	0.885	4.5	45	+ 1	33.5
Babassu oil	150	0.946	30.3	38 [a]	+20	33.4[a]
Methylester Babassu oil	127	0.879	3.6	63	+ 4	31.8
Palm oil (dendê)	267	0.918	39.6	42 [a]	+31	35.0[a]
Methylester of Palm oil	164	0.880	5.7	62	+13	33.3
Sunflower oil	232	0.924	37.1	35.5[a]	- 5	34.0
Methylester Sunflower oil	183	0.880	4.6	49	+ 1	33.5

[a] calculated [b] not specified since addition of naphtha

The low volatility of the plant oils, as expressed by their distillation curves, remains practically unaltered in the esters. Being pure compounds, the esters exhibit boiling points, rather than curves, in the temperature range of 310 to 325ºC. This is shown in Fig. 1 by the example of soybean oil ester in comparison to the distillation curve of gasoil. Due to this low volatility of the esters, evaporation of the fuel is more difficult and can, under certain engine operating conditions, result in an incomplete combustion. The part of the fuel thus left unburnt can cause high dilution of the lubricating oil, as was found mainly in DI engines (e.g. /3/, /4/).

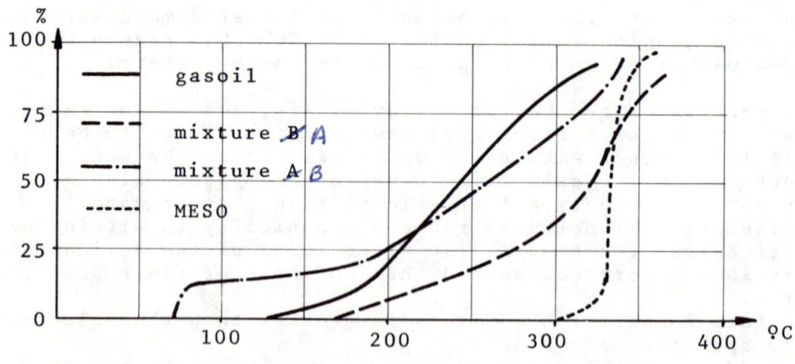

Fig. 1. Distillation curves of the test fuels

ENGINE AND VEHICLE TESTING ON STRAIGHT METHYLESTER OF SOYBEAN OIL

Because soybean oil is Brazil's main plant oil product (/5/), its methylester was selected for testing.

Dynamometer Testing of Engines

Performance Testing: The results of the comparative dynamometer tests on one given engine, at full load, fuelled alternatively on gasoil and MESO (methylester of soybean oil), are shown in Fig. 2.

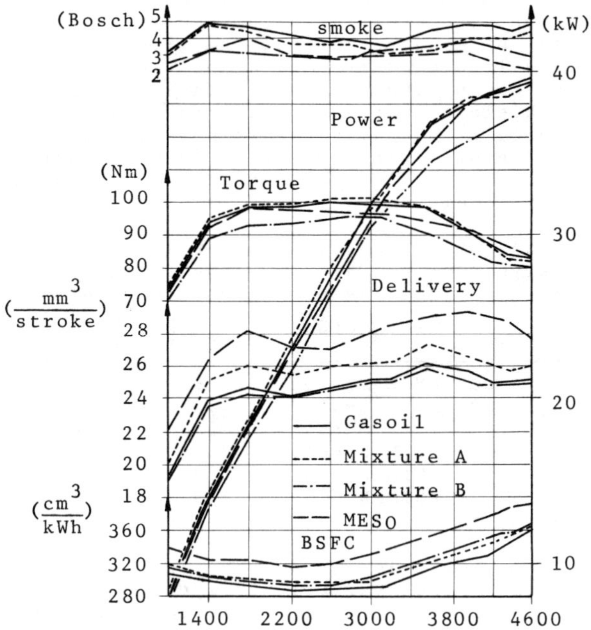

Fig. 2. Bench test results at full load of the same engine fuelled with the test fuels

The delivery of the injection pump is seen to be a little higher with the ester (on the average 7%), due to internal hydrodynamic variations in the pump. As the net volumetric calorific value of the methyl ester of soybean oil is about 6% lower than that of gasoil,this may explain the increase in volumetric specific consumption. Power and torque differ only marginally, but smoke emission is much lower with the methylester.

Durability Testing with MESO: In Volkswagen's rigorous durability test program for Diesel engines for light commercial vehicles, the engine is operated about 1,000 of the 1,400 hours at maximum power and about 300 at maximum torque. Throughout the durability test with straight ester, the same procedures set up for testing on gasoil have been followed strictly.

Specific attention was given to the investigation of whether the above-mentioned dilution of the lubricating oil was ocurring, with a consequent need to reduce the 75-hour oil drain period specified for testing on gasoil. To this end, a few preliminary tests were prepared with the lubricating oil of API CD characteristics (Shell Rimula CT-30), the engine being submitted to the same conditions established for the durability test. These preliminary tests showed that incomplete combustion, with consequent high lubricating-oil dilution, did not occur in our IDI Diesel engine with the relatively high compression ratio of 23. Because of this, the same lubricating-oil change period as stipulated for durability testing on gasoil was maintained. During the 1,418 hour durability test, 19 used-oil analyses were made. Table 2 shows the mean values of these and their respective standard deviations, together with the analysis of the new oil. Within the normal oil-change period, the viscosities, the insolubles and the alcaline capacity of the oil showed still acceptable levels, considering the extremely rigorous testing program. Under normal vehicle operating conditions, these viscosities would be considered too high. We know, nevertheless, from durability testing on gasoil, that oils with such viscosities at the end of the drain periods do not affect engine wear negatively under these testing conditions.

Table 2. Mean Value and standard deviation of 19 used oil analyses and analysis of the new oil (Shell Rimula CT-30)

Characteristics	Test Method	used oil mean	used oil stand.dev.	new oil
Viscosity Kin.cSt at 40ºC	ASTM D-445	187.8	15.4	114
Viscosity Kin.cSt at 100ºC	ASTM D-445	19.3	1.7	12.4
TAN	ASTM D-664	4.8	.8	3.1
TBN	ASTM D-664	3.25	.3	9.6
Insolubles Hep.	ASTM D-893	3.9	.7	
Insolubles Tol.	ASTM D-893	3.0	.6	

Values of torque, power, smoke levels and cylinder compression have remained within the normal variations expected in this type of test. After the durability run, the engine was torn down for examination and for measurement of wear. In the bearings, piston rings, cylinder bores and valve train, wear remained within the VW specifications, as shown in table 3 and 4. These data are considered to be representative of normal results in this test.

Table 3. Wear measurements after a 1,400 hours dynamometer durability test

	Clearance [a] gasoil (5 eng.) mean	gasoil (5 eng.) stand. dev.	MESO (1 eng.)
Crankshaft journal/ Main bearing	. 52	. 01	. 47
Big-end bearing (Connecting rod/Crankshaft)	. 59	. 05	. 53
Cylinder/piston	. 83	. 05	. 85
Exhaust valve stem/guide	0.74	0.07	0.69
Intake valve stem/guide	0.37	0.04	0.54

a Measured clearance divided by allowable maximum clearance (i.e., .47 = 47% of allowable maximum clearance)

Table 4. Piston ring wear measurements after the 1,400 hours durability test

	Clearance [a]		
	gasoil (5 eng.)		MESO (1 eng.)
	mean	stand. dev.	
Compression Ring I	.86	.11	.76
Compression Ring II	.79	.11	.71
Oil Ring	.89	.09	.70

[a] Measured ring gap increase divided by the maximum allowable ring gap increase

It should be remembered that the values shown in tables 3 and 4 are merely representations of the percentage of the maximum allowable wear. Thus, values of less than 1.0 indicate that the measured wear is less than the maximum allowable for the part involved.

Although the engine was not specifically rated for oil-related deposits, no unusual conditions were noted. In the combustion chamber and around the nozzles we found dry, soft, easily-removable deposits which were even considered slightly below normal. Our rotary injection pump (Bosch) is lubricated only by the fuel and thus might have been a critical component. Robert Bosch do Brasil analyzed the injection equipment and found the wear absolutely normal for this type of test. The injection tubing showed slight traces of incipient cavitation, which was also considered normal. The only change noted in parts subjected to contact with the fuel was a brown tone on the inner walls of the pump, except where polished by movement.

Vehicle Testing: At this writing, a Diesel Passat has run up 40,500 km, and a Diesel delivery van 5,200 km, on MESO. Drivability, performance,engine noise and cold starting are comparable to those with Diesel oil. Exhaust-gas smell is at a comparable level, although different in quality.

Table 5 gives consumption and CO and NO_x emissions as determined on the chassis dynamometer in the US Driving Cycles, with a Passat being fuelled either with methylester of soybean oil or with gasoil. We are not yet equipped to measure HC emissions from Diesel powered cars. Fuel consumption is, on the average, from 5 to 6% higher and falls in the range expected, as the net volumetric heat-content of the ester is 6% below that of gasoil. CO emissions are much lower; NO_x remains the same.

Table 5. Emissions and consumption of the same vehicle (Passat Diesel), fuelled on diesel oil and methylester of soybean oil (MESO), measured on the chassis dynamometer

Fuel	CDC [a] CO g/km	CDC [a] NO_x g/km	CDC [a] Consumption ℓ/100 km
Diesel oil	1.16	.68	7.0
Meso	.69	.69	7.4
Difference % Base: gasoil	- 40		~+ 6

[a] CVS City Driving Cycle started with warmed-up engine

Table 6 gives a typical used-oil analysis after testing the Passat Diesel under family driving conditions in São Paulo. Again, a lubricating oil of API CD characteristics was used (Texaco URSA LA-3). The sample was taken after the 7,500 km specified for the Passat on gasoil. The data show the same tendencies to variation as given in table 3, though in a less pronounced way, and thus indicate no problems.

Table 6. Typical used-oil analysis from testing the Passat Diesel (Texaco URSA LA-3)

Characteristics	Test Method	new oil	used oil
Viscosity Kin.cSt at 40ºC	ASTM D-445	114.6	127.9
Viscosity Kin.cSt at 100ºC	ASTM D-445	11.74	13.74
TBN	ASTM D-2896	12.82	12.36
Insolubles Pent.	ASTM D-893		.58
Insolubles Benz.	ASTM D-893		.29

Materials Compatibility with Methylester of Soybean Oil (see Table 7): Investigation into the behavior of the various materials in the fuel system that come into contact with MESO was conducted by submitting them to the routine tests prescribed by Volkswagen's internal standards for the approval or rejection of production parts.

For comparison of the behavior of plastic and rubber parts, separate immersions were made of each part in gasoil and MESO at ambient temperature, for times varying from 25 to 50 hours. Depending on the type of part, measurements were then made, and the changes noted, of such characteristics as diameter, thickness, volume, weight, hardness, breaking point and elongation.

Corrosion tests were also run on the metals in the injection-pump housing. Assemblies of three plates, aluminium, brass and steel, in that order were immersed respectively in MESO and gasoil at 50ºC for 6 days, after which changes in weight and aspect produced by electrochemical action were noted.

Two forms of MESO were used: normal and subjected to accelerated oxydation.

Fuel-tank behavior was analysed by partial filling with MESO for six months and then examining the appearance of the interior.

TESTS WITH METHYLESTER OF SOYBEAN OIL AS A GASOIL EXTENDER

With the objective obtaining a first indication of the technical feasibility of MESO as a gasoil extender, we resolved to submit to engine testing the following:

- a gasoil/MESO blend
- a ternary mixture gasoil/MESO/ethanol, in which the ester would act as extender and bridging agent.

For an initial evaluation, the determination of the ternary system was carried out at room temperature, using Brazilian gasoil, MESO and anhydrous ethanol. The fuel-specific properties of the basic components are given in Table 8. The various ternary mixtures were observed for 30 days for stability. Obviously, the phase diagram of the ternary system includes the gasoil-

MESO binary system.

Table 7. Effects of MESO on various materials in the fuel system compared with gasoil

Materials	Effect	
Nitrile rubber	Deterioration greater; not acceptable	
Viton	Comparable	
Polyformaldehyde (Polyacetal, Delrin)	Comparable	
Polyamide 12W	Worse but still acceptable	
Polyamide 6.6	Worse but still acceptable	
Asbestos fiber & nitrile rubber	Worse but still acceptable	
Granulated cork bonded with nitrile rubber	Worse but still acceptable	
Ebonite	Comparable	
Epichloridrine	Comparable	
White phenolic molding compound	Comparable	
Polyurethane foam	Noticeable drop in performance when immersed in engine lubricating oil contaminated with 10% MESO, as contrasted with the straight lubricant	
	CORROSION	
	MESO	MESO oxidated (ASTM D-525)
Aluminium	no appreciable increase	no appreciable increase
Brass	less	greater [a]
Steel	no appreciable increase	greater [a]
Fuel tank: phosphatized surface	Slight signs of corrosion at some points, but satisfactory.	

[a] further testing will have to decide about substitution

Table 8. Some fuel-relevant properties of the mixtures A & B and of their components

	MESO	Anhydrous Ethanol	Gasoil[b]	Mixture B	Mixture A
Kin.visc.,37,8°C ASTM D-445 cSt	5,22	1,18	2,13	2,32	2,67[c]
Flash point °C ASTM D-93	160	10	27,5	10	d
Cloud point °C ASTM D-97	- 1	<-40	-7	-6	d
Cetane Number ASTM D-613	45	8[a]	45	42	45[c]
Heating Value net MJ/ℓ	32,5	21,17	35,34	32,73	34,64[c]
Stoich. air/fuel ratio kg/kg	12,37	8,95	($C_{16}H_{34}$) 14,88	13,77	14,25[c]

[a] Third report on Diesel Blends - T.M. Letcher - Nov. 1980

[b] Measured data of Brazilian Diesel fuel containing naphtha

[c] calculated

[d] not determined

Figure 3 shows the phase diagram. It can be seen that an increasing proportion of ester in the gasoil also increases the proportion of ethanol tolerated. Furthermore, it can be seen from the diagram that, from a 25% ester content onwards, the tolerance to ethanol is total.

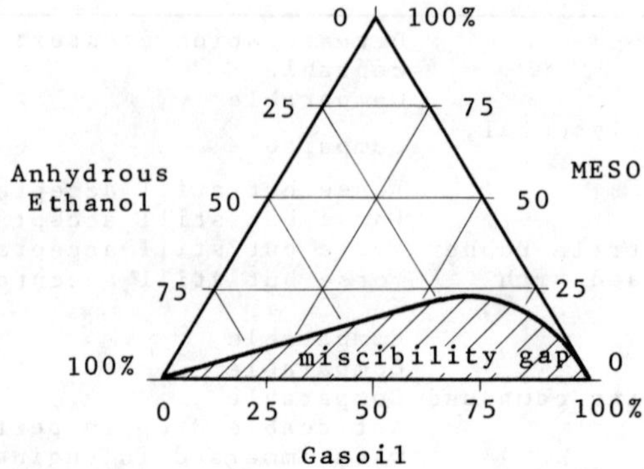

Fig. 3. Phase diagram of the ternary system gasoil/MESO/anhydrous ethanol at room temperature

Characterization of the fuels for testing

For these experiments, two mixtures were chosen which not only had to be totally homogeneous, but had also to satisfy the following criteria:

- the resulting fuel mixture had to be compatible with existing systems (Diesel engines and gasoil distribution system)
- the measured cetane numbers of the fuel mixtures should not be less than 42, a value often found in Brazil since the addition of naphtha to gasoil. The heating value should be similar to that of gasoil.
- because the alternative plant oils program, which would increase the availability both of soybean and other oils, has still to start up (/6/), we arbitrarily limited the amount of MESO in the mixtures to 25%.

The mixtures chosen were:

Mixture A: 75% gasoil + 25% MESO
Mixture B: 91% Mixture A + 9% anhydrous ethanol

The fuel-specific data more relevant to the technical performance of the blends are shown in Table 8. It is mainly because of the ethanol that mixture B has a heating value and a stoichiometric air/fuel ratio lower than those of gasoil. But, generally speaking, the fuels do not differ significantly one from another in the properties considered. The distillation curves of the mixtures are shown in Fig. 1. In the case of the ternary system (mixture B), we find a significantly different shape, in comparison with gasoil, caused by the high-boiling compounds that stem from the ester and the low-boiling characteristics caused by the ethanol.

Engine and Vehicle Tests with the Mixtures

Bench testing for Performance: The results of the comparative dynamometer tests at full load, with the same engine now fuelled alternatively with the two mixtures, are shown in Fig. 2, together with the results on gasoil and on MESO, which have already been discussed.

Power, torque and volumetric brake specific fuel consumption (BSFC) reflect the lower net volumetric heating values of the fuel mixtures under study. Smoke (BOSCH) decreased when the engine was running on the ternary mixture, which might be due to the presence of oxigenate compounds, mainly the ethanol.

Vehicle Testing on the Chassis Dynamometer: Table 9 gives fuel consumption and CO and NO_x emissions as determined on the chassis dynamometer, with one given VW Type II light-duty delivery van with the 1.6-liter Diesel engine, fuelled with mixture B and also with gasoil.

Volumetric fuel consumption is higher with the ternary blend and again reflects mainly the differences in the net volumetric heating value. NO_x emissions are roughly the same but CO is lower, which again might be due to the presence of the oxigenate compounds, mainly the ethanol.

Table 9. Emissions and consumption of the same vehicle (light delivery van - inertia weight 1360 kg), fuelled on gasoil and Mixture B, measured on the chassis dynamometer

	C D C [a]		
	CO g/km	NO_x g/km	Consumption ℓ/100 km
Gasoil	1.28	1.24	7.9
Mixture B	.36	1.26	8.36
Difference % Base: gasoil	- 72%	-	+ 5.8

[a] CVS City Driving Cycle started with warmed-up engine

CONCLUSION

Monoesters of plant oils, as typified by the methylester of soybean oil (MESO) used in the tests reported herein, reveal a high degree of compatibility with current production IDI-engined vehicles, whether as straight fuels, fuel extenders, or bridging agents that permit the introduction of ethanol into system-compatible gasoil blends.

REFERENCES

/1/ Pischinger, G.; Clymans, F.; Fernandes, F.: Consumption Data and Considerations on Three Approaches to Diesel Oil Substitution. Proceedings of 4th Int'l Symposium on Alcohol Fuels Technology, Oct. 1980, Guarujá - SP - Brazil.

/2/ Pischinger, G.; Siekmann, R.; Clymans, F.: Diesel-oil Substitution by Vegetable Oils - Fuel Requirements and Vehicle Experiments. Proceedings of the Conference on New Energy Conservation Technologies and their Commercialization, April 1981, Berlin, Germany.

/3/ Geoffroy, M.; Smith I.B.; Blackburn, J.H.; Pinchin, R.; V. da Rocha, A.: Esters as a Future Diesel Fuel: an Initial Assessment. Proceedings of 5th Int'l Symposium on Alcohol Fuel Technology, May 1982, Aukland, New Zealand.
/4/ Internal Progress Report Nr. 2/81: Test Program using Methyl-ester of Soybean Oil in 2 VW trucks (13 t GVW) with DI Diesel Engines, Dec. 1981, Research Division of Volkswagen do Brasil S.A., São Paulo, Brazil.
/5/ Production of Vegetable Oils, Statistical Year Book of Brazil, IBGE, 1980, Brasília, Brazil.
/6/ Vegetable Oil Production Plan for Energy Purposes; Ministry of Agriculture, submitted to the National Energy Council; Oct. 1980; Brasília, Brazil.

THE INFLUENCE OF LUBRICANT CONTAMINATION BY METHYLESTERS OF PLANT OILS ON OXIDATION STABILITY AND LIFE

R. W. Siekmann
D. Blackman

G. H. Pischinger
L. D. Carvalho

ABSTRACT

Lubricating oil deterioration due to contamination by methylester of plant oil is studied in laboratory tests in MacCoull apparatus and, for comparison,bench and mixed-driving tests. Methylester of soybean oil (MESO) is added intentionally to the lubricating oil. Laboratory tests show that the amount of double bonds introduced is significant in increasing viscosity and reducing TBN. Gas-chromatographic analysis of extracted ester shows degradation mainly in fatty acids with conjugated double bonds.

Tests in engines, both bench and mixed-driving, gave results similar to MacCoull but respectively and much less pronounced.

INTRODUCTION

The use of straight plant oils in Diesel engines is problematic, principally because of their high viscosities, the lower volatility expressed by their distillation curves, and - we assume - the glycerol. Deposits in the combustion chamber and fuel contamination of the lubricating oil,resulting in shortened drain periods, have been the main problems recurring in reports (1, 2).

In the transesterification of plant oils, the principal change is that glycerol is substituted by methanol or ethanol, whereas the fatty acids do not suffer major alterations in respect to their degree of unsaturation. The transesterification modifies most of the fuel-relevant properties of plant oils so that they meet the requirements of today's Diesel engines, especially the adequate viscosities and cetane numbers (3).

The distillation curves, however, remain practically unaltered (8). Engines fuelled with methylesters showed that deposits in the combustion chamber were practically eliminated. Fuel dilution of the lubricating oil, however, continues to occur under certain operating conditions of the engine, indicating an incomplete combustion due to the still low volatility of the fuel (4,5).

In direct-injection engines, under some operating conditions, the degree of dilution reached levels of up to 20%, resulting in accelerated deterioration of the lubricating oil and the consequent need for an earlier oil change.

EXPERIMENTS

In order to determine the causes and consequences of lubricating oil deterioration, we simulated oil dilution by preparing different mixtures of methylesters of vegetable oils in the lubricating oil(MIL-L-2104C, API CD, SAE 30). The mixtures were then subjected to a series of laboratory and engine tests.

The authors are: R. W. Siekmann, Dr.-Ing. Dipl.-Chem., Manager of Energy Technology and Alternative Fuels Dept.; G. H. Pischinger, Dr. techn.Dipl.-Ing., Manager of Research Division, Volkswagen do Brasil S.A.; D. Blackman, Technical Consultant, L. D. Carvalho, Chief Chemist, R&T Dept., Texaco Brasil S.A. Produtos Petróleo.

Laboratory Tests

With a view to submitting the mixtures to more critical conditions of temperature, oxidation by the air, and the presence of catalysts, a MacCoull apparatus was used. This apparatus simulates engine-bearing working conditions (except those of load) and permits evaluation of the lubricating oil resistance to oxidation (6).

The mixtures under test were circulated in the apparatus in the presence of air and with copper as a catalyst. In order to better determine the influence of the degree of unsaturation of the vegetable oils, the methylesters selected were those of soybean oil (MESO) and babassu oil (MEBO), these being respectively representative of a highly unsaturated oil (iodine number 128) and a highly saturated oil (iodine number 17).

The reference tests chosen for an initial evaluation of the lubricating oils were the kinematic viscosity at 40°C and the TBN (ASTM D-2896).

Influence of the Degree of Dilution: In order to determine the influence of the degree of the dilution on the stability of the lubricating oil, mixtures of 5, 10 and 20% MESO and MEBO were processed for 8 hours in the MacCoull apparatus.

Because preliminary tests on the influence of the temperature showed increased effect above 125°C, test temperatures were set at 150°C and 170°C.

In Figure 1, which presents the variation of TBN, the general picture is of a greater reduction in alkaline reserve with the increasing proportion of dilution, both for MESO and MEBO, with the values more critical at 170°C.

At 150°C the babassu methylester (MEBO) shows an insignificant drop in TBN for all the concentrations examined.

By contrast, MESO shows a greater drop and, at 20% dilution, a pronounced tendency to degradation in its unsaturated esters.

As for the behavior of the mixtures at 170°C, one sees, in the case of MESO at 20% dilution, a drop in TBN which would practically dictate an oil change, as the value is near the lower limits recommended by some engine manufacturers.

Note that, in the case of MEBO, even at a temperature of 170°C the drop in TBN is only about 8%.

The figure also shows that, in the test at 170°C, the 10% dilution by MESO produces about the same drop in alkaline reserve as does a 20% dilution at 150°C.

Influence of the Degree of Unsaturation: Because the 5 to 20% mixtures of soybean and babassu oil methylesters really amount to no more than the introduction into the lubricating oil of different degrees of unsaturation, the iodine numbers of these mixtures were calculated on the assumption of an iodine number zero for the lubricating oil.

Although the tests in the MacCoull apparatus were of short duration (8 hours), they already show a straight-line variation of TBN as a function of the double bonds introduced into the lubricating oil (Figure 2).

An increase in the content of double bonds of the monoesters in the lubricating oil results in the progressive deterioration of the oxy-thermic stability of the mixtures (7), permitting an increasing formation of acid compounds, with the consequent reduction in alkaline reserve.

The Influence of Test Duration: The variations in the viscosity and in the TBN as functions of the test duration were examined at 170°C and with 10% MESO as dilutant so as to increase the severity of the conditions.

The variation of the amount of ester artificially diluted into the lubricating oil was determined quantitatively during the test, by infra-red analysis,using a Perkin Elmer, Model 283 B IR-Spectrometer.

In order to obtain a standard, different mixtures of lube-oil with varying concentrations of MESO were prepared and analysed in the range of 1850 to 1700 cm^{-1}. It could be shown that results are in good agreement as the esters absorb at 1750 cm^{-1} and the vibration of products of oxidation at 1710 cm^{-1} do not disturb the determination.

Figure 3 shows the influence of reaction time on viscosity, TBN and the amount of ester as determined by infra-red.

Viscosity shows an increasing tendency with reaction time, doubling in about 16 hours of testing and finally reaches a levelling-off at an increase of 156%.

The values of TBN as determined by ASTM 2896 demonstrated a constant drop in alkaline reserve, the TBN reduction at 24 hours of testing reaching 44%.

The ester content drops slightly with reaction time which most presumably can be attributed to evaporation as would be shown by other experiments under similar operating conditions. Part of the ester also may be consumed in oxidation reactions.

As the foregoing experiments showed that premature degradation of the lubricating oil is heavily dependent on ester concentration (i.e., lubricant dilution) and especially on the amount of double bonds introduced, we studied the reactions of the methylesters of the various fatty acids involved.

For this purpose we employed methanol to extract the ester from the lubricating oil and then, by gas-chromatography, characterized it with respect to its fatty acid composition.

In our procedure, we used 75 ml of methanol to extract the esters from 10 ml of used oil and then drove off the methanol to leave the extract, thus obtained, in a form that could be submitted to GC analysis. GC analysis was carried out on a Perkin Elmer, Model Sigma 3 apparatus, using an FID detector and a temperature program in the range of 180°C to 230°C, with a rate of 2°C/min. A 0,14 µl sample was injected in each analysis and the determination of the fatty acid esters with Silar 10-C as column material. For quantitative analysis we used the methylester of lauric acid as an internal standard.

Blank tests showed that the extraction with methanol certainly picks up a part of the additives contained in the lubricating oil, but that this does not affect the subsequent GC analysis. Also observed was the preferential extraction of the unsaturated fatty acid esters, as was to be expected in view of the polarity of the solvent, this made it necessary to determine correction factors for each fatty acid ester.

Figure 4 shows the amount of the various methylesters of fatty acids which are contained in MESO as a function of reaction time in the MacCoull apparatus.

Because of the calculation method in GC analysis, a decrease in unsaturated fatty acids resulted indirectly in an unrealistic increase of the saturated compounds. Because the C_{16} and C_{18} saturated fatty acids can be considered more resistant to oxidation reactions, we assumed their concentration to be practically constant. The results thus obtained largely justify this procedure.

The discrimination of the different fatty acid esters shows that, with an increase in reaction time, it is mainly the unsaturated fatty acids that disappear. Most probably they are consumed in oxidation reactions as is also indicated by the drop in TBN. Obviously, due to the different amount of double bonds in the molecule, the fatty acids show different reactivity. While Oleic acid, C_{18-1}, suffers a reduction of only 25% during the reaction time, the consequences of the fatty acids with conjugated double bonds are significant. The linoleic acid, C_{18-2}, drops from 58.5% to 9.3% during reaction time, which represents a decrease of 84.1%. The influence of three conjugated double bonds increases still more the reactivity which results,for the amount of linolenic acid C_{18-3},in a drop of 97.6% during testing time.

It can be assumed that, for the case of a lubricating oil contaminated with MESO, it is the lower oxidation stability of the unsaturated fatty acids that will also finally influence the life of the lubricating oil. The methylester of babassu oil contains only 13.5% of oleic and 2.2% of linoleic acid, the rest of the fatty acids being totally saturated, which help to explain the results shown in Fig. 1, which revealed a lesser oxidation of the lube oil when contaminated with MEBO.

BENCH TESTING

Up to now, as far as we know, all experiments to determine the lubricating oil behaviour of DI engines when fuelled with esters of plant oils have shown excessive oil dilution, which forces a premature oil change. Because of this, an adequate evaluation of the lube oil and of different additive packages was rendered significantly difficult.

In order to carry out this type of study, the prechamber engine we used in the experiments is advantageous as no excessive lubricating oil dilution was experienced. It was therefore possible to simulate oil contamination as it occurs in DI engines and then study the effects with respect to increased lube oil degradation. The test engine data are given in Table 1.

For a first evaluation,two IDI engines, both fuelled with MESO were bench tested for 100h. We used the same lubricating oil as tested in the MacCoull apparatus, one engine having its lubricating oil intentionally contaminated with 5% of MESO. No higher dilutions were considered as they would have reduced the lube oil viscosity to values below the acceptable minimum. Both engines were examined and adjusted with respect to their comparability in power output, torque, characteristics of the injection system and blow-by.

The 100 h of bench testing were carried out under relatively severe conditions, the engine operating nearly all the time at maximum power or maximum torque. For a detailed discription of the test conditions, see (8).

The effects of oil contamination on viscosity and TBN are given in Figure 5 and are compared with the reference test.

In the case of this latter, a steady increase in viscosity is found which,in the severe bench test, practically doubles the viscosity in 100 hours. Significantly different is the change in viscosity when 5% MESO is added. The rate of increase is more pronounced and, although starting at a lower initial value, at about 65 hours of testing the viscosity catches up with, and passes, the increasing value of the reference test. Furthermore, the viscosity of the contaminated lube oil goes up in a rapidly steepening curve which,at 100 hours, reaches about 5 times the initial figure. As the analysis of the used oil shows, there is also a substantial increase in soot content over the testing period and it can be assumed that the steady thickening of the lube oil is mostly due to this and does not represent a break-down of the anti-oxidant capacity of the lube oil. This is also supported by the TBN analysis, as demonstrated in the following.

It can be seen that in the case of oil contamination the decrease in alkaline reserve is more accentuated and this can be related to a more intense formation of oxidation products with acid character. On an overall look, however, TBN values for both tests remain well within the acceptable range, which stipulates a minimum limit of 4 mg KOH/g for the method applied.

The infra-red analysis of the contaminated lube oil showed a constant decrease from the initial 5.2% down to 3.2% after 75 hours of testing. The determination of the ester content after 100 hours was not possible because of the diminished transparency of the sample. Considering these values and the results of the following GC-analysis, it must be kept in mind that, during the engine test, lube oil make-ups were carried out in all cases with lube oil containing 5% of MESO.

In the reference test, carried out with MESO as fuel but with normal lube oil, the IR analysis proved the fact that, for the IDI engine used, oil dilution practically does not occur. The values obtained were negligibly low at less than 1%.

In order to evaluate the extent to which the results of the MacCoull test with ester of plant oil are transferable to engine tests, the ester was extracted from the used lube oil and analysed by gas chromatography.

The results are given in Figure 6.

The methylesters of fatty acids examined show in general the same tendencies as observed in the MacCoull tests. The lower degradation of the fatty acids in the engine test reveals, however, that the MacCoull apparatus represents the more drastic oxidation conditions. Also in the engine test, while the stearic acid remains relatively stable, the unsaturated compounds C_{18-1}, C_{18-2} and C_{18-3} dropped about 31.6%, 59.6% and 78.0% respectively. Comparing their relative reduction from the MacCoull test with the bench test results, the decrease of C_{18-1} and C_{18-3} is greater in the engine test.

The significant reduction of the palmitic acid C_{16}, which was not observed in the MacCoull tests, can possibly be attributed to greater evaporation losses under engine operation conditions. In order to investigate this effect further separate tests are currently under way.

VEHICLE TESTING

We examined the lubricating oil behaviour under actual operating conditions in two VW Type II light-duty delivery vans over the normal drain period of 7500 km. Both vehicles, equipped with engines of the same performance as those studied in bench testing and operated under mixed-driving conditions with MESO as fuel had their lubricating oil contaminated with 5% of MESO. Both vehicles were operated mainly under full-load conditions. Within the drain period, used oil samples were taken every 1000 km and submitted to analysis. The used oil analysis of the two mixed-driving tests resulted in practically identical values.

In contrast with the results of bench testing, in which conditions are definitely more severe in terms of load and engine temperature, those of vehicle testing showed viscosities and TBN only slightly changed and therefore absolutely normal for within the drain period.

The lower rate of increase in viscosity can most probably be attributed to the significantly lower soot levels, detected by used oil analysis, which range between 0.3 and 0.9% whereas, in the bench test, soot levels at 100 hours came to more than 4%.

The fatty acid ester composition of MESO, extracted from the used oil samples, is given in Figure 7. The results confirm the less severe operating conditions when compared with those of bench testing, as already indicated by the TBN and viscosity data. Whereas in bench testing, oleic acid was partly consumed during the test period (-36.0%), under mixed-driving conditions it can be considered as stable. Only the more reactive compounds linoleic and linolenic acid, having conjugated double bonds, demonstrate a decrease of 16.3% and 42.0%, respectively. As already observed in bench testing, the palmitic acid tends to decrease during the oil drain period.

Normal lube oil make-ups during the test period, although carried out with 5% MESO, seem to have a stabilizing effect on lube oil behavior under mixed-driving conditions.

CONCLUSIONS

Laboratory testing with the MacCoull apparatus has shown that the degradation of lubricating oil contaminated with methylesters of plant oils is determined mainly by the amount of unsaturated fatty acids introduced. The reaction temperature is another important factor. Gas-chromatographic analysis has shown that it is mainly the linoleic and linolenic acids that are degraded in the test, being more reactive because of conjugated double bonds.

The most severe tests were the MacCoull, followed by bench-testing of the engines and, finally, mixed-driving vehicle tests; all the results were similar, but reflected the differences in severity.

Better lube oil behavior can be expected, according to these test results, to the extent that it becomes possible to reduce the contamination by conjugated double bonds; lines for investigation could be the reduction of lube oil dilution and the use of different plant oil esters.

REFERENCES

(1) Bacon, D.M., Bacon, N., Moncrieff, I.D., Walker, K.L.: The effects of Biomass Fuels on Diesel Engine Combustion Performance. Proceedings of the IV Int'l Simposium on Alcohol Fuels Technology - Vol. 1, Oct. 1980, Guarujá, SP, Brazil.

(2) Barsic, N.J., Humke, A.L.: Performance and Emissions Characteristics of a Naturally Aspirated Diesel engine with Vegetable Oils Fuels. SAE Paper Nr. 81 0262, Feb. 1981.

(3) Pischinger, G.H., Siekmann, R.W., Falcon, A.M., Fernandes, F.R.: Results of Engine and Vehicle Tests with Methylesters of Plant Oils as Alternative Diesel Fuels. Proceedings of the Fifth Int'l Alcohol Fuel Technology Simposium, May 1982, Auckland, New Zealand.

(4) Geoffroy, M., Smith, I.B., Blackburn, J.H., Pinchin, R., V. da Rocha, A.: Esters as a Future Diesel Fuel: an Initial Assessment. Proceedings of 5th Int'l Symposium on Alcohol Fuel Technology, May 1982, Auckland, New Zealand.

(5) Internal Progress Report Nr. 2/81: Test Program using Methylester of Soybean Oil in 2 VW trucks (13 t GVW) with DI Diesel Engines, Dec.1981, Research Division of Volkswagen do Brasil S.A., São Paulo, Brazil.

(6) MacCoull, N., Ryder, E.H. and Scholp, A.C., SAE Journal(Trans.)50,338(1942)

(7) Gunstone, F.D., Hilditch, T.P., J. Chem. Soc. 1022 - 1025 (1946)

(8) Siekmann, R.W., Pischinger, G.H.: Volkswagen of Brazil Activities and Results with Methylesters of Plant Oils as Diesel Fuel. Proceedings of Vegetable oils as Diesel Fuel, Seminar II, Oct. 1981, Peoria, USA.

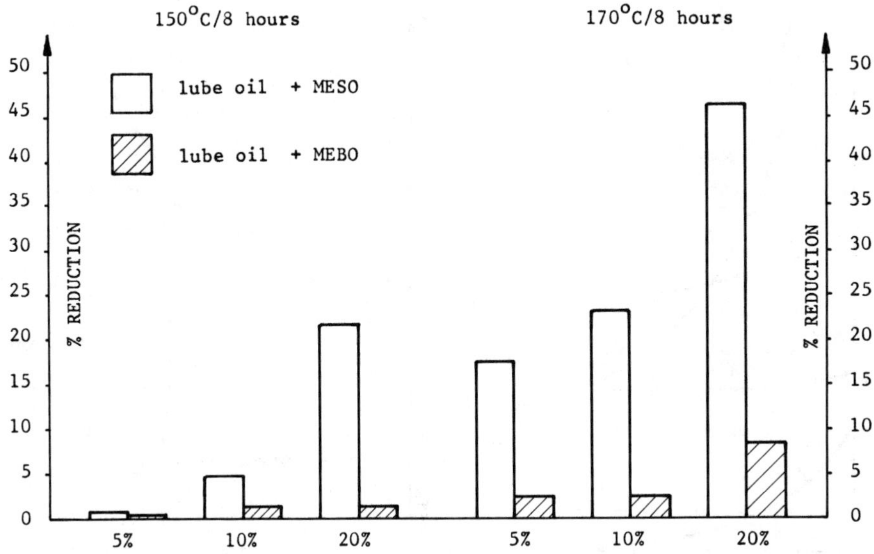

Figure 1 : MacCoull Test - Reduction of TBN as a function of the ester concentration in lube oil.

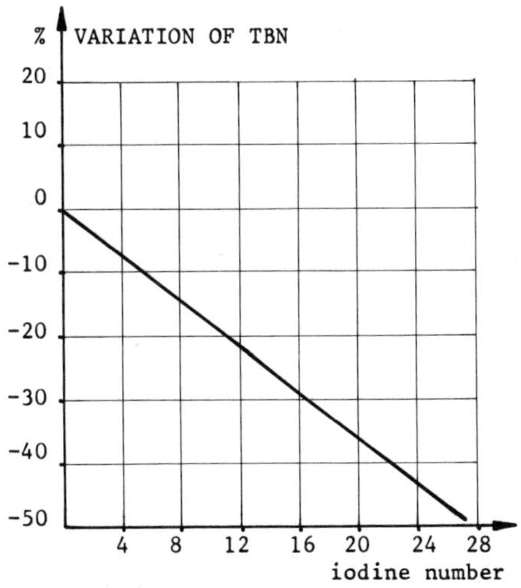

Figure 2 : MacCoull Test - Variation of TBN as a function of the amount of double bonds in lube oil (170°C/8 h).

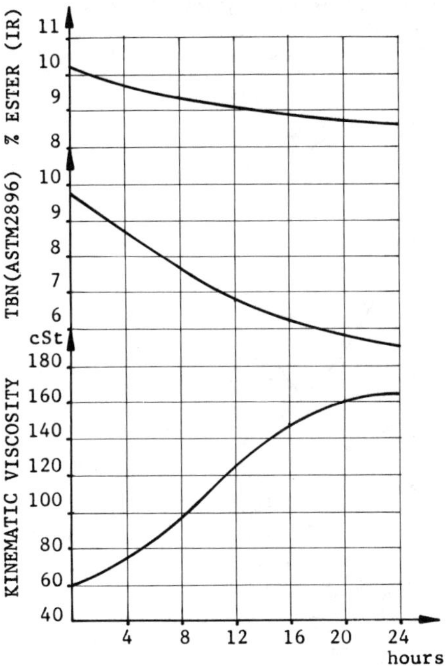

Figure 3 : MacCoull Test - Viscosity, TBN and ester content of lubricating oil varying with reaction time.

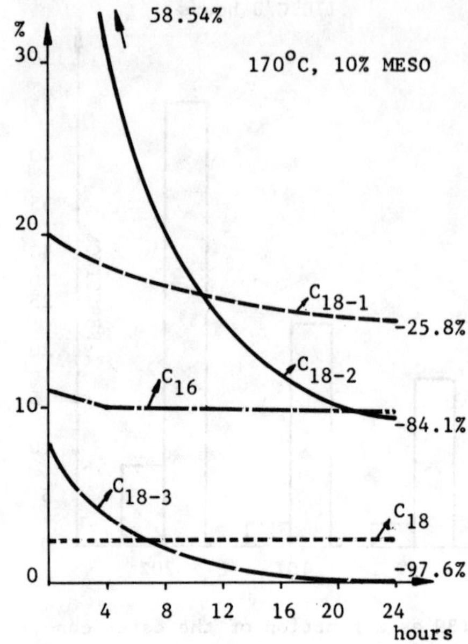

Figure 4: MacCoull test - Variation of fatty acid ester composition with reaction time.

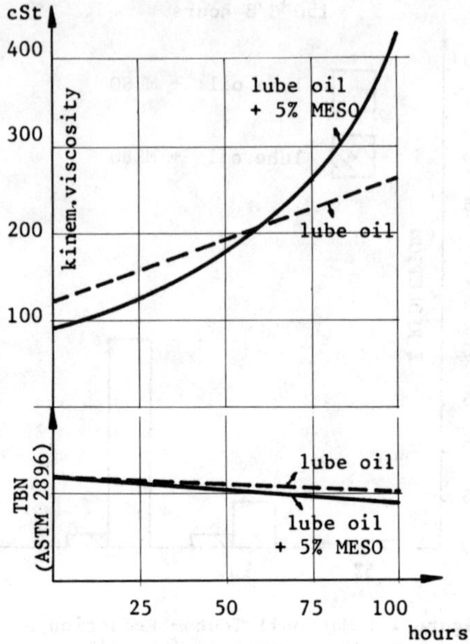

Figure 5: Variation of TBN and viscosity during bench testing.

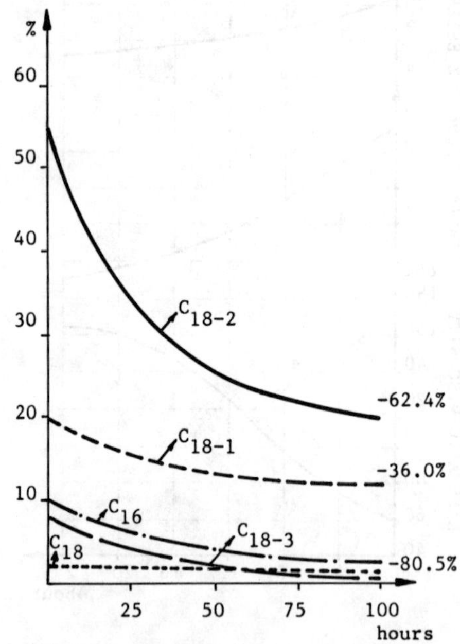

Figure 6: Variation of fatty acid ester composition during bench testing.

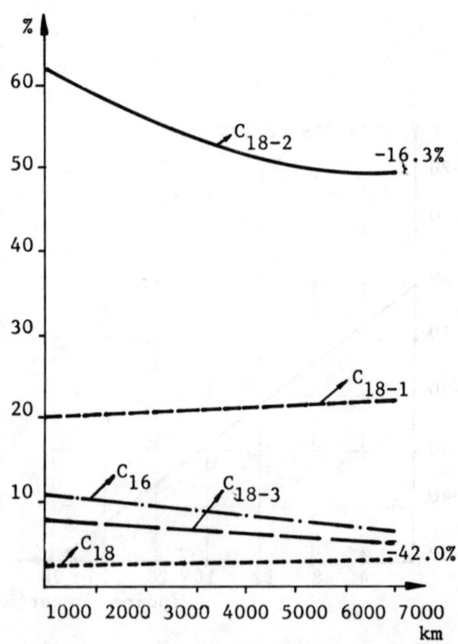

Figure 7: Variation of fatty acid ester composition during mixed-driving testing.

Number of cylinders	:	4 in line
Combustion process	:	swirl chamber
Displacement	:	1588 cm^3
Bore/Stroke	:	76.5 mm/86.4 mm
Compression ratio	:	23
Injection pump	:	rotary

Table 1: Data of the 1.6 liter IDI Diesel engine used for testing.

POLYMERIZATION OF VEGETABLE OILS

Roger A. Korus, Traci L. Mousetis, and Lawrence Lloyd

Vegetable oils are inherently less stable than commercial diesel fuels. The major chemical difference between vegetable oils and commercially acceptable diesel fuels is the high degree of unsaturation of the vegetable oils. Points of unsaturation are susceptible to oxidation, free radical formation, and the resultant oxidative polymerization leading to gum formation. Gum in a fuel will escape complete combustion as do heavier ends of diesel fuels. The resulting carbon residues contribute to deposits in the combustion chamber and around nozzle tips, coking on the piston rings, and carbon residue in the lubricating oil. The injection spray pattern is disrupted, the rings seize, and the crankcase oil viscosity increases. All these problems are common to the use of vegetable oils as a diesel fuel. Long term use of vegetable oils, especially under high engine load, is not possible unless these problems are eliminated.

Polymerization or gum formation occurs by an oxidative polymerization during storage and by a more complex oxidative and thermal polymerization at the high temperatures and pressures preceding combustion. In a compression ignition engine the conditions during and preceding combustion are very complex. The fuel-air mixture is very heterogeneous with regions of fuel alone and regions of fuel vapor, of air alone, and of vapor-air mixtures. Both oxidative and thermal polymerization can occur under these conditions. The performance of commercial fuels forms little carbon residue. However, the heavier ends of diesel fuel and fuels with a high gum content cause carbon residue formation similar to vegetable oils (Obert, 1973). To avoid this problem with diesel fuels, antioxidants and dispersants are added to prevent gum formation and maintain gum in solution.

LITERATURE REVIEW

Nearly all previous work on gum formation and deterioration of vegetable oils has been concerned with the drying oils used in paints, varnishes, and other coatings. For these applications highly unsaturated oils such as linseed, safflower, and soybean oils have been used. Linseed contains approximately 55% linolenic acid (18:3) and 15% linoleic acid (18:2). Safflower and soybean contain 55-85% linoleic acid. The mechanisms for gum formation by oxidative and thermal polymerization in drying oils have been summarized in several reviews (Formo 1979, Cowan 1979, Wexler 1966, Rheineck and Austin 1968, Wexler 1964). Oxidative polymerization occurs by the interaction of double bonds and oxygen to form peroxides. Conjugation of double bonds occurs to stabilize hydroperoxides. A free radical, chain reaction is inhibited as hydroperoxides decompose and the subsequent chain-growth polymerization reactions give high molecular weight crosslinked products.

Oxidative polymerization occurs by a free radical, hydroperoxide mechanism with a low activation energy. For most free radical polymerizations activa-

The authors are: Roger A. Korus, Associate Professor, Traci L. Mousetis, Graduate Assistant, and Lawrence Lloyd, Undergraduate Assistant, Chemical Engineering Department, University of Idaho, Moscow, Idaho 83843

tion energies for propogation are close to 5 kcal/mole and overall activation energies are 15-20 kcal/mole (Lenz 1967). Activation energies for thermal polymerizations are approximately twice that of free radical, chain-growth polymerizations (Rushman and Simpson 1955, Sims 1957). Thermal polymerization is negligible at temperatures below 250°C. So oxidative polymerization is the mechanism of storage deterioration. The mechanism of thermal polymerization has not been well characterized (Formo 1979, Cowan 1979, Wexler 1966, Rheineck and Austin 1968, Wexler 1964, Lenz 1967). From previous work it is impossible to predict the mechanism leading to gum formation at the high temperatures and pressures and heterogeneous conditions of diesel fuel combustion. The spray pattern, atomization, and droplet size will be affected by fuel viscosity and volatility with the more viscous and less volatile vegetable oils being more difficult to atomize. With vegetable oils conditions are favorable for gum formation and complete combustion is difficult to achieve.

A reduction in the rate of oxidative polymerization should also reduce the rate of thermal polymerization. So a vegetable oil selected for stability in storage would also reduce combustion problems. Thermal polymerization can occur by a Diels-Alder mechanism (Formo 1979) or a free radical mechanism (Cowan 1979), but both mechanisms show the same trend with vegetable oil combustion (Considine 1974). In both cases rates are increased as conjugation and unsaturation are increased. Although rates of oxidative polymerization increase with extent of conjugation and unsaturation, quantitative relationships have not been established. It has been shown that the rates of oxidation of films of triolein, (18:1), trilinolein (18:2), and trilinolenin (18:3) are 1:20:330 (Chipault et. al. 1951). Relative rates of oxidation of methyl oleate, methyl linoleate, and methyl linolenate were determined to be either in the order 1:12:25 or 1:2:4 (Wexler 1966, Kartha 1958). Decomposition of autoxized methyl esters of triglycerides and fatty acids has shown that oligomers as large as tetramers are formed as would be expected in a chain-growth polymerization reaction (Frankel et. al. 1960).

PROCEDURE

Vegetable oils were extracted with a CeCoCo expeller. The extracted oil was allowed to settle in 50 gallon drums and filtered through a three stage filtering system consisting of pre-filter, 20 μm filter, and 4-5 μm filter. Viscosity was measured with Cannon-Fenske viscometers in a water bath at 37.7°C. Fatty acid composition was determined by conversion of oils to methyl esters (Luddy et. al. 1968) and separation of esters in a 6 foot, 1/8 inch O.D. stainless steel column packed with 10% Silar 10C on 100/120 GAS CHROM Q at 180°C in a Varian 1400 gas chromatograph.

Thermal polymerization samples were sealed under vacuum in glass ampoules and heated in an electric furnace. Oxidative polymerization samples were heated in open glass beakers in a forced air convection oven. Lubrizol 560 and 565 were obtained from The Lubrizol Corporation. FOA 2 and FOA 3 were obtained from E. I. duPont de Nemours & Company, Petroleum Chemicals Division.

RESULTS AND DISCUSSION

Compositions of the vegetable oils used for determination of polymerization rates are shown in Table 1.

These vegetable oils had a wide range of vegetable oil composition ranging from the highly unsaturated linoleic varieties to the relatively saturated oleic and erucic varieties. The high linoleic acid safflower variety and the high oleic acid safflower variety represent the two extremes of unsaturation among the common vegetable oils resulting from alternation of a single gene. A comparison of these two safflower varieties demonstrates the effect of unsaturation without interference of other vegetable oil properties.

Table 1. Composition of Vegetable Oils

	Fatty Acids (%)					
	Palmitic 16:0	Oleic 18:1	Linoleic 18:2	Linolenic 18:3	Eicosenoic 20:1	Erucic 22:1
Rapeseed	3.01	18.56	14.84	7.08	10.46	45.08
Oleic Safflower	4.75	74.12	19.74			
Linoleic Safflower	5.87	8.84	83.76			

Relative viscosity measurements for the oxidative polymerization of safflower oil are shown in Figure 1. Rapeseed oil showed relative viscosity values similar to the high oleic safflower oil.

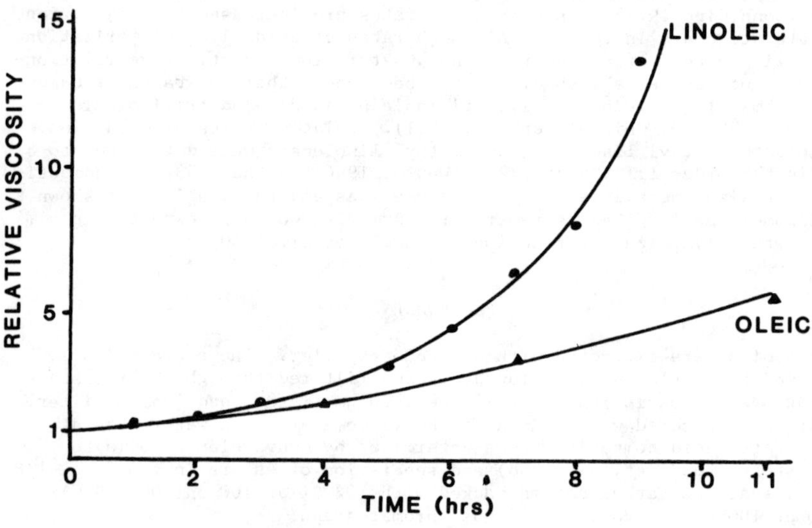

Fig. 1. Oxidative Polymerization of Safflower at 250°C.

Polymerization of rapeseed oil was measured at 240°C and compared to polymerization after the addition of fuel additives (Figure 2).

None of the additives tested reduced the rate of oxidative polymerization of rapeseed oil. Lubrizol 560 and 565 are multi-functional additives and were used at the maximum concentrations recommended for diesel fuels. FAO 3 is an antioxidant and was used at approximately three times the maximum concentration recommended for diesel fuels. All of these additives improve fuel stability in diesel fuels, but appeared to destabilize vegetable oils. FAO 2 is a poly(methacrylate) dispersant and had little effect on oxidative polymerization.

A comparison of oxidative vs. thermal polymerization of rapeseed oil is shown in Figure 3.

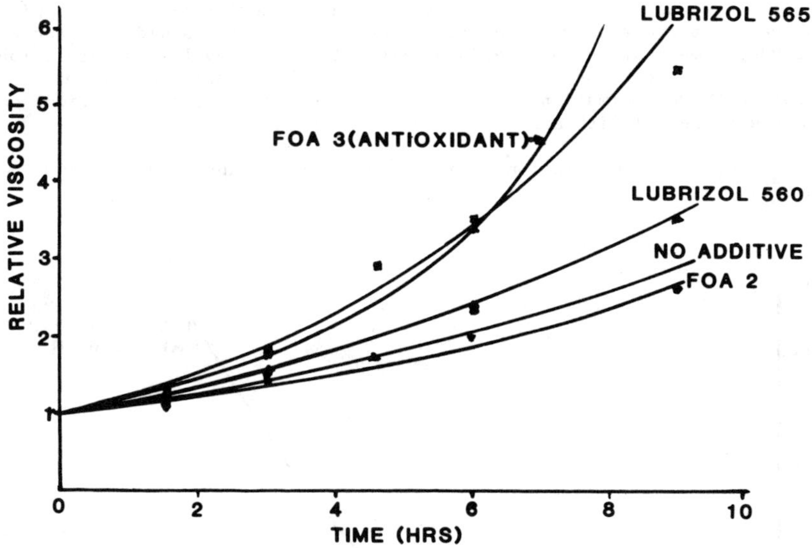

Fig. 2. Oxidative Polymerization of Rapeseed Oil With Diesel Fuel Additives at 240°C. The following additive concentrations were used: FAO 2 - 200 ppm, FAO 3 - 200 ppm, Lubrizol 560 - 75 ppm, and Lubrizol 565 - 3600 ppm.

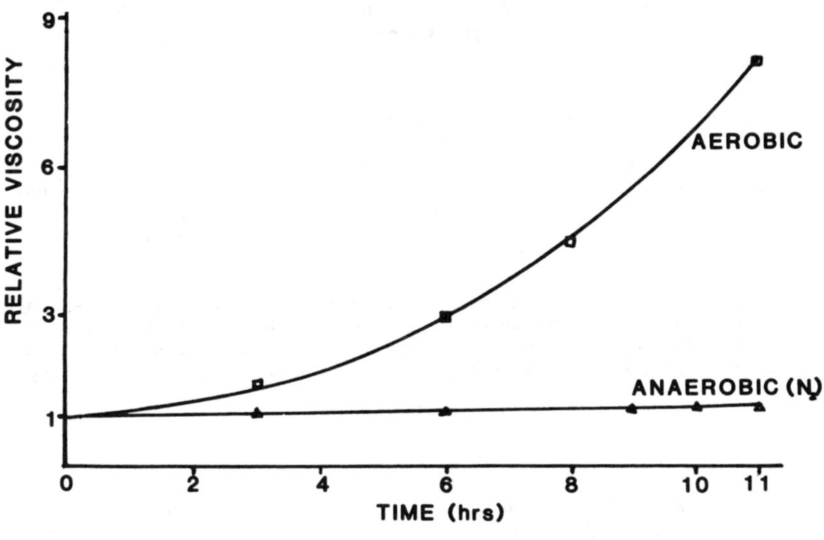

Fig. 3. Oxidative vs. Thermal Polymerization of Rapeseed Oil. Identical conditions were used for oxidative and for thermal measurements except the oven was continuously purged with nitrogen for the thermal (anaerobic) measurements.

At temperatures of 240°C and less the rate of thermal polymerization was negligible relative to oxidative polymerization. Therefore, oxidative polymerization has the greater potential for storage deterioration. The additives used to prevent storage deterioration in diesel fuels did not reduce oxidative polymerization in vegetable oils. The high level of unsaturation in vegetable oils overwhelmed any reasonable level of antioxidant. Oil deterioration by gum formation was reduced by anaerobic storage conditions and with a vegetable oil with a low level of unsaturation.

Rates of thermal polymerization became significant at approximately 300°C (Figure 4).

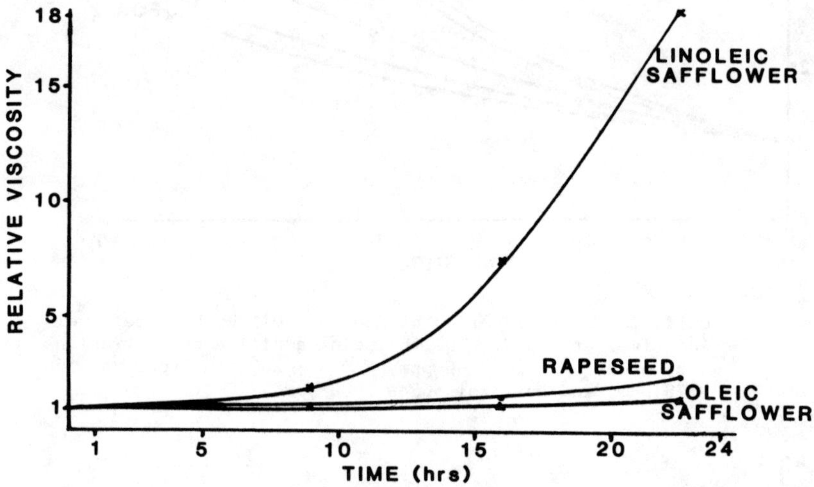

Fig. 4. Thermal Polymerization of Vegetable Oils at 300°C.

Rates of thermal polymerization showed a stronger dependence on degree of unsaturation than oxidative polymerization. Relative rates of oxidative polymerization of oleic and linoleic safflower were 1:3 (See Figure 1). Relative rates of thermal polymerization of oleic and linoleic safflower were approximately 1:30.

Thermal polymerization occuring by the Diels-Alder mechanism requires conjugated double bonds. Only linoleic and linolenic acids can isomerize to a conjugated double bond structure, so high linoleic safflower had a much higher rate of thermal polymerization than oleic safflower and high eruic rapeseed.

CONCLUSIONS

The addition of antioxidants and dispersants is not sufficient to eliminate gum formation in vegetable oils. Even with relatively unsaturated oils like rapeseed the extent of unsaturation overwhelms these additives. Fuel deterioration during storage will be minimized in an anaerobic storage environment and, to a lesser extent, with a lower degree of oil unsaturation.

Gum formation and carbon coking can also occur immediately preceding and during combustion. Thermal polymerization may be the dominant gum forming reaction under combustion conditions since thermal polymerization has a higher activation energy than oxidative polymerization and anaerobic conditions can occur within atomized fuel droplets. Carbon coking can be reduced with a lower degree of oil unsaturation and with better atomization of the fuel.

REFERENCES

1. Chipault, J.R., E.C. Nickell, and W.O. Lundberg. 1951. Official Digest Federation Soc. Paint Technol. 23:740.

2. Considine, D.M. (editor). 1974. Chemical and Process Technology Encyclopedia, pages 1129-1135.

3. Cowan, J.C. 1979. Encyclopedia of Chemical Technology (3rd Ed.), Volume 8. Wiley-Interscience. pages 130-150.

4. Formo, M.W. 1979. Bailey's Industrial Oil and Fat Products (4th Ed.), Volume 1. D. Swern (editor). Wiley-Interscience. pages 687-816.

5. Frankel, E.N., C.D. Evans, and J.C. Cowan, 1960. J. Am. Oil Chemists' Soc. 37:418.

6. Kartha, A.R.S. 1958. J. Sci. Ind. Res. 17B:135.

7. Lenz, R.W. 1967. Organic Chemistry of Synthetic High Polymers, Wiley-Interscience. pages 310-320.

8. Luddy, E.F., R.A. Barford, and R.W. Reimenshneider. 1968. J. Am. Oil Chemists' Soc. 45:549.

9. Obert, E.F. 1973. Internal Combustion Engines.

10. Rheineck, A.E. and R.O. Austin. 1968. Treatise on Coatings, Volume 1, Part 2. R.R. Myers and J.S. Long (editors). Marcel Dekker.

11. Rushman, D.F. and E.M.G. Simpson. 1955. Faraday Society Transactions. 51:230.

12. Sims, R.P.A. 1957. J. Am. Oil Chemists' Soc. 34:466.

13. Wexler, H. 1964. Chemical Reviews. 64:591.

14. Wexler, H. 1966. Encyclopedia of Polymer Science and Technology, Volume 5. Wiley-Interscience. pages 216-234.

FUEL ADDITIVES FOR VEGETABLE OIL-FUELED COMPRESSION IGNITION ENGINES

J.D.C. Baldwin
Assoc. Member ASAE

H. Klimkowski

M.A. Keesey

ABSTRACT

Three fuel additives were tested to determine their effect on fuel injector deposits in a 3-cylinder diesel engine when using a fuel blend containing 20% soy oil and 80% No.2 Diesel fuel oil. Fuel consumption was recorded and crankcase oil samples were analyzed for wear metals for comparison with results from a baseline test in which no fuel additive was used. Injector deposits were removed and their masses determined.

INTRODUCTION

A renewed interest in the use of vegetable oils as fuel substitutes or extenders began with re-evaluations of their performance in modern, unmodified compression ignition engines (Cochran et.al., 1980; Bruwer et.al., 1980). These and other studies have confirmed earlier findings that, when used in limited proportions, the addition of vegetable oil has little effect on short-term engine performance. Aside from any economic aspects, major drawbacks to the use of these fuels are increased viscosity and increased fuel injector deposits.

Although we have seen some reductions lately in world oil prices, the threat of an extended boycott remains a possibility. Being concerned about the need for an emergency fuel to ensure agricultural production during a period of fuel scarcity, our research efforts have been directed to improvements in the use of vegetable oil fuel extenders for emergency use with minimal fuel processing or engine modification. The possibility that a suitable fuel additive is currently available had to be considered. Therefore, while no problems have yet been attributed to the deposits found on injectors when using soy oil mixed, in limited amounts, with diesel fuel, the potential for reduced engine life, damage, or even engine failure has resulted in our study of fuel additives to reduce the formation of these deposits.

It is likely that deposit and viscosity associated problems cannot be solved entirely by the use of fuel additives. Minor alterations in engine operation may help, but major changes in engine design may be necessary to make vegetable oil fuel blends truly viable.

This paper attempts to address the deposit problem by presenting the results of tests using commercially available fuel additives to treat fuels containing 80% No.2 Diesel fuel oil and 20% refined soy oil.

The authors are: J.D.C.BALDWIN, Assistant Professor, H.KLIMKOWSKI, Research Associate, and M.A. KEESEY, Research Associate, Agricultural Engineering Dept., Louisiana State University.

INJECTOR DEPOSITS

In a compression ignition engine, combustion requires compression work to elevate the combustion air to sufficiently high temperatures to initiate combustion processes. These temperatures may be high enough to cause some de-hydrogenation of vegetable oil molecules, i.e., stripping off hydrogen molecules. If the ignition temperature of the resulting material is not achieved, smoke or soot will be produced. Similarly, if fuel is sprayed on the walls or piston, decomposition of the fuel may result in film-plating, and unburned oil may collect in the piston ring grooves. These accumulations may char or oxidize causing the rings to bind. Also, fuel residues on the injector tips may form and build with potential for clogging the openings or, at least, altering the fuel's spray pattern. As of this date, no such problems of this type have been encountered. Carbon deposits have been found to form within the injector itself, altering the spray pattern or even blocking the passage of fuel.

The higher viscosity of vegetable oil fuel blends contributes to these problems, as the effectiveness of fuel spray atomization and evaporation are reduced, leaving larger fuel particles with lower surface areas and greater probability of impinging on the walls and piston. Fuel droplets will then be either carried along, unburned, with the combustion products in the exhaust, or if thermally decomposed, appear as smoke particles.

Fuel additives may reduce injector deposits by several methods. For example, barium has been hypothesized to act as a de-hydrogenation inhibitor and tests have shown its beneficial effects as a smoke suppressant additive for diesel fuels (Miller, 1967).

FUEL ADDITIVE TESTS

A Ford 4600 tractor (3-cylinder, 201 CID) was connected to a portable, hydraulic brake dynamometer via the tractor's PTO (Power Take Off) shaft. Using a fuel blend containing, by volume, 20% soy oil (SBO) and 80% No. 2 Diesel fuel oil, a baseline for fuel consumption and injector deposits was established by operating the engine at its rated speed of 2200 RPM (656 RPM, PTO) and loaded to deliver 22 kW (30 hp). The engine was operated for 100 hours based on the tractor's hour-meter. At this speed, the metered time corresponds to an actual test period of 76 hours. Records were maintained on the volume of fuel consumed.

The injectors were removed at 50 hours, photographed, and the testing resumed. At the end of 100 hours the injectors were again removed and photographed. The deposits on each of the injectors were carefully collected and their masses determined. Samples of the crankcase oil were taken and analyzed for wear metal concentrations.

Three fuel additives were selected for testing. Additives A and B are often found in stores where diesel fuel additives are sold to the public. Additive C is a metal-based additive containing barium.

Each of the additives was added to the 20% SBO fuel blend according to the manufacturer's recommendations. Prior to each test the injectors were cleaned and adjusted. The tests were conducted in the same manner as in the baseline test.

RESULTS

The fuel additive dosage rate, average fuel consumption, injector deposit mass determinations, and crakcase oil analysis results are summarized in

Table 1. It should be noted that the tests with additive C were not for a full 100 hour test period because the tests were canceled after excessive deposits in one case and excessive smoking in the second.

Without the use of a fuel additive, an average of 20.7 mg of carbonaceous material was removed from each injector. The deposits, before removal, are shown in Fig. 1 and are representative of those found when using a 20% soy oil fuel blend in this engine.

Although the mass of the deposits is considerable, there has been no evidence from these tests showing that engine operation is impaired. The deposits are, for the most part, brittle and break easily. That portion of the deposits close to the injector surface, on the other hand, appears to be harder and are more difficult to remove.

Table 1. Fuel Additive Test Data Summary

Additive	Dose (mL/20L)	Fuel Consumption (L/h)	Inj. Deposit (mg) Cylinder #1	#2	#3	Wear Metal (ppm) Cu	Fe	Cr	Al	Si	Pb	Mo
---	--	6.13	17	23	22	3	20	2	5	5	7	1
A	62.5	5.80	20	25	26	2	21	2	9	6	10	-
B	93.8	5.68	15	20	24	2	20	1	10	5	10	-
C	100.0	6.13	1*	175	190	1	11	1	6	4	3	-
C	66.6	6.32	82**	132	100	1	9	1	4	7	2	1

* 50-hours
** 63-hours

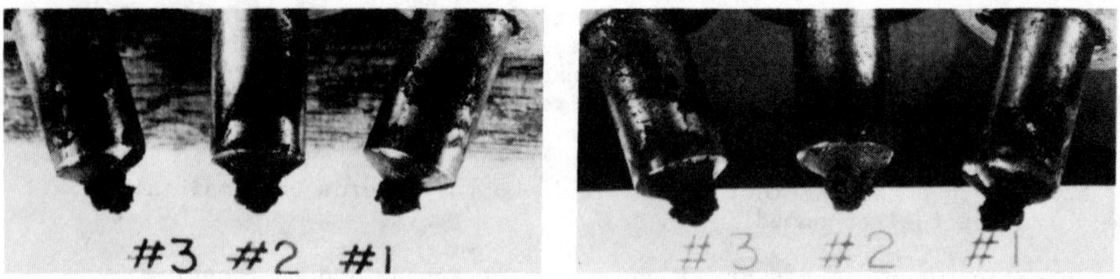

(a) 50 Hours (b) 100 Hours

Fig. 1. Injector Deposits with 20% SBO Fuel - Untreated

The injector deposits after 50 hours and 100 hours when using additives A and B are shown in Figs. 2 and 3. The average mass removed from each injector with either of these additives is comparable to, or greater than, those found when no additive is used: additive A, 23.7 mg; additive B, 26.3mg.

From the work by Miller (1967) it was our belief that additive C had the greatest probability of success because of its apparent success in reducing injector deposits in some standard diesel engines. The manufacturer of this additive, however, could not provide a dosage recommendation for our particular application but indicated that a 0.25 - 0.50% concentration was the general range when used with diesel fuel. Because the deposits produced when using vegetable oil fuel blends are greater than those that form when using

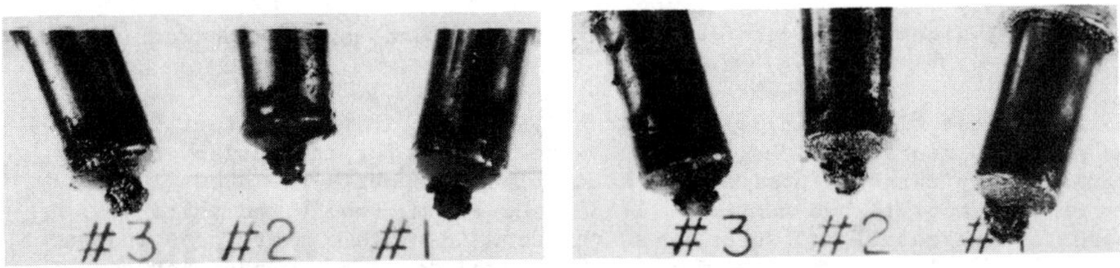

(a) 50 Hours (b) 100 Hours

Fig. 2. Injector Deposits with 20% SBO Fuel + Additive A

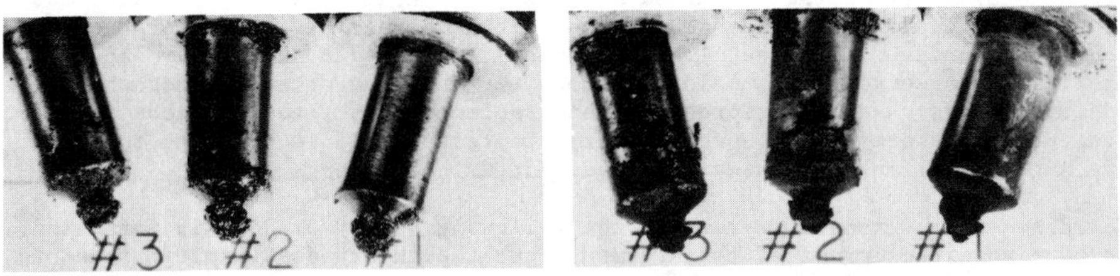

(a) 50 Hours (b) 100 Hours

Fig. 3. Injector Deposits with 20% SBO Fuel + Additive B

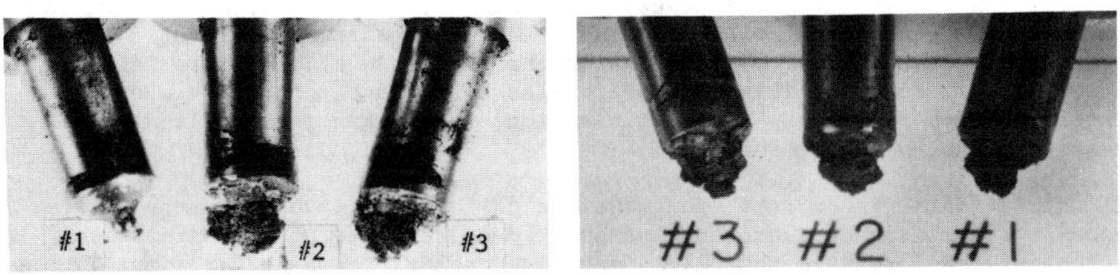

(a) 50 Hours, 100 mL/20L (b) 62.5 Hours, 66.6 mL/20L

Fig. 4. Injector Deposits with 20% SBO Fuel + Additive C

diesel fuel alone, we anticipated that a higher than normal treatment would be necessary.

The first test with additive C was at a 0.5% dosage rate. The test was stopped after 50 hours when the injectors were removed for inspection and significantly larger deposits were found. These deposits are shown in Fig.4a. Two of the deposits had masses of 175 mg and 190 mg, while the third was found to have only 1 mg. We observed the deposits with a microscope and were able to see a large quantity of whitish-colored deposits in addition to the normal carbonaceous material. The manufacturer suggested that the substance we observed was barium salts and that this was a common observation when too much additive is present. We presume that the lack of material on the No.1 injector had fallen off during the test, prior to the 50-hour inspection.

A second test with additive C at a 0.33% rate was made. The test was stopped at 62.5 hours because the engine had begun to smoke excessively. The deposits on the injectors are shown in Fig. 4b. The average mass was found to be 104.7 mg.

From the analyses of the crankcase oil samples taken at the end of each test, as shown in Table 1, there appears to be little effect of the use of these additives on the wear rates, i.e., no drastic changes in any metal concentration was observed. It should be recognized that the data obtained when additive C was used only represents results from shorter test periods so the concentrations are lower. Special wear rate tests would be required to evaluate the effects of these additives for extended usage.

For discussion purposes we suggest the following theory of deposit formation. We have no direct proof of the argument, but the theory does explain some of the observations made during this study.

At some point in the injection process, fuel wets the surface of the injector. While some of the fuel's more volatile components evaporate, the remaining fuel is exposed to the oxygen present and higher temperatures ititiate complex polymerization and partial oxidation, i.e., a drying effect, perhaps accelerated, by a catalytic effect of the injector surface. This film exhibits a highly adhesive character at this stage of deposit formation. In the combustion gases incomplete combustion of the fuel or thermal decomposition results in free, carbon-rich molecules or particles. Some of these particles impact the surface film on the injector and adhere to it or become trapped within the polymerizing material. As the process is repeated, injection after injection, in the initial stages, the initial material hardens and acts as an insulating material, reducing the temperature of the outer surface and reducing its adhesive character. Additional material accumulates around the injector openings as a result of dribbling when the injector valve closes. Subsequent injections force this material radially away from the opening.

If this scenario is correct, then the use of fuel additives with the intention of inhibiting de-hydrogenation may reduce the rate of polymerization, which may then preserve the film's adhesive quality for a longer time. Thus, a greater amount of mass could be captured by this surface. This would help to explain why the first two additives tested had little effect on deposits while the third additive, which should have proved effect, resulted in the greatest accumulation of material.

SUMMARY

Of the 3 additives selected for study, none provided an overall reduction in injector deposits, and instead, resulted in greater deposits. No adverse effects of the additives was apparent from examination of the wear metal analyses of the crankcase oil at the end of each test. Also, while the use of the additives indicated a fuel savings in some cases, the testing procedure at part load and constant speed are not typical of normal operation and any general conclusions drawn from the data would be misleading.

ACKNOWLEDGEMENTS

The authors wish to express their appreciation of the support for this project by the Louisiana Agricultural Experiment Station, CASRD, and Louisiana State University.

REFERENCES

1. Bruwer, J.J., B.v.D. Boshoff, F.J.C. Hugo, J. Fuls, C. Hawkins, A.N.v.d. Walt, and A. Engelbrecht. The utilization of sunflower seed oil for diesel engines. Agricultural Energy. 1980. 2:385-390.

2. Cochran, B.J., J.D.C. Baldwin, and L.R. Daniel. "DIESOL" - an alternative fuel for compression ignition engines. Agricultural Energy. 1980. 2:391-396.

3. Miller, C.O. Diesel smoke suppression by fuel additive treatment. SAE Paper - 670093. 12 p.

ATTEMPS TO PREVENT INJECTOR COKING WITH SUNFLOWER OIL BY ENGINE MODIFICATIONS AND FUEL ADDITIVES

A N van der Walt* and F J C Hugo*

ABSTRACT

The effect of injector tip temperature on coking propencity when sunflower oil is used as a fuel for direct injection engines, was tested. Partial retraction of the injector, the addition of a heat shield to the injector and cooling the injector with water was tried. Also, injector temperature was increased by reducing heat transferred to the cylinder head and preheating the sunflower oil. None of these measures could prevent coking of the injector tip. Coating the injector tip with Teflon and increasing the back leakage rate was also tried without success. Only a few of many additives tested, showed some promise of being able to prevent coking.

INTRODUCTION

When direct injection engines are run with neat sunflower oil as fuel, injectors will invariable coke up after 40 to 50 hours (Bruwer et al, 1980 and Galloway and Ward, 1980). Coking of the injectors leads to poor fuel atomization, less efficient fuel combustion, and dilution of lubricating oil by partially burnt sunflower oil which runs down the cylinder walls. Dilution of crankcase oil with sunflower oil leads to polymerization in the presence of oxygen and high engine temperatures. Dilution of the crankcase oil is enhanced at low loads and during idling of the engine. Even continuous running at 70% maximum power led to polymerization of the lubricating oil in a Perkins 4.248 engine.

Since injector coking is one main factor contributing to premature failure of direct injection engines running on neat sunflower oil, prevention of injector coking is one of the main concerns of researchers in the field of the application of vegetable oil as a diesel fuel replacement. Ethyl Ester of Sunflower oil proved to be one possible solution to the problem (Bruwer et al, 1980). Another possibility would be to manipulate the injector temperature to prevent coking from taking place. Since either too high an injector temperature or too low a temperature could be responsible for coking, the effect of both cooling and heating of the injector was tested. Also, if some coating for the injector tip could be found which prevented coke from sticking to the injector tip, coke would break off the injector tip as soon as it was formed. A Teflon coating was tested. Fuel additives might also play a significant part in the solution of the coking problem. Commercially available fuel additives were tried.

ENGINE MODIFICATIONS

Reduction Of Injector Tip Temperature

Retraction: If a smaller area of the injector was exposed to the flame during combustion it could be reasoned that the injector temperature should be lower. The quickest and easiest way to reduce injector tip area exposed to the flame is to re-

*Division of Agricultural Engineering, P/bag X515, Silverton, Republic of South Africa.

tract the injector tip from the combustion chamber, by inserting an extra washer on the sealing surface. Insertion of additional copper washers between injector body and cylinder head, resulted in a retraction of about 4 mm. This was the maximum possible retraction without seriously affecting the spray of the fuel in the combustion chamber. Two injectors on a Perkins 4.248 direct injection engine were retracted in this way. The engine was then run on a mixture of 60% sunflower oil, 40% illuminating paraffin at 70% maximum power for a period of 73 hours. No positive reduction of coking could be noticed on the retracted injector tips and the test was terminated.

Heat Shield: A heat shield, formed like a cylinder with one conical closed end and which could slide over the injector tip, was constructed.

Fig. 1. Fuel Injector And Heat Shield

Four holes were drilled in the conical end to correspond with the holes in the injector. The four holes in the heat shield were big enough not to interfere with the fuel spray. One such a heat shield was fitted to the injector of a Massey Ferguson 240 tractor. The tractor was then run on a mixture of 20% sunflower and 80% diesel fuel at 75% maximum power for a period of 13 hours. Removal of the injectors did not show a substantial reduction of coking from the use of the heat shield.

Cooling: A special shroud which permitted water cooling of one injector on a Massey Ferguson 240 tractor was constructed. A slot, running up one side of the injector, round the tip and back again down the other side, was ground to serve as a conduit for carrying the cooling water. A copper tube was brazed onto the feed end of the slot. A heat shield plus a length of tubing which could slide over the injector tip, formed a shroud round the injector and sealed off the open side of the slot. (Fig. 2)

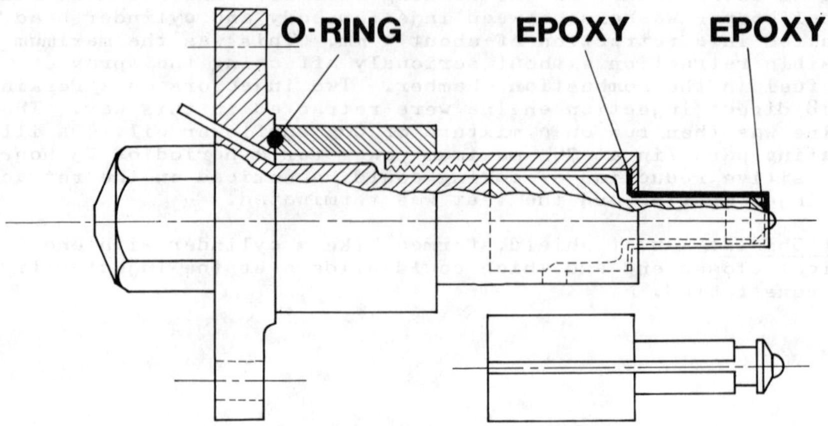

Fig. 2. Modified Injector For Water Cooling

Cold tap water was forced through the cooling shroud of the injector while the tractor ran on 20% sunflower, 80% diesel at 75% maximum power for a period of 13 hours. On removal, the cooled injector had less coke deposits on it than the unmodified injectors, but since coking could not be completely eliminated, it was decided that water cooling of the injector didn't produce a significant enough effect. This same test was repeated by running the tractor on neat sunflower oil. Again, water cooling of the injector did not prove to be a solution to the problem.

Teflon Coated Injector Tip: One injector was coated with Teflon at the tip, leaving the spray holes clear. This injector was fitted to a Massey Ferguson 240 tractor running on a 20% sunflower oil, 80% diesel blend. The test was started off with 3 hours of low idling. Inspection of injectors after this period showed no difference between the Teflon coated injector and the control injectors. When the load on the engine was increased, the higher operating temperatures caused the Teflon coating to come off the injector tip. It is suspected that the Teflon could not withstand the high surface temperature.

Increased Back Leakage Rate: By increasing injector needle clearance, back leakage rates could be changed. Back leakage was tested by applying a pressure of more than 150 bar, to the injector in an injector tester and noting the time it took to reduce the pressure gauge reading from 150 bar to 100 bar through back leakage. Three injectors were modified to achieve back leakage rates of respectively, 45 seconds, 30 seconds and 5 seconds on a 20% sunflower oil, 80% diesel blend. These injectors were fitted to a Massey Ferguson 240 tractor which was run at 80% maximum power for a period of 15 hours. On removal of the injectors it was found that there was very little difference in coking deposits between the 3 injector tips.

Increased Injector Tip Temperature: By replacing the copper washer with a fibre washer, heat transfer between injector body and cylinder head could be reduced. Further, by pre-heating the fuel line with exhaust gases, the cooling effect of the fuel on the injector was minimized. The injector tip temperature was increased by this method on one injector of a Massey Ferguson 240. The tractor was then run on a 20% sunflower, 80% diesel blend for 5 hours: 30 minutes at 70% maximum power followed by 30 minutes maximum power followed again by 4 hours of 70% maximum power. This treatment caused increased coking of the injector tip.

FUEL ADDITIVES

Quite a large number of commercially available fuel additives were tried in different tests to assess their ability to prevent injector coking in a direct-injection engine. The names of the additives tested, the fuel used, particulars of the test and the results are summarized in table 1.

One problem encountered in this research programme was to get a quantitative method for judging the amount of coke which had formed on the injector tip. Coking cannot be judged on a mass basis, since both the exact locality and the physical form of the coke formation must be taken into account. In most cases, the amount of coking was compared on the one hand to the coke which could be expected on an injector which was run on diesel under identical conditions, and on the other hand to the coke which could be expected on an injector which was run on neat sunflower oil. When the coke formation at the conclusion of a test was more or less comparable to that of neat sunflower oil (fig. 3), it was judged that no effect could be produced. When the coke formation was more or less comparable to that of diesel, or that a drastic reduction in coking had taken place, it was judged that the treatment had produced a promising effect worth further investigation.

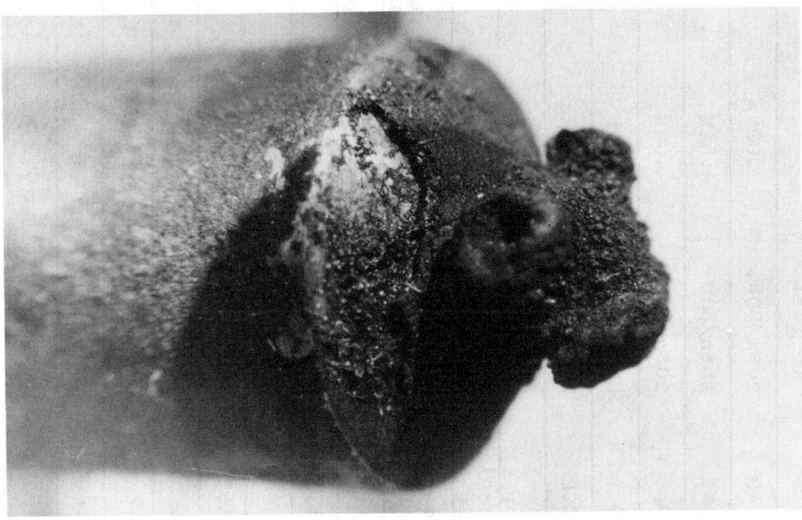

Fig. 3. Typical Injector Coking Produced by Neat Sunflower Oil

Table 1. Effect Of Different Additives On Injector Coking with Sunflower Oil

Name of Additive	Fuel used in test	% Additive	Test method	Result
Du Pont FOA2	20Sn/80D	.004%	MF 240 100 hrs at 80% Pmax	Coked injectors
Du Pont FOA2	20Sn/80D	.02%	MF 240 24 hrs at 80% Pmax	Coked injectors
Ethyl MPAD	20Sn/80D	.04%	MF 240 33 hrs at 80% Pmax	Coked injectors
Ethyl MPAD	20Sn/80D	.16%	MF 240 38 hrs at 80% Pmax	Coked injectors
Dieselex CC2	20Sn/80D	0.2%	Ford 7000 engine 24 hrs at 70% Pmax	Coked injectors
Wynn's Diesel Conditioner	20Sn/80D	5%	MF 275 Std 5 hr test*	Coked injectors
Wynn's Diesel Conditioner	Sn 100%	5%	MF 275 Std 5 hr test*	Promising results
Wynn's Diesel Conditioner	Sn 100%	5%	MF 275 1/2 hr Pmax 9 1/2 hr 70% Pmax	Coked injectors Phase separation problem
PSP Conditioner	Sn 100%	2%	MF 240 Std 5 hr test*	Coked injectors
PSP Conditioner	20Sn/80D	2%	MF 275 Std 5 hr test*	Coked injectors
Lubrizol 565	20Sn/80D	1%	MF 275 Std 5 hr test*	Coked injectors but exhaust pipe white
Lubrizol 565	Sn 100%	1%	MF 275 Std 5 hr test*	Promising reduction of coking Phase separation problem
Vasconol U	20Sn/80D	2%	MF 275 Std 5 hr test*	Coked injectors
Dieselol 166	20Sn/80D	0.1%	MF 275 Std 5 hr test*	Promising reduction of coking
Dieselol 166	20Sn/80D	1%	MF 275 Std 5 hr test*	Coked injectors
OFA 425 B	20Sn/80D	0.23%	MF 275 Std 5 hr test	Coked injectors

Table 1. (Continued)

Name of Additive	Fuel used in test	% Additive	Test method	Result
Lubrizol 8256A	20Sn/80D	1%	MF 275 Std 5 hr test	Coked injectors
Lubrizol 58	20Sn/80D	1%	MF 275 Std 5 hr test	Very promising reduction of coking
Frank Fehr FF4A	20Sn/80D	0.05%	MF 275 Std 5 hr test*	Very promising reduction of coking Phase seperation problem
Frank Fehr FF3	20Sn/80D	0.3%	MF 275 Std 5 hr test*	Promising reduction of coking
Carburol Autol Diesolite K	Sn 100%	0.25%	MF 275 Std 5 hr test*	Very promising reduction of coking No phase seperation problem
Vasconol U	20Sn/80D	0.2%	MF 275 Std 5 hr test*	Promising reduction of coking
Lubrizol 2153	20Sn/80D	1.4%	MF 275 Std 5 hr test*	Coked injectors
Fire Hart TD35	Sn 100%	0.2%	MF 275 Std 5 hr test*	Promising reduction of coking
Lubrizol 8005	Sn 100%	1%	MF 275 Std 5 hr test* without fuel filter	Very promising reduction of coking Phase seperation problem
Dieselol 167/2	Sn 100%	0.1%	MF 275 Std 5 hr test*	Very promising reduction of coking

Notes:

*Std 5 hr test : 1/2 hr at 70% max. power, 1/2 hr max. power, 4 hrs 70% max. power.

20 Sn/80D : 20% degummed Sunflower oil, 80% Diesel blend.

Sn 100% : Neat degummed Sunflower Oil as fuel.

Most of the tests were done on either a Massey Ferguson 240 or a Massey Ferguson 275 tractor. Since engine load conditions for practical farm work vary between very light loads and maximum power, any laboratory load that will produce coking on the injectors within a short period points to an unsatisfactory fuel. Initially, it was believed that a constant load of about 70% maximum power would be the quickest way to show up any coking problem. Practical experience with a Ford 7000 (without turbo charger) where the tractor was able to complete 1 000 hours' heavy farm work, but only a further 300 hours in a laboratory at 70% maximum power (Bruwer et al, 1980) pointed to this fact. Subsequent experience with a Massey Ferguson 240 proved that a significant, unacceptable amount of coking can be produced on injector tips within as little as 5 hours by running the engine on sunflower oil for 30 minutes at 70% maximum power, followed by 30 minutes at maximum power followed by 4 hours at 70% maximum power. This test procedure was subsequently adopted to facilitate a quick assessment of any additive. Any additive that could significantly reduce coking in this test, was considered promising enough to be evaluated on a cyclic load for a longer period of time. By noting power and smoke readings at eight hour intervals, coking of the injectors can be identified through power loss and increased smoke readings in a long-term test.

In the case of Lubrizol 8005, a severe phase seperation problem manifested itself after mixing of the additive and the degummed sunflower oil. This again resulted in severe fuel filter clogging. When the fuel filter was bypassed, a very promising reduction of coking on the injectors was noted. (Fig. 4)

Fig. 4. Very Promising Reduction of Coking Produced by Lubrizol 8005

It was felt however, that until phase seperation problems can be solved, those additives that do pose this kind of problem cannot present a practical solution.

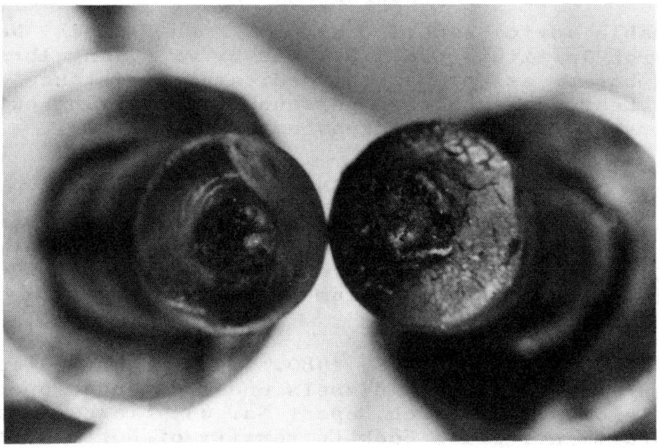

Fig. 5. Very Promising Reduction of Coking by Dieselol 167/2 (right) compared to that of Diesel itself (left)

The most promising additive tested to date was Carburol Autol Diesolite K. Its use resulted in a very marked decrease in coking, spray holes were unobstructed and no phase seperation problem was encountered. In a preliminary Perkins cyclic load test, the very promising results obtained in the 5 hour test were confirmed over a 24 hour period. A longer duration cyclic load test will now be undertaken.

CONCLUSIONS

Minor mechanical modifications to achieve reduction of the injector tip temperature by retraction, a fitted heat shield or water cooling did not prevent injectors from coking up, although the water cooled injector indicated a reduction of coking. Increasing the injector temperature increased the coke build-up.

Teflon does not seem to be a suitable material to prevent coke from sticking to the injector tip. Also, it does not seem to be able to withstand high engine temperatures. Some other materials which can be easily coated onto injector tips and which will be able to prevent coke adhesion and withstand engine temperatures, should be tested.

Increased injector back leakage rates could not solve the coking problem.

Only a few of the many additives tested showed some promise of being able to reduce coking. More work in this field will be done. The co-operation of companies supplying additives is of great importance, both in improving existing additives and in designing new additives specially for plant oil fuels.

ACKNOWLEDGEMENTS

The invaluable advice supplied by Dr L M du Plessis, National Food Research Institute C.S.I.R., and by mr A. Engelbrecht, on a personal basis, as well as the dedicated assistance of the Division's Learner Technician personnel in carrying out the engine tests is gratefully acknowledged.

REFERENCES

Bruwer, J.J., Boshoff, B.v.D., Hugo, F.J.C., du Plessis, L.M., Fuls, J., Hawkins, C., van der Walt, A.N., Engelbrecht, A. 1980. Sunflower Seed Oil as an extender for diesel fuel in agricultural tractors. Symposium of South African Institute of Agricultural Engineers.

Galloway, D.J., and Ward, J.F. 1980. Comprehensive testing of modern compression-ignition diesels under extended operations on vegetable oils. Research Report No. 64, Stage 1 and Stage 2, Physics Department, James Cook University of North Queensland, Australia.

INJECTOR-FOULING PROPENSITY OF CERTAIN VEGETABLE OILS AND DERIVATIVES AS FUELS FOR DIESEL ENGINES

Graeme R. Quick[1] Brian T. Wilson[2] Peter J. Woodmore[3]

ABSTRACT

The results of experiments specifically designed to study the propensity for certain vegetable oil candidates as fuels to foul the injectors of representative farm-type diesel engines are presented. The search for an expedient means of reproducing the injector-coking problem led to choice of linseed oil which has produced drastic effects in DI engines, tending to foul the injectors and cause ring-sticking in less than ten hours. Short and long-term engine test data are presented and the authors' current experiences in use of vegetable oils or their derivatives as diesel engine fuels are outlined. Viscosity is no longer indicted as the sole cause of coking. Esters are reconfirmed as good fuel candidates.

INTRODUCTION

This presentation is an extension of a paper presented at the 1980 Winter Meeting of the ASAE in which the situation in regard to the use of vegetable oils was reviewed and areas for further study were considered (Quick, 1980). In summary, vegetable oils were considered promising candidates as a fuel replacement or extender for diesel engines. Good short-term engine performance was overshadowed however by problems with injector coking, and ring-sticking in longer term operation in unmodified direct injection diesel engines. Means for overcoming the coking problems were discussed. This paper carries the subject a little further by identifying a means of rapidly exaggerating the injector-fouling problem.

Information is beginning to accumulate to show that indirect injected engines may be better able to cope with vegetable oils used as fuels (Bruwer, 1981). However, the great majority of diesel engines used in the rural community are direct injected.

COKING PHENOMENA

Compression-ignition engines have been tested on over 40 different 'bio-oils' since the first tests of such fuels were reported in 1900, and short-term performance has been very good on these candidate fuels. Sustained operation on vegetable oils as straight fuel replacements in unmodified direct injection engines eventually causes injector fouling which in turn leads to a train of events that could result in premature engine failure. For example, a Ford 2711E four cylinder direct injection engine was operated on straight crude degummed sunflower oil driving an industrial air compressor and reached only 87 hours before the engine had to be shut down due to misfiring, heavy

[1]Director of Agricultural Engineering, [2]Senior Technical Officer and [3]Technical Officer at the Agricultural Engineering Centre, Glenfield. N.S.W.

smoking and unsatisfactory performance. Examination of the dismantled engine showed all injectors to be seriously fouled, combustion chambers and ports heavily loaded with deposits, top rings stuck and the crankcase oil severely contaminated and partly gelled. It was surmised that the injectors had begun to foul first, interfering with injection, and probably causing 'overspray'. The subsequent fuel washing of the cylinder walls is then believed to have led to the ring sticking and lubricant contamination. This engine was reconditioned and then operated on a blend of 5 micron-filtered sunflower oil (20%) and distillate (80%). 1,000 hours of operation with the normal oil change intervals were then successfully completed without any major deleterious effects on the engine, although deposits were accumulating.

The use of blends with distillate is just one of a number of possible solutions to the coking problem when vegetable oils are used in diesel engines. These solutions fall under the usual two categories:

(1) Fuel Modifications: Blends, esterification, refinery cracking and possibly the use of additives.

(2) Engine Modifications: Dual-fuelling, fuel heating, injection system changes or other even more extensive engine modifications such as conversion to indirect injection.

The work reported here was based on the observations of the previous paper that: -

(1) Bosch smoke readings were lower for engines running on vegetable oils or esters than on diesel distillate in short-term tests; see for example, Table 1.

(2) There was a wide range in iodine number between different vegetable oils, see Figure 1, whereas the viscosity variation is minor by comparison. Iodine number relates to the reactivity of the vegetable oils, coconut oil being very low on the scale and linseed oil 17 times higher than coconut oil.

Table 1. Relative Performance of Kubota L 245DT 3 Cylinder NA, IDI Diesel Tractor Engine on Diesel Fuel and on Four Crude Degummed Vegetable Oils in Short-Term Dynamometer Tests

	Diesel Fuel	Sunflower	Rapeseed	Safflower	Linseed
Peak Power	100*	102	102	104	107
Maximum Torque	100	101	103	103	105
Bosch Smoke at Max. Torque	100	73	67	78	83

* Figures are rounded percentages of averaged test data, relative to diesel fuel

VISCOSITY

With the exception of castor oil, the viscosity range of the candidate oilseed fuels in Australia is narrow. Thus any viscosity-related problem with vegetable oils would be expected to be common to them all. Viscosity lowering by blending with distillate or by esterification has been shown to postpone

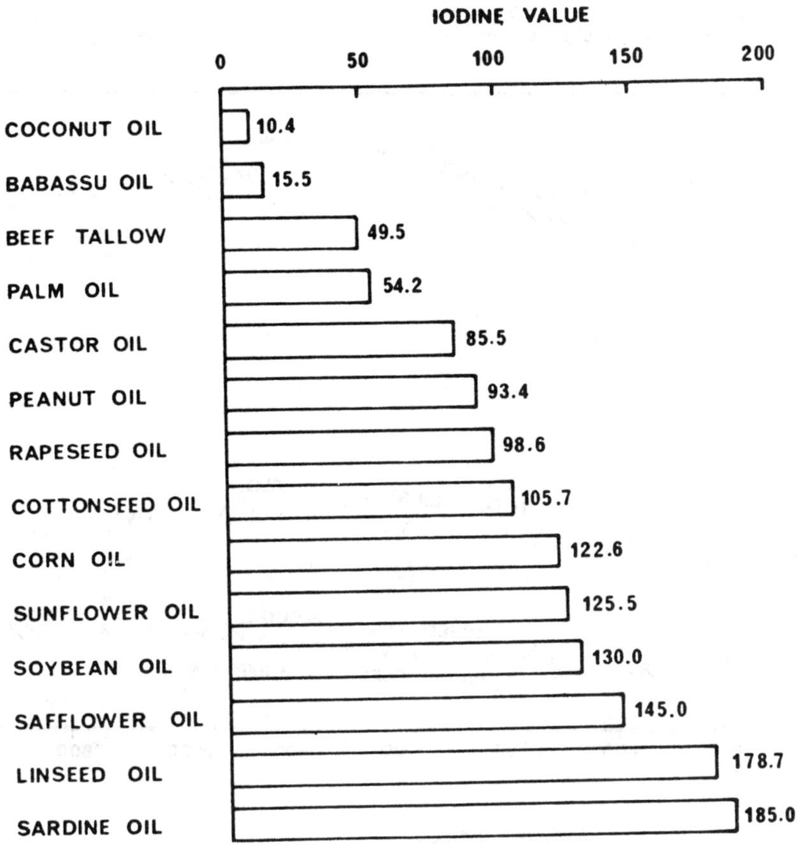

Fig. 1. Iodine Number Ranking for a Range of Bio-Oils. Iodine Number Relates to the Tendency of a Thin Film of the Oil to Form a Skin. (Major Source of Data in Graph, Swern, 1979).

coking problems (Siekmann & Pischinger, 1981). Heating the oils also lowers viscosity. Engine fuel heaters using coolant as the heating medium are commercially available and these could be readily employed to preheat oilseed fuels.

There has been some variability in engine hours accumulated before the onset of serious problems with different direct-injected engines (Hugo, 1982). Variability in short-term performance of diesel engines on these higher-viscosity fuels may be largely accounted for by the response of the injection pump/injector system. While some systems have in-built viscosity compensation, others may be sensitive to viscosity changes.

At Glenfield we have utilized a number of engines but more recently concentrated on a Lombardini Model 720 air cooled single cylinder direct injected engine. In Figure 2 the fuel demand characteristics for the Lombardini running on distillate and on linseed oil are compared. The engine exhibited some tendency to hunting or speed deviation on linseed oil but not on distillate. When operated on linseed oil heated to an average 85°C and compared with performance running on oil at the ambient temperature (average

20°C), there was no apparent effect on performance between the two fuels, nor did fuel heating to this extent make any difference to coking propensity.

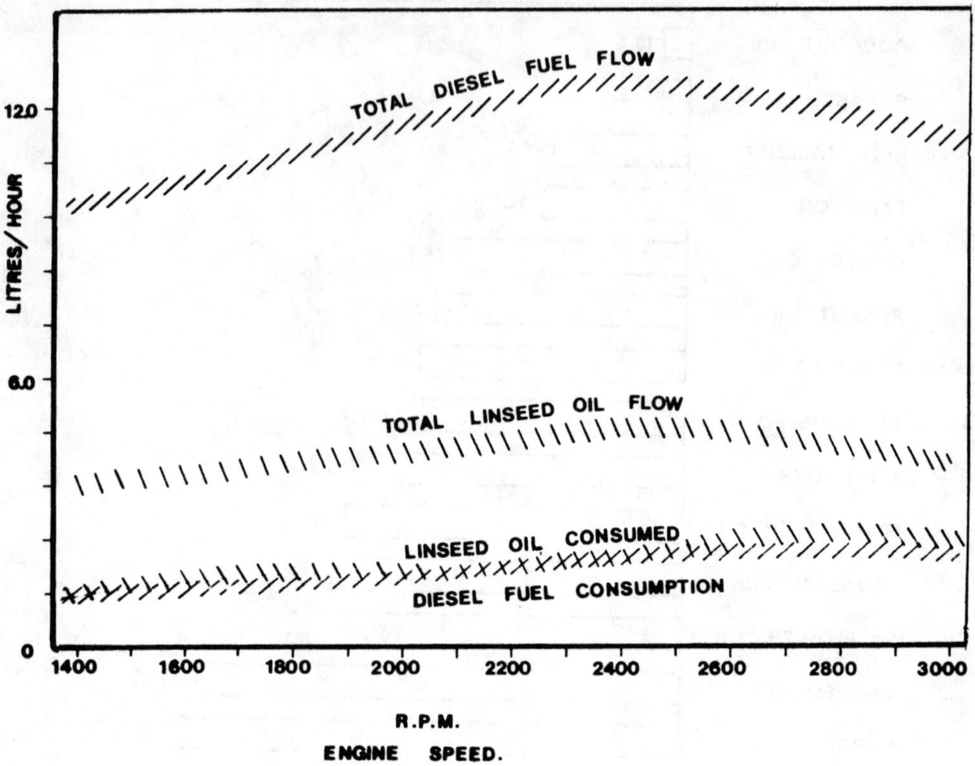

Fig. 2. Lombardini 720 NA, DI, Engine Fuel System Characteristics. Note Considerable Reduction in Total Fuel Flow on the Linseed Oil which at 40°C has a Viscosity of 7.5 Times That of Distillate. Total Fuel Flow was Actual Flow to the Fuel Pump, i.e. Fuel Consumption plus Total Bypass Flow.

'WORST OIL'

Linseed oil was selected as the candidate 'worst' oil to obtain baseline data as a standard of comparison for future work on injector fouling. The characteristic full governor characteristic curves for the Lombardini 720 engine runningon raw linseed oil filtered to 5 microns are shown in Figure 3. Linseed oil is the most readily available high iodine value bio-oil with characteristics (viscosity, energy density etc.) similar to the prime candidate oilseed fuels in Australia. The assumption was that if there was some additive or fuel modification anticipated to reduce injector fouling, it would become most readily apparent if it was evaluated first against or blended with linseed oil. A duration test was conducted on the Lombardini 720 engine powering a M & W dynamometer through a 3:1 reduction drive, and with the engine instrumented for torque, speed, exhaust gas temperature, fuel temperature and fuel flow (total and by-pass). In each duration test, which was run at constant speed, the engine was shut down after the deterioration in performance was deemed sufficient, usually at 50% to 80% of original power.

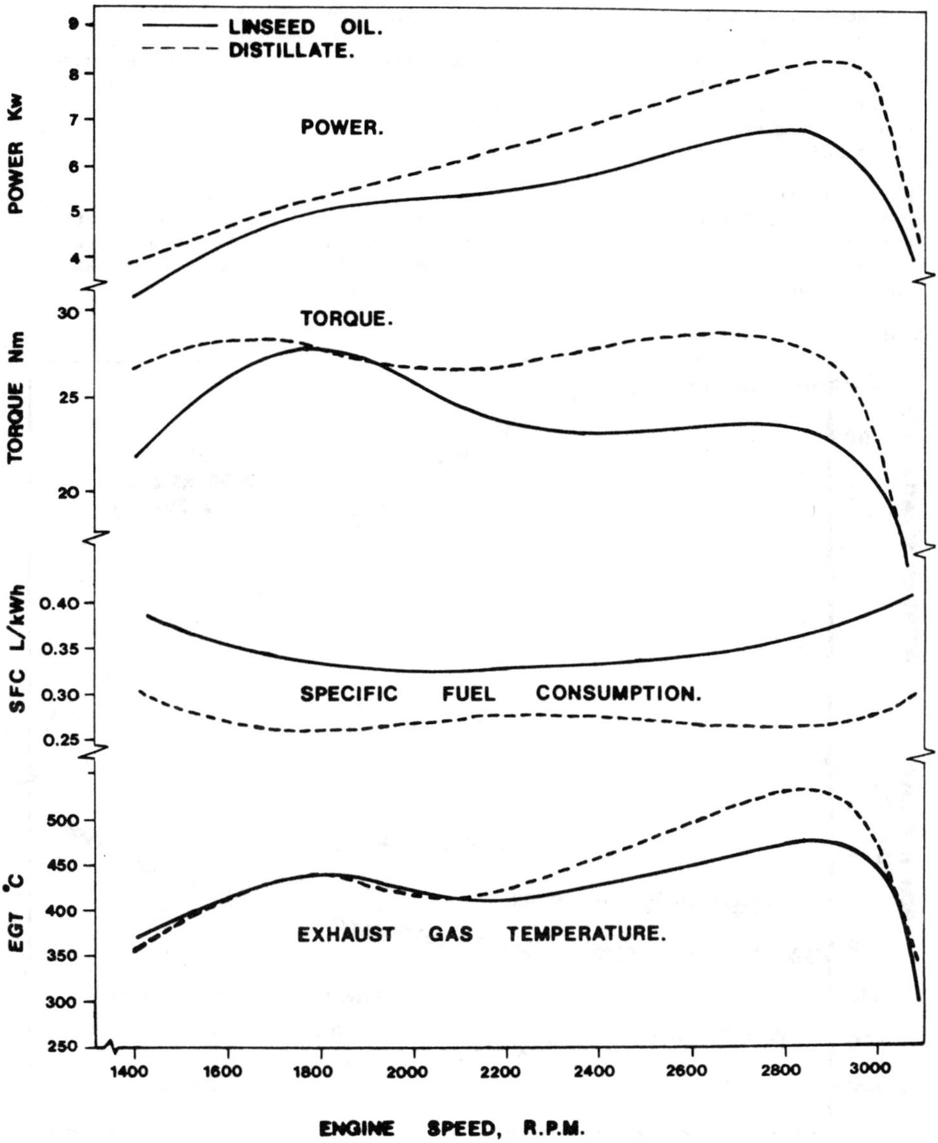

Fig. 3. Engine Characteristics on Distillate and on Filtered Linseed Oil Both at Ambient Temperature. Lombardini Model 720 NA, DI Single Cylinder Air-Cooled Diesel Engine.

See Figure 4 for example. This Model 720 lost 10% power after 7 hours operation and was down to below half start-up power after only 16 hours when operated at or near maximum torque setting on linseed oil.

It was observed that the Lombardini engine was difficult to re-start on linseed oil, whether the oil was cold or the fuel pre-heated, and that as the duration tests proceeded, smoke density increased, with the exhaust becoming more acrid and objectionable. Exhaust gas temperature declined with time and the load had to be reduced to maintain the predetermined speed. After each duration run the engine was pulled down and examined. Linseed oil tended to foul the four injector holes which were enshrouded by funnels of carbonaceous materials extending as far as 10 mm away from the injector tip. Hard deposits were beginning to form in the piston combustion space and around the valves and ports, although the rings were still free.

Between duration tests the engine was reconditioned and restored to operate at or near original power on distillate. The results from the on-going duration tests on this Lombardini engine are summarised in Table 2. The esterification technique used to produce the methyl ester was an improved version of that generally outlined by Fuls and Hugo (1981). The material produced at Glenfield was 95% methyl ester yield (5% L.S.O.) and viscosity the same as that of diesel fuel at 40°C.

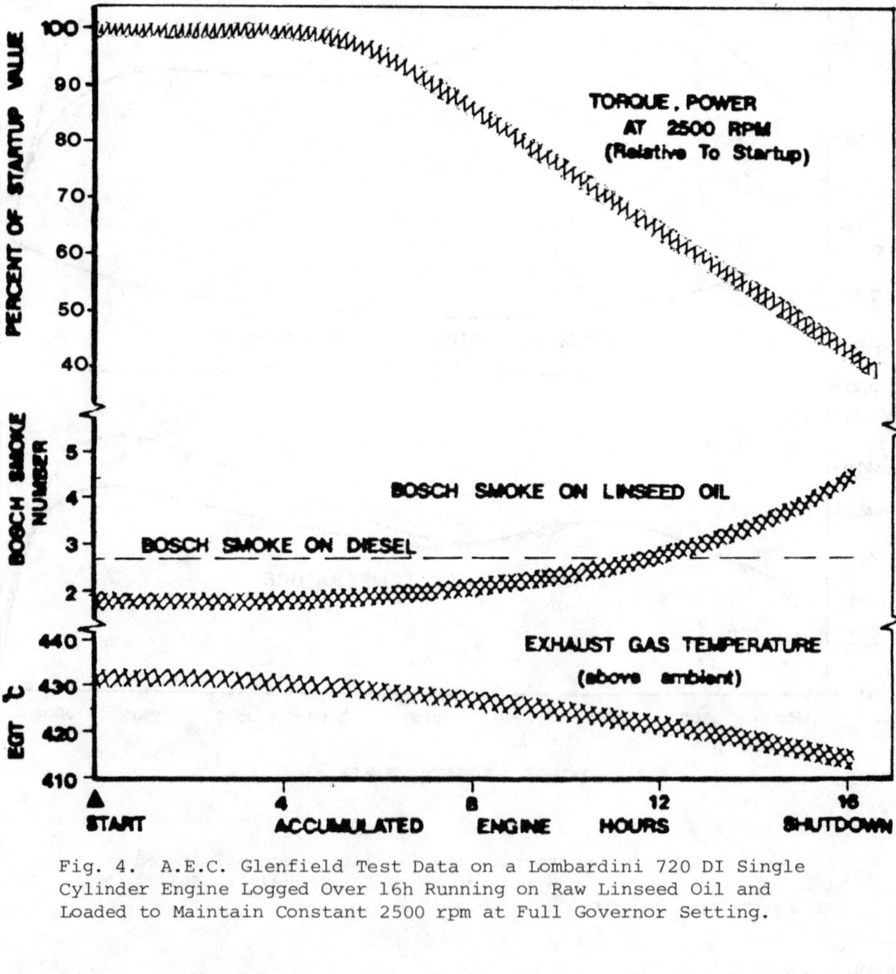

Fig. 4. A.E.C. Glenfield Test Data on a Lombardini 720 DI Single Cylinder Engine Logged Over 16h Running on Raw Linseed Oil and Loaded to Maintain Constant 2500 rpm at Full Governor Setting.

Table 2. Summary of Duration Tests on Lombardini Model 720 Operating on a Range of Non-Standard Fuels

Fuel	Fuel, Conditions	Comments	Hours to 90% of Initial Torque
Rapeseed Oil (R.S.O.)	Alkali Refined. Ambient Temperature.	Duration test terminated at 120 h. Rings siezed & gummed. Very little coking of injector.	86 (Test No. 15)
Linseed Oil (L.S.O.)	Raw Linseed Oil. Ambient Temperature.	Duration test to 16½ h. Severe fouling of injector, rings still free. Hard deposits.	7.0 (Test No. 11)
'Hot' L.S.O.	Raw Linseed heated to 85°C prior to injection	Duration test to 15 h. Coking as per 'cold' L.S.O.	8.5 (Test Nos 7&8)
L.S.O. + V100	V100 Fuel additive added to 'cold' L.S.O. in ratio of 1:1600.	Duration test to 9 h. L.S.O. possibly in degraded condition due to exposure to air and sunlight.	3.5 (Test No. 18)
L.S.O. + n-Butanol	10% n-Butanol in 'cold' L.S.O.	Duration test to 18 h.	10 (Test No. 10)
20% Distillate 80% L.S.O. CYCLING	Cycling test on fuels, alternately 1 minute on distillate, 4 minutes on L.S.O.	Duration test to 19½ h. Drop of 12% of power on the L.S.O. cycle from the distillate cycle at start of test. However distillate lost power at a faster rate.	Distillate Cycle 8 L.S.O. Cycle 14 (Test No. 19)
L.S.O. Methyl-Ester	Locally prepared methyl ester produced with sodium hydroxide catalyst.	Engine performance good on ester. Smoke density below distillate.	200 to date (Test No. 22)

CONCLUSIONS

1. In this paper baseline data on linseed oil as a diesel engine fuel is reported. The selection of linseed oil was made on the basis that this was the most readily-available high iodine value vegetable oil. This highly unsaturated oil provides a useful standard for assessments of fuel or engine modifications that might lead to the adoption of bio-oils as fuels for direct-injected diesel engines.

2. The engine of particular interest in this test series was a Lombardini single cylinder air-cooled direct injected unit which at maximum torque speed was down to 90% power after just 7 hours operation and below half of start-up

power after 16 hours on filtered raw linseed oil.

3. Heating the linseed oil to approximately 85°C in order to reduce viscosity made no appreciable difference to its performance.

4. The viscosity of linseed oil is close to that of the other widely-grown vegetable oils in Australian agriculture, such as sunflower, soybean and rapeseed. Yet the same engine running on rapeseed oil was capable of operating for 86 hours, or twelve times longer than on straight linseed oil. It was observed that rather than foul the injector, the rapeseed oil caused a decline in performance by its'varnishing' effect on the piston rings. Fuel viscosity cannot be indicted as the sole cause of injector fouling.

5. Chemical additives, claimed to reduce injector fouling when blended in diesel fuel, did not markedly improve the situation in linseed oil. There are indications that linseed oil, exposed to air and sunlight for extended periods, is even more likely to cause injector fouling.

6. A test was initiated using electronic timers and solenoid-actuated valves in which the engine was run alternately on linseed oil and then for a short period on diesel fuel. The theory was that perhaps the diesel would ameliorate the effects on the injector of the linseed oil. The results, for both 1 min/4 min and 4 min/12 min cycling routine, did not show significant improvement.

7. The only meaningful improvement found so far to enable linseed oil to be utilized as a longer-term fuel in this type of engine was to trans-esterify the linseed oil. The methyl ester was locally produced with comparative ease. The test was still continuing into the hundreds of hours at the time this paper was submitted and so far reconfirms that the ester is a good fuel candidate, even when derived from linseed oil - the 'worst' oil.

REFERENCES

1. Fuls, J. and Hugo, F.J.C. 1981. On-farm preparation of sunflower oil esters for fuel. Third I.C.M., Berlin, October.

2. Hugo, F.J.C. 1982. Plant oils as a fuel for diesel engines: experiences with sunflower oil. 10th Int. Sunflower Conf. Queensland, Australia.

3. Quick, G.R. 1980. Developments in use of vegetable oils as fuel for diesel engines. Paper No. 80-1525, Proc. ASAE 1980 Winter Meeting, Chicago, Illinois.

4. Siekmann, R. and Pischinger, G. 1981. Volkswagon do Brasil's activities and results with methyl esters of plant oils as diesel fuel. Vegetable Oils as Diesel Fuels - Seminar 2, Peoria, Illinois. October.

EVALUATION OF AN ON FARM PRESS

J. A. Blake
POS PILOT PLANT CORPORATION
U of S Campus
Saskatoon, Saskatchewan

ABSTRACT

POS has evaluated the Simon-Rosedowns mini-40 farm scale press in terms of throughput, oil removal, quality of oil, and effect of seed conditions on throughput and oil quality. Preliminary indications are that the press is very reliable and rugged. Oil output of 9-10 litres/hour has been attained for 6-7% moisture seed. Throughput drops off from 25 to 15 kg/hr and oil removal worsens considerably when 9-10% moisture seed is expelled.

EVALUATION OF AN ON FARM PRESS

POS Pilot Plant Corporation, Saskatchewan Research Council and the Department of Agricultural Engineering, University of Saskatchewan have been engaged in a joint project to evaluate the use of Canola oil as a diesel fuel extender/ replacer in diesel engines. One aspect of this project is the evaluation at POS of the Simon-Rosedowns Mini-40 screw press as a farm scale screw-press to expell the oil from Canola seed. (Canola is a low erucic acid, low glucosinolate type of rapeseed which has been developed in Canada for its nutritional qualities).

The Simon-Rosedowns press, (Figure 1) consists of a driven constant pitch worm inside a cage of twelve triangular shaped plates arranged perpendicularly to the axis of the worm. At the exit end of the cage, the worm has a smooth tapered section. The entire worm can be moved axially by a hand operated drive in order to narrow down the exit opening and thus adjust the exit cake thickness and the back pressure on the material between the worm and the cage. The cage section mounts between two heavy metal plates and is held in place by 3 bolts which pass through the apices of each triangular section and through each heavy plate. The distance between each cage is adjusted by placing metal washers of 0.13, 0.25, 0.51, or 0.76mm on the bolts between the sections. In addition, part of each side of each plate is hollowed out so that oil expelled outward between the plates does not have to travel the entire thickness of the plates but only a narrow region (land) surrounding the worm. The original cross-section of the centre hole in each plate was circular. Feed is by gravity into a hopper arranged just in front of the cage.

The initial arrangement had 0.76mm spacers on the first 6 plates at the feed end and 0.25mm spacers at the cake end. This configuration and a number of variations of spacing were of little use for the expelling of Canola. Feed rate was less than 15 kg/hr and the machine was prone to blockage, excessive fine material in the oil and puddling of oil inside the machine. This latter effect over lubricated the cage and throughput dropped drastically or ceased altogether.

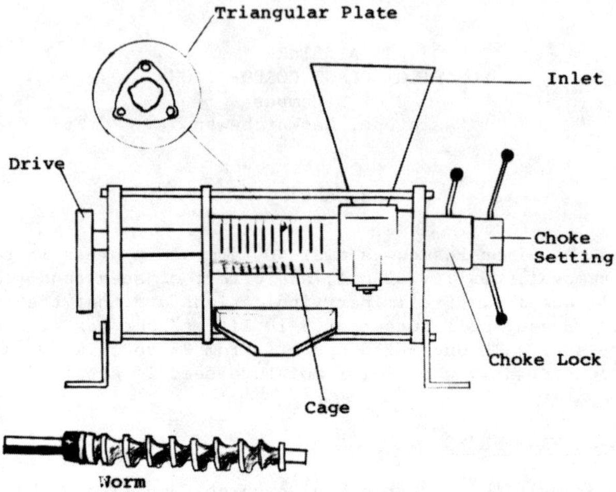

Figure 1.   Schematic Diagram of Mini-Press

Simon-Rosedowns provided a new cage. This cage has narrower lands and the central hole has 3 roughened cusps or lobes arranged equilaterally around the central hole. Simon-Rosedowns also recommended reducing the taper of the worm.

These changes improved the operation of the press but major problems were not eliminated until we switched from controlled flow to fully flooded input. The machine works best with a full hopper and no attempt to control feedrate.

Flooded input allowed us to operate continuously, but there were still problems with excessive fines and poor oil removal. After trying a variety of cage spacings, we carried out most of our work with the first 3 spacings at 0.76mm, the next 2 at 0.51mm and the last 6 at 0.25mm. This configuration has given oil with moderate fines level. Recently we have shown that changing the last 3 spacings to 0.13mm reduces the fines further. We have no exact value for fines as yet as we are developing a filtering system which will allow a more exact quantitation of fines.

Oil removal depends on choke setting, seed moisture, and characteristics of the seed. Choke setting is taken as a number engraved on a choke lock and varies from 0 fully open, to 30 fully closed. The practical range is from 20 to 26 or to cake thicknesses of 0.9 to 0.45mm (Figure 2). The good correlation exemplified between the cake thickness and choke setting is not carried over into the relationship of cake thickness to residual oil content or to feed rate when data for all seed moistures are included. A plot of percentage oil remaining vs. choke setting is approximately linear only if dry seed (less than 7% moisture) is pressed.

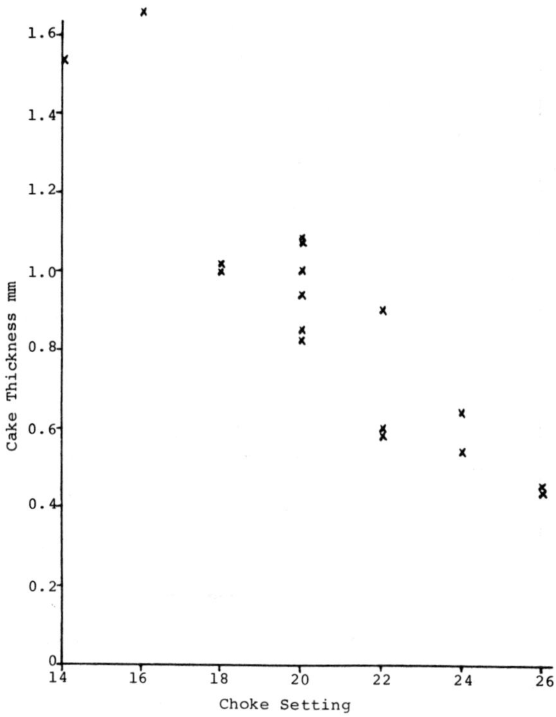

Figure 2. The Relationship of Cake Thickness to Choke Setting

A moisture content in excess of 9% adversely affects operation of the press. Throughput is reduced to about 60% of normal and it is difficult to obtain a residual oil content less than 25%. The cake also breaks into small soft pieces instead of the large leathery pieces obtainable at lower moistures. We have not ascertained as yet the upper limit of moisture for practical operation of the press but it appears to be in the 7-8% moisture range.

Feed rate is usually 25 kg/hr or less for residual oil less than 20% but in one run we attained 28.4 kg/hr and 17.3% residual oil. This was achieved with approximately 7% moisture seed. Our usual throughput is about 60-70% of rated throughput. Oil quality is excellent. Free fatty acid (ffa) is 0.13 to 0.38% on one lot of seed and 0.5 on another. The higher value on the first lot of seed occurred when we pressed 10% moisture seed. The free fatty acid values probably reflect the free fatty acid values in the seed. Expulsion of good seed by the normal cooking prepressing route would give 0.3 - 0.5% ffa. Peroxide values are excellent at 0.3 - 0.5 meq/kg. These values show that the minipress oil is of excellent edible quality and that it should not oxidize readily. Phosphorous content in the oil is in the range 8-50 ppm. This is much lower than the 200-300 ppm normal for oil produced by conventional expelling. This level of gums should give no problems in diesel fuel use. Sulphur levels at 6-7 ppm are high for edible use but negligible for diesel conditions.

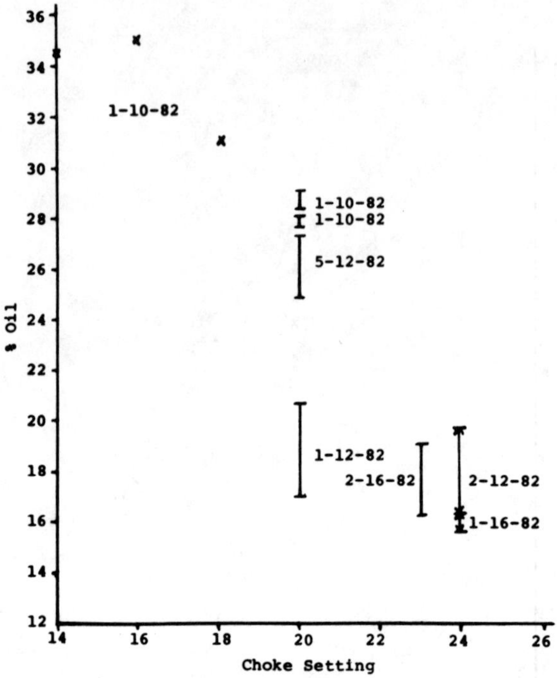

Figure 3. Percentage oil remaining in the seed vs. choke setting for Regent seed of less than 7% moisture. The numbers on the graph are run numbers.

We have also expelled two lots of cold seed. Unfortunately, the press is mounted in a heated room and seed at -12°C warmed to as much as 12°C while sitting in the input hopper. Even under these conditions we noticed a discharge of dry material from the front cage openings and a slightly higher residual oil content in the cake.

Cake temperatures vary from as low as 70°C to as high as 120°C. Oil temperature is only 50-60°C as most of the oil is expelled in the middle part of the press before the frictionally generated temperature rises very much.

The press has been very reliable in operation. Approximately 5 tonnes of seed have been pressed including 1 tonne of dehulled Canola pressed at very high pressure and low throughput. In this time only one thrust bearing has been replaced. Start-up is almost immediate. We have obtained good oil removal from as little as 4 kg of seed. There is no point in recycling cake, however, as oil removal from recycled cake is negligible.

Rate of oil production varies with conditions of operation but averaged 9 to 10 litres/hour (2.2 Imperial gallons/hour) for dry seed and residual oil in the cake of 16-20%.

Our major goal in this work has been to raise the capacity of the press while maintaining good oil removal.

We have not been able to reach the rated capacity with good oil removal. While capacity enhancement has been our main goal, we have also tried to determine relationships among choke settings and throughput, percentage oil in the cake, effect of moisture in the seed. This work is still ongoing and we hope for more definite correlations in the near future.

ON-FARM SOYBEAN OIL EXPRESSION

Robert W. Ramsey
Assoc. Member, ASAE

F. D. Harris
Member, ASAE

ABSTRACT

A mechanical screw press was used to determine the practicality of on-farm soybean oil expression. The press was integrated into a system designed to vary the input temperature of the soybeans. The system was simple and required minimal supervision making it attractive for on-farm use. Significant increases in oil production per bushel and oil production rate can be made by heating the soybeans.

INTRODUCTION

The power requirements of today's agricultural equipment is met almost exclusively by diesel fuel. Recently, the price of this fuel has increased and the supplies have become less reliable causing a search for an alternative source of internal combustion engine fuel. Vegetable oils offer a possible solution to this problem. They have many similar physical properties as diesel fuel and may be used as a substitute fuel.

The purpose of this research was to investigate the possibility of producing soybean oil on the farm. The objectives were to design a simple soybean oil expression system, measure the oil production and energy consumption of the system, determine the effect of heating the soybeans as a method of pretreatment and estimate the practicality of such a system.

Two methods of removing the oil from the oil seeds are currently available. The first is the extraction process in which the oil seeds are soaked in a solvent such as hexane. The meal is then removed and the hexane is evaporated, leaving the oil. The complexity of the solvent extraction process makes it undesirable for small on-farm processing plants (Pratt, 1980).

The second way to remove the oil is through a mechanical screw press, commonly called the expression process (Figure 1). The press consists of a worm shaft which rotates within a barrel and is constructed such that the volume displaced at the feed end is considerably greater than at the discharge end. Thus, as the material is conveyed from the feed end to the discharge end, it is subjected to increasing pressure, and oil is expelled through slots in the barrel. The screw press is the most practical approach to oil removal on the farm (Erickson, 1979).

In commercial screw press operations the oil seeds are rolled or crushed to rupture the oil cell walls and then they are cooked for 55 minutes until they reach a temperature of 116°C (240°F) and a moisture content of 2.2%

The authors are: Robert W. Ramsey, former Graduate Research Assistant, and F. D. Harris, Professor, Agricultural Engineering Department, University of Missouri-Columbia.

(Bederson, 1978). The soybeans in this study were heated to various temperatures between 27°C (80°F) and 116°C (240°F) before entering the screw press in order to improve the oil production of the system.

EQUIPMENT AND PROCEDURES

The expression device was a mechanical screw press, Model "Mini 40", manufactured by Simon-Rosedowns of Hull, England. The press has a nominal design capacity of 40 kg/hr (88 lb/hr) and features adjustable barrel shims and choke settings. The unit was powered by a 3-hp, 230 volt, 3 phase electric motor.

An electric grain roaster was used to heat the beans. The roaster was the "Roast-A-Tron" model manufactured by Mix-Mill, Inc. of Bluffton, Indiana. Rheostats were connected to the heating elements to control the temperature of the beans. A pulley reduction system was installed to reduce the output.

The roaster was installed above the press so that the beans would flow from the outlet of the roaster to the inlet of the press. The soybeans passed through a grain cleaner and over a magnet before entering the press in order to eliminate any rocks, metal or other debris. The electrical energy was recorded using standard kilowatt-hour meters. The temperature of the soybeans was measured using a thermocouple located in the inlet hopper, two inches above the worm screw. The oil production system is shown in Figure 2.

Approximately 30 bushels of soybeans were processed before the screw was fully polished and the oil production was up to normal levels. Before the screw was polished very little oil was produced and the press clogged frequently.

A final barrel shim spacing and choke setting were selected. The barrel shims control the size of the opening that the oil flows through as it is pressed from the soybeans. The barrel shims were arranged as follows; each box represents a space for shims between the barrel plates:

X = no shims
5 = 0.005 inch shims
10 = 0.010 inch shims
20 = 0.020 inch shims

X	X	X	X	X	X	5	5	10	20	20	20	X

↓ Outlet ↑ Inlet

The choke is the orifice located at the discharge end of the press which controls the pressure within the press. The choke was set at 24 which produced meal approximately .25 mm (.010") thick. This setting was determined experimentally to be the highest that did not cause clogging.

Certain procedures were necessary to prevent clogging of the press. In order to start the system the choke setting was reduced and the press was turned on. Soybeans were loaded into the hopper and allowed to flow through the press until the barrel became warm. Then the choke was moved to the final setting. To shut down the system, the hopper was allowed to empty and the choke setting was reduced. When meal stopped flowing from the choke the press could be turned off.

RESULTS

The oil output in liters per kilogram (gal/bu) reached a maximum at a temperature of approximately 77°C (170°F) as shown in Figure 3. As the temperature increased past this point the output dropped rapidly and problems with the operation of the press became more prevalent. At temperatures over 93°C (200°F) the meal had an increased tendency to clog the press and very high temperatures were created in the pressing operation causing scorching and burning.

The graph of liters per hour (gal/hr) (Figure 4) followed the same general trend. However, it reached a peak at approximately 54°C (130°F). The production rate dropped rapidly after this temperature.

The energy ratio is defined as the amount of electrical energy consumed by the heating and pressing operation divided into the energy contained in the oil that is produced. This ratio brings together the time used for oil production, the quantity of oil produced, and the amount of energy used. The graph of energy ratio versus temperature (Figure 5) shows that the most efficient point to operate the system is around 52°C (125°F).

The oil produced in each run was placed in jars and allowed to settle for one week. The percentage of total volume that was occupied by settled solids is shown in Figure 6. Oil produced around 82°C (180°F) contained the smallest amount of solids.

A comparison was made between the measured oil production and the theoretical value which was calculated using analysis of oil content before and after pressing. The following equation was used:

$$P = \frac{55.6 \text{ lb/bu}}{7.7 \text{ lb/gal}} \left[A - \frac{C\ (1-A)}{(1-C)} \right]$$

where, P = oil production in gallons per bushel
A = oil content before pressing
C = oil content after pressing

The measured oil production was reduced by the amount of solids that settled out. Table 1 shows the measured and calculated values as well as the percentage difference between the values.

The moisture content of the soybeans was between 10.3% w.b. and 10.6% w.b. with a specific weight of 3.48 kg/l (55.6 lb/bu). The oil content varied between 18.7% and 19.5%. The scatter in the graphs was probably due to differences in the soybeans as well as experimental errors.

OVERALL SYSTEM OPERATION

The screw press operated smoothly after the initial polishing of the screw and barrel was completed. The press could be easily started and stopped if certain procedures were followed.

Difficulty was encountered in keeping the output of the heater equal to the input of the screw press. The heater system would be improved if the rate of output was controlled by the level of soybeans in the inlet hopper of the screw press.

The control circuits for the auger and grain cleaner allowed the system to automatically maintain a constant supply of beans. Meal and oil were collected manually for data collection purposes, however, this would be easy to automate.

In general the oil production system performed very well with minimal supervision. No daily maintenance was needed and once the system was started it could be operated indefinitely given a supply of soybeans and storage for the meal and oil.

CONCLUSIONS

The oil production was found to be approximately..07 l/kg (.47 gal/bu) at a rate of 2.75 l/hr (.73 gal/hr) with no pretreatment of the soybeans. By increasing the temperature to 54°C (130°F) the rate of oil production was greatest. At 77°C (170°F) the rate of oil production per bushel was the greatest. The most efficient point to operate, based on energy ratio, was around 52°C (125°F). Thus, the best temperature to use depends on the particular situation.

Large differences were found in the amount of solids produced with the oil. Certainly a large part of the material that settled contained oil but further research must be done to determine if this oil can be effectively removed. If removal is a problem then the system be operated at approximately 82°C (180°F) where the solids are minimized.

The research to this point indicates that on-farm soybean oil production is practical. It is impossible to calculate the final cost of the oil since the cost of refining is unknown and could be significant.

REFERENCES

1. Bredeson, D. K. 1978. Mechanical extraction. Journal of the American Oil Chemists Society 55(11): 762-764.

2. Erickson, D. R. and P. Dixon. 1979. Soy oil as diesel fuel: economic and technical perspectives. American Soybean Association, St. Louis, Missouri.

3. Pratt, G. L. 1980. Sunflower oil for fuel. North Dakota Farm Research 38(2): 22-23.

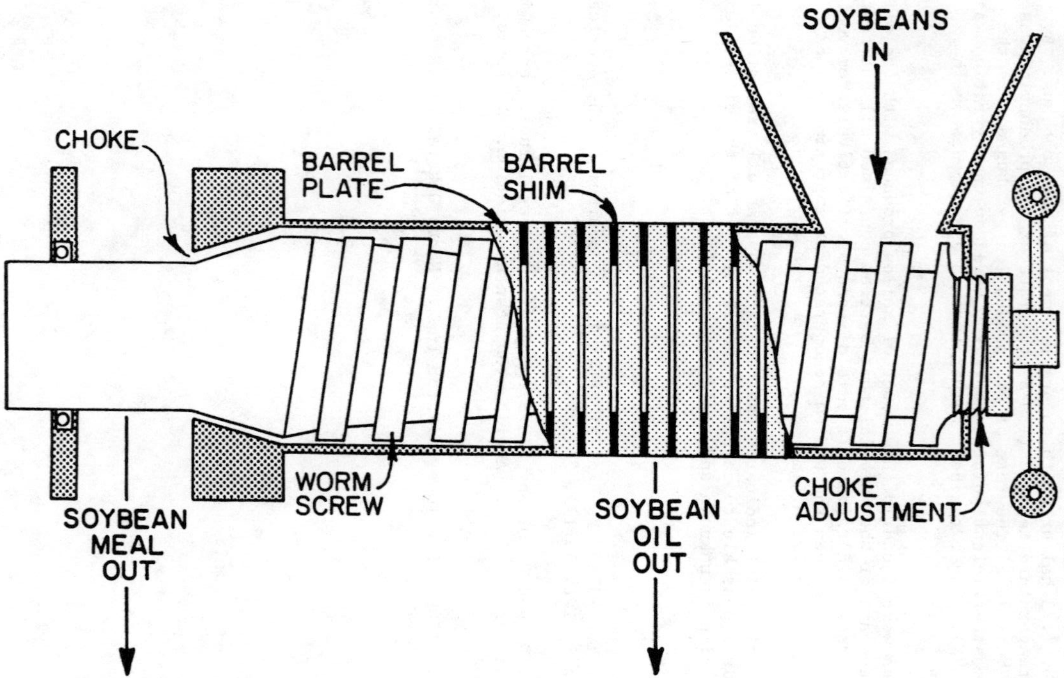

Figure 1. Mechanical Screw Press

Figure 2. Oil Production System

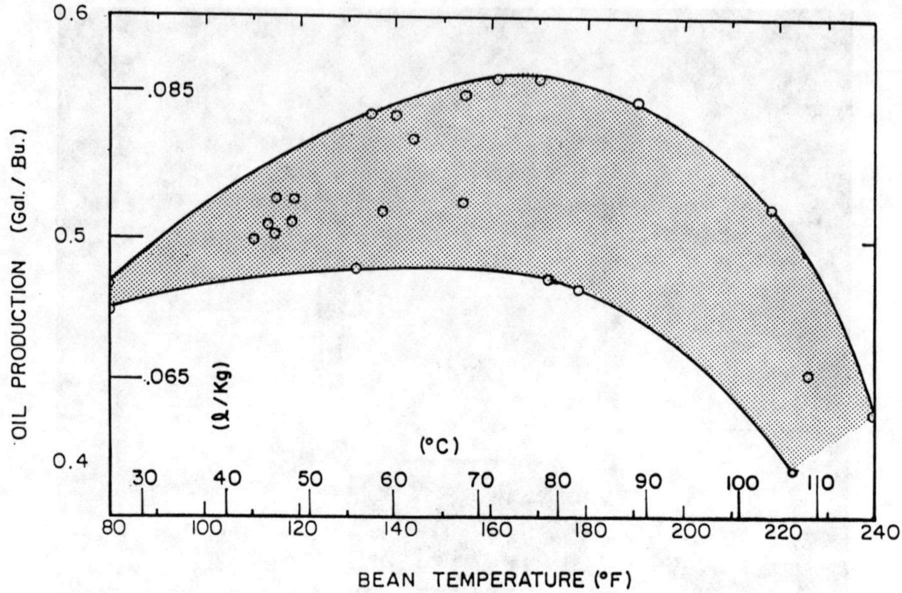

Figure 3. Oil Production Per Bushel

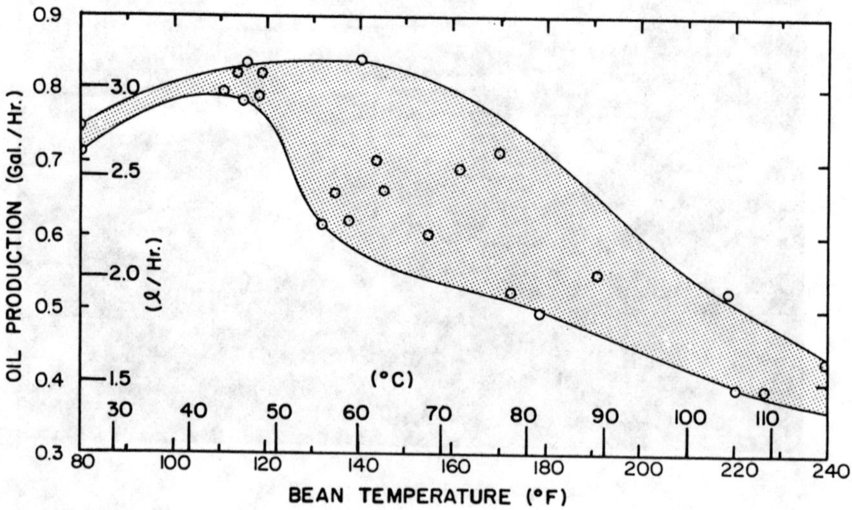

Figure 4. Oil Production Per Hour

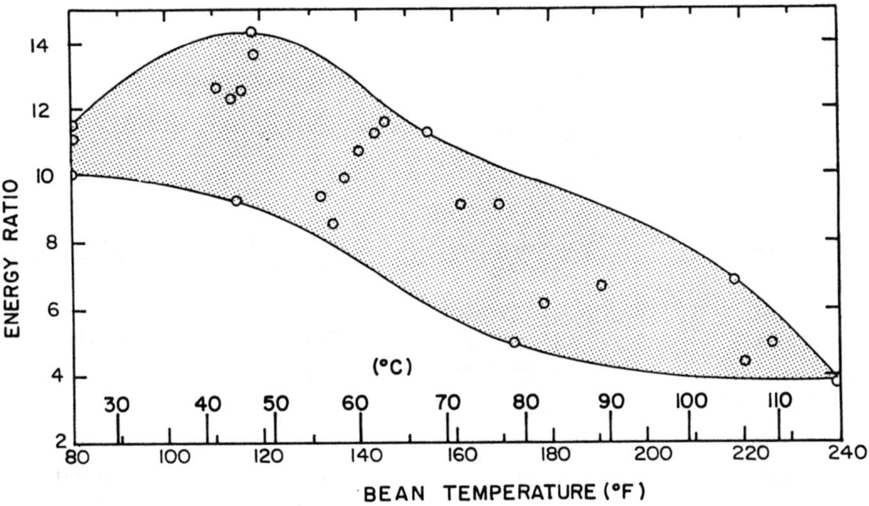

Figure 5. Energy Ratio

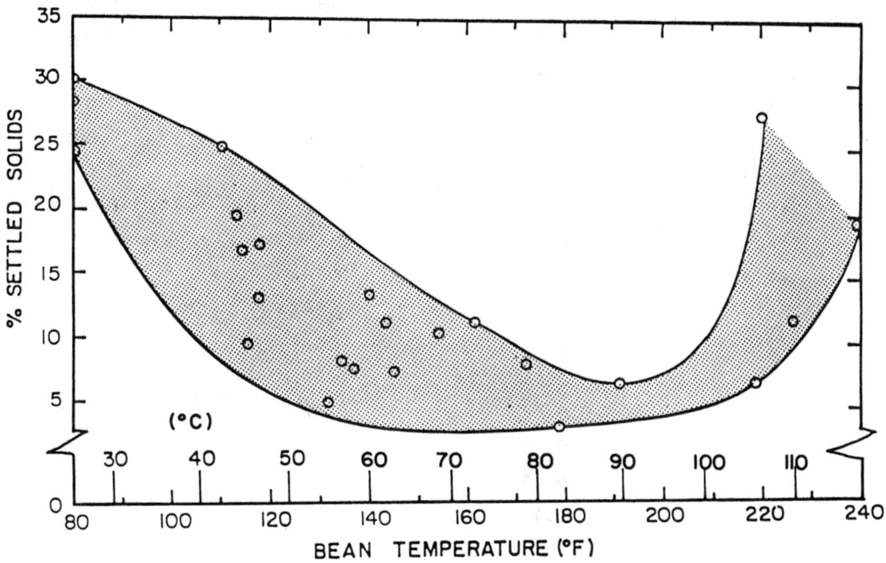

Figure 6. Percent Settled Solids

Table 1. Measured and Calculated Oil Production

Temperature (°F)	Measured (gal/bu)	Calculated (gal/bu)	Difference
80.0	0.29	0.33	-0.123
80.0	0.42	0.34	0.197
80.0	0.37	0.35	0.044
110.8	0.46	0.37	0.186
113.9	0.41	0.40	0.031
115.1	0.50	0.41	0.167
115.9	0.42	0.46	-0.111
118.4	0.33	0.41	-0.233
118.9	0.42	0.45	-0.066
132.4	0.46	0.46	0.001
134.9	0.42	0.50	-0.207
137.5	0.46	0.47	-0.024
140.5	0.48	0.48	0.009
144.0	0.46	0.48	-0.033
145.9	0.50	0.52	-0.044
154.8	0.43	0.46	-0.073
161.6	0.46	0.50	-0.097
169.9	0.41	0.54	-0.297
172.5	0.44	0.44	-0.005
178.9	0.45	0.46	-0.026
190.0	0.47	0.52	-0.099
218.6	0.45	0.47	-0.045
220.4	0.45	0.28	0.371
226.3	0.45	0.38	0.154
240.0		0.33	

AN AUTOMATED SMALL SCALE OIL SEED PROCESSING PLANT FOR PRODUCTION OF FUEL FOR DIESEL ENGINES

J.C. Thompson and C.L. Peterson*

Abstract

University of Idaho seed processing research is centered about a CeCoCo oil expeller. A seed preheater-auger, seed bin, meal auger, and oil pump have been constructed to complete the system, which is automated and instrumented. The press, preheater, cake removal auger, and oil transfer pump are tied into a central panel where energy use is measured and the process controlled. Extracted oil weight, meal weight, process temperature, and input energy are all recorded during operation. The oil is transferred to tanks where it settles for 48 hours or more. It is then pumped through a filtering system and stored ready to be used as an engine fuel. The plant has processed over 11,000 kg of seed with an average extraction efficiency of 78 percent.

Introduction

Extraction of oil from vegetable oil crops has been the subject of intense study by many researchers. Equipment and techniques for extracting the oil by both mechanical and solvent methods are well known. Equipment is commercially available from several sources. Recent interest in using vegetable oils such as from rape, safflower, sunflower and peanuts all of which have from 40 to 45 percent oil content has stimulated an interest in a small extraction plant which could be used on one farm or by individuals in a small cooperative. Not many of these small scale presses are known to be in operation, in the U.S., although, two particular machines have been actively promoted for this use. While the general techniques of operating a large plant are well known by engineers and scientists, the practical problems encountered with the efficient operation of a small press are not well documented.

The University of Idaho has developed an automated plant using a press of 40 kg/hr capacity. A seed preheater-auger, seed bin, meal auger, oil pump, oil storage, and oil filtration equipment complete the system. The plant is automated and instrumented. The press, preheater, oil filtering system, and auger are tied into a central panel where energy use is measured and the process controlled. Most of the equipment, with the exception of the screwpress, has been constructed locally.

This automated processing plant is an important part of the total vegetable oil program at the University of Idaho. In addition to providing valuable data on small scale screwpress extraction it also:

1. Provides all of the oil used in the engine testing program.
2. Provides all of the meal used in the animal industries nutrition studies.
3. Is used for long term storage studies with meal, oil, and oil seeds.
4. Provides data used by Agricultural Economics on cost budgets for on-farm processing plants.
5. Is used with the breeding program to supply oil samples, meal samples and extraction data on experimental varieties. A recent study compared 4 varieties of safflower, 2 winter rape varieties,

*The authors are Scientific Aide and Professor respectively, Department of Agricultural Engineering, University of Idaho, Moscow, Idaho, 83843.

and one sunflower variety. Significant differences existed in every category and demonstrated the value of the plant as a research tool.

6. Has been used to provide modest amounts of oil to other research programs not associated with the University of Idaho and to test for oil extraction for other products including peanuts, millet, and dog fennel (weedseed).

Review of Literature

Extraction of vegetable oils for food and industrial markets is a common practice, carried on successfully in many locations with a variety of vegetable oil sources. Nearly all previous extraction data reported has been concerned with large scale processing in plants operated by highly technical professionals. The concept of extracting oil by pressure with a screwpress is well known as are the various associated processes (Anjou, 1971; Beach, 1975; Norris, 1982).

Rapeseed, the most promosing fuel oil in the Pacific Northwest, is characterized by small seed size and high oil content. Anjou (1972) suggests the hulls are important to act as supporting material during extraction which increases oil yield but lowers feeding value. Some authors prefer crushing of the seed before pressing, whereas, Beach (1975) has the opinion that "cracking prior to flaking is unnecessary and redundant." Beach says the objective is to roll to a flake thickness of 0.2 mm with 0.25 to 0.30 mm the maximum thickness tolerable.

There is general agreement on the following discussion of screwpress operation (Anjou, 1972; Beach, 1975; Norris, 1982).

The mechanical screwpress has five elements: the main worm shaft and worm, drainage barrel, choke mechanism, motor transmission and bearings, and the loading system.

The press is designed to exceed a pressure of 1,000-1,400 kg/cm^2 on the seed and should reduce the oil content from 42-45 percent in whole seed to 14 or 15 percent in the processed meal. Power requirements are approximately 1 kw per metric ton of daily capacity. Cooking temperatures range from 77°C to 105°C with hold-up temperatures of 15-20 minutes. Longer times promote protein degradation. 80°C is the minimum temperature for myrosinase inactivation. Most plants today use a mild pressing operation in which approximately half or more of the oil is removed avoiding high pressure. The remainder of the oil is solvent extracted.

Crude rapeseed oil has small amounts of phosphatide called "gums" and free fatty acids which should be removed by hydrating with steam or hot water.

Filtration is ordinarily by settling and then forcing through a plate and frame filter system.

Cooking is done to breakdown oil cells, coagulate protein to facilitate oil separation, reduce affinity of oil for solid surfaces, insolubilize phosphatides, increase fluidity of oil, destruct molds and bacteria, and inactivate the enzyme, myrosinase.

The latter is essential for oil and meal to be free of hydrolysis products from the glucosinolates.

Anjou (1972) reports that the quality of both oil and meal are markedly affected by the method of cooking. The protein of the meal is often extensively denatured in the cooker, which improves feeding value, but too much heating can cause losses in amino acids. Lysine, an essential amino acid is the most heat sensitive. Overcooking darkens the oil and increases refining loss. Heat treating can also lower the oxidation stability of the oil, which would affect storage life and the amount of anti-oxident additives needed for both oil and meal. Free fatty acid content of the oil is also affected by temperature, moisture, and hold-up time.

Straight screwpressing has the advantages of simplicity and lower investment costs and the disadvantages of high power consumption, wear, and tear on equipment, residual oil in meal and high temperatures.

Information on the operation of small scale plants is much less available. While the basic theory is the same, more problems with quality control, efficiencies and handling and storage of seed, oil and cake are inevitable. Personnel have less training and equipment must be manufactured as inexpensively and simply as possible.

Simon-Rosedowns (1981), Anderson International (1981), and Chuo Boeki Goshi Kaisha (1981) all have bulletins describing processing equipment. The smallest is about 40 kg per hour capacity. Seed preheaters and some handling equipment are also described. Research papers or data on the operation of small units, the range of oil equality expected and economics of scale could not be found in the recent literature. Morgan and Schultz (1981) in a special report for the American Chemical Society state that "simple, low-cost extraction equipment is needed." They further state that "the literature in key technical areas - oil seed agronomy, oil extraction and conversion, and diesel utilization - is incomplete, especially for less well-known species."

University of Idaho Small Scale Processing Plant

The flow of materials and operation of the system are as follows: The raw product is screened as it is being dumped into a holding bin of 100 kg capacity. The holding bin sits directly over the pre-heater auger. On its way to the press, the seed is heated by conduction through the inner and outer walls of a special auger to temperatures ranging from 30 to 60°C. Heat is provided by 4 - 750 watt strip heaters in a closed forced air system controlled by two thermostats and protected by a memory limit switch. Before remove any metal scraps it might contain. This has proved to be a necessary precaution. The feed auger is controlled through the central panel utilizing a photo cell and related circuitry to keep the press hopper full. The press, a screw type, is presently adjusted by hand. The scroll moves in or out with respect to a tapered end bushing and is gaged by the oil flow and the thickness and consistency of the cake. The meal is carried away by a small 76 mm auger and collected in bags to be distributed to animal scientists for feeding purposes. The oil flows into a funnel shaped catch that feeds into a squeeze pump, controlled with a washing machine type liquid level switch. From there it is transferred to a holding tank and allowed to settle for a few days. The oil is forced through a 3-stage filter unit consisting of a recleanable pre-filter, 20 micron and then 4-5 micron throw away final filters. Finally the oil is stored in drums ready to be used in the engine testing phase of the program.

The existing filtration system is technically adequate to provide oil of high enough quality for use in diesel engine tests. However, the cost of the throw-away filters is 2.5 to 3 cents per L. Thus far, materials used in the cleanable pre-filters have not had a sufficiently small mesh size to extend significantly the life of the final 4-5 micron filter. Additional materials including metallic and plastic materials of smaller sieve sizes will be tested. Final filter life as determined by pressure drop through the filter and total cost of filtration per liter of oil are the evaluation criteria.

Extracted oil weight, meal weight, process temperatures, and input energy are all recorded during operation. The system requires a very minimum of supervision, however, a person collecting data has been with the system during all of its operation thus far. The plant has processed 11,370 kg of seed and produced over 4,000 liters of oil in 250 hours of operation. An average extraction efficiency of 78 percent has been obtained to date. Sediment in the oil has averaged only 1 percent, the oil left in the meal has been 15.3 percent. The power required for the total system has been 0.16 kw-hr/L. A summary of the data is shown in Table 1.

Table 1. Total Seed Processed 7/8/81 to 4/9/82
(Winter Rape Represents 90% of Total)

Seed In (kg)	Oil Out (kg)	Oil Out (L)	Oil Ext. (%)	Press Eff. (%)	Oil in Seed (%)	Oil in Meal (%)	Solids in Oil (%)	Power Require. (kw-hr/L)	Total Processing Time (hrs)
11372	4226	43306	37.2	80.0	45.0	15.0	1.1	0.16	250

Maintenance

The operation of the "New Type 52" CeCoCo press is quite simple and requires very little routine maintenance. However, the screw arrangement is vulnerable to foreign material in the seed such as metal or rock. If the tapered end of the scroll becomes scored the efficiency of the press is drastically reduced. After about 14,000 L of oil had been produced this problem was encountered. Precautions were taken, such as screening the seed and incorporating a magnetic metal trap. These measures were felt to have eliminated particles contaminating the seed, but the problem persisted. Further investigation revealed a worn bushing on the drive end of the scroll allowing it to score against the tapered end ring when adjusted for optimal oil expression. The bushing and shaft were then refitted to true alignment. During the interim, various methods of rebuilding the screw and tapered end ring were explored. The ring was first rebuilt with brass, then a steel insert and finally a spray powder hard facing technique. None of these methods worked satisfactorily. The auger, made of manganese steel, responded well to the spray powder build-up and a turning and polishing operation. The end ring was then bored out to the base cast iron and the resulting combination has worked well. Since these problems have been corrected over 5.5 Mg of seed has been processed with good success, giving an overall efficiency of nearly 85 percent. New parts will be evaluated against the rebuilt parts to gage the effectiveness of the technique.

It should be noted that if the screw fails during the processing of one Mg of seed $0.20 - 0.30 extra cost per liter of oil is incurred to replace or rebuild the screw.

Data Collection and Analysis

In order to obtain an accurate materials balance, the weight of the seed input is measured against the weights of the products out. An electronic scale was used to take accurate weight measurements. The oil weight was determined with a calibrated digital counter which records revolutions of a squeeze tube transfer pump. Seed and meal samples were taken at regular intervals and oil content measured on a Newport Nuclear Magnetic Resonance Analyzer (NMR). These values were used to calculate press efficiency. Oil samples were also evaluated for sediment using filtration and drying methods and for fatty acid content with a gas chromatograph. Temperature at various locations on the preheater and press were measured using thermocouples. Power requirements were tabulated with amp and watt meters.The meal was evaluated for suitability as a livestock feed by the Animal Sciences Department.

Seed Types Processed

Three oil seed crops selected because of their adaptability to the Pacific Northwest, have been processed in the press. In addition, peanuts from Georgia at the request of USDA personnel, and a few other minor crops have also been processed.

Rape. Winter Rape is currently produced on less than 3,000 ha in northern Idaho. Seed yields in excess of 5,600 kg/ha have been achieved in experimental trials. It's production benefits other crops in a cereal-legume rotation by reducing the build-up of pathogens and insects. It has been grown in the Palouse of Northern Idaho since about 1938. The majority of the seed processed here has been Winter Rape seed of the Dwarf Essex variety. It is easily processed, has a high oil content, and is locally available. The extraction percentage has been very good (80-88%). The oil flows out clean (1% solids) and the cake is dry and easy to handle. The disadvantages of rape are its high content of erucic acid and glucosinolates leaving the meal somewhat toxic. The oil is also quite viscous causing flowability problems under cold conditions.

Safflower. An oilseed crop grown in Arizona, California, Montana, and North Dakota safflower was found to be adapted to the dryland areas of northern Idaho and could be raised commercially if suitable markets can be developed. The safflower has a tough hull and requires more power than the other seeds as can been seen in Table 2. The majority of it has been fairly trashy and leaves a high sediment percentage in the oil. The cake is very coarse but high in protein and fiber making it a good feed supplement.

Sunflower. Native to North America, most of the sunflower raised in the U.S. are produced in the Dakotas, Minnesota and Texas. However, two oil-types were evaluated in Northern Idaho with favorable results. Traditionally sunflower seed is hulled prior to a pressing operation, however due to availability and cost, the seeds could not be hulled and were processed as is. Some difficulty was experienced in pressing as the seed would back up in the chamber. The slick hulls seemed to be causing the problem. Extraction efficiency was the lowest of the seeds evaluated, averaging about 70%, see Table 2.

Table 2. Crop/Cultivar Experiment (Summary)

Seed Type	Seed In (kg)	Oil (kg)	Out (L)	Oil Eff. (%)	Press Eff. (%)	Oil In Seed (%)	Oil In Meal (%)	Solids In Oil (%)	Power Require. (kw-hr/L)	Proc Time/Mg (hrs)
Saffl.	364	100	106	27	75	36.2	13.2	2.7	0.17	16.5
Sunfl.	91	32	36	35	69	44.4	16.1	2.9	0.13	16.8
W. Rape	182	67	74	37	85	43.7	13.7	.95	0.08	22.5

Peanut. The Southern Agricultural Energy center at Tifton, Georgia provided 45 kg of raw, shelled peanuts to be evaluated in the plant. Tests were made under three temperature conditions, and extraction was found to be very temperature dependent. The seeds were run at room temperature, preheated to 40°C and finally preheated to 60°C. Results appear in Table 3. Preheating the nuts increased the extraction efficiency from about 11 percent up to 88 percent compared to room temperature processing. The cold seeds being fairly soft were forced out through the cage bars when close adjustment was made. Backing off the screw then caused all of the material to be forced past the end ring with little or no oil extracted. When the seeds were preheated to 60°C extraction proceeded smoothly with high efficiency and a hard dry cake.

Other. Millet and dog fennel have also been tried in small quantities. The oil content of the millet was very low and would not flow through the press without clogging. The dog fennel did yield about 17 percent oil, but a very disagreeable odor made extraction quite objectionable.

Table 3. Georgia Peanut Experiment Trial.

Temp °C	Seed In (kg)	Oil Out (kg)	Oil Out (L)	Oil Ext. (%)	Press Eff. (%)	Oil In Seed (%)	Oil In Meal (%)	Power Require. (kw-hr/l)
22	9.1	0.4	7.27	4.7	11	41.0	40.5	0.22
40	9.1	2.3	0.45	25.0	61	41.0	21.3	0.24
60	18.2	6.5	1.17	36.0	88	41.0	8.5	0.36

An attempt was also made to rerun some wet (high oil content) rape seed meal obtained from operating the press when the scroll was damaged. The consistency of the meal did not lend itself to oil expulsion. The soft meal was forced through the cage bars similar to the trial with cold peanuts.

Discussion

Several types of information have been obtained with the press. These include an oil extraction versus temperature experiment with winter rape, data on power requirements and extraction efficiency at various time periods in the life of the plant, and a crop and cultivar processing experiment.

Oil Extraction Versus Temperature.

Most sources advocate heating the seed to 75-120°C and holding it for 30-60 minutes to increase efficiency. Early studies with the University of Idaho press concentrated on drawing a correlation between oil extraction and temperature, however the preheater was found to be inadequate to attain temperatures above 85°C. It was found that temperature had very little affect on oil extraction. In fact, a slightly negative correlation was found. There does seem to be an optimum temperature within the present working range. If seed temperature is held at 40°C the press temperature will stabilize at 55-60°C. Oil flow and cake consistency are very good at this point. As the press temperature approaches 70°C the oil flow drops off and is reflected in the cake. Similarily until the press reaches a temperature of about 50°C operation is somewhat sluggish. Plans for the immediate future call for building a cooker capable of reaching the 120°C range in order to experiment with the recommended temperature.

Data at Various Time Intervals.

Table 4 gives data collected during four time periods during the life of the press. The initial set was taken when everything was fairly new including the operator, and experiments were being designed and methodology worked out. Second was a period of mechanical problems discussed earlier. The efficiency dropped and much oil was lost in the cake. After the problems were corrected, trials were conducted with favorable results. The latest period reflects good operation during production of over 2500 L of rape oil for use in the upcoming engine endurance trials.

Table 4. Data Representing Various Time Periods of Plant Operation.

Dates	Seed In (kg)	Oil Out (kg)	Oil Out (L)	Oil Ext. (%)	Press Eff. (%)	Comments
7/8/81-8/20/81	3610	1249	1369	34.0	75.2	Initial Break
8/20/81-11/23/81	1385	366	401.6	26.5	57.5	Mech. Problems
11/23/81-1/12/81	636	191	212.0	39.9	77.2	Crop/Cultivar Exp.
1/12/82-4/9/82	5739	2356	2584.4	41.1	86.3	Mech. Problems Solved

Seed Type and Cultivar Processing Experiment.

A recent study in cooperation with the Plant and Soil Science Department was performed on four varities of safflower, two varities of winter rape, and one varity of sunflower. A summary of the data appears in Table 2. The experiment involved running 5 - 18 kg samples of each variety through the press and collecting the data for a materials balance analysis as well as time, temperature, and power data. This information appears on Table 5. Note that significant differences exist between crops and between cultivars within crops showing the usefulness of the system in agronomic studies.

Table 5. Crop/Cultivar Processing Experiment.

OIL CONTENT

	Whole Seed (%)	Meal (%)	Extraction (%)[1]
Safflower			
S-208	35.9 e*	14.0 bc*	73.5 bc*
S-112	36.4 e	12.5 c	76.5 b
UC-1	36.7 e	10.3 d	75.0 b
S-541	38.8 d	12.7 c	75.5 b
Winter Rape			
Dwarf Essex	45.9 a	14.4 b	83.4 a
Sidal	41.5 c	13.0 bc	86.5 a
Sunflower			
Hybrid 894	44.4 b	16.1 a	69.9 c

OIL EXTRACTION

	kg/Mg	L/Mg	time/Mg (hrs/Mg)	sediment (%)
Safflower				
S-208	262 f*	282.5	16.6 bc*	6.95 a*
S-112	275 e	295.8	16.2 bc	2.91 bc
UC-1	272 e	296.7	16.0 bc	2.52 bcd
S-541	290 d	311.7	17.3 bc	2.58 bcd
Winter Rape				
Dwarf Essex	380 a	417.7	18.4 b	1.01 cd
Sidal	358 b	388.9	15.3 c	0.89 d
Sunflower				
Hybrid 894	308 c	334.6	22.5 a	2.91 b

POWER REQUIREMENT

	kw-hr/hr	kw-hr/Mg	kw-hr/L
Safflower			
S-208	2.98 a*	49.6 a*	0.174
S-112	3.06 a	49.4 ab	0.166
UC-1	3.00 a	49.5 ab	
			0.166
S-541	2.92 a	50.6 a	0.161
Winter Rape			
Dwarf Essex	1.82 c	33.3 c	0.079
Sidal	2.20 b	33.7 c	0.087
Sunflower			
Hybrid 894	2.02 bc	43.4 b	0.129

*Means within a column not followed by the same letter differ at the 0.05 level of probability by Duncan's multiple range test.
Each data point represents the average of five replications, 18 kg per sample.

[1]Percent of oil in whole seed that was extracted.

Conclusions and Recommendations

One of the objectives of this phase of the energy project at the University of Idaho was to demonstrate the feasibility of a farmer owned and operated vegetable oil plant that would provide a high percentage of liquid fuels needed for crop production. The plant described in this paper has proved to be a workable system readily adaptable to most farm operations in the Pacific Northwest. Initial investment costs are estimated at $10,000, and production costs per liter of oil range from 0.5 to 1.00 dollars. The economic balance of using the oil for fuel vs. selling it on the open market is obviously unfavorable at current diesel fuel prices, however as the availability of fossil fuels diminish and if prices continue to rise, the economics may swing in favor of vegetable oil to power U.S. Agriculture. It has also been suggested that if plants were established they could provide oil for existing oil markets with the option of switching the production to fuel use in time of emergency.

Acknowledgements

Work on this project has been conducted under state project H787 and a DOE-USDA grant number 59-2161-1-6-040-0.

The authors express appreciation to George Brocke of Kendrick, Idaho for supplying the CeCoCo press used in the plant, and to Dick Auld and Bill Bettis, University of Idaho Plant Scientists, for help with the crop and cultivar experiment, and to Roger Korus of the University of Idaho Chemical Engineering Department for help with chemical analysis.

Literature Cited

Anderson International Corp. 1980. The Anderson Expeller Press DUO. Bulletin A-151-R. Strongsville, Ohio, 44136.

Anjou, K. 1972. Manufacture of Rapeseed Oil and Meal. Chapter 9 in Rapeseed edited by L.A. Applequist and R. Ohlson. Elsevier Publishing Company, New York, N.Y. 10017.

Auld, D., B. Bettis, and N. Porter. 1981. Development of Oilseeds as Alternative Crops for the Pacific Northwest. Yearly Report of the University of Idaho, Washington State University, and Oregon State University Agricultural Experiment Stations submitted to the Pacific Northwest Regional Commission.

Beach, D. 1975. Primary Processing of Vegetable Oils. Chapter 23 in Oilseed and Pulse Crops in Western Canada - A symposium edited by J.T. Harapiak. Western Co-operative Fertilizers Limited, Calgary, Alberta, T2P 2N1.

Bruwer, J.J., B.D. Boshoff, F.J.C. Hugo, L.M. duPlessis, J.Fuls, C. Hawkins, A.N. VanderWalt, and A. Engelbrecht. 1980. Sunflower Seed Oil as an Extender for Diesel Fuel in Agricultural Tractors. Presented at the 1980 Symposium of the South African Institute of Agricultural Engineers on 11 June 1980.

Chuo Boeki Goshi Kaisha. 1981. CeCoCo Oil Expeller. Advertising Literature and Communication. "CECOCO" Exhibition and Demonstration Farm Center, Ibaraki City, Osaka-Fu, Japan.

Engelman, H.W., D.A. Guenther, and T.W. Silvis. 1978. Vegetable Oil as a Diesel Fuel. Diesel & Gas Engine Power Division of ASME. Paper 78-DG-P-19 presented November 5-9, 1978.

Goering, C.E. and M.J. Daugherty. 1981. Energy Inputs and Outputs of Eleven Vegetable Oil Fuels. ASAE Paper 81-3586 presented at the 1981 Winter Meeting of ASAE, Chicago, Illinois.

Hofman, V., K. Kaufman, D. Helgeson, and W.E. Dinusson. 1981. Sunflower for Power. NDSU Cooperative Extension Service Circular AE-735. Fargo, North Dakota.

Hugo, F. 1981. Sunflower Oil as a Diesel Fuel Replacement: The South African Research Program. Proceedings of Vegetable Oil as a Fuel Seminar II, NAEC-NRRC, Peoria, Illinois.

Morgan, R.P. and E.B. Shultz,Jr. 1981. Fuels and Chemicals from Novel Seed Oils. C & EN Special Report. American Chemical Society, Washington, D.C. 20036.

Norris, F.A. 1981. Extraction of Fats and Oils. Chapter 3 of Bailey's Industrial Oil and Fat Products. John Wiley and Sons, New York, N.Y.

Peterson, C.L., G.L. Wagner, and D.L. Auld. 1981. Performance Testing of Vegetable Oil Substitues for Diesel Fuel. ASAE Paper No. 81-3578 presented at the 1981 Winter Meeting of ASAE, Chicago, Illinois.

Peterson, C.L., D.L. Auld, V.M. Thomas, R.V. Withers, S.M. Smith, B.L. Bettis. 1981. Vegetable Oil as an Agricultural Fuel for the Pacific Northwest. University of Idaho, Agricultural Experiment Station. Bulletin No. 598.

Pischinger, G., F.C. Clymans, and R.S. Sickman. 1981. Diesel Oil Substitution by Vegetable Oils - Fuel Requirements and Vehicle Experiments. Volkswagen do Brazil S/A, 04217. Sao Paulo, Brazil.

Quick, G.R. 1980. Developments in Use of Vegetable Oils as a Fuel for Diesel Engines. ASAE Technical Paper No. 80-1525. ASAE, St. Joseph, Michigan 49085.

Simon-Rosedowns Limited. 1981. Turn Seed into Oil and Cattle Cake. Advertising Circular. Hull, England HU2 OAD.

EXPELLOR EXTRACTED RAPE AND SAFFLOWER OILSEED MEALS FOR POULTRY AND SHEEP

V. M. Thomas*, R. J. Katz, D. A. Auld,
C. F. Petersen and E. A. Sauter

The objective of these studies was to evaluate the feeding value of on-the-farm expellor extracted rape (RSM) and safflower (SM) oilseed meals for poultry and sheep. Rapeseed meal and SM contained 30.7 and 25.8% crude protein (CP) and 21.7 and 8.7% fat, respectively. Rapeseed meal contained a total glucosinolate concentration of 78.3 µmoles/g. A 22-day feeding trial was conducted with 6-day-old chicks. Rapeseed meal and SM replaced 25 or 50% of the soybean meal (SBM) protein in isonitrogenous (23% CP), isocaloric (3250 kcal ME/kg) diets. Birds fed SBM and 25 or 50% SM consumed more ($P<.01$) daily feed and gained more ($P<.01$) per day than those fed 25 or 50% RSM. Birds fed RSM had enlarged thyroid glands in comparison to those fed SBM. Two lamb digestion trials were conducted to evaluate the effect of replacing cottonseed meal (CSM) protein with either RSM or SM on nitrogen utilization and DM digestibility. Replacing 100% of the CSM protein with RSM had no effect ($P>.05$) on dry matter digestibility and N utilization. Nitrogen balance studies indicate that expellor extracted SM may replace up to 75% of the CSM protein in diets for wethers.

INTRODUCTION

The feeding value of oilseed meals varies with the species of seed and method of oil extraction used. In general, meals produced by mechanical extraction (expellor) of oil from seed contain more fat and a lower percentage of protein than those produced by solvent extraction using hexane.

Most rapeseed meal (RSM) currently fed to livestock is called "canola" meal and is a Canadian product. The term "canola" identifies rapeseed and its products of recently developed varieties that contain 5% or less erucic acid in the oil and 3 mg or less glucosinolate per gram in the meal. No high glucosinolate rapeseed meal is produced in Canada today. Intact glucosinolates are not toxic, but are hydrolysed in the intestines of livestock to form toxic by-products which cause growth rate depression, organ pathology and hypothyroidism (Bell 1965, Van Etten et al. 1969). Poultry, when fed high levels of rapeseed meal (but not canola meal), suffer from growth depression and enlarged thyroid glands (Tookey et al 1969, Wetter and Craig 1959). Ruminants are less susceptible to glucosinolates than other animals. Their resistance is probably associated with the degradation of the toxic breakdown products by microflora in the rumen. In Northern Idaho growers raise a winter variety of rape that contains a high concentration of glucosinolate. Currently, the majority of rapeseed meal fed world wide is canola meal, although some high glucosinolate meal is fed to ruminants. Canola meal produced from a mixture of canola varieties contains from 37 to 40% crude

---

*The authors are: V. M. THOMAS, Assistant Professor, R. J. KATZ, Research Technician, D. A. AULD, Associate Professor, C. F. PETERSEN, Professor and E. A. SAUTER, Professor, Dept. Animal Sciences and Plant and Soil Sciences, University of Idaho, Moscow.

protein and has an amino acid profile that compares favorably with that of soybean meal.

Safflower meal (SM) produced by prepressed solvent extraction contains approximately 24% crude protein (CP), 30-35% crude fiber and 1% fat. Safflower meal has been used in poultry diets with good results (Petersen et al. 1957, Kuzicky and Kohler 1968, and Wylie et al. 1972). Its major limitations are a high fiber content (Valadez et al. 1965) and low lysine content (Wylie et al. 1972). Safflower is used in ruminant diets when available at a competitive price.

A small scale expeller manufactured by CeCoCo was used for oil extraction in our studies. The press has a 40 kg/hr capacity. A seed preheater-auger, seed bin, meal auger, oil pump, oil storage and oil filtration equipment complete the system. An average extraction efficiency of 77% has been obtained to date. Becasue the efficiency of oil extraction is only 77%, meal produced as a by-product has slightly less protein and more fat than meals produced by pre-processing and solvent extraction. The objective of these studies was to evaluate the feeding value of on-the-farm expeller-extracted RSM and SM for poultry and sheep.

EXPERIMENTAL PROCEDURE

Poultry Study: One hundred and fifty male broiler chicks of uniform body size six days of age were randomly allocated into five groups. Each group of chicks was divided into three replicates of ten chicks each and placed in electrically-heated wire chick brooders with automatic water cups. The basal diet contained corn and soybean meal (SBM) and was formulated to contain 23% CP and 3250 kcal ME/kg diet. Rapeseed meal and SM replaced either 25 or 50% of the SBM protein in isonitrogenous, isocaloric diets (table 1). The chicks were fed the experimental diets ad libitum for a 22-day feeding period and chick feed consumption/day, chick weight gain/day, and chick feed/gain were determined. On day 25, six birds from the control, 50% SM, 25% RSM and 50% RSM diets were weighed and killed and the thyroid gland removed. A thyroid length to live weight ratio was calculated.

Standard A.O.A.C. (1975) procedures were used for CP and ether extract (EE) determination. Glucosinolate content of the RSM was determined by the method of Wetter and Young (1976). The amino acid composition was determined with a Beckman amino acid analyzer following protein hydrolysis with 6 N HCl. Statistical analysis of the data was according to the procedures of Anderson and McLean (1974) with significant differences between means determined by Newman-Keuls sequential range test.

Lamb Digestion Trials: Two digestion trials were conducted with 15 crossbred wethers with three animals per treatment group. In both trials, treatments consisted of a cottonseed meal positive control and a urea negative control diet, and three test diets in which RSM or SM replaced 33, 66 or 100% (RSM) or 25, 50 and 75% (SM) of the CSM protein. Diets were formulated to contain 13% CP on a DM basis. Ingredient composition of the control and test diets are shown in tables 2 and 3.

Animals were housed in metabolism cages and fed twice daily at 12-hour intervals. The digestion trial consisted of a 14-day adjustment period and a seven-day total collection period. Lambs were fed ad libitum during the first ten days and then at 90% ad libitum for the remainder of the trial. Daily samples of feed and orts were collected at each feeding and composited for analysis. Fecal samples were collected daily and a subsample was taken for analysis. Total daily urine (acidified) was collected and a 10% aliquot was taken for analysis.

Feces were dried in a forced-air oven at 50 C for 48 hours. Both feed and fecal samples were ground in a Wiley mill using a one-mm screen. Nitrogen content of feeds, feces and urine were determined by A.O.A.C. (1975) procedures. Data in the digestion trials were analyzed by procedures described previously in the poultry feeding trial.

RESULTS AND DISCUSSION

Chemical Composition: Expeller extracted RSM and SM contained 30.7 and 25.8% CP, 21.7 and 8.7% fat and 25.2 and 42.3% acid detergent fiber, respectively (table 4). RSM contained a total glucosinolate concentration of 78.3 µmoles/g. Amino acid composition of RSM was similar to that of SBM; however, SM had lower concentrations of most amino acids in comparison to SBM.

The high fiber content and poorer amino acid profile of expeller extracted SM are two factors limiting its use in ration formulation. The high concentration of glucosinolates of the RSM limits its usefulness as a feed for livestock. It is important to emphasize that the RSM fed in this study was not from a canola variety known for having a low glucosinolate concentration.

Poultry Study: Birds fed SBM and 25 or 50% SM consumed more ($P<.01$) daily feed than those fed 25 or 50% RSM (65.9, 64.9, 63.0 vs 52.8, 37.9 g/day) (table 5). Feed consumption was greater ($P<.01$) for 25% RSM birds than those fed 50% RSM. Growth rate was greater ($P<.01$) for birds fed SBM and 25 or 50% SM vs 25 or 50% RSM (37.6, 36.85, 36.55 vs 30.0, 18.5 g/day). Daily gain was greater ($P<.01$) for 25% RSM versus 50% RSM. Birds fed SBM, 25 or 50% SM and 25% RSM converted feed to gain more efficiently ($P<.01$) than those fed 50% RSM (1.72, 1.73, 1.68 1.74 vs 2.02 g feed/g gain). Birds fed 25 and 50% RSM had greater thyroid length to weight ratios than those fed SBM or SM, indicating that the glucosinolate content of RSM impaired thyroid metabolism (table 6). Expeller extracted SM can effectively replace up to 50% of the SBM protein in diets for broiler chicks. The replacement of 25 or 50% of SBM protein with RSM decreased bird performance. There is a need for the development of a low glucosinolate variety of winter rape if the meal is to be fed to poultry.

Lamb Digestion Trials

Rapeseed Meal: Nitrogen intake was similar across treatment groups (24.5 g/day, table 7). Fecal and urinary N excretion tended to be lower for lambs fed CSM in comparison to those fed urea and RSM diets. Because of the lower fecal and urinary N losses, lambs fed CSM tended to retain more g of N daily than the other treatment groups. Percentage urinary N excreted was lower ($P<.05$) with the CSM diets than with the urea and 66% RSM diets. Lambs fed the CSM diet retained more ($P<.05$) absorbed N than those fed the urea or 66% RSM diets. These data indicate that rapeseed meal may be a satisfactory source of supplemental protein for sheep. It is interesting to note that the high glucosinolate content of the RSM diets did not impair digestibility of dry matter or N utilization which is in contrast to the data reported for poultry. Glucosinolates may be converted to non-toxic compounds in the rumen.

Safflower Meal: Nitrogen intake was similar across treatment groups (36.1 g, table 8). No differences ($P>.05$) in nitrogen balance were found. Lambs fed the 75% SM diet tended ($P>.05$) to retain a higher percentage of absorbed N than the other treatment groups. Dry matter digestibility was higher ($P<.05$) for lambs fed the urea diet versus the CSM or SM diets. This response may have been due to the higher proportion of grain in the urea diets; however, calculated total digestibility nutrient content between all diets did not differ by more than 5%. It appears that expeller extracted SM may replace up to 75% of the CSM protein in diets for wethers.

REFERENCES

1. Anderson, V.L. and R.L. McLean. 1974. Design of Experiments (1st Ed.). Marcel Dekker, Inc., New York.

2. A.O.A.C. 1975. Official Methods of Analysis (12th Ed.). Association of Official Analytical Chemists, Washington, DC.

3. Bell, J.M. 1965. Growth depressing factors in rapeseed meal. VI. Feeding value for growing finishing swine of myrosinage-free solvent-extracted meal. J. Anim. Sci. 24:1147-1151.

4. Kuzmicky, D.D. and G.O. Kohler. 1968. Safflower meal-utilization as a protein source for broiler rations. J. Poultry Sci. 47:1266-1270.

5. Petersen, C.F., A.C. Wiese, G.J. Anderson and C.E. Lampman. 1957. The use of safflower oil meal in poultry rations. Poultry Sci. 36:3-8.

6. Tookey, H.L., C.H. VanEtten and M.E. Daxenbichler. 1969. Glucosinolates. In Toxic Constituents of Plants Foodstuffs. 2nd ed. J.E. Liener (ed.). Academic Press, New York, pp. 103-142.

7. Valdez, S., W.R. Featherston and R.A. Pickett. 1965. Utilization of safflower meal by the chick and its effect on plasma lysine and methionine concentration. Poultry Sci. 44:909-915.

8. VanEtten, C.H., M.E. Daxenbichler and J.A. Wolff. 1969. Natural glucosinolates (thioglucosides) in foods and feeds. J. Agr. Food Chem. 17:403-491.

9. Wetter, L.R. and B.M. Craig. 1959. Varietal and environmental effects on rapeseed. I. Isotiosynate and thiooxazolidone content. Can. J. Plant Sci. 39:395-399.

10. Wetter, L.R. and C.G. Craig. 1976. A thiourea-UV assay for total glucosinolate content in rapeseed meals. J. Am. Oil Chem. Soc. 53:162-164.

11. Wylie, P.W., S.M. Talley and J.N. Freeman. 1972. Substitution of linseed meal nad safflower meal for soybean meal in diets of growing pullets. Poultry Sci. 51:1701.

Table 1. Ingredients and Chemical Composition of Chick Feeding Trial Diets

Item	Source of supplemental protein				
	100% SBM control	25% RSM	50% RSM	25% Saff	50% Saff
Ingredients, %					
Dwarf Essex rapeseed meal	...	13.95	27.89	...	...
Safflowerseed meal	...	...	...	16.60	33.19
Yellow dent corn	62.00	54.18	45.76	43.94	25.57
Soybean meal	34.10	27.01	20.04	29.05	24.07
Animal fat	0.00	1.00	2.50	6.50	13.25
Limestone	1.00	1.00	1.00	1.00	1.00
Dicalcium phosphate	2.00	2.00	2.00	2.00	2.00
DL-methionine	.15	.11	.06	.16	.17
Tm salt	.25	.25	.25	.25	.25
Vitamin mix[a]	.50	.50	.50	.50	.50
	100.00	100.00	100.00	100.00	100.00
Chemical composition					
Crude protein[b], %	23.9	23.7	22.4	23.7	23.1
Crude fat, %	2.6	7.3	11.3	8.9	15.5
Acid detergent fiber, %	9.6	10.6	12.5	13.6	18.8
Total glucosinolates (μmoles/g)[c]	...	10.9	21.8	...	...
Amino acids[d]					
Cystine	1.54	1.35	1.37	1.22	1.43
Lysine	5.18	4.98	5.33	4.51	4.58
Methionine	2.00	1.98	2.08	1.77	1.88

[a] Vitamin mix contains per pound: 1,200,000 USPU-vitamin A; 300,000 ICU-vitamin D_3; 400 IU vitamin E; 1.6 mg vitamin B_{12}; 800 mg riboflavin; 5000 mg niacin; 1100 mg d-pantothenic acid; 50,000 mg choline chloride; 400 mg menadione sodium bisulfite complex

[b] Nitrogen x 6.25

[c] Isothiocyanates plus oxazoliolinethiones

[d] Percent of protein

Table 2. Composition of Lamb Digestion Trial Diets (Rapeseed)[a]

Item	Protein supplement: Cottonseed meal	Urea	33% RSM	66% RSM	100% RSM
Barley	65.0	70.5	63.3	61.9	60.1
Wheat straw	20.0	20.0	20.0	20.0	20.0
Cottonseed meal	7.0	...	5.2	3.2	...
Rapeseed meal	...	...	3.5	6.9	11.9
Safflower meal	...	...	...	...	...
Urea	...	1.0	...	...	...
Molasses	5.0	5.0	5.0	5.0	5.0
Limestone	1.0	1.0	1.0	1.0	1.0
Monosodium phosphate	.5	.5	.5	.5	.5
Dynamate[b]	...	.5	...	...	...
Ammonium chloride	.5	.5	.5	.5	.5
Trace mineral salt	1.0	1.0	1.0	1.0	1.0

[a]Expressed on a dry matter basis

[b]Contains not less than 22% sulfur, 18% potassium and 11% magnesium

Table 3. Composition of Lamb Digestion Trial Diets (Safflower)[a]

Item	Protein supplement: Cottonseed meal	Urea	25% SM	50% SM	75% SM
Barley	57.1	68.7	52.0	47.0	42.0
Wheat straw	20.0	20.0	20.0	20.0	20.0
Cottonseed meal	13.9	...	11.7	9.5	7.2
Rapeseed meal	...	...	...	...	...
Safflower meal	...	...	7.3	14.5	21.8
Urea	...	1.7	...	...	...
Molasses	4.0	4.0	4.0	4.0	4.0
Limestone	1.7	1.7	1.7	1.7	1.7
Monosodium phosphate	.8	.8	.8	.8	.8
Dynamate[b]	...	.6	...	...	...
Ammonium chloride	.8	.8	.8	.8	.8
Trace mineral salt	1.7	1.7	1.7	1.7	1.7

[a]Expressed on a dry matter basis

[b]Contains not less than 22% sulfur, 18% potassium and 11% magnesium

Table 4. Amino Acid and Chemical Composition of Rapeseed, Safflowerseed and Solvent Extracted Soybean Meal[a]

Item	Rapeseed meal	Safflowerseed meal	Soybean meal
Amino acid[b]			
Aspartic acid	5.9	6.9	10.9
Threonine	3.6	2.2	3.7
Serine	3.6	3.1	5.0
Glutamic acid	13.8	13.4	17.2
Proline	5.3	3.4	5.1
Glycine	4.1	3.8	4.0
Alanine	3.6	3.1	4.2
Cystine	1.2	1.1	1.5
Valine	3.8	3.0	4.2
Methionine	1.6	1.0	1.4
Isoleucine	2.9	2.2	3.8
Leucine	5.4	4.4	7.2
Tyrosine	2.3	1.7	3.5
Phenylalanine	2.9	2.7	4.6
Lysine	5.2	2.4	6.0
Histidine	2.1	1.5	2.4
Arginine	4.6	5.9	6.7
Chemical composition			
Crude protein, %	30.7	25.8	49.4
Crude fat, %	21.7	8.7	0.9
Acid detergent fiber, %	25.2	42.3	9.1
Total glucosinolates, μmoles/g[c]	78.3	...	...

[a]Data expressed on a dry matter basis

[b]Percent of protein

[c]Isothiocyanates plus oxazolidinethiones

Table 5. Performance of Male Broiler Chicks Fed Rapeseed and Safflowerseed Meal Diets in a 22-day Feeding Trial

Item	Diets				
	Control	25% RSM	50% RSM	25% SM	50% SM
Gain/chick/day (g)	37.60[a]	30.00[b]	18.50[c]	36.85[a]	36.55[a]
Feed consumption/chick/day (g)	65.97[a]	52.76[b]	37.93[c]	64.93[a]	63.03[a]
Feed/gain	1.72[a]	1.74[a]	2.02[b]	1.73[a]	1.68[a]

[a,b,c]Means within same row with different superscripts differ ($P<.01$)

Table 6. Effect of Glucosinolate Intake on Thyroid Length:Body Weight Ratios in Hubbard Cockerel Chicks After 25-day Feeding of Rapeseed and Safflowerseed Meal Diets

Diet	Total glucosinolate intake[d]	Thyroid length[e]	Weight of bird (kg)	T:W[f]
Control	...	6.38 (10)	1.15	5.57[a]
50% Saff	...	6.18 (9)	1.13	5.46[a]
25% RSM	0.014	7.89 (12)	.94	8.42[b]
50% RSM	0.020	6.46 (12)	.61	10.54[c]

[a,b,c]Means within the same column with different superscripts differ ($P<.05$)

[d]Total amount of isothiocyanates plus oxazolidinethiones ingested per bird in moles

[e]Mean value in millimeters. Values in brackets refer to the number of thyroids examined

[f]Thyroid length, (mm)/bird wieght (kg)

Table 7. Effect of Protein Source on Dry Matter Digestibility and Nitrogen Utilization (Rapeseed Meal)

	Protein supplement				
Item	CSM	Urea	33% RSM	66% RSM	100% RSM
Apparent digestibility, %					
Dry matter	69.1	68.2	68.3	66.0	66.8
N balance, g/day					
Intake	21.5	26.9	25.1	24.0	25.3
Fecal	6.8	9.2	8.4	8.6	8.0
Urine	8.1	12.0	10.7	11.8	11.2
Absorbed	14.7	17.7	16.8	15.5	17.3
Retained	6.7	5.7	6.0	3.6	6.0
% of N intake					
Fecal	31.9	34.2	33.3	35.4	31.8
Urine	37.5[a]	44.6[b]	42.6[ab]	49.5[b]	44.7[ab]
Absorbed	68.1	65.8	66.7	64.6	68.2
Retained	30.6[a]	21.05[ab]	24.1[ab]	15.1[b]	23.5[ab]
% absorbed N retained	44.8[a]	31.9[b]	36.1[ab]	23.4[b]	34.5[ab]

[a,b]Means within the same row with different superscripts differ ($P<.05$)

Table 8. Effect of Protein Source of Dry Matter Digestibility and Nitrogen Utilization (Safflower)

Item	Protein supplement				
	CSM	Urea	25% SM	50% SM	75% SM
Apparent digestibility, %					
Dry matter	59.1[a]	68.3[b]	56.5[a]	58.6[a]	56.2[a]
N balance, g/day					
Intake	34.3	34.2	35.9	35.2	41.0
Fecal	10.5	8.8	11.1	10.5	11.2
Urine	19.3	19.4	20.9	20.6	21.6
Absorbed	23.7	25.4	24.8	24.8	29.8
Retained	4.5	6.0	3.9	4.2	8.2
% of N intake					
Fecal	30.6	25.8	31.2	29.7	27.5
Urine	56.1	56.6	58.5	58.1	52.4
Absorbed	69.4	74.2	68.8	70.3	72.6
Retained	13.3	17.8	10.3	12.3	20.2
% absorbed N retained	19.2	23.6	14.5	17.3	27.8

[a,b]Means within the same row with different superscripts differ ($P<.05$)

EVALUATION OF SOYBEAN OIL-AQUEOUS ETHANOL MICROEMULSIONS FOR DIESEL ENGINES

C. E. Goering
Sr. Member ASAE

A. W. Schwab

R. M. Campion
Student Member ASAE

E. H. Pryde
Exec. Affiliate ASAE

ABSTRACT

Hybrid fuels, formed by microemulsifying aqueous ethanol in soybean oil, were evaluated by burning them in a diesel engine. No. 2 diesel fuel was also burned to provide baseline data. The hybrid fuels performed nearly as well as No. 2 diesel despite having lower cetane numbers and less energy content. At present, the hybrid fuels are more expensive than No. 2 diesel fuel and their effect on engine durability is unknown.

INTRODUCTION

Dr. Rudolph Diesel used peanut oil to fuel one of his engines at the Paris exposition of 1900 (Nitscke and Wilson, 1965). Intermittently since then, there have been other studies on the use of vegetable oils as fuel for diesel engines. Recently, the foreseeable depletion of world petroleum reserves and the instability of conventional petroleum sources has generated renewed interest in such studies.

Neat vegetable oils are too viscous for prolonged use in direct-injected diesel engines, and several techniques are being used to reduce fuel viscosity. Viscosity can be reduced by blending the vegetable oil with other, less viscous liquid fuels and the resulting blends have been termed hybrid fuels. The most popular hybrid diesel fuels have resulted from the blending of vegetable oils with conventional diesel fuels. Nonpetroleum hybrid fuels would be preferable in the longer term to completely eliminate dependence on petroleum. Several organic solvents are available for blending into nonpetroleum, hybrid fuels for diesel engines.

Ethanol is an organic solvent that can be produced in rural communities from agricultural feedstocks. However, neither anhydrous nor aqueous ethanol are

---

This study was a part of Project No. 390 of the Agricultural Experiment Station, College of Agriculture, University of Illinois at Urbana-Champaign. It was supported in part by grants to the University of Illinois from Cooperative Research, USDA, and from The Illinois Soybean Program Operating Board.

The authors are: C. E. Goering, Professor, and R. M. Campion, Student, Department of Agricultural Engineering, University of Illinois, Urbana, IL; and A. W. Schwab, Research Chemist, and E. H. Pryde, Research Leader, Northern Regional Research Center, Agricultural Research Service, U.S. Department of Agriculture, Peoria, IL 61604.

directly miscible with most vegetable oils. Immiscible liquids can be joined into macroemulsions (i.e., ordinary emulsions), but continued agitation is required to sustain such emulsions. Use of microemulsions eliminates the need for agitation.

Microemulsions are transparent, thermodynamically stable colloidal dispersions in which the diameter of the dispersed-phase particles is less than one-fourth the wavelength of visible light (Boruff, et al., 1982). The microemulsions form spontaneously when suitable surfactants are used to sufficiently reduce the interfacial tension between the dispersed and continuous phases. Hybrid fuels of special interest are water-in-oil type microemulsions in which the vegetable oil is the continuous phase and aqueous ethanol forms the dispersed phase.

In the research reported herein, organically derived surfactants were used to microemulsify aqueous ethanol (190 proof) with soybean oil. Soybean oil was chosen because it is least expensive of the domestically grown vegetable oils and is available in the largest quantities (Goering and Daugherty, 1981). Two series of hybrid fuels that were investigated were ionic and nonionic microemulsions of aqueous ethanol in soybean oil. The objective of the study was to evaluate the hybrid fuels by burning them in a diesel engine.

FUEL PROPERTIES

The compositions (by volume) of the hybrid fuels are given in Table 1. SAE Standard J313C (SAE, 1979) establishes limits for certain key properties of

Table 1. Composition of the Hybrid Fuels

Fuel Component	Chemical Formula	Fuel: Ionic Hybrid	Fuel: Nonionic Hybrid
Soybean oil	--	52.3	53.3
190-proof ethanol	C_2H_6O	17.4	13.3
1-butanol	$C_4H_{10}O$	20.5	33.4
Linoleic acid	$C_{18}H_{32}O_2$	6.54	0
Triethyl amine	$C_6H_{15}N$	3.27	0

No. 2 diesel fuel. Table 2 compares certain of those limits with measured properties of the nonionic hybrid fuel. Properties were measured using standard ASTM procedures (ASTM, 1981). The hybrid fuels were not tested for water and sediment, ash, and total and active sulphur because earlier research had shown that the components of the hybrids were very low in these contaminants (Goering, et al., 1981a; Goering, et al., 1981b). Distillation temperatures were not measured because earlier research had shown that vegetable oil cracks during distillation (Goering, et al., 1981a). While they are not shown in Table 2, the fuel properties of the ionic microemulsions would differ little from those of the nonionic microemulsion, except that the viscosity of the former was 8.77 mm^2/s.

Calculated properties of the hybrid fuels and of diesel fuel are shown in Table 3. The higher heating values (HHV) of the hybrid fuels were calculated from component HHV's that were measured by the ASTM standard bomb calorimeter method (ASTM, 1981). The HHV's of the hybrid fuels were approximately 81% of that of No. 2 diesel fuel. The stoichiometric air-fuel ratios of the hybrids were approximately 79% of that of No. 2 diesel. Fuel properties differed

Table 2. Comparison of Fuel Properties

Property	ASTM Method	Fuel: Nonionic Hybrid	Fuel: Limits for No. 2 diesel
Flash point, °C	D93	27.8	51.7 min
Cloud point, °C	D2500	a	b
Pour point, °C	D97	-65	c
Water & Sediment, %	D1796	-	0.05 max
Carbon residue, %	D524	0.18	0.35 max
Ash, %	D482	-	0.01 max
Dist. temp. at 90% point, °C	D86	-	282-338
Viscosity at 37.8°C, mm^2/s	D445	6.77	1.9-4.1
Total sulphur, %	D129	-	0.05 max
Active sulphur, %	D130	-	d
Cetane No.	D613	25.1	40 min

[a]Cloud points were not attempted because fuel separated into 2 layers at 0°C.

[b]Specified at 6°C above the tenth percentile minimum ambient temperature.

[c]Not specified by SAE but usually 4 to 6°C below cloud point.

[d]Test interpreted by comparison of immersed copper strip with standard immersed strips.

Table 3. Calculated Properties of the Fuels

Property	Fuel: Ionic Hybrid	Fuel: Nonionic Hybrid	Fuel: No. 2 diesel
Higher heating value, kJ/kg	36687	37045	45343[a]
Stoichiometric A/F ratio	11.60	11.57	14.55

[a]Measured.

little between the hybrids except that the ionic hybrid was more viscous.

The cetane rating of the hybrids was well below the SAE minimum for No. 2 diesel fuel (Table 2) and so use of a cetane improver was investigated. Figure 1 shows the effect of primary akyl nitrate (PAN) on the cetane number of the nonionic hybrid fuel. A 10% concentration is a massive amount of PAN and this was the amount required to bring the cetane rating of the nonionic microemulsion to the minimum of 40 for No. 2 diesel fuel.

EQUIPMENT

A John Deere* Model 152 power unit was used for testing the fuels. John Deere Model 830 tractors use the same engine. The three-cylinder, naturally aspirated direct-injection diesel engine displaced 2.491 liters and was rated at 26.3 kW at a speed of 2400 rev/min. The engine used a Roosa Master distributo type injection pump with normal injection advanced 26° before head

*The mention of firm names or trade products does not imply that they are endorsed or recommended by the University of Illinois or the U.S. Department of Agriculture over other firms or similar products not mentioned.

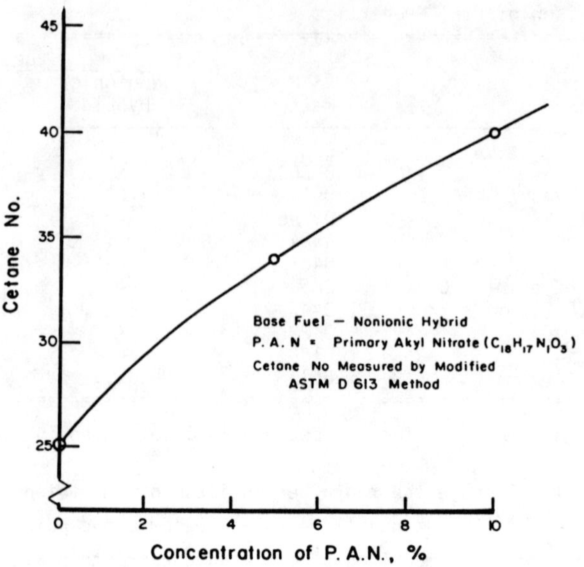

Figure 1. Cetane Enhancement of a Hybrid Fuel

dead center. The engine was connected through an overcentering clutch to a Midwest Dynamic Type-768 eddy current dynamometer. Fuel consumption and engine speed were measured through use of an automatic weighing system and a standard chronotachometer that measured elapsed time and engine revolutions while 100 grams of fuel were being burned. Chromel-alumel thermocouples and an Omega model 199 digital indicator were used to monitor exhaust and coolant temperatures. A counter flow heat exchanger with automatic control of secondary water was used to regulate the temperature of the engine coolant. Air was supplied to the engine through an orifice meter connected to a double surge tank. A calibrated, inclined manometer permitted measurement of the pressure drop across the orifice. The double surge tank included a boost fan to maintain atmospheric pressure at the inlet of the engine.

PROCEDURE

The engine was started and run on No. 2 diesel fuel until the coolant reached the controlled temperature of 89$\pm$3°C. A baseline test was then run on No. 2 diesel fuel. Loading began at 2527 rev/min high idle speed and increased until governor's maximum speed was reached. At each load, the load, speed, fuel and air consumption and exhaust and coolant temperatures were measured.

After completion of the baseline run, the engine was switched to the nonionic hybrid fuel, the fuel return line was diverted to a waste container, and the engine was run until the diesel fuel was flushed from the system. The same test procedure used in the baseline tests was then repeated for the nonionic hybrid fuel.

The ionic hybrid fuel was then tested using a procedure similar to that for the nonionic hybrid. Finally, a second baseline test was run on No. 2 diesel fuel.

RESULTS

Figure 2 and Table 4 show the key results of the engine tests. Although the hybrids contained 19% less energy per kilogram than No. 2 diesel fuel (Table 3), the nonionic microemulsions was able to produce almost the same peak power as diesel fuel (Table 4). The ionic microemulsion produced 5% less

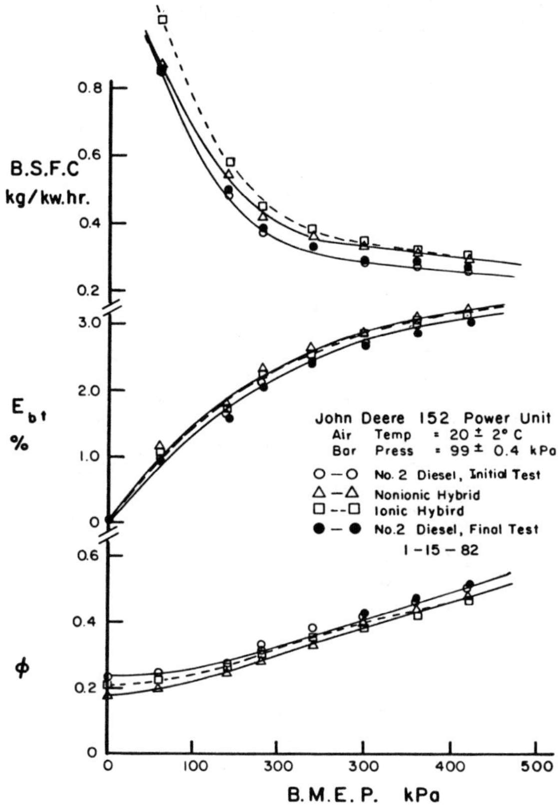

Figure 2. Performance of a Diesel Engine or Diesel and Hybrid Fuels.

Table 4. Engine Performance at Maximum Power

Test fuel	Max Power kW	Fuel Supplied mg/injection	Energy Supplied kJ/injection	Brake Thermal Efficiency %
No. 2 diesel	24.1	86.1	3.91	30.5
Ionic hybrid	22.9	101.2	3.71	32.2
Nonionic hybrid	23.7	99.9	3.70	32.3
No. 2 diesel	23.9	86.9	3.94	30.3

power than No. 2 diesel. The higher viscosity of the hybrids increased the size of the fuel injections (Table 4), but not enough to achieve equal energy per injection. The approximately 6% less energy per injection of the hybrids was offset by a corresponding gain in brake thermal efficiency, and so enabled the engine to produce more power than was expected.

The engine speeds (and thus the mechanical efficiencies) were essentially the same for all of the fuels tested. As indicated by the equivalence ratio (Ø, or the actual fuel-air ratio divided by the stoichiometric fuel-air ratio) in Figure 2, the hybrid fuels burned leaner than diesel fuel because of the oxygen contained in the hybrids. The leaner burn gave better thermal efficiency for the hybrids. Although the increased efficiency and increased fueling rate were able to maintain high power output of the engine on the hybrid fuels, the brake specific fuel consumption (BSFC) was higher with the hybrids as shown in Figure 2. All of the data in Figure 2 are plotted against brake mean effective pressure (BMEP), or specific torque. For the 2.491 L test engine, the torque in Newton meters would be 0.198 times the BMEP.

DISCUSSION

The low cetane numbers of the hybrid fuels had no adverse effect on their short-term performances in the engine. The engine ran well on the hybrid fuels after warmup, and the audible diesel knock was the same as for No. 2 diesel fuel. Cold starting with the engine at room temperature was easily achieved on the hybrid fuels by using an ether assist. The cetane number of the hybrids does not seem indicative of their performance. A study is needed to evaluate the energy release patterns of the hybrid fuels during combustion, and such a study is now underway at the University of Illinois.

The safety aspects of hybrid fuels must be considered. The most volatile component of the hybrid fuels was ethanol, with a flash point of 14.4°C. The flash point of the nonionic hybrid fuel was 27.8°C and, based on similarity of composition, the flashpoint of the ionic hybrid would be close to 27.8°C. Thus, the hybrid fuels are less volatile than ethanol and could be handled with procedures considered to be safe for handling ethanol.

The effect of the hybrid fuels on engine durability has not yet been determined. The short-term performance tests were accomplished in 3.5 hours. There was no degradation in engine performance during that time because, as shown in Figure 2, the final run on No. 2 diesel fuel substantially reproduced the initial run. Exhaust temperatures averaged from 15 to 37°C lower with the hybrid fuels as the load increased from idle to governor's maximum. The cooler burning would be helpful to exhaust valves and other temperature-stressed parts. However, many more hours of testing will be required to evaluate the effect of the hybrid fuels on engine durability.

The cost per liter of fuel is estimated to be $0.54 for the ionic hybrid, $0.48 for the nonionic hybrid and $0.24 for No. 2 diesel fuel at the wholesale level. The price of diesel fuel is currently depressed because of the current economic recession and the glut of petroleum on the market. Until petroleum prices rise and the price gap narrows, the hybrid fuels are not competitive with petroleum. However, theprice of petroleum could rise quickly if oil imports were interrupted, and the hybrid fuels could then compete as emergency fuels. In the longer term, world reserves of petroleum will be depleted and all fuels will have to be renewable.

SUMMARY AND CONCLUSIONS

Hybrid fuels were formed by creating microemlusions of aqueous ethanol in soybean oil. A nonionic microemulsion was formed by using 1·butanol as a surfactant, and an ionic microemulsion was formed by using a mixed surfactant. The objective of this study was to evaluate the hybrid fuels by burning them in a diesel engine and to compare them to the No. 2 diesel fuel that was burned in baseline tests. The following conclusions were drawn from the study:

1. The nonionic microemulsion produced nearly as much engine power as No. 2 diesel fuel despite having a 19% lower heating value.

2. Increased viscosity of the hybrid fuels gave a 16% increase in the mass of each fuel injection at maximum power, but the injections contained 6% less energy than those of No. 2 diesel fuel.

3. Oxygen in the hybrid fuels produced leaner combustion and yielded a 6% gain in thermal efficiency over No. 2 diesel fuel at full power.

4. Brake specific fuel consumption was 16% higher with the hybrids than with No. 2 diesel fuel at full power.

5. The nonionic fuel gave slightly higher thermal efficiency, higher engine power and lower specific fuel consumption than the ionic hybrid and thus was slightly more effective as a fuel.

6. Diesel knock was no worse for the hybrid fuels than for No. 2 diesel fuel and thus the low cetane numbers of the hybrid fuels was not reflected in engine performance.

7. The hybrid fuels were less volatile than ethanol and thus could be handled safely with procedures considered safe for ethanol.

8. The effect of the hybrid fuels on engine durability is unknown and should be determined before the fuels are commended for general use.

9. The hybrid fuels are currently too expensive to compete with No. 2 diesel fuel but could serve as an emergency fuel if petroleum supplies were interrupted.

REFERENCES

1. ASTM 1981. Annual Book of Standards, part 23, petroleum products and lubricants. Am. Soc. for Testing Materials, Philadelphia, PA.

2. Boruff, P. A., A. W. Schwab, C. E. Goering and E. H. Pryde. 1982. Evaluation of diesel fuel-ethanol microemulsions. Transactions of the ASAE 25(1):p. 47-53.

3. Goering, C. E. and M. J. Daugherty. 1981. Energy inputs and outputs of eleven vegetable oil fuels. ASAE Paper No. 81-3586, ASAE, St. Joseph, MI 49085.

4. Goering, C. E., D.R. Hunt, S. Furcich and M. J. Daugherty. 1981a. Vegetable oil-solvent blends as diesel fuel. Proceedings of Regional Workshops on Alcohol and Vegetable Oil as Alternative Fuels, Northern Agr. Energy Center, Peoria, IL.

5. Goering, C. E., A. W. Schwab, M. J. Daugherty, E. H. Pryde and A. J. Heakin. 1981b. Fuel properties of eleven vegetable oils. ASAE Paper 81-3579, ASAE, St. Joseph, MI 49085.

6. Nitscke, W. R. and C. M. Wilson. 1965. Rudolph Diesel, pioneer of the age of power. The University of Oklahoma Press, Norman, OK.

7. SAE, 1979. SAE Handbook. Society of Automotive Engineers, Warrendale, PA.

PROPERTIES AND PERFORMANCE TESTING WITH BLENDS OF BIOMASS ALCOHOLS, VEGETABLE OILS AND DIESEL FUEL

S. Vinyard
Member ASAE

E. S. Renoll
Member ASAE

J. S. Goodling
Member ASME

L. Hawkins
Undergraduate Student, ME

R. C. Bunt
Graduate Student, ME

ABSTRACT

This paper is a presentation of results from three related efforts to determine the technical feasibility of using alcohols and vegetable oils blended with Diesel oil as fuel for unmodified compression ignition engines. Several different vegetable oils were successfully tested in a single cylinder engine. Sunflower oil was blended from 50% to 80% by volume with Diesel fuel and used in a multicylinder engine. Thermophysical property data were gathered on pure and blended fuels and are reported. A spray parameter, ε, was found which would predict the necessary change in valve opening pressure to render the atomization of the new fuel similar to that for which the injection system was designed. Engine testing showed that fuel consumption was substantially reduced upon setting the injectors at the new VOP.

INTRODUCTION

The apparent predicted dearth of fossil fuels has spurred an interest in biomass-derived substitutes for the internal combustion engine. The majority of investigations have concentrated on vegetable oils in diesel engines in pure or blended forms (Engleman 1978*; Barsic and Humke 1981; Peterson, et al. 1981) and alcohol blends in spark ignition engines (API 1976; Pefley 1979). Some work has been done on alcohol blended fuels as Diesel substitutes, but the ultimate applications have not been satisfactory (Wrage and Goering 1979; Pratt 1980; Rao 1981). Most performance testing reported was done by comparison of engine characteristics with Diesel fuel under similar operating conditions. Only recently has extensive investigation of applicable thermophysical properties been reported for vegetable oils (Peterson, et al. 1981; Goering, et al. 1981). Little work is found which reports results of engine modifications necessary to optimize engine performance based upon the properties of these alternative fuels or blends.

The authors are: S. Vinyard, U. S. Army Tank Command, Warren, MI; E. S. Renoll, Professor, Agricultural Engineering Department; J. S. Goodling, Professor, Mechanical Engineering Department; L. Hawkins, undergraduate student; R. C. Bunt, graduate student; Mechanical Engineering Department, Auburn University, Auburn, AL.

---

*A complete bibliography is not intended here; only a few representative authors are referenced.

This paper shows the results of three systematic steps of a two-year effort to determine the technical feasibility of using alcohols and vegetable oils blended with diesel oil as fuel in unmodified compression ignition engines. The first phase explored the use of six different agricultural oils in a single cylinder Diesel engine. The purpose of these tests was simply to see if the engine would function properly and to gain experience in alternative fuel engine testing. One of these oils was chosen as a prime target for the second phase which consisted of blending from 50 to 100% sunflower oil with diesel fuel and determining engine performance as a function of percent blend on a multicylinder engine. The third phase was to find thermophysical properties of two pure alcohols, pure vegetable oils and diesel fuel, to blend them to a mixture which had kinematic viscosity properties near those of diesel fuel, to modify the engine externally for accommodation of the new blend, and finally to observe engine performance with the new blends.

EXPERIMENTS

Phase One

Four different vegetable oils were tested under identical conditions in a small, single-cylinder compression-ignition engine. They were crude sunflower, crude and refined cottonseed, soybean, and peanut oils. Filtered, used cooking oil and Number 2 diesel fuel were also included in this evaluation.

The engine used was a Peter Model AA1 diesel. It is a single-cylinder, air-cooled, overhead valve, four-stroke engine with a Lanora combustion chamber, with bore and stroke of 69.9 and 57.2 mm and displacement of 219.6 cm^3. With a compression ratio of 17:1, the engine is rated at 2.8 kW at 3,600 rpm. The dynamometer used was a GO-Power Model DY-7DD water brake dynamometer designed for small, single-cylinder engines. Engine speed was measured by a strobe light tachometer and engine torque by a hydraulic load cell. Fuel, oil, and cylinder head temperatures were measured with thermocouples. A burette and stopwatch were used to measure fuel flow volumetrically. Testing and data reduction were done following the procedures set forth in SAE Standard 816b, Engine Test Code.

Full power curves were obtained and the results showed that the vegetable oils ranged from 86% (refined soybean oil) to 97% (refined peanut oil) of that obtained with diesel fuel. Refined SFO showed a peak power decrease of only 7% from that for diesel fuel. Due to differing properties of the various fuels, consumption was compared on mass and volume bases. In all cases, specific fuel consumption decreased, with refined peanut oil showing the best improvement (27% and 19% respectively) and crude cottonseed oil the least (13% and 4% respectively). Refined SFO displayed improvements of 27% and 11% on a mass and volume basis. Thermal efficiency (the inverse of OBSFC) showed similar trends.

Though all the vegetable oils showed improvement over diesel fuel in fuel efficiency, the results are somewhat in question. Improvements in thermal efficiency are much higher than expected. Since power was lower with each of the oils than with diesel fuel, only a significant improvement in combustion efficiency could bring about the improvements in OBSFC and thermal efficiency. The high viscosity of the oils may provide better penetration into the air cell of the Lanora combustion chamber causing a change in the combustion process. The high viscosity could also change the advance in injector timing due to less leakage in the injector pump. Cylinder pressure and heat release data would be useful in further investigations to determine the nature of vegetable oil combustion in this engine.

Phase Two

Based on the fact the refined sunflower oil is a readily available agricultural product and the results in Phase One showing that fuel consumption was decreased, it was chosen as the alternative fuel for the second part of this effort. SFO has higher density, much higher viscosity, lower Cetane Number and lower specific energy than diesel fuel. These differences will have a definite effect on the performance of a diesel engine with an injection system designed specifically for diesel fuel. This phase was devised to determine how seriously an engine's performance is affected by increasing amounts of blended substance (SFO). Blends from 50 to 100% by volume of refined sunflower oil were made with diesel fuel and run in a 4 cylinder, 4 stroke, 2.82 liter (172CID), Ford Diesel tractor engine with bore 9.9cm. (3.9 in.) and stroke 9.14 cm (3.6 in.). A Roosa Master injection pump supplied fuel to the direct injector nozzles, each with four 0.27 mm orifices. A water brake dynamometer (Go Power D-502) was used to load and sense torque. All procedures and measurements specified by SAE StandardJ816b were followed. The engine was first run at full power using diesel fuel and data were taken to find if the engine met manufacturer's performance curves. Having confirmed this, variable speed, continuous power runs were made at 85% of full power torque for all fuels used. Individual data sets are not presented here. Rather Figure 1 shows a comparison of smoothed fuel consumption curves for the various blends as a function of engine speed. The engine failed to operate for mixtures above 90% SFO.

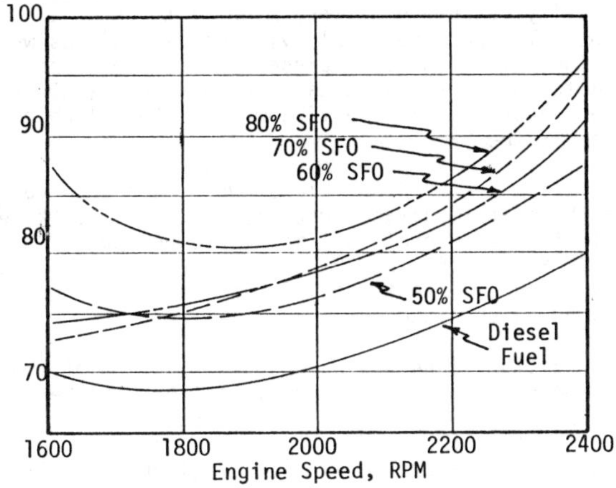

Fig. 1. Fuel Consumption Compared for Diesel Fuel with Volumetric Blends of Sunflower Oil

The trend is an increase in fuel consumption approximately proportional to the amount of SFO added to the diesel fuel except at the lowest speeds. This could be explained as follows: OBSFC is an indication of the mass fuel consumption, but fuel is injected by volume in this Diesel engine. The higher density of SFO would cause higher mass injection for the same volume; because of high injection pressures, it is normal to have intentional leakage past the injector needle, to be pumped back into the fuel system (for cooling and lubrication.). The greater viscosity of SFO would reduce this leakage, again resulting in more fuel charge entering the combustion chamber.

Phase Three

Based on the experiences from above, it was decided to gather extensive thermophysical property of the fuels, in pure and blended forms. These properties included higher heating value, kinematic viscosity, density and surface tension. This was in hopes of finding a blend of either of two new alcohols (Dioxalane and MEK)*, vegetable oil and diesel fuel that would have properties similar to 100% diesel fuel. The next step was to compare the spray pattern of the alternative combination to that of diesel fuel. It was suspected that a change of the valve opening pressure (VOP) or injection timing would alter engine performance in a positive fashion with the new fuel. Finally engine performance data were taken.

Fuel Properties: Properties were measured using the applicable ASTM Tests or Standards. These included higher heating value at constant volume for liquid fuels (isothermal and adibatic bomb calorimeters using ASTM D240), kinematic viscosity (Cannon-Fensky Routine viscometers using ASTM D445), and surface tension (Harkins and Brown 1918). Density was measured at room temperature by use of an accurate pycnometer and beam balance scales. Most properties were measured two or three times independently by different experimenters. The averaged results are shown in Table 1. Some data are not reported due to unreproducibility or inconsistency in the experimental results. As can be seen from Table 1, the heating values of the alcohols are quite low while those for the vegetable oils are only several percent lower in comparison with Diesel fuel. Viscosity measurements were made at the nominal temperatures of 40, 60, 80 and 100C. The pure vegetable yielded viscosities which are an order of magnitude higher than those of Diesel fuel. Dioxalane alcohol was much lower by a factor of four. No data were obtained at 80 and 100C because of excessive evaporation in the viscometer. No reliable data were gotten for MEK because of the same reason.

Mixtures of MEK or Dioxalane, SFO and Diesel were made on a volume basis in anticipation that the most important property of the blend (kinematic viscosity) would resemble that of diesel fuel at all temperatures tested. Such was the case for 10% MEK, 5% SFO and 85% diesel. Surface tension was nearly the same for all substances investigated. Heating values for the pure and blended fuels behaved as predicted.

Spray Studies: Atomization, penetration and distribution of fuel in a combustion chamber are dependent upon nozzle configuration, chamber turbulence pressure, and fuel properties. The former two are not influenced by the fuel for a given engine and to alter either would be a major and expensive modification of the engine. However, for a fuel of given properties, the VOP will influence the velocity of injection (primarily through the Reynolds number) and hence the entire spray. It is known from semi-empirical studies (Harmon, 1955) that a spray pattern is a function of Reynolds number, Weber number, ratio of kinematic viscosities and ratio of densities for the two substances involved (fuel and air).

These two fuels were recently found from biomass conversion processes, which are the property of Auburn University. Dioxalane and MEK were derived from waste or low grade hardwood residues which were rich in hemi-cellulose. MEK and Dioxalane were also derived from a pulp and paper plant waste stream which had a high xylan content. The hemicellulose and xylans were converted to xylose and then fermented to butaneidiol (Butylene glycol) using a microorganism strain specifically mutated to efficiently convert hardwood wastes. The butanediol was then converted to MEK by an energy efficient liquid phase process. These processes for the transformation of hardwood wastes to MEK are proprietary processes of Auburn University. Butanediol extracted from the fermenter product, was converted to Dioxalane by the Mathias process.

Table 1. Measured Thermophysical Properties of Petroleum and Biomass Fuels

Fuel	Higher Heating Value	Kinematic Viscosity, cSt at Nominal Temperature, C				Surface Tension at T Dyne/cm/T(C)
	KJ/Kg	40	60	80	100	
Diesel Oil No. 2	45282	2.91	2.12	1.59	1.22	27.26/25.0
Dioxalane	34552	0.752	0.573			27.35/24.0
MEK	33758					
Butanediol 2, 3	27421	16.64	6.85	3.45	2.04	
n-Butanol		2.22	1.45	1.005		
ME(10%)SFO(15%)DF2(75%)	43054	3.09	2.20	1.685	1.35	26.88/25.0
DIOX(10%)DF2(90%)	44003	2.57	1.81			26.84/25.0
MEK(10%)DF2(90%)	44216	2.48	1.77	1.37	1.067	
DIOX(10%)SFO(5%)DF2		2.54	1.78	1.36	1.09	26.90/24.6
SFO (CRD)	41449	18.60	11.21	6.21	4.90	40.95/22.5
SFO (REF)	39417	31.31	17.53	9.59	7.49	31.97/21.2
SFO(%)/DF2(%)						
10/90	44998					27.81/22.0
20/80	44651	5.53	3.66	2.63	1.99	28.72/27.0
25/75		5.51	3.52	2.55	1.92	28.33/21.5
30/70	43723	7.26	15.21	4.08	3.25	25.20/20.0
40/60						
50/50	42473	9.43	6.17	4.77	3.19	26.84/21.7
60/40	41415	11.57	7.21	5.16	3.94	28.89/24.0
70/30	40755	16.16	9.32	6.24	4.43	30.30/25.6
80/20	40103	17.84	13.09	6.69	4.89	29.70/23.2
90/10	39463	23.05	14.02	8.72	5.89	
Peanut Oil (REF)	40630	34.82	19.24	12.34	8.28	32.00/24.0
PNO(%)/DF2(%)						
10/90	44785					30.16/24.5
20/80	43815					29.32/23.1
30/70	43848					28.64/23.0
Soybean Oil (REF)	40922	32.27	17.8	10.85	7.48	30.90/21.0
Cottonseed Oil (REF)	38201	32.85	18.05	11.03	7.60	32.20/20.0
Corn Oil (REF)		31.46	19.99	11.39	7.60	36.62/20.4

A theoretical drop size and spray angle were calculated using Harmon's equations as a function of measured properties. Measured valves for the actual spray angle and area were obtained. This was done by removing one injector from the engine, forcing a single pulse of fuel through it, and collecting the fuel sprayed on absorbent paper. The discolored area (A) was measured with a planimeter and the angle of dispersion (Θ) calculated from that. A spray pattern parameter, ε, defined as

$$\varepsilon = \frac{(A/\Theta)\text{measured}}{(A/\Theta)\text{theoretical}}$$

was then determined for each fuel. For diesel, the value was found to be 15.4. Other fuels were tested similarly until a VOP was found which yielded ε nearly that for diesel fuel. Table 2 shows the results for all fuels tested. It can be seen that as much as 18% increase in VOP was required to obtain the same value of ε.

Table 2. VOP for Various Fuels required for ε = 15.4

Fuel Type	VOP, kPa (psi)	Change
Diesel Fuel No. 2	18960 (2750)	0
10% Diox/90% DF2	19480 (2825)	+2.7%
10% MEK/90% DF2	18750 (2720)	+1.1%
25% SFO/75% DF2	22270 (3230)	+17.5%
10% Diox/5% SFO/85% DF2	18270 (2650)	-3.6%
10% MEK/15% SFO/85% DF2	21990 (3190)	+16.0%

Engine Testing: All testing was done in the manner described in Phase Two above with several additions. Water and oil temperatures were controlled external to the engine and fuel filters were changed with each new fuel. The reams of experimental data gathered would probably be of no interest to the reader and are therefore not shown here. However, Figure 2 shows dramatically the effect of changing the VOP in a predictable fashion on OBSFC.

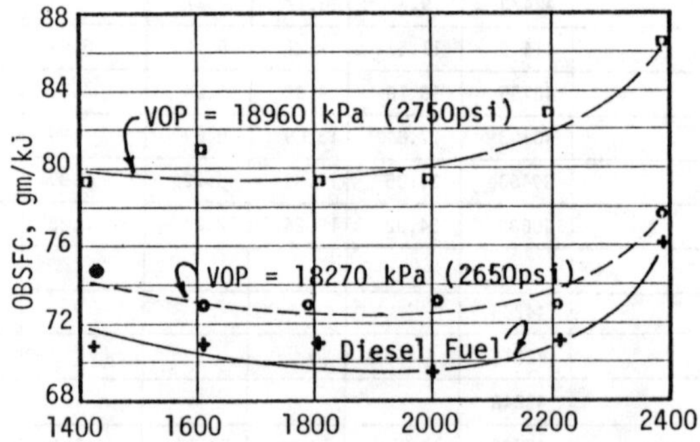

Fig. 2. Results of 10% Dioxalane, 5% Sunflower Oil and 85% Diesel Fuel at two VOP compared to 100% Diesel Fuel

Three curves are shown in Figure 2. The lowest is OBSFC as a function of RPM for diesel fuel. The highest is for 10% Dioxalane, 5% SFO and 85% DF2 at the manufacturer's specified VOP of 18960 kPa (2750 psi). The middle curve is for the same fuel at a VOP of 18270 kPa (2650 psi). As can be seen, the BSFC is reduced drastically from 14% above that for diesel to only 4% above.

All other fuels investigated showed similar trends but not quite as pronounced an improvement.

CONCLUSIONS

A simple external test was devised which indicates that VOP for a new fuel can be predetermined to yield a spray pattern similar to that for the standard fuel. Without major fuel injection or engine modification, the machine can be adjusted accordingly to accept the new fuel and it will perform nearly as well with the new fuel as with diesel fuel. Alcohol up to 10% by volume, have been successfully run in a common multicylinder Diesel engine without a separate fueling system. No studies on either lubrication or emissions were performed for the alcohol/oil/diesel fuel combinations and thus, no statement can be made regarding the total acceptability of these fuels.

ACKNOWLEDGEMENTS

The author gratefully acknowledges the support of the Mechanical and Agricultural Departments and the Engineering Experiment Station of Auburn University for their support of this work.

REFERENCES

1. American Petroleum Institute. 1976. Alcohols, A Technical Assessment of Their Application as Fuels. July 1976

2. Barsic, N.J. and A.L. Humke. 1981. Vegetable Oils: Diesel Fuel Supplements? (Automotive Engineering 89(4) 89(4):37-41

3. Engleman. H.W., D.A. Guenter, and T.W. Silvis. 1978. Vegetable Oils as a diesel fuel. Diesel and Gas Engine Power Division of ASME, Paper No. 78-DG-P-19, ASME WAM, Nov 1978

4. Goering, C.E., A.W. Schwabe, M.J. Daugherty, E.H. Pryde, A.J. Heakin. 1981. Paper No. 81-3579. 1981 Winter Meeting ASAE, Chicago, Illinois, Dec. 1981

5. Harkins, W.D., F.E. Brown. 1918. The Determination of Surface Tension (Free Surface Energy), and the Weight of Falling Drops. Journal American Chemical Society. 37:499-524, 1918

6. Harmon, D.B. 1955. An Equation for Predicting a Mean Drop Size in High Speed Spray. Univ. of Cal. Pub., 5. 1955

7. Pefley, R.K. 1979. Alcohol Fueled Automobile Performance - Emissions - Environment, Univ. of Santa Clara, CA, Oct. 1979

8. Peterson, C.L., G.L. Wagner, D.L. Auld. 1981 Performance Testing of Vegetable Oil Substitutes for Diesel Fuel. Paper No. 81-3578. 1981 Winter Meeting ASAE, Chicago, Illinois, Dec 1981

9. Pratt, T.N. 1980 Alternate Fuels Requirements for Medium Speed Diesel Engines. 7th Energy Technology Conference, Washington DC, March 1980

10. Roa, B.H. 1981. Combustion of a Wide Range of Fuels in a Diesel Engine. FUEL 60: 554-556

11. Wrage, K.E., C.E. Goering. 1979. Technical Feasibility of Diesohol. Paper No. 79-1052. 1979 Summer Meeting ASAE, CSAE, Winnipeg, Canada, June 1979

ALTERNATIVE FUEL BLENDS AND DIESEL ENGINE TESTS

David E. Braun and K. Q. Stephenson
Department of Agricuiltural Engineering
The Pennsylvania State University

The U.S. dependence on uncertain sources of imported petroleum warrants expeditious and continuing research for the development and testing of alternative fuels. Particularly vulnerable to shortages and high costs of liquid fuels are the transportation, agricultural, and industrial users of mobile machinery and equipment. The major power sources for these industries will continue to be diesel engines. Coupled with the uncertain, diminishing supply, and predicted exhaustion in this century of non-renewable, petroleum based fuels, the development of alternative renewable sources is urgent. Agriculturally based sources are renewable, would reduce U.S. dependence on imports, and provide needed new markets for farm products.

OBJECTIVES

The objectives of this work were to determine the effects of vegetable oil, ethanol and diesel fuel blends on diesel engine performance, and evaluate their combustion characteristics. Specifically, the principal objectives were as follows:

1. Determine diesel engine performance with various blends of degummed soybean oil, neat ethanol, and diesel fuel.
2. Determine fuel blend viscosities and compare their P-T diagram combustion characteristics.
3. Inspect injectors, valves, lubrication oil, and system for fuel blend effect on coking, contamination, and fouling.

REVIEW OF LITERATURE

Diesel tractor performance and engine durability, including long term field tests, were conducted with a variety of ethanol-based, multi-component fuel blends as reported by Meiring et al. (1981). Satisfactory results were obtained by co-solvent additives for blend stability and octyl nitrate to adjust the cetane rating to the equivalent of diesel fuel.

Peterson et al. (1981), found that engine life in long term tests of five different vegetable oils tested as substitutes for diesel fuel, was related to fatty acid content and viscosity of the oils. In short term tests, vegetable oils were found to give performance, power output, and fuel consumption equivalent to diesel fuel.

Goering et al. (1981), evaluated and compared the fuel properties of eleven vegetable oils. Their results showed that all the vegetable oils were much more viscous and more reactive to oxygen than No. 2 diesel fuel. Vegetable oils with the most favorable combinations of properties for fuel oils were soybean, cottonseed, corn, rapeseed, and sesame. Quick (1980), reported that coking of fuel injectors was a problem in unmodified diesel engines with long term operation on four kinds of vegetable oils. Short term tests with various vegetable oils as fuel gave results equivalent to diesel fuel.

The authors are: DAVID E. BRAUN, Graduate Assistant, and K. Q STEPHENSON, Professor, Agricultural Engineering Department, The Pennsylvania State University, University Park, PA. 16802.

Yarbrough et al. (1981) found that the gums in sunflower seed oil were detrimental to diesel engine performance, and even degummed and dewaxed sunflower seed oil seriously contaminated the lubricating oil during engine operation. Kaufman et al. (1980) compared sunflower oil and blends of sunflower oil with diesel fuel, to baseline results of three diesel engines. Overall, they found that engine performance on sunflower oil was comparable to operation on diesel fuel in short term tests.

The Division of Agricultural Engineering, Republic of South Africa (1981) issued a news release reporting success in operating a diesel engine tractor for 2,300 hours on pure sunflower oil. Their results were obtained with an indirect injection engine. Inspection after 1,000 hours of operation showed the engine was free of abnormal wear and injector coking.

A study on the production, marketing, and utilization of degummed soybean oil as a diesel fuel extender was conducted and reported by Adams et al. (1981) for Gold Kist, Inc. Their results showed a favorable energy balance for soybean oil as fuel; an adquate supply and a present, 1981, world-wide surplus; and it was the lowest cost alternative diesel fuel available in fuel quantities for the immediate future.

Ryan et al. (1981), reported on the qualitative aspects of alternate diesel fuels composed of vegetable oil components in solution with petroleum-derived fuels. Vegetable oil mixtures of soybean, sunflower, peanut, and corn with diesel fuel have been satisfactory in four-stroke cycle engines though vegetable oils are much higher in viscosity and slightly lower in heating value. Studies of ethanol microemulsions with diesel fuel indicated potential for reducing the quantity of petroleum-derived fuel, and neat ethanol was reported to be usable as diesel engine fuel if surface ignition was provided.

A summary of investigator's results on vegetable oils for diesel fuel was published in Appendix I of Vegetable Oil as Diesel Fuel Seminar II (1981). The wrap-up summary of that seminar stated: 1) certain pre-combustion chamber engines reportedly operate more satisfactorily than direct-injection engines on both neat vegetable oils and mixtures with diesel fuel; 2) most problems with vegetable oil fuels in diesel engines resulted from improper combustion; 3) lubrication oil contamination did not occur under conditions of proper combustion; 4) soybean oil for fuel should be degummed and sunflower oil dewaxed, and 5) blended or hybrid fuels appear to be the most probable substitutes for diesel fuel in unrestricted use.

EQUIPMENT

An Allis-Chalmers D-17 tractor with a Model D-262, 4 stroke, 6 cylinder energy cell diesel engine was used in these tests. The energy cell diesel combines the precombustion chamber and air cell into a hybrid design. The energy cell consists of a major and minor chamber, which opens into the main combustion chamber. The cell is on the opposite side from the injector and contains approximately 10-15% of the top dead center volume of the combustion chamber. The cell induces a secondary turbulence which aids more complete combustion. The engine was rated at 38 Kw (51.14 hp) at 1650 rpm. The injection system contained a Roosa Master injector pump and Allis Chalmers pintle type injectors with an opening pressure of 13.8 MPa (2000 psi). Static injection timing was 16° BTDC; displacement was 4.3L and compression ratio was 15.7:1.

An M & W Gear PTO Dynamometer was used to load the engine. Thermocouples were installed to monitor intake and exhaust manifold temperatures. An electronic digital top loading balance was used to measure consumed fuel weight for each run. A 60 tooth gear and magnetic pickup produced a time base pulse for P-T diagrams with pulse-to-pulse time intervals equal to six degrees of crankshaft rotation. Combustion chamber pressure was measured with a piezoelectric pressure transducer mounted in the number one cylinder energy cell port.

PROCEDURE

The engine was disassembled to determine its initial condition and make needed repairs and adjustments. A new injector tip was installed on cylinder number one to check wear during subsequent tests. Baseline tests were conducted with No. 2 diesel fuel. A fuel blend of 40 percent diesel fuel, 40 percent degummed, dewaxed soybean oil, and 20 percent ethanol (40:40:20) by volume was used in the engine to compare to diesel fuel baseline data. The blends were agitated by aeration utilizing a low pressure air source. The engine was tested by the following running time sequence procedure:

30 minutes	low idle, no load
30 minutes	rated speed, rated load
60 minutes	60% speed, 60% load
60 minutes	80% speed, 20% load

Maximum power output (full rated load) data was taken following each time sequence.

After 25 hours of operation, the engine was disassembled, inspected, and serviced. A new fuel blend of 30 percent diesel, 40 percent degummed, dewaxed soybean oil, and 30 percent ethanol (30:40:30) by volume was used. The test sequence was continued for another 25 hours (50 total) of operation. The engine was again disassembled and inspected. Crankcase oil was drained and inspected following each 25 hour period.

FUEL CHARACTERISTICS

Diesel injector systems are designed for a certain fuel viscosity which provides the most favorable spray pattern for combustion. Soybean and other vegetable oils have a very high viscosity compared to diesel fuel. The higher viscosity will change operating characteristics of the injection system, such as pump performance, system pressures, and spray patterns. Since ethanol has a very low viscosity combining it with the vegetable oil reduces the viscosity of the blends.

The fuel blends used in these tests were not completely miscible. The 40:40:20 blend remained mixed at room temperature ($20^{o}C$) for a long period of time. The 30:40:30 blend separated after a few seconds at the same temperature. Heating this blend to a temperature of $48^{o}C$ corrected the separation problem. A low pressure air source was used to agitate the blend in the test contanier to prevent separation during test runs. The fuel bypass line was returned to the fuel container which increased the temperature of the blend. Observations of fuel flow at the return line proved that no separation of components occured in the injection system.

A Cannon-Fenske calibrated viscometer was used to determine the viscosities of the fuel blends. It was found that the blends were about twice as viscous as diesel fuel and much less than soybean oil (Table 1). All readings were taken at 37.78°C (100°F) except that of the 30:40:30 blend. This blend was tested at 48.89°C, 54.44°C,, and 60°C (120°F, 130°F, and 140°F) and extrapolated to a 37.78°C (100°F) value. The ethanol blends significantly reduced the high viscosity of soybean oil and diesel fuel blends and the injector spray patterns were not noticeably different from diesel fuel as observed during injector tests.

Table 1. Viscosities of Fuels at 37.78°C (100°F)

Fuel	Specific Gravity	Viscosity (Centistokes)	Viscosity (Centipoises
No. 2 Diesel Fuel	0.8400	2.85	2.39
Soybean Oil	0.9118	38.09	34.73
Ethanol	0.7800	1.39	1.08
40-40-20	0.8562	5.78	4.95
30-40-30	0.8462	5.29	4.48

DISCUSSION

The full load, full governor power output of the engine declined about 1.5 percent for the 40:40:20 blend and 4 percent for the 30:40:30 blend (Graph 1). Thermal efficiencies tended to increase compared to diesel fuel (Graph 2). The power reduction was anticipated because of the lower energy content of the blends (Table 2). The higher thermal efficiencies recovered some of the power loss due to lower Btu contents of the blends. Engine performance was not measurably affected by the lower energy content.

Table 2. Energy and Power Output.

Fuel	Energy (KJ/GM)	Energy (%)	PTO Brake KW	Power (%)
Diesel	46.21	100	44.34	100
40-40-20	39.64	85.8	43.66	98.5
30-40-30	37.72	81.6	42.52	95.9

Following fifty hours of operation, inspection revealed no excess carbon buildup. The cylinder walls were smooth and well lubricated. The combustion chamber ends of the energy cells were coated with slightly more carbon than the other parts of the combustion chamber, possibly due to the injector position, spraying fuel directly at that point. The cells themselves may have reduced carbon accumulations by creating a secondary turbulence that provided more complete combustion of the fuel. The lower viscosity of the blends, compared to soybean oil, maintained a more normal spray pattern and reduced carbon deposits and coking of the injectors.

After seven hours of operation on the 40:40:20 blend, a black tar-like substance was observed at the connection between the exhaust manifold and muffler. The residue accumulated up to the 25 hour mark at all low speed cycles of the time sequence. During the full speed, full load period, the residue burned off on the exhaust manifold. When the fuel was changed to a 30:40:30 blend, larger amounts of the residue accumulated in the exhaust. The substance was not found inside the exhaust manifold or near the valves. It was related to the combustion of the soybean oil since no similar residue was observed in operation on diesel fuel or previous ethanol fumigation tests.

The pressure transducer with charge amplifier produced pressure-time diagrams of the combustion process. The recorded pressures may be slightly higher than those in the main combustion chamber because of the position of the transducer in the energy cell (6, pg. 606). The main chamber pressures may be lower and smoother. The traces show a slight ignition delay at full governor, full load conditions (Figure 3). At idle speeds, clear differences were evident in the pressure-time diagrams. The ignition delay caused by ethanol was clearly shown, but no audible knock was detected in these tests (Figure 4).

CONCLUSIONS

Engine performance was normal with the fuel blends used. A slight decrease in fuel energy supplied gave a decrease in available horsepower, but higher' thermal efficiences partially offset the decrease. No unusual problems were found in the 50 hours of short term tests. Continued tests are required to determine long term effects of the blends.

The fuel blend viscosities were approximately double that of diesel fuel but much less than that of soybean oil. Observed injector spray patterns showed no appreciable differences on a test stand when comparing the blends to diesel fuel. Some contamination of the lubrication oil was indicated, but results of used oil analyses were not available at this time. The combustion pressure-time diagrams showed slight ignition delay due to the ethanol component, however, no knock was audible while tests were in progress.

Inspection of the engine after each 25 hour period of operation revealed no excess carbon buildup. The injector tips and energy cells were free of coking. Longer term tests will be conducted to determine the extent of carbon buildup problems. The indications from these tests showed that carbon buildup and injector coking were not problems with the combustion of the blends used.

The blends showed little change in engine performance by substitution of 70% petroleum based No. 2 diesel fuel with renewable sources of energy. The use of these fuel blends were promising and continued research will determine what problems may occur in long term tests.

LITERATURE CITED

1. Adams, Curtis H., W. S. Luttges, Michael Mullen, M. C. Rand, Joseph F. Peters, Bernard J. Schorr, and M. Carl Ziemke. (1981). Study on Production, Marketing, and Utilization of Degummed Soybean Oil as a Diesel Fuel Extender. Prepard for Gold Kist Inc., P.O. Box 2210, Atlanta, GA 30301.

2. Division of Agricultural Engineering, Department of Agriculture and Fisheries, Rep. of South Africa. Sunflower Oil as Tractor Fuel: Research Report. News Release (1981).

3. Goering, C. E., A. W. Schwab, M. J. Daugherty, E. H. Pryde, and A. J. Heakin. (1981). Fuel Properties of Eleven Vegetable Oils. ASAE Paper No. 81-7019. ASAE, St. Joseph, MI 49085.

4. Kaufman, K. R., M. Ziejewski, M. Marohl, and H. L. Kucera. (1980). Performance of Diesel Oil and Sunflower Oil Mixtures in Diesel Farm Tractors. ASAE Paper No. 80-1054. St. Joseph, MI 49085.

5. Meiring, P., R. S. Allan, and P. W. L. Lyne. (1981). Ethanol-Based Multiple Component Fuels for Diesel Tractors. ASAE Paper No. 81-1055. ASAE, St. Joseph, MI 49085.

6. Obert, Edward F. <u>Internal Combustion Engines</u>. 3rd Ed. International Textbook Co., Scranton, PA. 1968.

7. Quick, Graeme R. (1980). Developments in Use of Vegetable Oil as Fuel for Diesel Engines. ASAE Paper No. 80-1525. ASAE, St. Joseph, MI 49085.

8. Peterson, C. L., G. L. Wagner, and D. L. Auld. (1981). Performance Testing of Vegetable Oil Substitutes for Diesel Fuel. ASAE Paper No. 81-3578. ASAE, St. Joseph, MI 49085.

9. Ryan, T. W. III, W. E. Likos, and C. A. Moses. (1981). Future Diesels May Use Hybrid Fuels. <u>Automotive Engineering</u>. Vol. 89(1) pp. 63-65.

10 Vegetable Oil as Diesel Fuel Seminar II. (1981). Northern Agricultural Energy Center, Northern Regional Research Center, Peoria, IL. Oct. 21 and 22, 1981.

11. Yarbrough, C. M., W. A. LePori, and C. R. Eugler. (1981). Compression Ignition Performance Using Sunflower Seed Oil. ASAE Paper No. 81-3576. St. Joseph, MI 49085.

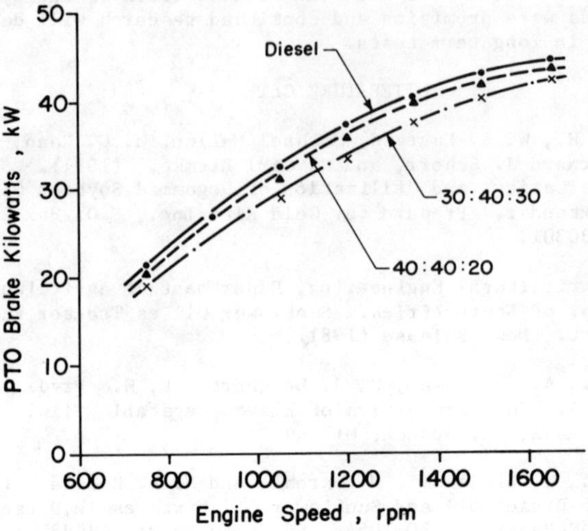

Graph 1. Brake PTO Power vs Speed at Full Governor and Full Load.

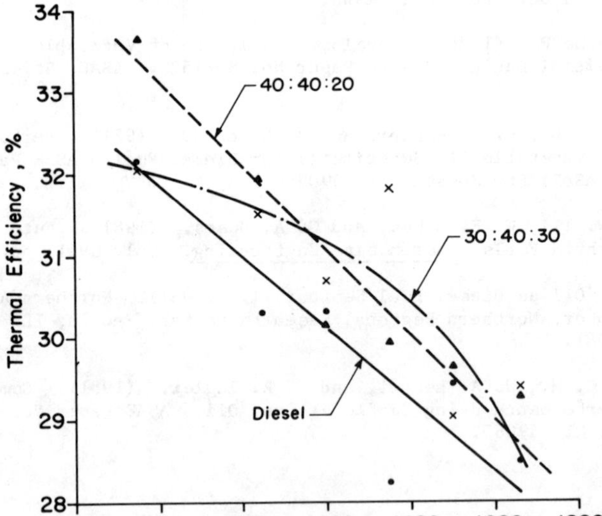

Graph 2. Engine Thermal Efficiencies vs Speed at Full Governor and Full Load.

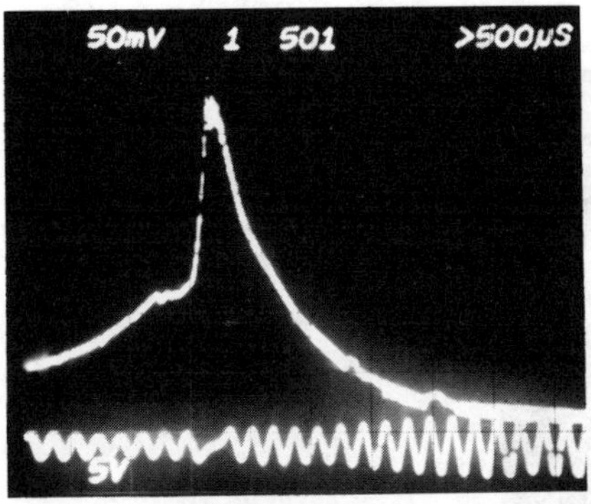

(a.) Diesel Fuel at Full Load, Full Governor and 1650 RPM.

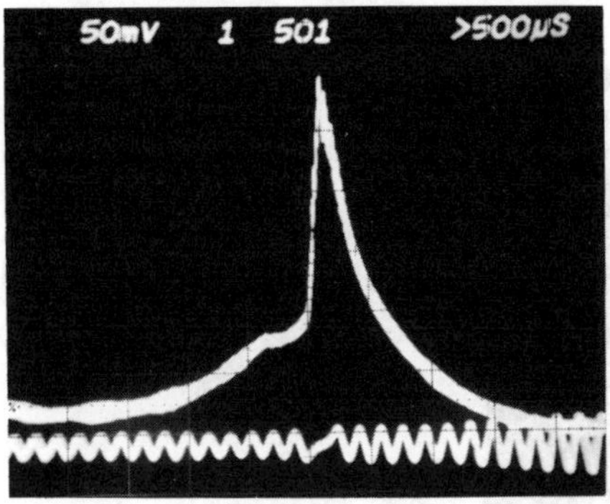

(b.) 40:40:20 Blend at Full Load, Full Governor and 1650 RPM.

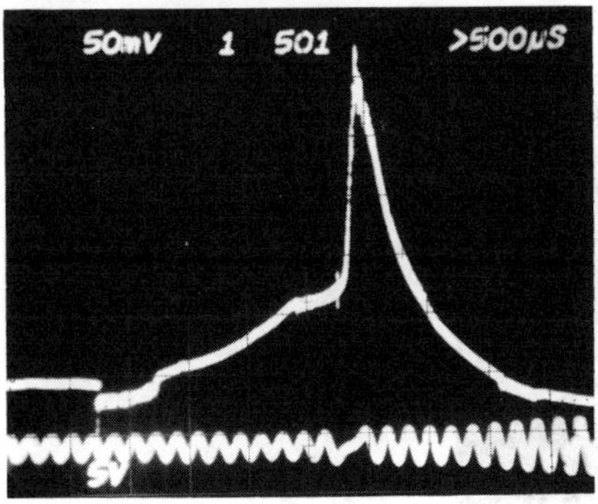

(c.) 30:40:30 Blend at Full Load, Full Governor and 1650 RPM.

Figure 3. P-T Diagrams of Blends Combustion compared to Diesel Fuel.

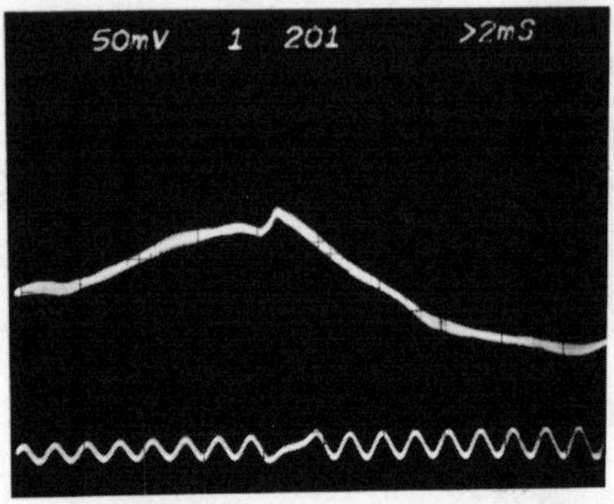

(a.) Diesel Fuel at Idle.

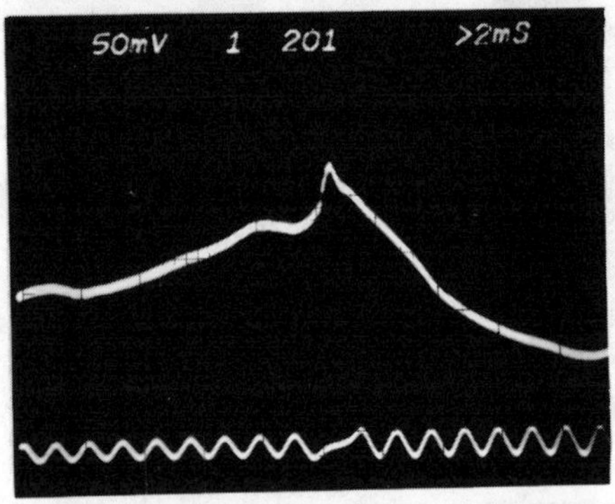

(b.) 40:40:20 Blend at Idle.

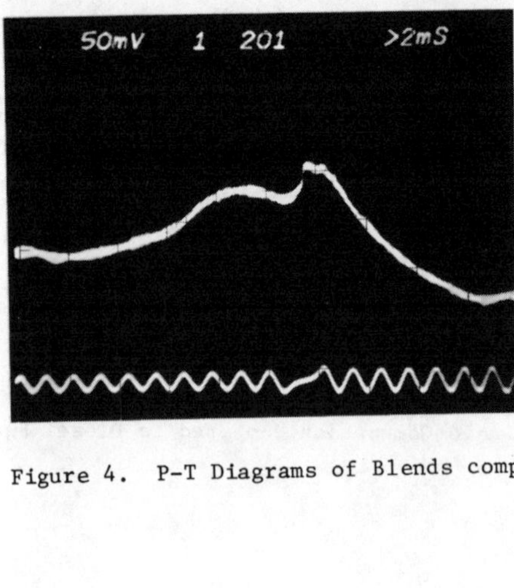

(c.) 30:40:30 Blend at Idle.

Figure 4. P-T Diagrams of Blends compared to Diesel Fuel.

SOME CORRELATION OF DIESEL ENGINE PERFORMANCE WITH INJECTION CHARACTERISTICS USING VEGETABLE OIL AS FUEL

K. S. VARDE*

ABSTRACT

It is known that the performance of a DI diesel engine is greatly affected by the injection and spray characteristics. These are influenced by the injection system design and by the properties of the fuel used. The effect of some of the physical properties of diesel fuel and soybean oil on the injection and spray parameters were measured and compared in the study. The injection system used in this investigation was the same as the one used in the engine performance evaluation using the two fuels. The performance of the engine is correlated with some of the spray parameters but it seems that the chemical composition and properties of the oil play important role in the emissions area of the engine.

INTRODUCTION

Oils derived from farm-products have long been considered as potential fuels for combustion engines but until recently these farm derived fuels attracted very little attention. The possibility that the conventional petroleum base engine fuels may not be available in sufficient quantities nor at a reasonable price has revived interest in exploring vegetable oils as fuels for diesel engines. Literature search shows a number of recent studies performed to evaluate vegetable oils as diesel substitute (Barsic and Humke, 1981; Engelman, et al, 1978; Forgeil and Varde, 1981; Hofman, et al, 1980; Quick, 1980; Ryan, et al, 1980; etc.)

The performance of a direct injection (DI) diesel engine is uniquely determined by the fuel properties and by the manner in which the fuel is introduced in the combustion chamber which, in turn, depends on the spray characteristics of the fuel emerging through the injector holes. These characteristics are strongly dependent on the injection system design, fuel properties and the underlying hydraulics of the system. Current diesel engines have been developed to their level of performance primarily by using diesel fuel with controlled properties. Since vegetable oils have different physical and chemical properties than those in diesel fuel, it is expected that a DI diesel engine designed for diesel operation will perform differently when fueled by vegetable oils.

A number of earlier studies investigated the engine performance when operated on diesel fuel and vegetable oils. But not much has been done to correlate how the relative differences in some of the physical properties of vegetable oils and diesel influence spray and hence engine performance. The present study looks at the effects of some of the properties of diesel fuel #2 and soybean oil on the injection pressure and other spray parameters. Attempt is made to correlate these with the performance of a naturally aspirated, single cylinder DI diesel engine, which was equipped with two different spray nozzles during the tests.

*The author is: K.S. Varde, Associate Professor, Department of Mechanical Engineering, The University of Michigan-Dearborn.

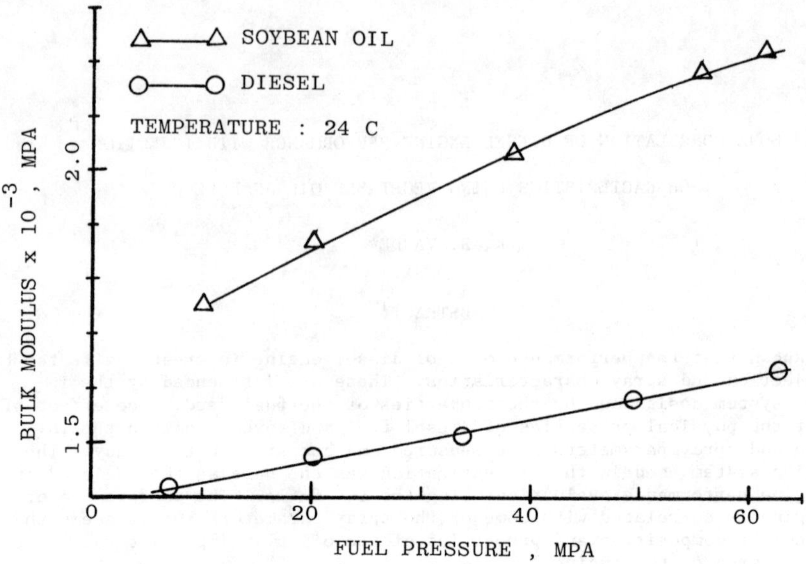

FIG. 1. VARIATION OF BULK MODULUS FOR DIESEL AND OIL

EXPERIMENTAL ENGINE SET-UP

The test engine was a single cylinder, naturally aspirated, DI diesel engine with a bore and a stroke of about 95 mm and a compression ratio of 17. The engine was coupled to a water brake dynamometer for loading purposes. The fuel injection system of the engine comprised of a plunger type pump and an injector with a needle opening pressure of about 19 MPa. The original injector had four orifices, each 0.32 mm in diameter. The tests were conducted with two different injection nozzles: the original nozzle (nozzle A) and another one having identical hole pattern as nozzle A but with orifice diameter of 0.36 mm (nozzle B).

The engine was run on two different fuels: diesel #2 and hydrogenated soybean oil. The latter was refined and had a very low wax content. Table 1 shows some of the properties of the fuels used in the program; some of these

Table 1. Engine Specifications.

Bore, mm	95
Stroke, mm	95
Compression Ratio	17
Cooling	Air-cooled
Injection Timing, °BTDC	24°
Injection Pump	Bosch plunger pump
Needle Opening Pressure, MPa	19
Orifice Spray Cone, deg.	150

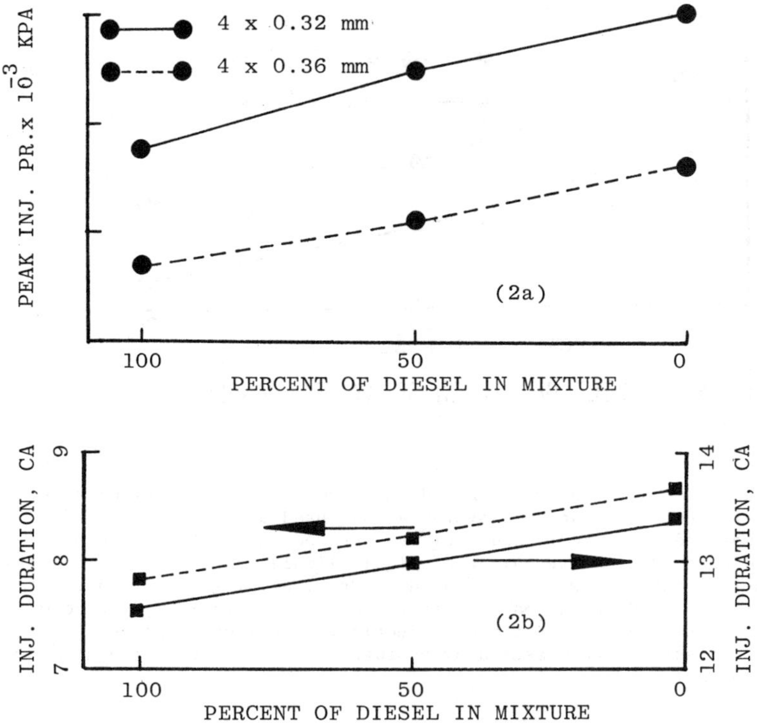

FIG. 2. EFFECT OF OIL ON INJECTION PRESSURE AND INJECTION DURATION.

properties were measured in the laboratory while the others were obtained from the suppliers. Each test comprised about 8 hours of time duration during which the engine was run up to full load at a constant speed of 35 rev/s. The engine was always started and switched off on diesel fuel to avoid soybean oil from staying in the injection system for prolonged time. The fuel was filtered between the supply tank and the injection pump by using dual filters of 10μ size.

Table 2. Fuel Properties

Property	Diesel	Soybean Oil
Density, kg/m^3	830	899
Heating Value, KJ/kg	42580	37064
Cetane No.	46	34
Viscosity, mm^2/s, 18C	3.5	71.8
Heat of Vaporization KJ/kg	255	218
Flash Point, C	60	330
Sp. Heat (Liquid), KJ/kg.k	1.84	2.0
Surface Tension, dyne/cm	28	34

Exhaust gas samples were withdrawn to analyse NOx and HC content using a chemiluminescent analyser and a flame ionization detector, respectively. Smoke levels were measured by a Bosch smoke meter; at least two samples were withdrawn during each test point and the average of these two was considered

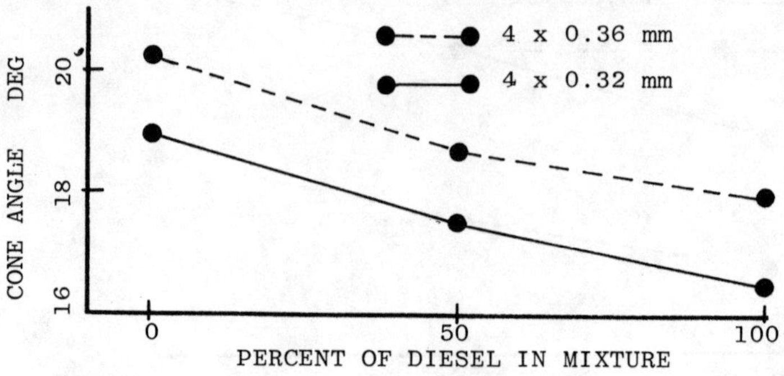

FIG. 3. EFFECT OF OIL ON SPRAY CONE ANGLE.

to be the smoke level.

To measure actual injection duration and spray cone angle, experiments were conducted with the same injection system as was used on the engine. The engine was motored at 35 rev/s and fuel was sprayed into a liquid chamber containing diesel #2. The technique of liquid-liquid injection for measuring spray parameters has been well documented by Gray, et al (1966). Injection duration for full load operation was determined by synchronizing strobe with the start of injection and the end of injection. Spray cone angles of the fully developed spray were measured by photographing the spray with high speed framing camera.

RESULTS OF INJECTION AND SPRAY STUDY

Due to differences in some of the physical properties of the two fuels, the values of some of the injection and spray parameters for the soybean oil are expected to be different than the corresponding values for the diesel fuel. For a given speed peak injection pressure and injection pressure history depends, among other things, on the fuel bulk modulus. Since information on the bulk modulus of oils is very meager, experiments were conducted to measure the isothermal bulk modulus of the oil using the P-V-T method. The apparatus and the method used in the experiment have been described elsewhere (Varde, 1982). The results, shown in Fig. 1, shows the bulk modulus of the soybean oil to be about 35% higher than that for diesel #2 at 24C. The bulk modulus

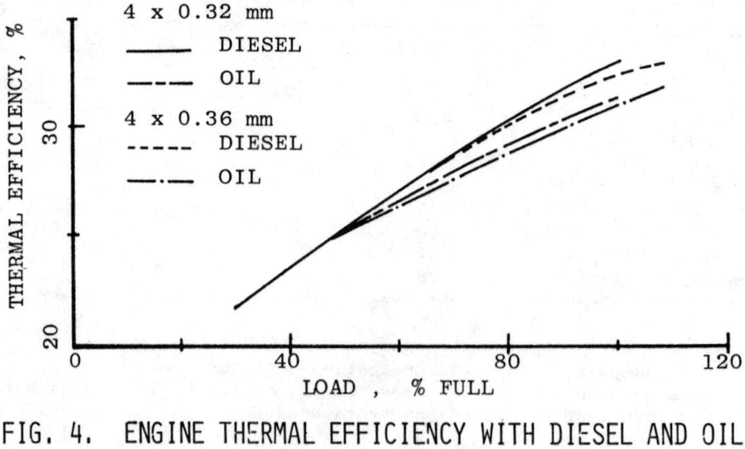

FIG. 4. ENGINE THERMAL EFFICIENCY WITH DIESEL AND OIL

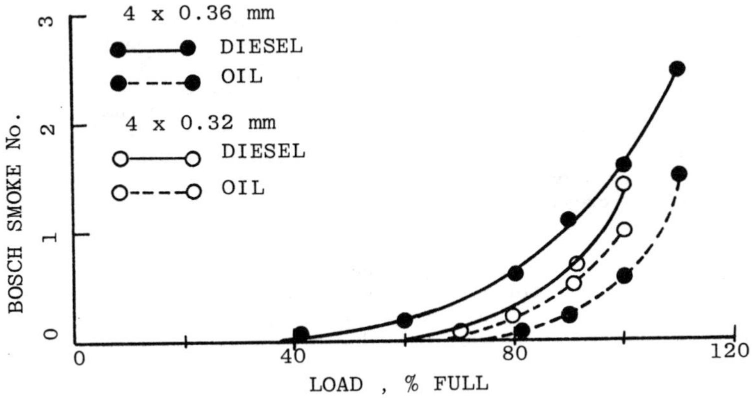

FIG. 5. SMOKE LEVELS WITH DIESEL AND OIL

was also found to increase with an increase in injection pressure.

The effect of increased bulk modulus on peak injection pressure for the full load operation can be seen in Fig. 2a. For both the nozzles injection pressure increased as the composition of the soybean oil in the mixture was increased. Peak pressure for the 4 x 0.36 mm nozzle was lower due to the larger orifice area of the nozzle. Fig. 2b shows the effective injection duration for the full load operation. Injection duration increased as the fuel viscosity increased, a phenomenon previously observed by Gray, et al (1966) in a jerk pump type of injection system. It is interesting to note that the injection duration with nozzle B decreased substantially indicating a high injection rate with this nozzle.

It is generally recognized that the spray cone angle is one of the parameters that determine the entrainment of air in the fuel spray. Cone angles are strongly dependent on the density of the medium into which fuel is sprayed as well as on the characteristic dimensions of the nozzle (Reitz and Bracco,

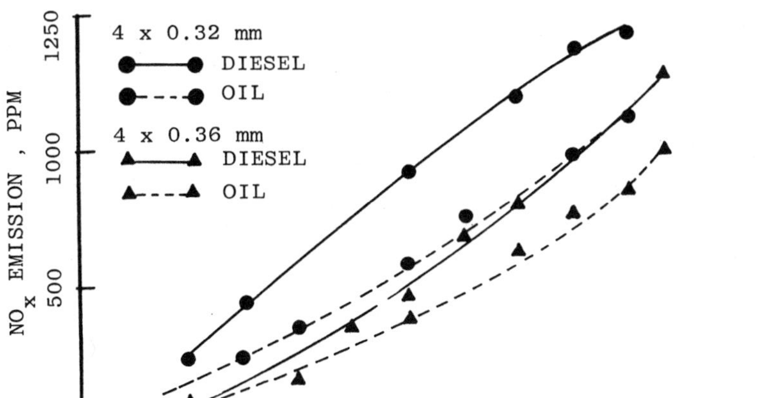

FIG. 6. NOx EMISSION WITH DIFFERENT FUELS

1980). High speed photographs of the fuel spray were used to measure the cone angles of the fully developed spray and Fig. 3 shows a comparison for the full load operation. These cone angles are not the actual angles that would be found in the combustion chamber of the engine since the fuel spray was injected into diesel which has much higher density than the compressed air in the combustion chamber prior to fuel injection. Nevertheless, a relative comparison shows spray cone angle to decrease as the oil content increases. An earlier study (DeJuhasz, 1932) had shown cone angle to depend on fuel discosity; a decrease in fuel viscosity increased the spray cone angle and vice-versa, similar to the findings of the present study.

For a given injection velocity, spray cone angle can be correlated with the entrainment of air and it is believed that a decrease in cone angle results in reduction of the amount of air entrained in the spray. This is expected to affect emissions generated during the combustion process.

ENGINE RESULTS AND CORRELATION

Using either deisel #2 or soybean oil, the engine was run at a constant speed of 35 rev/s using either of the two nozzles. The corrected maximum power output with nozzle B was about 9% more than that with the 4 x 0.32 mm nozzle. It appears that the larger orifices help to inject more fuel in the cylinder probably due to lower leakage past the plunger; this may occur because of relatively lower injection pressures experienced with nozzle B than nozzle A. For a given nozzle, the engine developed almost identical maximum power with diesel and oil, inspite of the fact that the oil had about 12% lower energy content on mass basis. Since soybean oil has higher viscosity, the internal leakage in the injection system is expected to be lower than with the diesel. This was verified by measuring the fuel flow rate with diesel and the oil. Thus, the lower energy content of the oil is compensated by higher volume flow rate at full load operation. Similar trends have been experienced in a recent study (Baranescu and Lusco, 1982) dealing with sunflower oil as a diesel fuel.

Figure 4 shows the engine thermal efficiency with the two fuels. At part load operation, there is not much difference in the efficiency but as the engine load increases, some effect of the fuel can be noticed. Both the nozzles showed lower efficiency with the oil. It appears that the overall combustion time for the oil was somewhat longer than for the diesel fuel; in fact, for a given load and the injection nozzle, the exhaust temperature for the oil was consistently higher than for the diesel by about 10 to 15 C indicating relatively more energy loss in the exhaust. Nozzle B yielded slightly lower efficiency than nozzle A probably due to larger droplets formed with larger orifice nozzle (Sass, 1929). The exhaust temperature for nozzle B was found to be about 5 C higher than for nozzle A using oil as the fuel inspite of the fact that the injection duration for the full load operation, shown in Fig. 2b, was substantially lower for the nozzle B. It reinforces the hypothesis that the soybean oil requires somewhat longer time for the chemical reactions to complete than the diesel fuel.

The effects of vegetable oil on the exhaust smoke levels are shown in Fig. 5. For any given nozzle, smoke levels were found to be consistently lower for the oil than for the diesel fuel. The chemical kinetics of the formation and decomposition of soot in the combustion chamber of a diesel engine is highly complex. And inspite of the advances made is understanding soot and its intermediates in flames, there is no complete agreement on the elementary soot forming and soot burning chemistry. If diesel fuel is considered by itself, one can expect higher smoke levels with nozzle B as shown in Fig. 5. It has lower injection pressure and larger orifices which would result in larger droplets and relatively poorer atomization than nozzle A. However, nozzle B has larger spray cone angle but this may be conpensated by the lower injection velocity thereby keeping air entrainment almost unchanged.

Based on the properties of oil, particularly the visocisty and the cetane number, one might expect increased smoke with the oil for a given nozzle. It is known that an increase in viscosity of the liquid fuel has an adverse effect on the fineness of atomization (Lubbock and Bowen, 1948). The results shown in Fig. 5 contradicts this and they are in agreement with some of the earlier studies (Forgeil and Varde, 1981; Baranescu and Lusco, 1982) but not others (Humke and Barsic, 1981). Various mechanisms for soot formation and consumption have been proposed but it is generally agreed that excess oxygen tends to suppress soot formation, which is why smoke levels are negligible at part load operation. It is also known that OH radicals are responsible for removing soot. The oil used in the study was hydrogenated while the soybean oil used in other study (Humke and Barsic, 1981) was either crude or degummed. It is generally believed that dehydrogenation steps during a combustion process can lead to soot formation; if this is true then a hydrogenated oil will lead to lower smoke levels even though OH radicals play some role in dehydrogenation. Thus one can expect smoke levels with oils either to increase or decrease with respect to diesel fuel depending on the chemical composition of the oil used as a fuel.

Figure 6 shows the variation of NOx with engine load, injection nozzles and the fuels used in the investigation. For a given nozzle, NOx was found to be higher for diesel fuel. In a diesel engine, NOx formation is related to

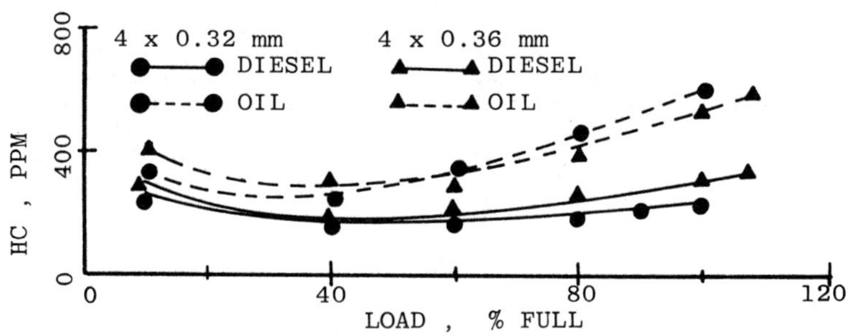

FIG. 7. HYDROCARBON EMISSION WITH DIFFERENT FUELS

the local concentrations of exygen molecules and the temperature in the local zone. As discussed earlier, soybean oil is expected to have relatively larger droplet sizes and poorer atomization. This will lower local temperature history resulting in lower NOx. In addition, the chemical composition of the oil can affect local concentration of the oxygen atoms. For example, the fatty acid composition of the soybean oil contain large proportion of oleic and linoleic acids (Ziejewski and Kaufman, 1982). During combustion process the chemical composition of the oils can alter NOx formation through changes in local concentration of O-atoms.

For a given fuel, NOx emission was lower for nozzle B even though the injection rate of the nozzle was higher than for the nozzle A. The larger orifices and lower injection pressure is expected to increase mean droplet sizes and reduce mixing in the combustion chamber. Both of these effects are known to disfavor NOx formation mechanism. The results shown in Fig. 6 do agree with the expected findings based on the spray parameters and NOx formation mechanism.

Emissions of hydrocarbons did not change much for a given fuel but, on the overall, soybean oil generated more HC emission than diesel fuel, as shown in Fig. 7. Unburned hydrocarbons in a diesel engine exhaust generally consist of either the original or the decomposed fuel molecules or recombined intermediate compounds. The spray characteristics had shown spray cone angle for oil to be lower than the corresponding angle for the diesel fuel. Thus, the center core of the oil spray plume will be denser with fuel drops; in addition, higher oil viscosity is expected to adversely affect atomization. Both of these factors will lead to an extended oxidation reaction in the combustion chamber, thereby leading to increased emissions of HC.

Based on the engine performance results and the spray parameter study, both the plysical and chemical properties of the soybean oil appear to play important role in the combustion process of the oil. The power output, thermal efficiency, emissions of NOx and HC can be correlated with the physical properties such as heating value, viscosity, bulk modulus, etc. and the resultant spray characteristics such as peak injection pressure, spray cone angle, quality of atomization, etc. However, soot formation mechansim appear to be more dependent on the chemical composition of the oil rather than the physical properties.

CONCLUSIONS

On the overall, DI diesel engine performance has been shown to depend on the properties of soybean and diesel fuel through variations in injection and spray parameters. Fuel properties such as bulk modulus, viscosity, etc., affect peak injection pressure, injection duration, injection pressure-time history, spray cone angle, quality of atomization, etc. These variations together with other fuel properties have influence on the maximum power developed by the engine, engine thermal efficiency and some of the gaseous emissions. Smoke levels in the exhaust appear to depend more on the chemical nature of the oil and it seems that physical parameters have secondary influence.

REFERENCES

1. Baranescu, R.A. and J.J. Lusco, 1982. Sunflower Oil as a Fuel Extender in D.I. Turbocharged Diesel Engines. SAE Paper 820260.

2. Barsic, N.J. and A.L. Humke, 1981. Performance and Emissions Characteristics of Naturally Aspirated Diesel Engines with Vegetable Oil Fuels II. SAE Paper 81095.

3. DeJuhasz, K.J., O.F. Zahn and P.H. Schweitzer, 1932. On the Formation and Dispersion of Oil Sprays. Penn State College Bulletin No. 40.

4. Engelman, H.W., D.A. Guenther and T.A. Silvis, 1978. Vegetable Oil as a Diesel Fuel. ASME Paper 78-DGP-19.

5. Forgiel, R. and K.S. Varde, 1981. Experimental Investigation of Vegetable Oils Utilization in a D.I. Diesel Engine. SAE Paper 811214.

6. Gray, J.T., N.T. Meckel and R.J. Mannheimer, 1966. Some Observations on the Liquid Injection Technique as a Research Tool. SAE Paper 660748.

7. Hofman, V., D. Zimmerman, D. Helgeson, W. Dinusson and C. Fanning, 1980. Sunflower Oil as a Fuel Alternative. Paper presented at the Agricultural Energy Alternatives Workshop, North Dakota State University, Fargo, N.D.

8. Lubbock, I. and I.G. Bowen, 1948. The Gas Turbine and its Combustion Problems. Joint Conference on Modern Applications of Liquid Fuels, Inst. of Petroleum and Inst. of Fuels, London: 21-23.

9. Quick, G.R., 1980. Developments in Use of Vegetable Oils as Fuel for Diesel Engines. Paper 80-1525 presented at the ASAE meeting, Chicago.

10. Reitz, R.D. and F.V. Bracco, 1979. On the Dependence of Spray Angle and Other Spray Parameters on Nozzle Design and Operating Conditions. SAE Paper 790494.

11. Ryan, T.W., W.E. Likos and C.A. Moses, 1980. The Use of Hybrid Fuel in a Single Cylinder Diesel Engine. Paper presented at the Central States Section meeting of the Combustion Institute, Baton Rouge, Louisiana.

12. Saas, F., 1929. Compressorless Diesel Engines. J. Springer Publishers, Berlin, Germany.

13. Varde, K.S., 1982. Dependence of Diesel Fuel Bulk Modulus on Pressure and Temperature. To be published in Can. J. Chem. Engr.

14. Ziekewski, M. and K.R. Kaufman, 1982. Endurance Test of a Sunflower Oil/ Diesel Fuel Blend. SAE Paper 820257.

PERFORMANCE, DURABILITY AND LOW TEMPERATURE EVALUATION OF SUNFLOWER OIL AS A DIESEL FUEL EXTENDER

Rodica A. Baranescu and Joseph J. Lusco
International Harvester Company - Engine Engineering

ABSTRACT

The paper presents the results of a research project to evaluate performance and durability of direct injection turbocharged diesel engines using sunflower oil and blends thereof. Alcaline refined sunflower oil and three different blends of sunflower oil and diesel fuel were comparatively tested against #2 diesel fuel for: physical and chemical characteristics, fuel injection system performance, short term engine performance, propensity to nozzle deposits buildup, limited durability operation and low temperature starting capability. Results are presented for the various phases of the project and correlations between the fuel characteristics and engine acceptability are discussed.

The interest in vegetable oils as fuels for diesel engines is not new. It may seem surprising today that in the late 1920's, engineers and researchers could think of oils derived from plants as potential fuels for diesel engines. The idea of using a product that could be obtained from renewable sources with characteristics that in appearance were similar to diesel fuels, was very attractive especially in countries rich in oleaginous plants and poor in mineral resources. Experiments were reported about engine performance potential using vegetable oils. Durability and serviceability issues were addressed or inventoried for further research [20], [9], [17].* Even the approach of chemically changing the vegetable oil composition (approach embodied in what today is called transesterification of oils) was mentioned in the published literature of that period. However, research was mostly academical and never materialized in practical solutions. Competition of petroleum based fuels was too strong, diesel fuels were in plentiful supply, at reasonable cost and with excellent characteristics for engine performance and life.

The last decade has witnessed our growing concern over the exhaustion of conventional energy sources in an environment that continuously increases its demand on liquid fuels. The conscientious search of alternative sources for petroleum base liquid fuels, the specific need of becoming energy self-sufficient and therefore less vulnerable to shortages, embargos or geopolitical considerations has brought the vegetable oils again into focus. In only a few years, information has started to flow-in again. Engine manufacturers, academic communities and research establishments have active programs pertaining to vegetable oil subjects; variety of sources, physical and chemical characterization, engine performance, and - as time went by - incipient results on engine life, serviceability and durability. In parallel, research is carried out to understand the availability potential of vegetable oil plants, the cost of processing the oil, the impact in the energy scenario, the practicality of oil fuels in normal or emergency situations, the price, the right balance between food and fuel, etc.

---

*Numbers in brackets indicate references at the end of the paper.

Issues are numerous; attitudes cover all the range from overoptimistic to overpessimistic. But as information accumulates we seem to rediscover what was learned from the long period of engine experience. Any new fuel different from the ones we already know with narrow established specifications for good performance and long life, changes the complicated balance between the fuel and the engine. Intensive research and testing are required before a new match can be established and the practicality can be demonstrated.

In spite of some similarities, vegetable oils and diesel fuels are different in physical properties and in chemical structure. From strictly engine performance standpoint, these differences materialize in peculiarities of fuel injection process, air/fuel mixing, combustion initiation and development. The degree of performance alteration will depend on the fuel used and on the engine sensitivity to fuel injection and combustion characteristics.

From a durability viewpoint, the characteristics of vegetable oils are likely to affect the amount of engine and fuel system deposits, the rate of engine wear and the low temperature operation.

The paper presents the results of a research project carried out at the International Harvester Corporation-Engine Engineering to evaluate performance and durability of direct injection turbocharged engines using vegetable oils and oil blends. Correlations of the physical and chemical properties of the fuel with the injection process and with the combustion development were established.

The test engine was a DT-436B, a direct injection, turbocharged six cylinder engine, representative of the IH line of diesel engines used in farm tractors and combines. The engine main features are:

- bore/stroke	109.2mm/135.9mm
- compression ratio	16.3:1
- maximum output	134 KW @ 2500 rpm
- maximum torque	616.9 Nm @ 1750 rpm

The fuel injection system comprised a single plunger rotary injection pump (American Bosch Model 100) high pressure lines (1.80mm I.D. x 900 mm), and hydraulic differential 17 mm nozzles. Nozzle opening pressure: 25.85 MPa; static injection timing: 15 degrees BTDC. The injection pump was equipped with an internal hydraulic actuated speed advance device; it provided a total of 14 degrees advance between low idle speed and 1800 rpm.

As alternative fuel, sunflower seed oil was chosen due to its potential availability in many parts of the country where IH agricultural equipment is used.

The specific objectives of this project were aimed at:

- Determing the performance of an engine dedicated to conventional diesel fuel when running on sunflower oil and blends thereof.
- Evaluating engine or fuel system changes to allow satisfactory operation on alternative fuels or blends containing alternative fuels.
- Determining longer term influence of fuel blends upon engine durability and performance.

PHYSICAL AND CHEMICAL CHARACTERIZATION OF THE FUELS TESTED

As a prelude to engine and injection system testing, the test fuels and the blends were analyzed according to ASTM standard procedure established for petroleum base fuels.

Sunflower oil was obtained from commercial suppliers. It was alkaline refined and degummed, practically the edible type. The chemical characterization of this oil given by the supplier specified a low wax content, which should be advantageous when operating at lower temperatures.

As diesel fuel, our own laboratory fuel was used, a blend intended to meet the midpoint of the API gravity limits specified for EPA Type 2 exhaust emissions test fuel; it also meets IH internal specifications.

Three blends were obtained by mixing the above fuels in the following proportions, by volume:

- 25% sunflower oil; 75% diesel fuel
- 50% sunflower oil; 50% diesel fuel
- 75% sunflower oil; 25% diesel fuel

Table 1 lists the properties of the test fuels and blends, determined by analysis or calculated by mass averaging data of the two constituents.

Kinematic viscosity measured at 40°C and 100°C for the blends and the two constituents is presented in Figure 1. The range of 1.9 to 4.1 cST viscosity (ASTM D975) recommended for #2 diesel fuel was not met by any of the above blends tested. According to these specifications - at 40°C - a blend of 20% sunflower oil and 80% diesel fuel would be the maximum admissible limit as far as fuel viscosity is concerned. During the engine tests carried out under the present project the average temperature at the injection pump inlet was approximately 40°C. Figure 1 shows also to what extent fuel heating lowers viscosity; at 100°C blends with up to 60% sunflower oil can be brought within the specification limits.

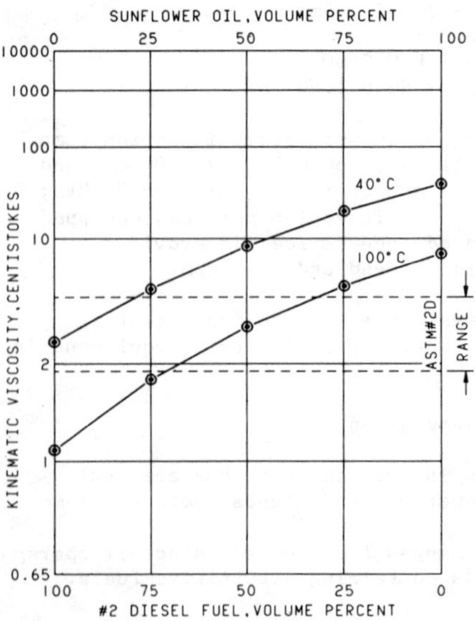

Fig. 1. Kinematic Viscosity of Sunflower Oil and Diesel Fuel Blends

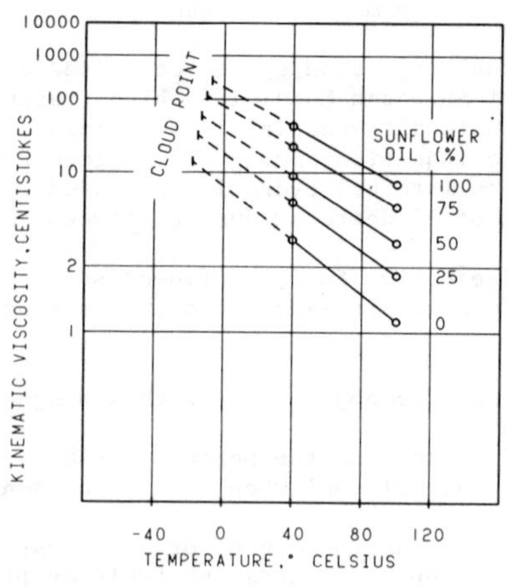

Fig. 2. Viscosity Temperature Characteristics of Sunflower Oil and Diesel Fuel Blends

Cloud points measurements emphasized the significant difference among the fuels and the blends at lower ambient temperatures. Subsequent fuel system and engine starting tests were run at temperatures at least 6°C higher than the cloud point for the respective fuel or blend to prevent fuel filter plugging. Approximate viscosity at cloud point was calculated based on extrapolation from values measured at 40°C and 100°C. (See Fig. 2 - broken lines). Figure 2 and Table 1 show the dramatic increase in viscosity that the sunflower oil and its blends exhibit at cloud point. As an example viscosity of sunflower oil is 250 cST while the viscosity of #2 diesel fuel at the same temperature is under 10 cST. This increased viscosity is likely to affect flow through the fuel lines, to lower the internal leakage that provides injection pump lubrication and to cause serious injection system failure.

The increase in mass density with the percentage of sunflower oil in the blend (see Table 1) provides more mass flow through the injection system and compensates to same extent the lower heat content of the fuel blends on a volume basis.

The lower cetane numbers measured for sunflower oil and its blends account for the longer ignition delay which is always experienced in engine tests. However, for vegetable oils the cetane number does not give the usual information on fuel combustion one would expect. A brief explanation on this matter is given below.

Table 1. Properties of The Fuels Tested

Fuel Properties		Sunflower Oil	75/25 Sunfl./ Diesel	50/50 Sunfl./ Diesel	25/75 Sunfl./ Diesel	#2 Diesel
Mass density kg/m^3		0.9224	0.9036	0.8860	0.8670	0.8478
Kinematic viscosity (cSt)	40°C	32.54	17.41	8.96	4.66	2.47
	100°C	7.79	4.88	2.93	1.77	1.08
Cloud point, °C		-7	-10	-13	-15	-18
Extrapolated viscosity at cloud point		250	125	60	28	13
Heating value kJ/kg	gross	39,700	41,225	42,600	44,275	45,800
	net	37,000	38,475	39,800	41,425	42,900
Carbon residue % by weight		0.34	-----	0.19	-----	0.35[a]
Cetane number		31	-----	40.7	-----	48.1

[a] measured on 10% residuum according to ASTM, D524-Special Procedure

The cetane number is a measure of self-ignition ability of the tested fuel. Under given engine conditions, the lower the cetane number of a fuel, the longer the ignition delay and the longer the amount of fuel injected until self-ignition occurs. During the ignition delay, a chain of processes leading to self-ignition develop - some physical in nature (fuel atomization, ignition and mixing with air), other chemical (fuel oxidation reactions precursors of self-ignition).

For petroleum derived fuels that are blends of hydrocarbons of various molecular weights - the physical and chemical processes develop quickly as the fuel quantities are injected. When self-ignition occurs, there is enough fuel already prepared to sustain combustion, therefore rates of pressure rise in the cylinder are high. Thus, for all petroleum derived fuels that meet the ASTM specifications for cetane number method, a low cetane number is associated with longer ignition delay and rough combustion.

Vegetable oils have a different chemical structure closer to that of a pure substance. Viscosity is high and volatility is low; therefore, the physical ignition delay should be much longer. When self-ignition occurs, a smaller amount of fuel is ready for combustion; consequently, during the first phase of combustion the pressure rise rate should be significantly lower. It is probable that the fuel accumulated in the cylinder during ignition delay burns initially slower and later faster. However, if this fast combustion occurs after TDC the pressure rise is moderated by the increasing volume of the combustion chamber.

Most experiments as well as our own have shown that vegetable oils exhibit longer combustion duration with moderate rates of pressure rise unlike the petroleum derived diesel fuels of low cetane numbers.

FUEL SYSTEM PERFORMANCE

The fuel injection system set to production specification for the farm tractor application was tested on a Friedmann-Maier test stand for comparative performance with #2 diesel fuel, the three blends and sunflower oil.

The set-up simulated the configuration of the injection system on the engine using the original nozzles, injection lines and fuel filters. Fuel temperature was kept at 40°C throughout the test. Injection line pressure was measured with AVL strain gage transducer mounted at the inlet of #1 injection nozzle. A variable inductance transducer was used to record needle lift of the same nozzle. A Robert Bosch Injection Rate Indicator

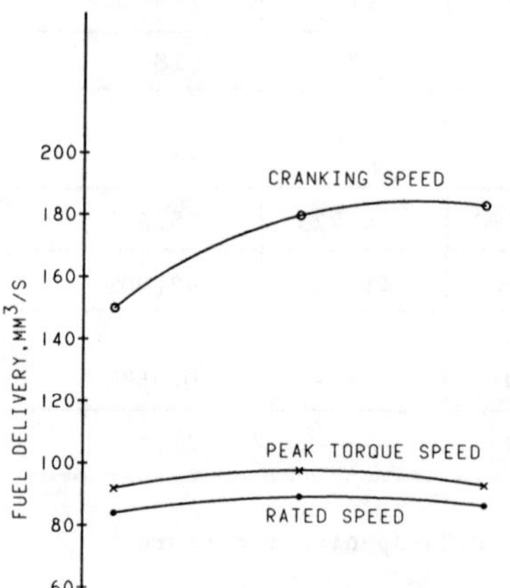

Fig. 3. Injection Pump Delivery for Sunflower Oil and Diesel Fuel Blends

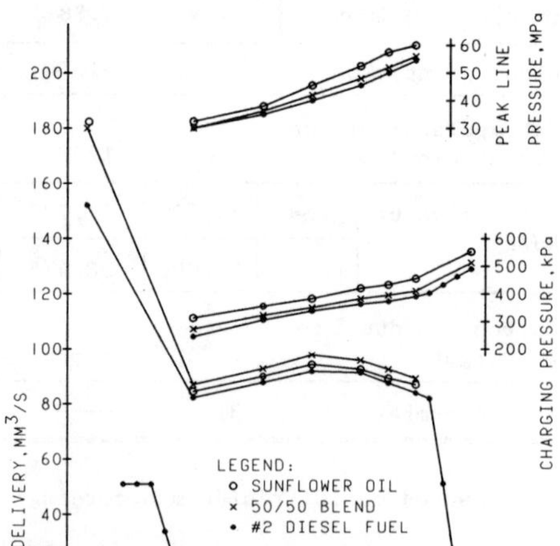

Fig. 4. Comparative Results of Fuel System Testing With Sunflower Oil and Diesel Fuel Blends

was used to obtain rate of injection. The fuel spray was visualized in an atmospheric chamber and photographs were taken under various conditions. Timing marks were generated in one degree increments with transducers mounted at the test bench flywheel. Line pressure at the pump exit and the nozzle entrance as well as needle lift were recorded with a four channel Tektronics oscilloscope and back-up photographs were taken. Pressure and motion signals were recorded, digitized and analyzed using a Norland 3001 wave analyzer and oscilloscope.

Figure 3 shows the change in pump delivery at various conditions as a function of sunflower oil percentage, in the blend. On a volume basis delivery increased when more viscous fuels were pumped. This was due to the fact that as viscosity increased the amount of fuel that escaped during the pumping stroke between the plunger and the barrel diminished and the net flow to the high pressure lines increased.

At peak torque speed and rated speed conditions, the fuel delivery increase was less than 10% and it peaked for the 50/50 blend; further increase in sunflower oil percentage slightly diminished the fuel delivery. This may be attributed to insufficient pump charging as blends became increasingly viscous. At cranking speed condition the fuel delivery increased as the blends had relatively more sunflower oil.

The comparative results of fuel system testing with diesel fuel, pure sunflower oil and a 50/50 blend under normal operating fuel temperature - 40°C - are summarized in Figure 4. Fuel delivery history on a volume basis is presented as a function of engine speed. The relative position of the curves shows that the difference between #2 diesel fuel operation and sunflower oil operation is not as dramatic as one might expect.

Final filter pressure (charging pressure) was somewhat higher with pure sunflower oil, but did not significantly change when the 50/50 blend was used.

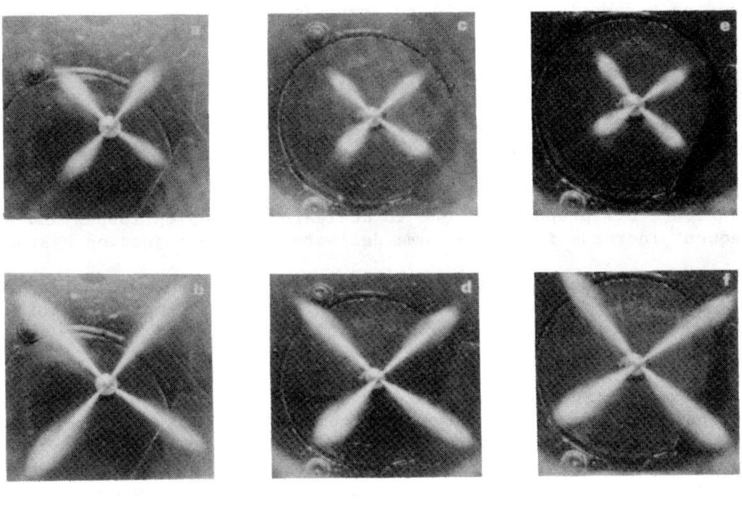

Fig. 5. Comparative Fuel Spray Photographs at Two Constant Plume Lengths - a), b) - Diesel Fuel; c), d) - 50/50 Blend; e), f) - Sunflower Oil

Peak pressures at the nozzle were higher with the alternative fuels due to the greater fuel delivery rate at the pump and to the lower nozzle discharge rate. Pressure differential at the final filter was not sensibly affected by blend operation and therefore it is not presented in Fig. 4.

Spray pattern photographs were taken for the diesel fuel and the blends tested in an attempt to emphasize eventual gross differences in spray geometry. Two constant plume lengths were chosen and under various working conditions atmospheric sprays were photographed. Although quantative evaluation was not possible with this method, some trends were visible which indicated that blends with sunflower oil provided a heavier, more dense core of larger size particles that reflected light and made the plumes appear brighter. Figure 5 shows the comparative photographs taken at rated speed and load conditions with diesel fuel, the 50/50 blend and with pure sunflower oil.

ENGINE PERFORMANCE

The engine was tested in comparative performance testing with all three blends - 25/75, 50/50, 65/25, and with sunflower oil.

Baseline data was obtained first using diesel fuel after which the injection system setting was not altered. This simulated the practical situation when a dedicated engine for diesel fuel use would operate instead with a fuel blend containing sunflower oil or with pure sunflower oil.

Performance testing was carried out according to SAE recommended procedures and comparative performance curves at full load and part load were produced. For purpose of emphasizing the difference in various parameters of engine performance as a function of blend composition relative curves were used throughout this part of engine analysis.

Engine Performance At Full Load Fuel Delivery

Figure 6 shows the relative variation of some magnitudes pertaining to engine performance as a function of the blend composition. Three constant values of engine speed were chosen as parameter of the curves: rated speed-2500 rpm, peak torque speed-1750 rpm and 900 rpm. The pump setting corresponded to full load fuel delivery for #2 diesel operation.

Relative fuel delivery variation is presented in Figure 6a. Two features were evident during the tests, namely:

- at constant engine speed, the mass fuel delivery increased with the sunflower oil proportion, due to decrease in pumping losses and subsequent increase in fuel volume delivered by the injection system.

- at constant blend composition, the mass fuel delivery increased with engine speed, due to the effect of the latter on pumping duration. Indeed at higher speed, pumping duration - expressed in time units - is shorter and less fuel flows back past the plunger.

Relative brake power output variation is recorded in Figure 6b. At rated speed the increase of fuel delivery with the percentage of sunflower oil generated power increase. At peak torque speed though, a peak of power increase was reached for an intermediate blend; further increase in sunflower oil percentage brought about a decrease in power. At the low speed tested, 900 rpm, the power output decreased with the percentage of sunflower oil in the blend. The explanation for this behavior lies in the lower energy content of the blends as they contained a higher percentage of sunflower oil. Likewise the fuel injection process, the combustion process and the power output were affected by the blend composition more so when the engine operated at lower speed.

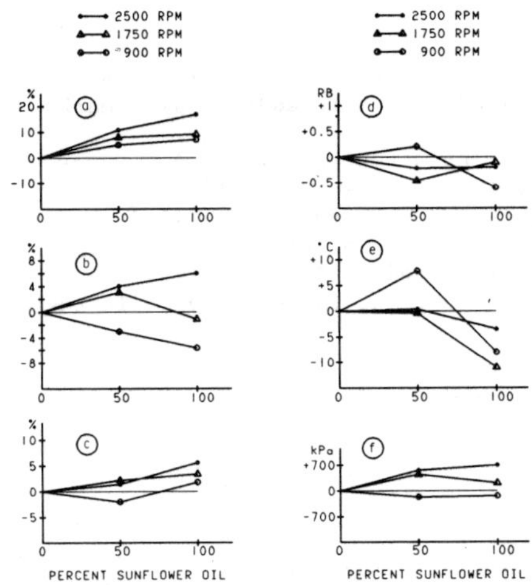

Fig. 6. Relative Variation of Performance characteristics at Full Load Fuel Delivery of Sunflower Oil and Diesel Fuel Blends: a) Fuel Mass Flow, by Weight; b) Power Output; c) Thermal Efficiency; d) Smoke Level; e) Exhaust Temperature; f) Peak Cylinder Pressure

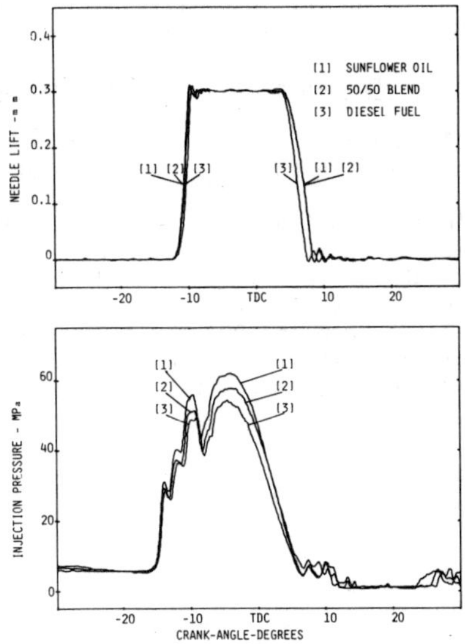

Fig. 7. Needle Lift and Injection Pressure at Full Load Fuel Delivery and 2500 rpm

Thermal efficiency was calculated using the measured values of specific fuel consumption and the fuels and blends tested (Fig. 6c). One arrived thus at comparable means of assessing how efficient the energy in the fuel was converted to output. Higher speeds provided somewhat improved thermal efficiency even with high percentage of sunflower oil in the blends. At low speed, however, energy conversion was worse. Both gain and loss in thermal efficiency were relatively modest in absolute value.

Some gain in thermal efficiency when operating with sunflower oil - or other vegetable oil - was previously reported by other researchers; [10], [13]. Lacking more in-depth information about the combustion mechanism of oils and oil blends, one must speculate at this point that the slower fuel preparation for combustion and the corresponding longer mixing time with air may help complete the chemical reactions of oxidation and improve conversion efficiency. At the same time, however, the slower combustion process and the delayed energy release rate are conducive to lower thermal efficiency. If this effect prevails over the gain due to more complete combustion, the net result is a decrease in thermal efficiency. The variation in smoke level (Fig. 6) seems to be consistent with the variation in thermal efficiency.

Exhaust temperature variations (Fig. 6d) were insignificant; in-cylinder peak pressures increased with the percentage of sunflower oil at higher speeds but at 900 rpm, lower peak pressures were attained due to poor injection system and combustion performance.

Figure 7 shows needle lift and pressure traces taken at full load delivery for rated speed . Throughout almost the entire injection duration fuel pressure at the nozzle was higher when more viscous blends were used, due in part to the higher fuel delivery, but also to the effect of fuel viscosity on the injection process; the needle lifted sooner and closed later. This leads to poorer combustion, especially at the end of the process; likewise the nozzle dribble is adversely affected.

Engine Performance At Half Load Fuel Delivery

Half load curves are presented in Figure 8. They generally illustrate similar trends with the ones presented at full load with some additional remarks.

Increase in fuel delivery occurred with all blends tested (Fig. 8a). Power increase occurred only at the highest speed tested (fig. 8b) while at 900 rpm power output steadily decreased when the percentage of sunflower oil increased. Thermal efficiency improved somewhat at the highest speed tested, but at 1750 rpm and 900 rpm blend operation resulted in lower thermal efficiency. The loss in efficiency was more important (about 15%) while the gain was modest (about 5%).

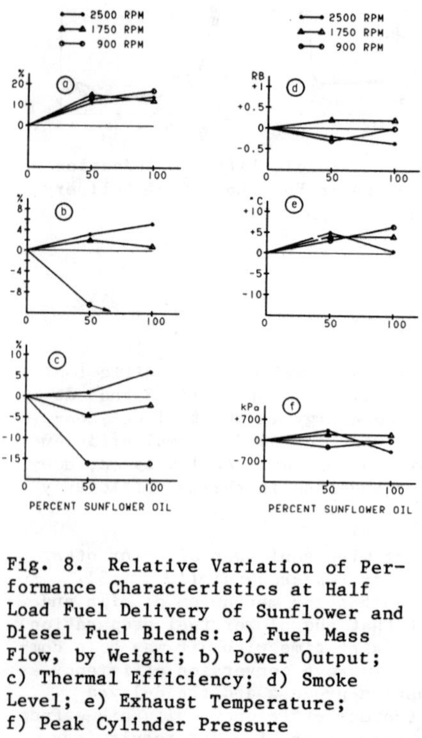

Fig. 8. Relative Variation of Performance Characteristics at Half Load Fuel Delivery of Sunflower and Diesel Fuel Blends: a) Fuel Mass Flow, by Weight; b) Power Output; c) Thermal Efficiency; d) Smoke Level; e) Exhaust Temperature; f) Peak Cylinder Pressure

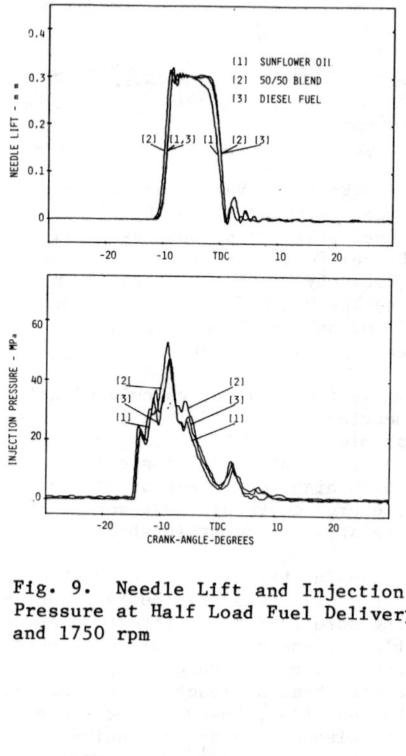

Fig. 9. Needle Lift and Injection Pressure at Half Load Fuel Delivery and 1750 rpm

Other data regarding smoke level, exhaust temperature, cylinder peak pressure, are given in Figure 8d), e), f).

Injection traces recorded for the three blends at half load fuel delivery are shown in Figure 9. Injection pressure was higher for the more viscous blends only during the first part of the injection process until the peak pressure was reached at the nozzle. After that, injection pressure for the more viscous blends was equal to or lower than the corresponding variation in diesel operation. Needle closing was slightly earlier with fuel blends. The injection pressure level and the slower decay rate at half fuel delivery indicate that spray characteristics are less satisfactory under this condition, even more so when a more viscous blend is injected instead of diesel fuel.

Figures 10 and 11 record injection traces at high idle condition - 2800 rpm and at low idle - 700 rpm. The difference between the blends appeared even more significant under these conditions, where the amount of fuel injected was small and the effect of the fuel and injection system reaction were relatively more important. The needle never reached the stop, the pressure level during injection was of the same order of magnitude as the pressure after injection, the residual pressure was sensibly different from one fuel to the other. Nozzle dribble should have been very significant under these conditions, according to pressure variation traces.

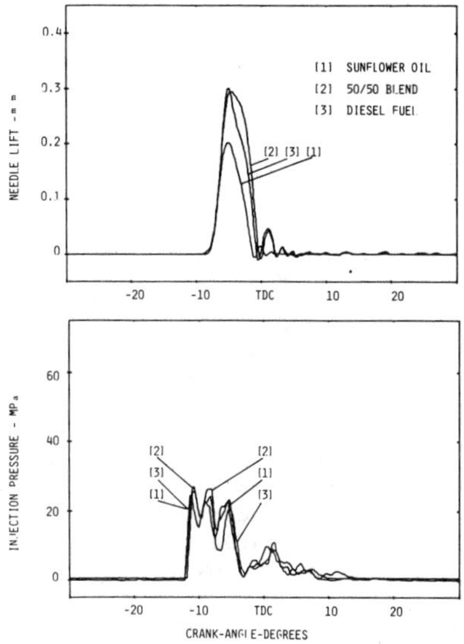

Fig. 10. Needle Lift and Injection Pressure at High Idle Condition

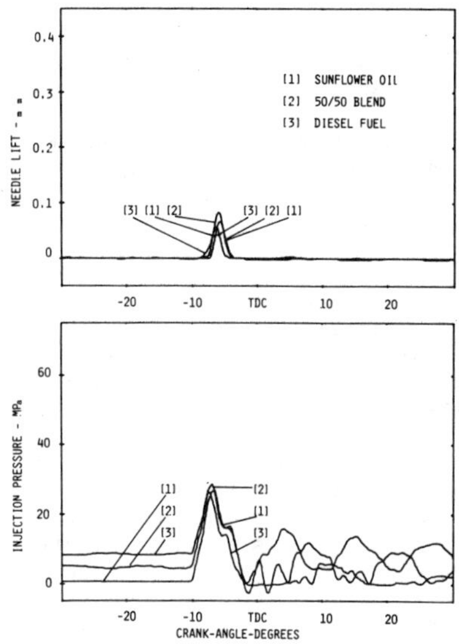

Fig. 11. Needle Lift and Injection Pressure at Low Idle Condition

<u>Nozzle Deposits During Performance Testing</u>

It is well known (and researchers have confirmed during their testing) that vegetable oils tend to obstruct nozzles with heavy carbon deposits [6], [20], [18]. During our experiments, we gave considerable attention to nozzle deposits, trying to understand which circumstances are conducive to their occurrence.

After each set of performance tests with a particular blend, the nozzles were removed, inspected, and cleaned prior to installing them back in the engine. Figure 12 shows the nozzle buildup for two of the blends and for pure sunflower oil. As a general trend, the deposits increased in amount with the percentage of oil in the fuel blend. They were harder than the usual deposits formed with diesel fuels. They did not appear to affect engine performance and did not affect the fuel spray. Pictures taken under higher magnification revealed the fact that deposits formed craters around the nozzle orifices but the orifices were not obstructed.

The deposit formation presented in Fig. 12 corresponded in each case to several hours of operation, but due to the schedule of performance testing, it was not possible to correlate the severity of buildups with specific engine

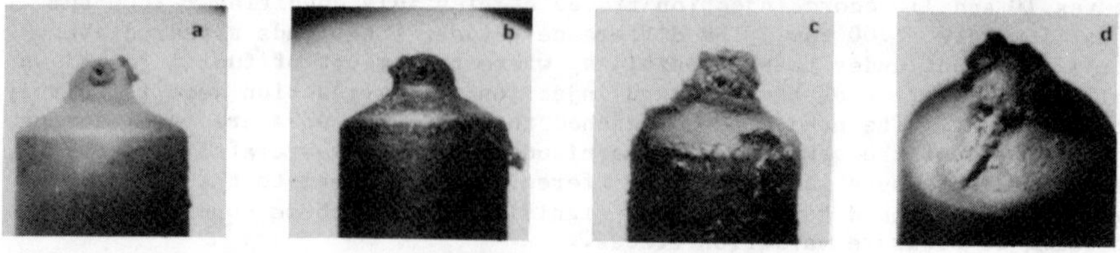

Fig. 12. Nozzle Deposit Buildup During Performance Testing (Several Hours): a) 50/50 Blend; b) 75/25 Blend; c) and d) Sunflower Oil

working conditions. At first, one may speculate that low regimes (low load and speed) must account for deposit formation due to poor spray patterns, reduced pressure at the nozzle during the last part of injection and fuel dribble after injection end.

A one-hour test was conducted at low idle condition with a 50/50 blend after which nozzles were inspected; no deposits were found around the orifices, instead the nozzles were wet and oily. This led us to believe that a combination of high nozzle tip temperatures - characteristic to high regimes, and poor injection cutoff-characteristic to low regimes - is likely to allow deposits to build up.

To verify this assumption, a one-hour test was run at full load, rated speed with a 50/50 blend starting with clean nozzles. According to recorded injection traces, this regime should produce the best spray characteristics. After the test, the nozzles exhibited quite heavy deposits (Fig. 13); two nozzles were cleaned and the engine was operated for another hour at low idle. The clean nozzles exhibited the same clean but wet oily appearance while the rest of the nozzles had the same deposits previously built covered with a wet oily layer. After these tests, we believe that deposits on the nozzle are formed even under high load and speed conditions due to the high viscosity of the fuel that lengthens the injection duration provides relative poor injection cut-off and nozzle dribble after injection. The high temperatures of the nozzle tip and the increased coking tendency of the blends containing sunflower oil contribute to the accelerated deposit buildup reported whenever vegetable oils are used alone or in blends with diesel fuel.

ENGINE DURABILITY TESTING

The engine was subjected to a limited durability testing of 200 hours to evaluate longer term influence of fuel blend operation on performance and durability of the engine and the injection system.

The 50/50 fuel blend was selected for the test, considered to be the optimum choice from the standpoint of both a more substantial replacement of diesel fuel and a more practical field application.

The test cycle was the one recommended by the Alternate Fuels Committee of EMA for vegetable oils and oil blends. It consisted of:

- 30 minutes at low idle
- 30 minutes at high idle
- 60 minutes at rated power speed
- 60 minutes at peak torque speed
- run five of the above cycles (15 hours) and shut down for the remainder

of the day (overnight)
- repeat the full cycle above until 200 hours of running are completed.

As lubricant, our own laboratory oil was used, meeting API service category CD. During the test, power and fuel consumptions were continuously monitored in accordance with SAE test procedures and with our internal specifications.

Oil samples were taken every fifty hours for inspection of wear metals and oil contaminants. Oil and filter were changed twice during the test, after each hundred hours of operation.

The engine completed the test without any performance problems. Daily starting was normal under the test cell environment. Fuel system operation recording did not show any deterioration that implied engine shutdown and inspection. Pressure drop across fuel filters did not change throughout the test period. Wear metals did not increase beyond the usual values encountered in durability tests. Used lubricant viscosity was within normal limits during the test and degradation of lubricant additives was not encountered.

ENGINE TEARDOWN AND INSPECTION

Upon completion of the durability testing, the engine was dismantled and subjected to inspection.

All nozzles exhibited heavy deposits on the outside of the tip and especially around the orifices in a crater like manner. Figure 14 shows a typical nozzle tip as taken out from the engine in a comparative view with a clean nozzle and one that operated through a durability test with #2 diesel fuel. The deposits were hard and their removal required scraping. Light colored particles were visibly imbedded in the carbonaceous deposits and this led to further investigation. Photographs were taken under high magnification which revealed

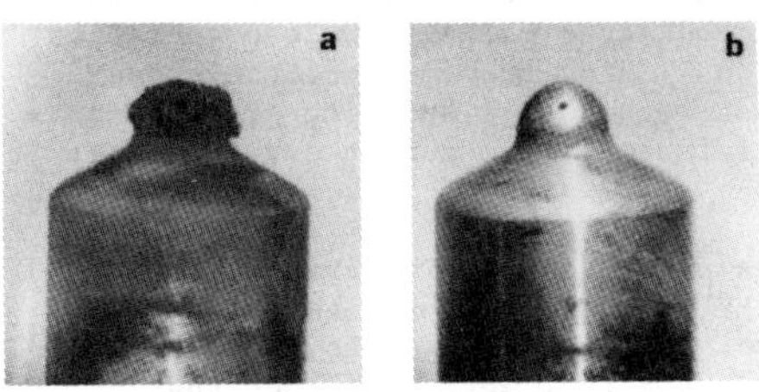

Fig. 13. Nozzle Deposit Buildup After One Hour Operation at: a) Full Load, Rated Speed; b) Low Idle

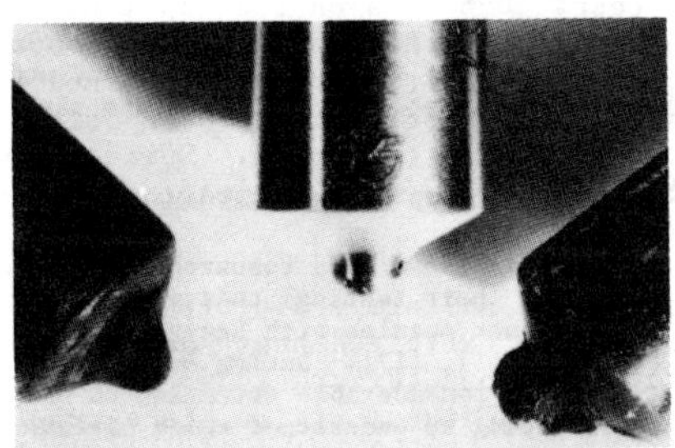

Fig. 14. Nozzle Tip: Right - After 200 Hours Durability cycle Operation With 50/50 Blend; Center - Clean; Left - After 200 Hours Durability Cycle Operation With Diesel Fuel

plate-like shapes and striations resembling the cutting wear chips. Energy dispersive x-ray spectrum was used to analyze the composition of the particles. Iron and typical low alloy steel elements were found which was an indication of premature wear occurring in the fuel system.

Injection system wear was judged normal during the test hours. Fuel filter clogging was not encountered, but sticky deposits were found on the filter cartridge cap. Injection pump plunger looked normal, however the distributor body had sticky deposits on areas exposed to stagnant fuel. The same was true for the pump body, the delivery valve seat, the injector body, etc., in general, where there was no continuous fuel flow or where clearances were rather small and the fuel was in longer contact with the metal. This condition that may become more severe during longer priods of engine shutdown or endurance operation is likely to impair the operation of the pump device, increase friction, affect sealings.

Fig. 15. Typical Deposits Buildup on The Piston After 200 Hours Durability Test With a 50/50 Blend

Fig. 16. Carbon Deposit Buildup on The Combustion Chamber Walls After 200 Hours Durability Test With a 50/50 Blend

All pistons exhibited quite heavy deposits on the ring belt area on the crown and on the surface of the combustion chamber. All deposits were of a wet, oily appearance and did not flake off in time as usually dry deposits generated during diesel operation do. Figure 15 illustrates a typical piston after the test. Deposits on the top land were slightly less than on the intermediate land where the accumulation groove of the balanced pressure design* was almost filled with carbon. Unburned fuel collected during low load operation most probably contributed to deposit buildup. Ring grooves were not excessively filled with carbon. During the test hours, ring and ring groove wear patterns were normal, indicating that satisfactory ring motion and ring seal were achieved.

Besides the uniformly distributed deposits on the combustion chamber walls, all pistons exhibited heavy deposits on the areas corresponding to the fuel spray positions (Figure 16). This feature, that is not seen to this extent in an optimized diesel engine, is a proof that the fuel jets were such that a liquid core touched the wall and liquid fuel inpingement occurred under most operating conditions. Since vaporization of such liquid jets is slow anyway - and even slower when blends with sunflower oil are injected, most of the fuel burns on the chamber walls leaving the heavy residue behind. The fuel overpenetration that occurred was due in part to the increase in injection pressure, but mostly to the larger droplet size of the fuel blend which prevented good atomization of the liquid jet.

The engine liners did not show signs of wear; honing patterns were still visible, but the same heavy oily deposits were present in the top land area. Likewise, the flange surface of the liner was covered with deposits. Figure 17 shows one typical liner after the test.

Cylinder head deposits were heavy, oily, especially around the fire dam area and on the walls of the exhaust port (Figure 18).

*Balanced Pressure is a patented design of the piston that provides improved oil consumption during the engine life.

Fig. 17. Engine Liner After 200 Hours Durability Test With a 50/50 Blend

Fig. 18. cylinder Head and Exhaust Port Deposits After 200 Hours Durability Test With a 50/50 Blend

Turbine blades were coated with oily deposits that covered the entire blade length. Although during the test hours they did not unduly affect the turbine performance, a prolonged time operation is likely to cause uneven accumulation and result in turbocharger unbalance and damage.

LOW TEMPERATURE TESTS

To evaluate the influence of low ambient temperature upon the fuel injection system and engine operation, two series of tests were conducted in our laboratory cold room; fuel injection characteristics on the injection test stand and engine starting tests.

The fuel injection test stand was relocated in the cold room and all the tests previously described were repeated for straight diesel fuel, 50/50 fuel blend and straight sunflower oil. For the first two fuels test temperature was -4°C; however, for sunflower oil the test was carried out at -1°C due to the high cloud point value of this fuel. In all cases cold temperature increased the fuel delivery throughout the whole speed range, the fuel pressure downstream the final fuel filter as well as the peak pressure values in the injection lines.

At the end of the test, while operating with straight sunflower oil at high idle condition, an injection pump failure occurred. The ensuing inspection revealed that the plunger had seized in the distributor head. Due to the uniqueness of this occurrence, it was difficult to establish all the factors that contributed to this failure; however, the insufficient lubrication in the plunger area due to high viscosity of the sunflower oil at the test temperature has undoubtedly played a major role.

For evaluation of cold starting behavior the 50/50 blend was compared against diesel fuel according to International Harvester's internal specifications at two temperature conditions: -1°C (20°F) and -7°C (20°F). The base engine was cold soaked at the test temperature for at least 18 hours previous to the test, and no starting aids were used. Cranking speed was 100 rpm. Engine lubricant SAE-10W was used.

Acceptability of engine starting was evaluated by measuring time to first firing, total cranking time and total acceleration time up to 2000 rpm. Several starting tests were conducted to ensure repeatability and reliability of the data.

Table 2 shows the comparative results of the test. All data indicated unacceptalbe increase in starting time. According to our criteria for good reliable engine starting with conventional fuels, the fuel blend has obviously failed the starting test. The long cranking time elapsing from cranking start to the first firing and the subsequent long time to self-sustained operation has brought about accumulation of fuel blend in the cylinders, in the exhaust ports and in the exhaust manifold. Right after starting considerable amount of raw fuel was coming out the exhaust line, and the exhaust odor was unusual.

In conventional operation, engine starting is associated with a high level of hydrocarbon emissions due to the fact that the excess fuel does not react to completion or reactions are temperature quenched. As the engine idles after starting, the hydrocarbon level gradually decays.

Table 2. Comparative cold starting data for diesel fuel and 50/50 diesel fuel/sunflower oil blend.

Ambient Temp. °C		Time To First Firing, sec	Total Crank Time sec	Total Accel. Time To 2000 rpm sec
-1°C	#2D fuel	1	3	4
	50/50 blend	10	43	47
-7°C	#2D fuel	6	14	18
	50/50 blend	29	50	52

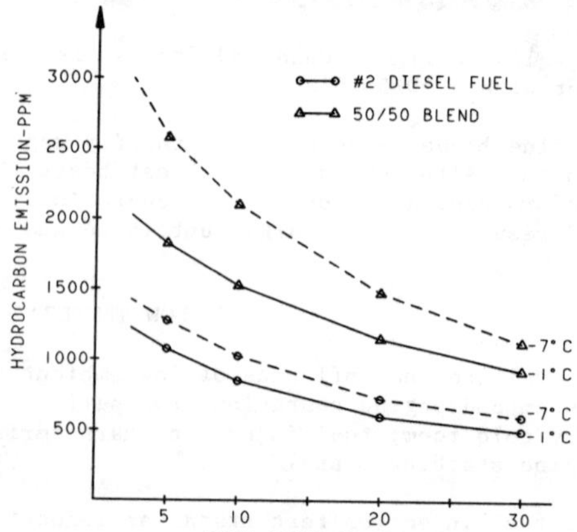

Fig. 19 Hydrocarbon Emission During and After Engine Starting

The hydrocarbon emission and the hydrocarbon clean-up characteristics were comparatively evaluated for the two operations. Continuous exhaust sampling and a hot flame ionization detector were used for this test according to the specifications of the SAE J215 - Recommended Procedure. Figure 19 shows the comparison between the two operations. At both test temperatures the hydrocarbon level was objectionably high in blend operation and stayed high throughout the testing time. According to our internal specifications, such level is not acceptable in normal operation.

CONCLUSIONS

In general context, our work shows that vegetable oils are candidates for alternative fuels due to their similarities with conventional fuels. However, due to their differences, vegetable oils and their blends are not yet a viable practical solution since they affect the performance, the durability and the serviceability of the engines, especially the more fuel sensitive direct injection engines. At the same time, the on-farm processing technology has not been developed to make vegetable oils economically feasible alternatives for fueling diesel engines of agricultural applications.

The activity carried out under this project has led us to the following conclusions:

1. The use of sunflower oil blended with diesel fuel in a direct injection engine brings about modifications in the fuel injection process characteristic to injection of viscous fuels namely; increase in injection pressure, longer injection duration, delayed needle closing, early needle opening. All these effects are conducive to longer combustion duration.
2. The modified injection characteristics and the blend properties generate a peculiar spray pattern with larger drops and fuel over-penetration. The subsequent combustion follows a different (still unknown in its basic features) mechanism of energy release.
3. The increased propensity to leave residues behind, make the blends with sunflower oil prone to deposit buildups after only short periods of operation. Our tests have shown that nozzle deposits tend to be blown off to some extent during operation, while piston and liner deposits are more stable.
4. The deposit buildup occurs even at high load and speed when blends exhibit satisfactory operation. Low speed and load operation increases the tendency to build up deposits by addition of unburned fuel that includes the carbonaceous deposits in a sticky oily matter that probably hardens on the hot engine parts.
5. Durability testing of 200 hours is insufficient for revealing long-term performance of sunflower oil blends. A new good engine, in contolled fuel conditions with a controlled fuel supply, may not fail during the first 200 hours like the case was with the engine tested under this project.
6. Engine shutdown for longer periods of time accelerates gum formation where the fuel contacts the bare metal and this is likely to impair further engine and injection system function.
7. Cold temperature operation is a very critical issue related to sunflower oil and the blends thereof. High viscosity can cause fuel system problems, failed starting, unacceptable emission level and injection pump failures due to lack of lubrication.

Opportunities are open for further research of vegetable oils. In fact, only now did we start to accept that oils may be more than an ingredient in an artist's paint or a seasoning in everybody's salad. We are in the process now of accumulating data about how oils behave as fuels or as fuel extenders. As we encounter troubles and emphasize limitations, we realize that more research will be needed to understand how and where to use at best the potential of vegetable oils.

REFERENCES

1. "Alcohol and Vegetable Oil as Alternative Fuels", Proceedings of Regional Workshops, April, 1981, Organized by USDA-SEA Extension and Purdue University, West Lafayette, IN

2. ASTM D524, "Ramsbottom Carbon Residue of Petroleum Products"

3. ASTM D613, "Ignition Quality of Diesel Fuels by the Cetane Method"

4. ASTM D975, "Standard Specification for Diesel Fuel Oils"

5. Baranescu, R. A. and Lusco, J. J., "Sunflower Oil as a Fuel Extender in Direct Injection Turbocharged Diesel Engines", SAE Paper #820260, February, 1982

6. Barsic, N. J. and Humke, A. L., "Performance and Emission Charcteristics of Naturally Aspirated Diesel Engines With Vegetable Oil Fuels", SAE Paper #810262, February, 1981 and #81095, September, 1981

7. Bruwer, J. J. el al, "The Utilization of Sunflower Seed Oil as a Renewable Fuel for Diesel Engines", Presentation at the 1980 Symposium of the South African Institute

8. Cardis Fishinger, M. K. el al, "Service Trial of Waste Vegetable Oil as Diesel Fuel Supplement", SAE Paper #811215, October 1981

9. Chowhuny, D. H. el al, "Indian Vegetable Fuel Oils for Diesel Engines" Bulletin No. 19, Department of Scientific and Industrial Research Bureau, India

10. "Diesel Fuels", SAE 313C - Information Report

11. Engelman, H. W. and Guenther, D. A., "Vegetable Oil as a Diesel fuel", ASME Publication 78DGP19, 1978

12. Forgiel, R. and Varde, K. S., "Experimental Investigation of Vegetable Oils Utilization in a Direct Injection Diesel Engine", SAE Paper #811214, October, 1981

13. Fort, E. F. el al, "Evaluation of Cottonseed Oils as a Diesel Fuel", SAE Paper #820317, February, 1982

14. Goering, C. E. el al, "Vegetable Oil - solvent Blends as Diesel Fuel", Proceedings of Regional Workshops, April, 1981, Organized by USDA-SEA Extension and Purdue University, West Lafayette, IN, pp 256-266

15. Hardenberg, H. O. and Ehnert, E. R., "Ignition Quality Determination Problems With Alternative Fuels for Compression Ignition Engines", SAE Paper 811212, October, 1981

16. Kaufman, D. R. el al, "Performance of Diesel Oil and Sunflower Oil Mixtures in Diesel Farm Tractors", Summer Meeting of American Society of Agricultural Engineers, Orlando, Florida, 21-24 June, 1981

17. Seddon, R. H., "Vegetable Oils in commercial Vehicles - Gas and Oil Power", August, 1942, pp 135-142

18. "Sunflower Oil as Tractor Fuel", Research Report, June, 1981, Department of Agriculture and Fisheries, Republic of South Africa

19. Ziejewski, M. and K. R. Kaufman, "Endurance Test of a Sunflower Oil/ Diesel Fuel Blend", SAE Paper #820257

20. Walton, John, "The Fuel Possibilities of Vegetable Oils - Gas and Oil Power", 33, July, 1938, pp 167-168

PERFORMANCE OF WINTER RAPE (BRASSICA NAPUS) BASED FUEL MIXTURES IN DIESEL ENGINES

Gary L. Wagner and Charles L. Peterson*

Abstract

Winter rape is well adapted to the Palouse region of Northern Idaho and Eastern Washington. Nearly all of the current U.S. production is grown in this region. Yields of 2200 to 2700 kg/ha with 45 percent oil content are common. Even though present production only 2000 to 2500 ha per year, the long history of production and good yields of oil make winter rape the best potential fuel vegetable oil crop for the region. Winter rape oil is more viscous than sunflower oil (50 cSt at 40°C for winter rape and 35 cSt at 40°C for sunflower oil) and about 17 times more viscous than diesel. The viscosity of the pure oil has been found too high for operation in typical diesel injector systems. Mixtures and/or additives are essential if the oil is to be a satisfactory fuel. Conversely, the fatty acid composition of winter rape oils is such that it is potentially a more favorable fuel because of reduced rates of oxidation and thermal polymerization.

This paper will report on results of short and long term engine tests using winter rape, diesel, and commercial additives as the components. Selection of mixtures for long term screening tests was based on laboratory studies which included high temperature oxidation studies and temperature-viscosity data. Fuel temperature has been monitored at the outlet of the injector nozzle on operating engines so that vicosity comparisons at the actual injector temperature can be made.

Introduction

Agricultural production in the United States requires 8.3 billion liters of diesel fuel to produce crops from 138 million hectares of crop land (ERS-USDA, 1976). In Idaho 155 million liters of diesel fuel are required to produce crops from 2.7 million hectares of cropland. Between 55 and 60 liters of diesel fuel are required for each hectare of crop production.

Idaho farmers, like those in the rest of the country, are concerned about the availability and increasing cost of their energy supplies. Agricultural enterprizes are especially dependent on a plentiful supply of fuel for the diesel engines used for off-road power. The possibility of using vegetable oils as a direct substitute for diesel fuel is one of several concepts for on-farm production of fuel. Vegetable oils, however, show promise of providing all the liquid fuel needed on a typical farm by diverting 10 percent or less of the total acreage to fuel production.

Anything that sounds this good must have some negative aspects and vegetable oil based fuels are no exception. Engine life, in certain cases, has been shown to be drastically reduced and the cost of vegetable oil based fuels still exceeds that of commercially available diesel fuel.

Data from a regional crop introduction project (Auld et al., 1981) indicates that sunflowers, safflowers, and winter rape produce the highest yields of

---

*The authors are Scientific Aide and Professor, respectively, Department of Agricultural Engineering, University of Idaho, Moscow, Idaho, 83843.

vegetable oil in the Pacific Northwest. Based on this research, five types of oil have been considered: oleic safflower, linoleic safflower, high erucic acid rapeseed (winter rape), low erucic acid rapeseed (spring rape), and sunflower. This paper will discuss the short and long term engine tests which have been conducted to the present time using vegetable oil fuels.

Procedure

The vegetable oils were obtained from commercial processors in all cases except for the winter rape oil used in the endurance tests. This oil was processed with a CeCoCo expeller operated by the University of Idaho.

The oils were evaluated in short term engine tests to determine their effect on engine performance. Three cycles of long term (830 hours) tests have been conducted to evaluate the potential effects on engine life.

Short term engine tests were conducted on a Ford 4-cylinder, 2.8 L, direct injection diesel engine with hole type injectors connected to a General Electric cradled dynamometer. The engine fuel system was modified by adding an additional spin-on fuel filter and a 3-way, hand operated, two position, directional control valve which allowed rapid switching between the No. 2 diesel used as a standard and the test fuel. Each vegetable oil fuel powered the engine for 20 minutes. An identical diesel fuel test was conducted both before and after each vegetable oil test. Switching between tests was done under load in all cases. The fuel filter on the test oil line was changed for each new vegetable oil blend. Data collected includes: 1. dynamometer load and rpm, 2. fuel consumption at five minute intervals, and 3. temperature of the incoming air, oil, engine coolant, fuel and exhaust. Each oil was tested three times and the data analyzed using a completely randomized design. Similar procedures were used for varying power and torque performance tests.

The Ford engine was replaced with a Lister two-cylinder, 1.27 L, direct injection engine. This engine was used in additive screening tests, the results from which are not yet available.

In January 1982, a John Deere model 4239, 3.92 L turbo charged, direct injection engine was installed on the dynamometer. The engine break-in has now been completed and short term testing is under way.

Two Yanmar model TS70C, 376 cc, single cylinder, water cooled, diesel engines with precombustion chambers and pintile type injectors were used for long term tests.

Test cycle number one used 100 percent safflower in one engine and No. 2 diesel in the second engine. Test cycle number two used the same fuels only the engines were reversed. Test cycle number three used 70 percent winter rape - 30 percent diesel in one engine and 70 percent winter rape - 30 percent diesel with a DuPont FOA-2 (DuPont, 1980) additive in the second engine. Test cycle number four is currently under way using the same fuels as test cycle three but the engines are reversed. Before and after each test the engines were completely disassembled to measure clearances and weigh critical components such as bearings and rings.

Safflower oil was chosen for the first tests because of its low viscosity. The winter rape blend was chosen because erucic acid has only one double bond which reduces the oxidation and polymerization rates. The diesel fuel was added to the winter rape to reduce the viscosity to acceptable levels.

The engines were connected to electric generators with resistor banks for load units and were operated at wide open throttle with the load cycled on and off at 15-minute intervals. Once each day the engines were stopped for general maintenance and were stopped for longer periods at irregular intervals for convenience of running the tests.

The total time of operation was determined during the first test and was simply the time at which the safflower fueled engine would no longer start which occurred at 830 hours. This time has been repeated for the other three tests.

A three cylinder Ford model 4600 tractor was also operated on 100% linoleic safflower oil for 150 hours over a 15 month period. The tractor was first operated on a 50-50 mixture of sunflower oil and diesel beginning December 27, 1979. Then on June 6, 1980 the tractor's main tank was filled with 100 percent safflower oil, which was used as the principle fuel until August 1981 when the engine was disassembled for inspection. The tractor was used for approximately 120 hours for general farm work and the remainder on short term demonstrations in many areas of Idaho. Engine inspection included an evaluation of the diesel injector pump by a commercial repair firm.

A Satoh Bison model S670D, 1.30 L tractor equipped with swirl chambers and pintile injectors is currently operating on a 50% winter rape - 50% No. 2 diesel blend with a Lubrisol 565 additive. The tractor is used by the University of Idaho Plant Science farm in their spring and summer tillage work. Operator observations, cylinder compression, and pto power tests are performed every 50 hours to monitor tractor condition and performance.

Three hydraulic load stands equipped with Wisconsin 1.0 L air cooled, direct injection engines have been constructed and are being interfaced with a Hewlett-Packard 85F micro-computer and 3054 DL data logger. The computer will control both the test cycle, according to Engine Manufactures Guidelines, and analyze and collect data on-line. Six engines are available which will make for continuous operation of the test stands since dissassembly and inspection of the engine requires about the same amount of time as running the tests.

Physical Properties

Viscosity-temperature relationships for the vegetable oil - diesel blends have been developed (Peterson et al., 1981). Actual measurement of the fuel temperature at the injector on two engines, one air-cooled and one water-cooled, has shown temperatures of 85°C to be typical. In selecting a blend of oil for the long term engine tests, temperatures of the fuel at the injector were used to compare viscosity of the blend with diesel fuel recommended viscosities. Diesel fuel was added to bring the viscosity of the blend close to the upper limit (13.1 cSt) set as the emergency fuel specification published by the Cummins Engine Company (1980). A temperature of 140°C would need to be reached by 100 percent winter rape to equal the viscosity of the 70 percent winter rape - 30 percent diesel mixtures at 85°C (actual injector temperature); whereas, comparing the fuels at 40°C would indicate a fuel temperature of the 100 percent winter rape of only 82°C to obtain equal viscosities. Heating to raise the temperature of the fuel at the injector pump did not result in a corresponding increase in injector fuel temperature. It was found that the rapid transfer of heat away from the injectors to the engine head and away from the fuel lines prevented the fuel temperature at the injector from increasing significantly. A 77°C increase in fuel temperature at the injector pump resulted in only 25°C increase at the injector.

Short Term Performance Tests

During the maximum power and fuel consumption tests, (Table 1) power output of the engines did not vary significantly for the different fuel types although there was a general tendency for power to increase 2-3 percent at the 50-50 blends.

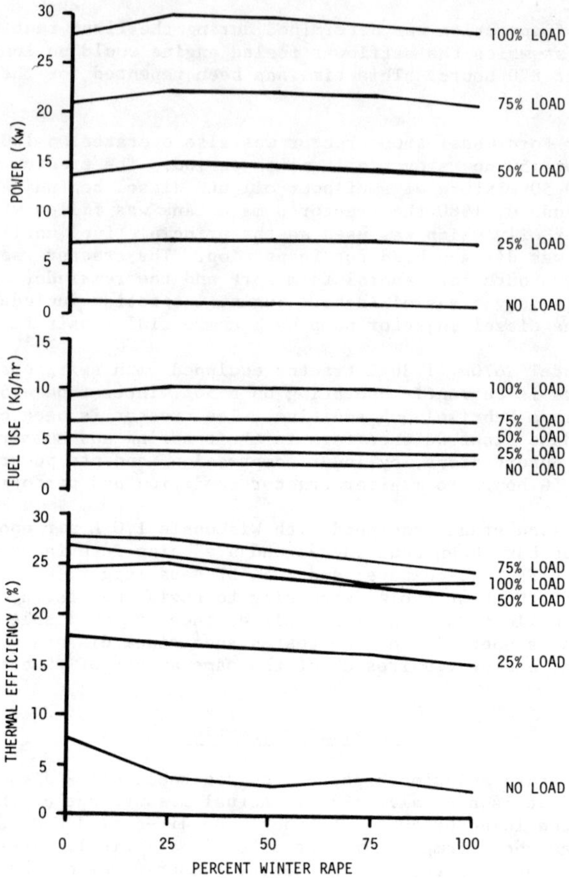

Figure 1. Power, fuel consumption, and thermal efficiency measured in a varying load test with winter rape-diesel blends as fuel. Engine used was a 2.8 L direct injection 4-cylinder diesel with hole-type injectors manufactured by Ford (the data is typical of the other vegetable oils tested.

Table 1. Diesel engine* performance using vegetable oil and vegetable oil-diesel mixtures as a fuel.

Fuel Type	kW	Fuel (kg/h)	Fuel (L/h)	BSFC** (kg/kwh)	BSFC** (kwh/L)	Thermal Efficiency (%)
100% Winter Rape	28.9	9.7	10.6	0.33	2.72	26.8
75% Winter Rape	29.4	10.4	11.7	0.35	2.54	24.7
50% Winter Rape	29.6	10.3	11.7	0.35	2.56	24.3
25% Winter Rape	29.2	9.9	11.4	0.34	2.56	24.2
100% Diesel	28.2	9.3	11.0	0.33	2.58	24.1

*Information in this table was obtained on a 172 cid Ford 4-cylinder, 2,200 rated rpm diesel engine.

**BSFC is Brake Specific Fuel Consumption and is a measure of work output per unit of fuel used.

In the interest of space, only the data for winter rape blends are shown for the varying power and fuel consumption tests (Figure 1). The data for the other fuels is very similar.

The test engine was severely damaged during the torque tests. Operating the engine with 100 percent sunflower oil under load at reduced engine RPM severely gummed the piston rings causing an almost immediate loss of power, and an increase in engine blow-by. This experience causes the authors to believe that maximum torque tests are an essential part of endurance test cycles and should be included in test procedures.

Long Term Engine Screening Tests

Long term testing thus far has included 1660 hours on 100 percent linoleic safflower oil (two 830 hours tests), 1660 hours on 100 percent diesel (two 830 hours tests), and 850 hours on 70 percent winter rape - 30 percent No. 1 diesel and 850 hours on 70 percent winter rape - 30 percent No. 1 diesel with DuPont FOA-2 dispersant additive added at the rate of 150 ppm. Power, fuel consumption, and wear data for the three long term test cycles are shown in Table 2.

Table 2. Power, fuel consumption, and wear data from long term screening tests using single cylinder, 376 cc, pre-combustion chamber Yanmar TS706 test engines.

	Test #1		Test #2		Test #3	
Engine No.	11	22	11	22	11	22
Fuel	100% Saff.	100% Diesel	100% Diesel	100% Saff.	70% W. Rape 30% Diesel	70% W. Rape 30% Diesel
Additive	None	None	None	None	None	DuPont FOA-2
Hours of Test	830	830	883.4	837.8	853.8	851.0
Comp. Change (%)	N.A.	N.A.	-6	-10	-6.2	-3.1
Power Drop (%)	N.A.	N.A.	+5[1]	+2[1]	4.5	2.7
Fuel Cons. L/hr	0.83	0.79	0.79	1.36	0.87	1.06
Stuck Rings	2	0	0	2	0	0
Ring #1, g[2]	0.173	0.124	0.123	0.150	0.169	0.260
Ring #2, g	0.078	0.039	0.054	0.262	0.086	0.188
Ring #3, g	0.037	0.029	0.042	0.380	0.047	0.098
Oil Ring, g	0.022	0.034	0.038	0.107	0.048	0.121
Rod Bearing, g	0.088	0.026	0.059	0.138	0.014	0.019
Piston, g	0.00	0.09	0.32	0.00	0.10	0.40
Upper Cyl. Wear, mm	0.025	0.025	0.010	0.028	0.013	0.025

[1]These data are not directly comparable because of changes in the fuel setting.
[2]Numbers are change in weight of component between start and finish of test in grams.

Measurements show about twice the wear rate for the safflower fueled engine compared to the diesel engine. The safflower fueled engine also showed more carbon build-up on the injector nozzle. At the conclusion of the first test, the injector in the safflower engine had seized and the engine would no longer start. A broken fan belt at about 700 hours allowing the engine to overheat and insufficient oiling of the valve train caused by a blocked oil line both undoubtedly affected the results of the first test. The second test, however, without those problems, still resulted in the piston ring gumming problems, a loss of compression, and increased blow-by.

For the test durations indicated and for the particular engine used, the winter rape blend is clearly superior to the linoleic safflower as a fuel. In all cases where the safflower oil has been used, the engines have suffered severe degradation primarily a result of ring gumming, lubrication oil thickening, and some injector gumming. None of these problems were found in the engines fueled with winter rape oil. Some gumming was noted on the upper piston land although it had not progressed to a point where it had any effect on engine performance.

Oil analysis data confirmed the results of engine measurements. A tendency for the oil to become more viscous was noted in the 100 percent safflower tests. The lubricating oil was analyzed at each 100 hour change interval and the viscosity compared with that of the new oil. An average ratio of beginning to final lubricating oil viscosity of 2.6 was noted for the safflower engine, 1.6 for the diesel engines, and an increase of only 1.4 and 1.6 for winter rape blend and winter rape blend with additive, respectively (Table 3).

Table 3. Oil analysis data for long term endurance tests (850 hours) with 376 cc displacement single cylinder precombustion chamber, Yanmar Diesel engines (data by Montech, Inc., Spokane, Washington). Figures reported are the average of data collected at the end of each oil change or approximately 100 hours.

Fuel Used	Ave. Hours On Oil	Viscosity (cSt @ 38°C)	Fuel Dilution[3]	Oxidation[3]
100% Safflower	104.7	291.5[1]	N	M-E
70% Winter Rape 30% Diesel	105.4	143.6[2]	N	N
70% Winter Rape 30% Diesel with FOA-2 Additive	104.2	166.2[2]	N	M
100% Diesel	110.4	182.8[1]	N	N

Fuel Used	Silicon[4] (Dirt)	Iron	Copper	Aluminum	Chromium
100% Safflower	21.6	131.5	5.25	34.0	5.49
70% Winter Rape 30% Diesel	13.8	37.4	4.4	8.4	2.2
70% Winter Rape 30% Diesel with FOA-2 Additive	20.8	70.6	6.00	16.2	3.9
100% Diesel	8.0	37.9	3.2	7.9	2.5

Fuel Used	Magnesium[4]	Silver	Moly	Tin	Lead
100% Safflower	19.75	0.75	4.25	19.1	18.3
70% Winter Rape 30% Diesel	20.2	0.2	1.00	4.4	8.2
70% Winter Rape 30% Diesel with FOA-2 Additive	20.2	0.5	2.2	9.6	9.8
100% Diesel	18.5	0.03	0.5	4.0	10.1

[1]Oil used was Delo 400 SAE30 with initial viscosity of 113.0 cSt @ 40°C.
[2]Oil used was Delo 300 SAE15w-40 with initial viscosity of 101 cSt @ 40°C.
[3]N = Normal
M = Marginal
E = Extreme
[4]Values given are in parts per million.

The oil analysis also indicates excessive oxidation, high iron, aluminum, chromium, silver, molybdenum, tin, and lead for the engine operated on linoleic safflower. The engine fueled with winter rape blend was equivalent to diesel in nearly all of these factors. No engine maintenance problems would have been detected by the oil analysis data during the long term endurance cycle using winter rape as a fuel.

It would appear from the oil analysis, wear measurements, and engine performance that the fuel additive was detrimental to the engine. This engine had more oil consumption throughout the tests which may have been a function of the seating of the rings. Replication of the test is underway to see if a trend develops on the affect of the additive. The additive did appear to drastically decrease fuel filter plugging.

While the test data shows that use of a winter rape blend may be feasible, additional testing will be required with many types of engines before it can be recommended for general use. Anyone contemplating the use of vegetable oil in a diesel engine should be aware of the possible consequences and be prepared to assume the risks.

Tractor Demonstration

A Ford 4600, 3-cylinder tractor has been used for demonstration purposes fueled by 100 percent alkali refined safflower oil. The engine was first operated on a 50-50 sunflower-oil diesel fuel mixture December 27, 1979 and demonstrated on this mixture until June, 1980 when it was switched to 100 percent safflower oil. The engine continued on 100 percent safflower oil for 15 months and accumulated 154 engine hours; and 120 hours doing field work and 35 hours in demonstration work. At the end of the test period the power had declined by 30 percent and the tractor was quite hard to start. Engine inspection showed severe piston ring seizing, sluffing off the piston onto the cylinder walls, and pitting the rod bearings. This engine has now been rebuilt and a new fuel will be selected based on results of the laboratory tests and the long term screening tests.

The injector pump from the Ford 4600 tractor was taken to Spokane Diesel Pump Repair in Spokane, Washington for inspection and testing. The pump is manufactured by CAV and is the rotary jerk type. During the pumps disassembly and inspection, a gold colored sediment accumulation was discovered under the pumps top cover. This was the only abnormal condition found. The head, barrel, and charge pump condition are good indicators of the fuel's effects on the pump. All of these areas showed no wear or abnormal markings. Also, no fuel gumming was found inside the pump. The injector pump was then placed on a pump test stand where high idle, low idle, fuel delivery, charge pump pressure, pump advance, and high and low idle speed settings were all checked and found to be normal. A slight adjustment in fuel cut-off setting was made.

Conclusions

While vegetable oil cannot yet be recommended as a fuel for general use, considerable progress in utilization has been made. The following specific conclusions result from the study.

1. High viscosity and tendency to polymerize within the cylinder are major chemical and physical problems.

2. Attempts to reduce the viscosity of the vegetable oil by pre-heating the fuel were not successful in increasing the temperature of the fuel at the injector sufficiently to be of value.

3. Short term engine performance with vegetable oils as a fuel in any proportion show power output and fuel consumption to be equivalent to the diesel fueled engine.

4. Severe engine damage occurred in a very short time period in tests of maximum power with varying engine RPM. Additional torque tests with all blends need to be conducted.

5. A blend of 70 percent winter rape - 30 percent No. 1 diesel has been successfully used to power a small single cylinder diesel engine for 850 hours. No adverse wear, effect on lubricating oil or effect on power output was noted.

6. A diesel injector pump used for 15 months and with 154 hours of operation with linoleic safflower had no abnormal wear, gumming, or corrosion.

Literature Cited

Auld, D., B. Bettis, and N. Porter. 1981. Development of oilseeds as alternative crops for the Pacific Northwest. Yearly Report of the University of Idaho, Washington State University, and Oregon State University Agricultural Experiment Stations submitted to the Pacific Northwest Regional Commission.

Cummins Engine Company. 1980. Fuel for Cummins Engines. Service bulletin No. 3379001-03. Columbus, Indiana, 47201.

DuPont. 1981. Fuel oil additive No. 2. Specification sheet A040531. E.J. duPont, deNemours, and Company, Petroleum Chemicals Division, Wilmington, Delaware, 19898.

Economic Research Service - USDA. 1976. Energy and U.S. Agriculture: 1974 data base. U.S. Government Printing Office.

Peterson, C.L., D.L. Auld, V.M. Thomas, R.V. Withers, S.M. Smith, and B.L. Bettis. 1981. Vegetable oil as an agricultural fuel for the Pacific Northwest. University of Idaho Agricultural Experiment Station Bulletin No. 598.

Peterson, C.L., G.L. Wagner, D.L. Auld. 1981. Performance testing of vegetable oil substitutes for diesel fuel. ASAE Technical Paper No. 81-3578, ASAE, St. Joseph, Michigan, 49085.

SWEDISH TESTS ON RAPE-SEED OIL AS AN ALTERNATIVE TO DIESEL FUEL

Eric Johansson Olle Nordström

SUMMARY

The cheapest version of Swedish rape-seed oil was chosen. First the rape-seed oil was mixed in different proportions with regular diesel fuel. A mixture of 1/3 rape-seed oil and 2/3 regular diesel fuel (R 33) was then selected for a long-term test.

A Perkins 4.248 diesel engine was used for laboratory tests.

Four regular farm tractors, owned and operated by farmers, and two tractors belonging to the Institute have been running on R 33. Each tractor was calibrated on a dynamometer according to Swedish and ISO-standards before they were operated on R 33. Since then the tractors have been regularly recalibrated.

The test tractors have been operated on R 33 for more than 3 400 hrs.

An additional 1 200 hrs have been covered by the laboratory test engine.

None of the test tractors have hitherto required repairs due to the use of R 33, but some fuel filters have been replaced. Some fuel injectors have been cleaned due to deposits on the nozzles.

INTRODUCTION

The National Machinery Testing Institute at Uppsala started a research programme on rape-seed oil as an alternative fuel for diesel engines in May 1980. The cheapest version of Swedish rape-seed oil (crude Lobra oil) was selected. The programme started by mixing rape-seed oil and regular diesel fuel in a common type of graduated tube with a volume of 500 cubic centimeters. The following proportions were selected:

Rape-seed oil	Regular diesel fuel
30 %	70 %
40 %	60 %
50 %	50 %
75 %	25 %

During the mixing procedure we observed that a homogeneous mixture was more easily obtained by adding rape-seed oil to regular diesel fuel than vice versa. The two fuels mix well at room temperatures and the mixtures remain homogeneous for a long period of time (more than 12 months at room temperatures) when 30 or 40 % of rape-seed oil was used. In the graduated tube containing 30 % rape-seed oil no deposits were found, but the tube containing 40 % rape-

The authors are: Eric Johansson, Director, and Olle Nordström, Officer in Charge of Tractor Testing, National Machinery Testing Institute, Uppsala Sweden

seed oil showed a very small amount of deposits when the mixture had been stored at room temperature for more than a week. The deposits did not increase with time. The fifty-fifty mixture had considerable deposits already after a week.

A pilot study on a VOLVO BM T 650 tractor using 1/3 rape-seed oil and 2/3 regular diesel oil (R 33) showed no problems of separation of the two oils or deposits at the bottom of the fuel tank. The dynamometer tests indicated the same power performance for the two fuels.

Regular checks of the density of the R 33 blend at three different heights in a regular farm fuel tank with a volume of three cubic meters during the winter 1981/82 showed no difference in density at the bottom, middle or top of the storage tank.

The tests were then enlarged and since early 1981 the National Machinery Testing Institute has been operating six different tractors in regular farm work. An International Harvester 1086 tractor has been operated on R 33 for more than 1240 hrs from April '81 to March '82. Besides the six tractors, a Perkins 4.248 diesel engine has been operated in a dynamometer stand for approximately 1200 hrs during the same period of time.

LABORATORY TESTS ON A PERKINS 4.248 DIESEL ENGINE

The engine, Perkins 4.248, is a 4-cylinder diesel engine with direct fuel injection. The displacement is 4.1 l och the compression ration 16:1. The engine is naturally aspirated and has a distributor type fuel injection pump.

The engine was connected to a hydraulic dynamometer and equipped according to ISO 1585 (Net power).

Fuel and lubricating oil

Fuel:	One-third rape-seed oil and two-thirds diesel fuel, R 33	
	Viscosity at 20°C mm^2/s	9.1
	Carbon residue, % by weight	0.30
	Ash content, " " "	0.036
	Density at 15°C	0.865
Lubricating oil:	Period 1 and 2, Shell Tractor Universal oil, SAE 15W/30 API CD/SE	
	Period 3, Shell Myrina oil, SAE 15W/40	

Running-in period

Running-in was done on pure diesel fuel during a period of 30 hrs at varying load.

Performance tests on diesel fuel and on R 33

Full load performance is shown in figure 1, giving the corrected power figures. The power was the same on both fuels. The specific fuel consumption was slightly higher on R 33 depending on the lower specific heat for this fuel (41.1 MJ/kg and 43.0 MJ/kg respectively). The efficiency was higher for R 33 at speeds under 2200 rev/min.

Fuel consumption on part load was measured at two different governor settings, 2200 and 1760 rev/min. Figure 2 shows that the consumption was similar with both fuels.

Endurance test on R 33

Period one: The endurance tests started after the running-in period and the performance test. The engine operated under a load cycle at two governor settings, 2200 and 1760 rev/min respectively. The duration of the cycle was one hour.

The engine kept its full power for about 10 hrs. After 14 hrs the power had decreased by slightly more than 1 kW (2 %). At the end of the period the total power drop was 2.6 kW (5 %). The exhaust smoke was 4 Bosch-units at the start of the period and 5.8 units at the end.

The period ended after 66 hrs running. The total time, including running-in and performance tests, was 106 hrs.

Inspection of the injectors showed rather severe deposits on the injector nozzles, figure 3.

Period two: Apart from changing of the injectors, no adjustments were made to the engine.

The engine operated at cycled load during the first 35 hrs, and subsequently at a constant load of 36 kW at 2240 rev/min, interrupted by short periods when measurements of maximum power were made.

The power at the beginning of the period was about 52.5 kW at 2200 rev/min. The power decreased very little during the first 30 hrs. From 30 hrs to 75 hrs the power dropped by, on average, 90 W/hr. After 75 hrs the power decreased at a fairly constant rate of about 205 W/hr. At the end of the period the power was 38.7 kW, the total power drop during the period thus being 13.7 kW (26 %).

Measurement of the compression-pressure showed normal pressure on cylinders 2 - 4 (2.8 MPa), but only 0.7 MPa on cylinder 1. Inspection of the engine revealed that the piston rings were sticking in cylinder 1 and that there were severe deposits on all nozzles.

Period three: The piston ring grooves in cylinder 1 were cleaned and new piston rings fitted. Renovated injectors were fitted in all cylinders.

The crude rape-seed oil used during this period was filtered through a filter with a nominal filter grade of 5 um (.005 mm). The lubricating oil used was Shell Myrina SAE 15W/40.

The engine operated during the first 10 hrs mostly at cycled load, after which it operated mainly continuously day and night. During about two-thirds of the time the power was 36 kW at 2240 rev/min, amd for the rest of the time the engine operated at full load at 2200 rev/min.

After a running-in period of about 10 hrs the max. power was measured to about 52 kW. From 30 hrs to 100 hrs the power drop was 6 kW (85 W/hr). The power increased after 100 hrs from 45.7 kW to 49.6 kW at 130 hrs and was rather constant to 200 hrs.

At 260 hrs there was another power drop to 41.5 kW followed by increasing power to 52 kW at 285 hrs. The engine worked fairly satisfactorily until about 500 hrs. Between 500 and 700 hrs disturbances in fuel delivery occurred, which were located to a plunger sticking in the injection pump (distributor type). When the plungers had been polished, the power increased to 52 kW. The engine has run totally 1200 hrs during period No. 3. No power drops were noted during the latest 400 test hrs.

GROUP TESTING ON POWER-TAKE-OFF PERFORMANCE

Dynamometer tests were made of the pto performances on a number of tractors (14) on regular diesel fuel or R 33. The injection setting and timing was the same on both fuels. The tests were made at three different governor settings: maximum speed, 80 % of maximum speed and 60 % of maximum speed.

The results given in table 1 show the relative power and fuel consumption by mass (g/MJ). The power ratio (power at R 33/power on diesel fuel) varied from 92 to 105. The average value was 98.8 at maximum speed. The power was thus 1.2 % lower when using R 33. At the other governor settings the power loss was 0.5 % and 0.7 respectively on average.

The fuel consumption measured on a mass basis was 6 - 7 % higher on R 33. Calculated to a volume basis this corresponds to a 2.5 % higher value on R 33.

Table 1. Power and fuel consumption ratios on diesel fuel and R 33 respectively

Tractor	Power ratio*			Fuel ratio**		
	max. speed	80 % of max.	60 % of max.	max. speed	80 % of max.	60 % of max.
Case 1490	100.0	100.0	100.0	107.2	108.4	107.8
Deutz 7207 C	98.6	98.0	100.0	107.4	106.8	108.9
Fendt 309 LSA	100.1	98.0	99.0	108.5	107.7	105.4
Fiat 880 DT-5	95.8	95.6	97.0	108.3	108.0	107.8
Ford 6610	100.1	99.1	99.4	104.8	105.6	106.6
Ford 7710	99.7	101.1	102.0	108.5	105.0	106.6
IH 844 XL4	98.1	101.7	102.2	104.6	105.7	106.1
John Deere 2140	100.0	104.0	104.8	108.3	109.9	107.6
MF 595	100.5	100.0	99.5	104.6	105.8	107.0
Same 95 DT	98.7	97.5	98.4	105.9	107.0	107.5
Valmet 802	99.3	101.5	101.6	104.2	104.2	104.0
VOLVO BM T 700	92.0	95.6	94.3	108.7	106.2	105.0
Zetor 10045	100.0	96.7	95.2	109.8	110.1	104.7
Zetor 8045	100.2	101.5	100.0	104.8	105.4	103.9
Mean value	98.8	99.3	99.5	106.8	106.8	106.4

$$\text{* Power ratio} = \frac{\text{Power at R 33} \times 100}{\text{Power at diesel fuel}}$$

$$\text{** Fuel ratio (mass basis)} = \frac{\text{spec consumption g/MJ R 33} \times 100}{\text{spec consumption g/MJ diesel fuel}}$$

TESTING UNDER ORDINARY FARM WORK CONDITIONS

The tests were intended to study the problems which might occur during long-term operation in ordinary farm work on the R 33 fuel without any alterations of the engine.

Before the testing period the performances of the engines were checked by dynamometer tests on the power-take-off. During the test the running time, type of work, fuel consumption and actual disturbances were noted. The tractors will regularly be checked on the pto dynamometer with respect to power performance and fuel consumption.

Tractors used in tests

At the Alnarp station (southern Sweden)

Ford TW-10	operated for	700	hrs	on	R 33
VOLVO BM T 650	" "	450	"	"	"
VOLVO BM T 600	" "	300	"	"	"

At the Ultuna station (central Sweden)

International 1086	operated for	1240	hrs	on	R 33
VOLVO BM T 650	" "	270	"	"	"

At the Röbäcksdalen station (northern Sweden)

VOLVO BM T 650 operated for 500 hrs on R 33

Ford TW-10: The tractor has a naturally aspirated diesel engine with direct fuel injection, cylinder displacement 6.57 l and rated engine speed 2300 rev/ /min.

The tractor was used for a variety of implements including 5-furrow semi-mounted plough, harrows, sugarbeet harvester, and for transport work.

Disregarding the changes of fuel filters four times, the engine worked satisfactorily. The injectors were cleaned in connection with a power check after 500 hrs of operation.

The total running time on R 33 was 700 hrs and total fuel burned was 9.6 m^3.

VOLVO BM T 650 (Alnarp): The tractor has a naturally aspirated diesel engine with direct fuel injection, cylinder displacement 4.2 l and rated engine speed 2300 rev/min. The fuel injection pump is of in-line type.

A power check after the test period showed the same power on diesel fuel and on R 33.

The tractor has been used for 450 hrs by the testing Institute on different types of work. Eleven fuel filters were changed, but no other repairs or adjustments were made during the test period. The engine has worked satisfactorily.

VOLVO BM T 600: The tractor has a naturally aspirated diesel engine with direct fuel injection, cylinder displacement 3.7 l and rated engine speed 1950 rev/min. The fuel injection pump is of in-line type.

A power check before the test period showed the same power on diesel fuel and on R 33.

The tractor has been used at a farming school on different types of farm work for 300 hrs. The engine has worked satisfactorily. The total consumption of R 33 was 1.9 m^3.

International 1086: The tractor has a turbo-charged diesel engine with direct fuel injection, cylinder displacement 6.8 l and rated speed 2400 rev/min.

Before the test period the pto power performance was 101.5 kW on diesel fuel and 104 kW at R 33 at rated engine speed, 2400 rev/min. The fuel consumption on a mass basis was higher on R 33, but the efficiency was slightly better on R 33 at speeds under 2400 rev/min. The smoke density was 2.5 Bosch units for both fuels at 2400 rev/min. At 2070 rev/min the smoke was 3.4 Bosch units for R 33 and 3.7 for diesel fuel. At 1600 rev/min the smoke figures were 5.2

and 5.9 respectively.

A power check was made after a running time of 350 hrs. The power at rated speed, 2400 rev/min, was about 4 % lower than at the start of the test. After about 1000 hrs operation on ordinary work an observable power loss occurred and a power check showed that the power was down to 80 kW. Figure 4 shows that the power increased to 96 kW when the injectors and the air filter were changed.

The engine was inspected after a running time of 1240 hrs on R 33. The underside of the cylinder head was covered by a rather thick layer of heavy carbon. Similar deposits were also observed on the pistons above the upper piston rings. Carbon deposits were observed on the inlet valves and in the inlet pipe. The deposits were hard.

The exhaust valves were fairly clean, as was the entire exhaust system. The cylinder wear was measured to about 5 µm (.005 mm).

No abnormal observations were made from any other points of view.

The tractor worked during the spring, summer and autumn of 1981, and the winter och spring of 1982. Disregarding the changes of fuel filters some times, the engine worked without disturbances. Cold weather starting was possible down to about -15°C. The total fuel consumption was 14 m^3 and total time 1240 hrs.

<u>VOLVO BM T 650 (central Sweden)</u>: The same engine as described for southern Sweden.

The tractor was used by the Institute on various types of work for 270 hrs. The engine operated satisfactorily. Three fuel filters were changed.

<u>VOLVO BM T 650 (northern Sweden)</u>: The samne engine as described for southern Sweden.

The tractor was used on various types of practical farm work for 485 hrs and consumed 3.9 m^3 fuel (R 33).

Disregarding the change of fuel filters twice, the engine worked satisfacorily. The injectors were cleaned in connection with a power check after 400 hrs of operation.

RESULTS OF THE FIELD TESTING

Six tractors have operated under ordinary farm work conditions for more than 3400 hrs on the fuel R 33. The total fuel consumption has been over 40 m^3.

The engines have worked without disturbances. The fuel filters have been replaced at shorter intervals than on regular diesel fuel. Power loss due to injector coking have occurred. One tractor has operated 1000 hrs before cleaning of the injector. The wear of the cylinder liners has been measured to 5 µm (.005 mm) on one tractor after 1240 hrs operation on R 33.

The field testing will continue with the intention to gain more experience of service life on tractors using R 33 as fuel.

PRODUCTION OF RAPE-SEED OIL IN SWEDEN

A normal Swedish rape-crop will yield 2800 kgs per hectar (2460 lbs per acre). Approximately 1000 liters of rape-seed oil (crude Lobra oil) will be extracted per ha. If a Swedish farmer devotes 10 % of his arable land for rape-seed production he will get a quantity of rape-seed oil that corresponds to his need of liquid fuel on a liter basis.

A cubic meter diesel fuel cost 2142 Sw Crowns May 10, 1982. The corresponding price of a cubic meter of crude Lobra oil was 2465 at the same time.

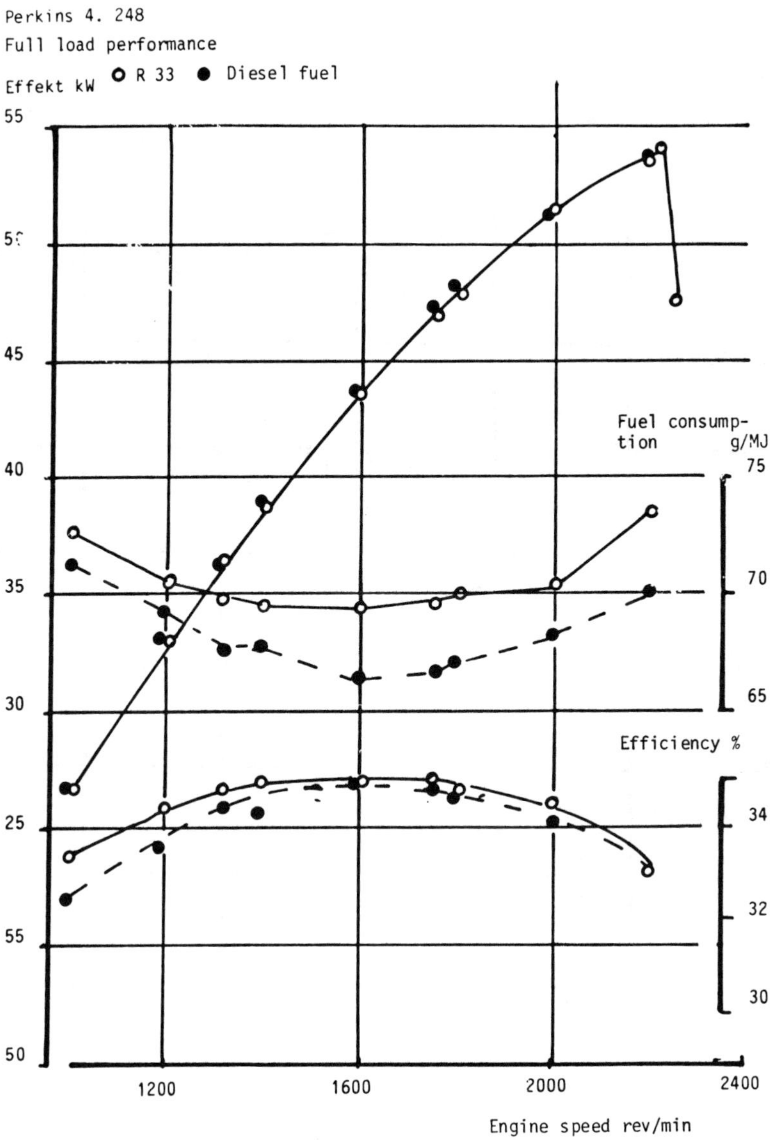

Figure 1.

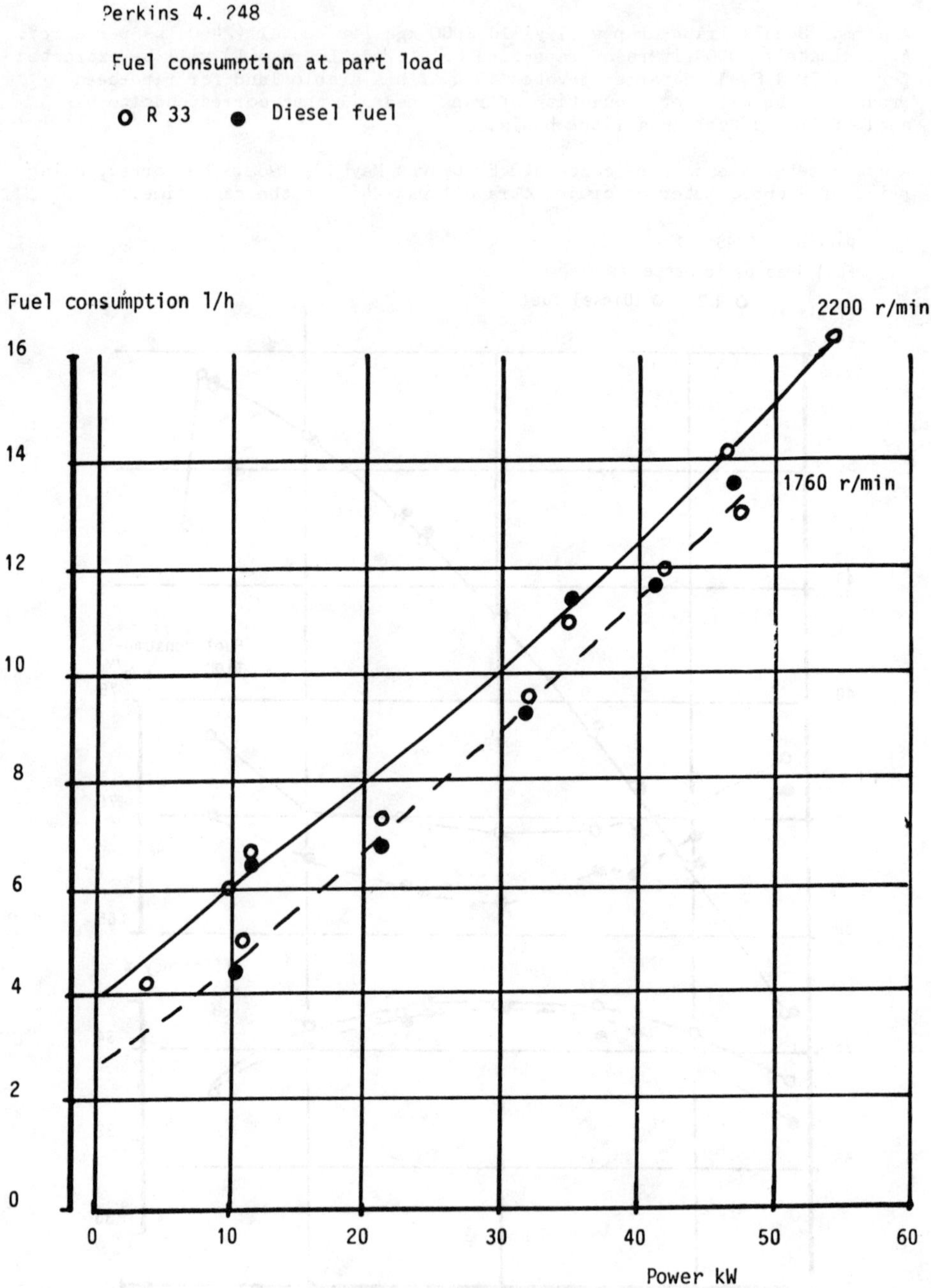
Perkins 4. 248
Fuel consumption at part load
R 33
Diesel fuel
Fuel consumption l/h
2200 r/min
1760 r/min
16
14
12
10
8
6
4
2
0
0
10
20
30
40
50
60
Power kW

Figure 2.

PERKINS 4.248 Nozzle coking

Injector nozzle after period 1.

Figure 3.

Figure 4. Pto power performance at 2400 rev/min engine speed.

Stage 1. After 350 hrs in field work, R 33

Stage 2. According to 1 plus 2 hrs full load dynamometer operation, R 33

Stage 3. After 1025 hrs in field work, R 33

Stage 4. According to 3 plus 0,5 hrs full load dynamometer operation, R 33

Stage 5. According to 4 plus exchange of injectors, R 33

Stage 6. According to 5 plus exchange of air cleaner filter, R 33.

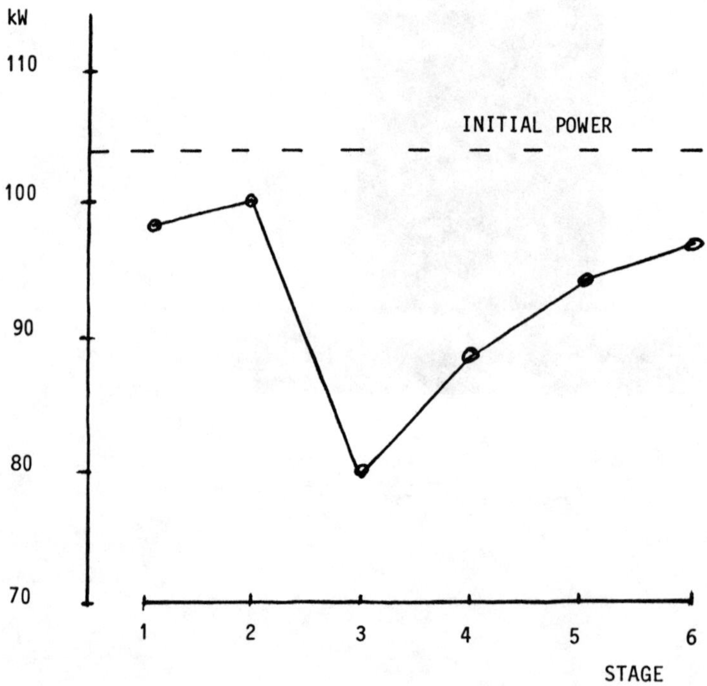

ENGINE DEPOSIT AND POUR POINT STUDIES USING CANOLA OIL AS A DIESEL FUEL

R. C. Strayer W. K. Craig G. C. Zoerb
Member ASAE

ABSTRACT

Engine tests conducted during previous investigations have established the viability of using canola oil as a substitute for diesel fuel on a short term basis, but also revealed the need to assess possible combustion chamber deposits from long range testing. Low temperature problems in handling vegetable oils has also been recognized as posing a threat to their use in winter operation. This paper reports a procedure involving a direct comparison of running two different fuels in an engine simultaneously to study deposit problems, and also reports on three attempted methods - fuel blending, fuel heating and fuel additives to reduce the pour point of canola oil.

Introduction

Research involving the use of vegetable oils as diesel fuel extenders or substitutes now appears to be world wide, with every conceivable oil being tested. Work at the University of Saskatchewan, Department of Agricultural Engineering being conducted jointly with the Saskatchewan Research Council and the Protein, Oil and Starch Pilot Plant has been restricted to low erucic acid rapeseed (now known as canola) oil and rapeseed (high erucic) oil.

Phase one of the research program established the viability of using canola oil as a fuel substitute for short periods of time and in relatively warm weather. (Strayer et al, 1980). However, it also revealed problems that required further investigation - in particular, the need to assess both combustion chamber deposits and lubricating oil contamination, and the problem of handling and using the fuel at low temperature. Thus the second phase of the program, as it relates to the Agricultural Engineering Department, is concentrating on these problems.

Cylinder Deposit Analysis

A review of other research reveals that a considerable amount of study has been done on this problem. Bruwer (1980), Peterson (1981) and others have all reported varying degrees of problems with combustion chamber deposits, injector fouling and piston ring sticking associated with the use of various vegetable oils when used as a fuel. It appears that in general the problem was least troublesome when running air cooled, precombustion chamber engines, and greatest with direct injection engines. It also is evident that some vegetable oils, such as linseed oil, are far worse than others, while methyl and ethyl esters of some oils create the least problem. Blending vegetable oils with diesel fuel appears to be one way of reducing the deposit problem.

The authors are: R. C. Strayer, Associate Professor, Agricultural Engineering Dept., W. K. Craig, Saskatchewan Research Council, and G. C. Zoerb, Professor and Head, Agricultural Engineering Dept., University of Saskatchewan, Saskatoon, Sask.

Canola is the major oil seed crop grown in Saskatchewan, and to date there is little apparent research on its use as a fuel source, or the problems of engine deposits associated with its use. It was therefore decided to pursue an investigation with canola oil and its esters in comparison with diesel fuel. It also seemed wise to select the type of engine which others have found to create the most adverse deposit problems. Accordingly, a two cylinder, 12 kW Petter water cooled engine was selected. This engine is direct injected with a medium-to-high swirl open combustion chamber. In order to get a valid comparison, each cylinder was operated with an independent fuel supply and system, such that two fuels or blends could be run simultaneously, and the results in each cylinder compared. The engine was operated at about 90% of governed speed and under a load which varied from 70% to 90% of full load, so that over each test period the power output would average about 80% of the engine's rated power. This, it was felt, would simulate an engine under normal loading conditions. Prior to starting each test, the injection pumps were adjusted such that each cylinder was delivering the same power.

Observations - The first test to be run using this procedure involved a comparison of summer diesel fuel running in number two cylinder, and filtered unrefined extractor oil (low erucic canola oil) running in number one cylinder. A total of 150 hours was put on the engine in six operating days, after which the injectors, exhaust manifold and cylinder heads were removed. The following comparisons were noted:

Fuel Injectors - Both had carbon deposits, but there was no significant difference.

Cylinder Heads - The cylinder operating on canola oil showed a slightly greater amount of deposit, but neither cylinder exhibited what would be considered excessive fouling.

Pistons - Slight deposit on both, but little difference.

Exhaust Manifold - This was the only part where there was a significant difference in deposits. The side of the manifold connected to number one cylinder contained substantially more material, and unlike the soft soot from cylinder number two, was quite hard and somewhat sticky.

At this point, it was concluded that a test period of 150 hours was insufficient, so without any cleaning of deposits the engine was reassembled and the test continued.

Low Temperature Problems

Viscosity: Everyone involved with the use of vegetable oils as a diesel fuel substitute has discovered that these oils have viscosities many times that of diesel fuel, and in addition have a low viscosity index which gives rise to very high viscosities at low temperature (Fig. 1).

The effect is twofold; flowability becomes a problem as temperatures approach 0°C. (Strayer, et al, 1981). It has also been established (Bruwer et al, 1980) that the spray pattern from the injection nozzle is adversely affected as viscosity increases, resulting in inferior combustion and contributing to cylinder deposits, particularly in direct injection engines. There is therefore a definite need to reduce viscosity, and as can be observed from Fig. 1, at least three methods are effective - heating the oil, mixing it with diesel fuel, or converting the oil to an ester.

Pour Point: Of considerable concern if an engine is to be run under low temperature conditions is the relatively high pour point of canola oil. To assess this problem, a test apparatus was constructed (Meister 1980) which

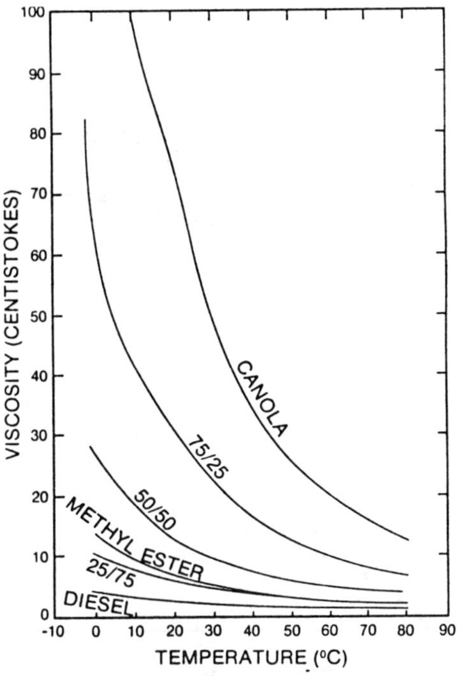

Fig. 1

VISCOSITY VS TEMPERATURE
FOR CANOLA
AND DIESEL BLENDS

pumped canola oil through a five micron diesel fuel filter at 55 kPa. The apparatus was then placed in a cold room and the flow rate measured as the temperature dropped. From Fig. 2 it can be observed that the flow started to drop drastically at about +10°C, and was effectively zero at -4°C. Since any engine in Saskatchewan can be exposed to temperatures below +10°C virtually any month of the year, this problem is even more serious than that caused by high viscosity.

Three studies were undertaken to find a solution - the use of fuel blends, fuel system heaters, and various pour point depressants were all investigated.

Fuel Blends

Four different blends with canola oil or methyl ester as the base were tested in an attempt to find a practical way of reducing pour point. It was the original intention to use ASTM D97-66 as the test procedure, but later it was decided that a valid comparison with diesel fuel would not necessarily be obtained, since the ASTM test is specifically intended for petroleum based fuels only (article 1.1, ASTM test D97-66). A somewhat different procedure was then devised, which gave a direct visual comparison, but did not establish an exact pour point.

Samples to be tested were mixed, and placed together on a rack in a refrigerated room. Observations were made continuously, and a recording of these observations taken at 3°C increments as cooling of the samples proceeded. In every sample some evidence of cloudiness first appeared, which occurs

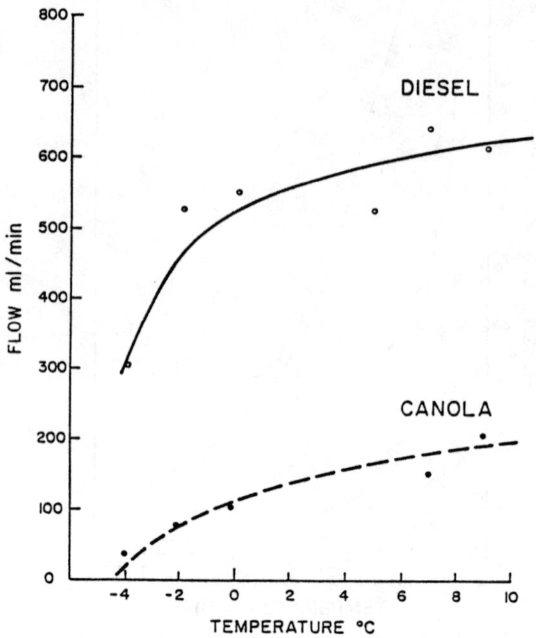

GRAPH OF DIESEL VS CANOLA FLOW AT 55 KPa, MEASURED IN FRONT OF FIRST FILTER (GAUGE 1)

Fig. 2

also in petroleum fuels as the wax particles start to separate. Further cooling started to produce solidification, and an estimate was made of the amount solidified in each sample. The procedure was continued until substantial solidification was reached with most of the samples. The procedure was repeated three times, and the figures shown in Table 1 are averages of the four runs.

Since the flowability of any of the mixtures or fuel samples will cease when any solidification appears, the only one of the fuels tested that would have any value as a fuel below -18°C would be the 90% Methyl Ester - 10% Ethanol mixture, and it was so cloudy that it would flow only very slowly.

In a separate test, a mixture of 50% methyl ester - 50% ethanol appeared to be reasonably fluid down to -30°C. However, this was not pursued as it had previously been established that such a mixture would not perform satisfactorily in an engine, nor was it a stable mixture.

Fuel System Heaters

The first method used to heat the fuel system was to construct a water-to-fuel heat exchanger (Youzwa, 1982). This required the construction of a new fuel tank into which a copper tube coil was incorporated, and this was connected to the water line on the discharge side of the cooling system. The water temperature in this line was regulated between 70°C and 90°C by a thermostatically controlled valve in the engine cooling system. Two more thermostatically controlled valves controlled the flow through the heat

Table 1. Pour point results with methyl ester and canola blends

Temp °C	90% Canola 10% Diesel	50% Canola 50% Diesel	100% Canola	90% Canola 10% Ethanol	Methyl Ester	90% M.E. 10% Ethanol
- 6	Fluid	Fluid	Fluid	Fluid	Fluid	Fluid
- 9	"	"	"	"	Cloudy, Fluid	"
-12	"	"	Cloudy, Fluid	"	Very Cloudy	"
-15	"	"	"	10% Solid	Cloudy	
-18	50% Solid	"	50% Solid	30% Solid	Solid	Cloudy
-21	70% Solid	20% Solid	70% Solid	50% Solid	Solid	Very Cloudy
-24	90% Solid	40% Solid	90% Solid	Solid	Solid	Very Cloudy

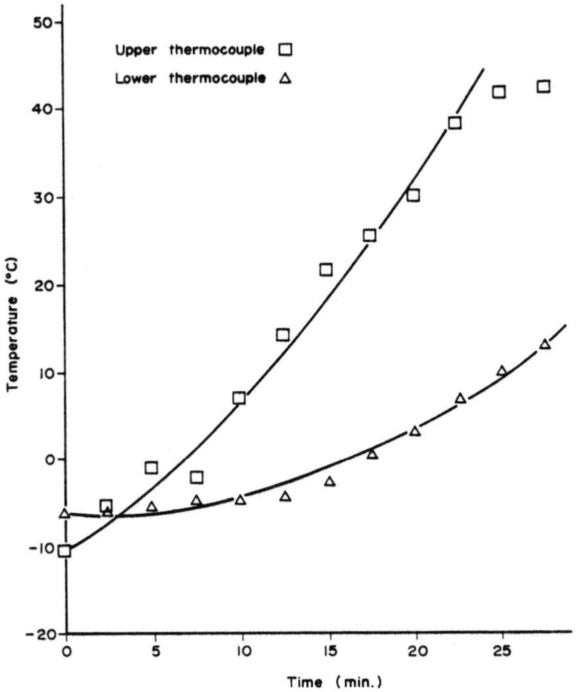

Fig. 3

Rate of heating of canola oil in heat exchanger tank (Youzwa, 1982)

exchanger, allowing full flow until the fuel temperature reached +15°C, at which time the flow would by-pass.

To conduct a test, the fuel tank, filled with canola oil, was placed in a freezer and cooled to about -10°C. It was then wrapped with insulation, connected to the water system, and the engine was started. Thermocouples placed in the bottom of the tank, and close to the top, measured the temperature, which was continuously recorded.

As can be observed from Fig. 3, the rate of heating was much higher at the top of the tank than at the bottom, and it became evident that because of the slow rate of circulation of the canola oil in the tank this would not be a very practical type of fuel heater to use. Future attempts will include a heater under the tank, or located as close to the bottom as possible.

An in-line heater was also connected into the fuel system just ahead of the injection pump. No data were recorded, but the heater will also be given further testing, as it may be a desirable component in the system to reheat the fuel just prior to injection.

Pour Point Depressants

A variety of chemicals, known as pour point depressants, have been used in the petroleum industry to reduce the pour point of lubricating oils and diesel fuels which are to be used at low temperature. Both petroleum based oils and vegetable oils exhibit a "cloud point" and a "pour point" which is related to a precipitation, separation and congealing of wax particles in the oil as the temperature is reduced. It appears that the probable mechanism of a pour point depressant is to coat the wax particles and prevent crystal growth and solidification. The compounds commonly used for this are methacrylate polymers, alkylated naphthalene or phenols (Obert, 1968), and in general all, by themselves, have a high viscosity and pour point.

To study the effect of some commercial pour point depressants on canola oil and methyl ester, five different products commonly used for engine lubricating oil and/or diesel fuel were obtained from an oil refinery, and mixed in varying proportions. The five used were:

1. Paradyne 25 (sold by Imperial Oil)
2. Paraflow 149 (sold by Imperial Oil)
3. Acryloid 1505-C
4. Acryloid 155-C
5. Tolad T-35 (sold by Petrolite)

Attempts were made to determine the chemical formula, but the information was not available.

The amount of depressant needed with petroleum oils is apparently quite variable, and can be effective in concentrations as low as 50 ppm, or may require several times that amount. As a first attempt, three mixtures of each depressant - 100 ppm, 200 ppm and 300 ppm were made with canola oil and with methyl ester. These were placed in a freezer and observed as the temperature dropped. For comparison purposes, samples of straight canola oil and methyl ester were also refrigerated.

It was evident very soon that none of the pour point depressants tested had any beneficial effect on either canola oil or methyl ester at any concentration tested. In fact, Paraflow 149 appeared to raise both the cloud point and pour point of both oils by about 5°C.

A second series of tests was run at a concentration of 1000 ppm, and a slight improvement was noted with both Acryloid additives.

Conclusions

Of the three methods employed to find a solution to low temperature engine operation when using canola oil, fuel heating appears to be the most practical when thinking of engine operation only. However, it must be appreciated that this will involve much more than just heating the fuel system of the engine itself. Everyone involved in storing, transporting or transferring the vegetable oil would have to be equipped with devices in tanks and pumps so that heat could be provided when needed. This could be a very costly and complex system where large storage tanks or truck tanks were involved. Thus while it may be very practical for the fuel system on an engine, it may be completely impractical when applied to the handling and distribution of the fuel.

It would appear, then, that the better solution would be to follow the lead of the petroleum industry and find chemical additives which would achieve the desired result of reducing cloud and pour points. There is no reason to believe that such chemicals are not available, and continued research will be required to develop this solution to the problem.

References

1. Bruwer, J. J., B. Van D. Boshoff, F.J.C. Hugo, L. M. duPlessis, J. Fuls, C. Hawkins, A. N. van der Walt and A. Engelbrecht. 1980. Sunflower seed oil as an extender for diesel fuel in agricultural tractors. Presented at the 1980 Symposium of the South African Institute of Agricultural Engineers.

2. Meister, S. and A. Cole. 1981. The potential of canola oil as a diesel fuel. Undergraduate thesis, Department of Agricultural Engineering, University of Saskatchewan, Saskatoon.

3. Obert, Edward F. 1968. Internal Combustion Engines, Third Edition. International Textbook Company, Scranton, Penn.

4. Peterson, C. L., G. L. Wagner and D. L. Auld. 1981. Performance testing of vegetable oil substitute for diesel fuel. Presented at the ASAE Winter Meeting, Paper No. 81-3578.

5. Strayer, R. C., E. Coxsworth and R. G. Chopiuk. 1980. Canola oil as a fuel for diesel engines. Presented at the Pacific North West Region, ASAE Annual Meeting.

6. Strayer, R. C. and W. K. Craig. 1981. Low temperature problems using canola oil as a diesel fuel. Presented at the North Central Region, ASAE/CSAE Joint Meeting, October 1981.

7. Youzwa, T. 1982. The design and testing of a heated fuel system for canola oil. Undergraduate thesis, Department of Agricultural Engineering, University of Saskatchewan, Saskatoon.

Acknowledgement

This project is being funded by Agriculture Canada.

LABORATORY ENDURANCE TEST OF A SUNFLOWER OIL BLEND IN A DIESEL ENGINE

M. ZIEJEWSKI K.R. KAUFMAN
Assoc. Member ASAE

ABSTRACT

This paper compares the effects of using a 25-75 blend (v/v) of alkali refined sunflower oil and diesel fuel in a diesel engine as compared to a baseline test on diesel fuel. There were no significant problems with engine operation during the baseline test. However, problems were experienced while using the blended fuel. The major problems were (1) abnormal buildup on the injection nozzle tips, (2) injector needle sticking, (3) secondary injection, (4) carbon buildup in the intake ports, (5) carbon deposits on the exhaust valve stems, (6) carbon filling of the compression ring grooves, and (7) abnormal lacquer and varnish buildup on the third piston land.

FUEL PROPERTIES AND CHARACTERISTICS

The fuels utilized in these tests were D-2 diesel control fuel (Phillips Reference Fuel) and a 25-75 blend (v/v) of alkali refined sunflower oil and D-2 diesel control fuel. Fuels were obtained from commercial sources. No additives were used. All fuels were filtered through a five micron filter. Samples of the test fuels were analyzed using American Society of Testing and Materials (ASTM) standard procedures. Test fuel properties are in Tables 1 and 2.

Table 1. Diesel Control Fuel Properties

Cetane Number	46.3
Distillation Range	
IBP, ^{o}C	200
10% Point, ^{o}C	224
50% Point, ^{o}C	257
90% Point, ^{o}C	296
End Point, ^{o}C	323
Gravity, ^{o}API	35.7
Sulphur, %	0.25
Hydrocarbon Composition	
Aromatics, %	29.8
Flash Point, ^{o}C	75
Viscosity, @ $40^{o}C$; mm^2/s	2.44

Table 2. Sunflower Oil Properties

Specific Gravity		0.9174
Viscosity @ $38^{o}C$, mm^2/s		33.9
Flash Point, ^{o}C		274
Peroxide Value, meq.		14
Acid Value, mg/g		0.1
Iodine Value, g/g		130
Moisture, %		0.03
Phosphatide Content, %		0.01
Fatty Acid Distribution, %		
Palmitic	16:0	6.0
Stearic	18:0	4.2
Oleic	18:1	18.7
Linoleic	18:2	69.3
Linolenic	18:3	0.3
Arachidic	20:0	0.4
Eicosenoic	20:1	0.1
Behenic	22:0	1.0

APPARATUS AND PROCEDURE

These tests were conducted at the Allis-Chalmers Engine Division in Harvey, Illinois. A direct injection, intercooled and turbocharged, four cylinder Allis-Chalmers diesel engine, model 433I, was selected because of its typical current design, relatively small size, and low fuel consumption. The dis-

---

The authors are: M. Ziejewski, Research Associate, and K.R. Kaufman, Assistant Professor, Agricultural Engineering Dept., North Dakota State University.

displacement of this engine is 3.28L with a 9.84 cm bore and 10.80 cm stroke. It has 14.5:1 compression ratio. A Stanadyne Roosa Master Distributor Pump Type DB2 was used with two 9.40 mm pumping plungers. The injection nozzle used was a Robert Bosch nozzle with four 0.32 mm diameter orifices, a 1.1 mm sac length, and a 1.0 mm sac diameter. Prior to the test run the engine was completely disassembled, all critical parts were measured, and the engine was rebuilt in strict accordance with all furnished specifications. Following this preparation, the engine was installed on a dynamometer test stand equipped with appropriate accessories for controlling speed, load and other engine operating conditions. The engine was operated on a cycle recommended by the Alternate Fuels Committee of the Engine Manufacturer's Association (EMA). The EMA cycle for the AC 433I engine is presented in Table 3. The cycle was repeated five times (15 hours). Average cycle power was maintained at about 70%. After 15 hours on the cycle, the engine was shut down for nine hours. This procedure was repeated until 200 hours of operation on the test cycle had been completed. At the conclusion of the test, the engine was disassembled, visually inspected, measured, and rated to determine the extent of varnish and carbon formation.

Table 3. EMA Test Cycle for AC 433I Engine

Step	Speed (RPM)	Torque (kNm)	Brake Mean Effective Pressure (kPa)	Time Min.
1	2300	3.36	1196	60
2	1955	3.72	1327	60
3	2070	0.93	333	30
4	750	0	0	30
				180

Fuel consumption was measured on a weight basis with a Cox Instrument Fuel Consumption Weight System, Type 402. A Bosch Model EFAW 68A smokemeter was used to analyze exhaust smoke. Three samples were taken and the average reading was recorded. The dynamometer used to load the engine was a Dynamic Asorbing Dynamometer, Model 1014 D.G. Fuel injection line pressures were measured at the nozzle with a Kistler Model No. 607F122 piezoelectric pressure transducer. The pressure output signals were conditioned with a charge amplifier. A magnetic inductance sensor manufactured by DALEC, Data Electronics Corporation sensed the location of a top dead center (TDC) indicator affixed to the flywheel. Outputs from the pressure tranducer and magnetic sensor were displayed on a Nicolet Instrument Corporation Explorer III digital oscilloscope and stored in a Tektronix 4052 Graphic Computer System. Fifteen line pressure reading were collected for each cylinder every 50 hours of the test cycle. The data was taken for all four cylinders at rated load (2300 RPM) and at peak torque (1800 RPM). Additionally, pressure traces for cylinder #4 were taken at high (2530 RPM) and at low idle (750 RPM).

RESULTS AND CONCLUSIONS

General Test Observations

The EMA recommend 200 hours of operation on their test cycle. However, in both tests the engine logged more total hours of operation because of time needed to take performance data and pressure readings throughout the test. The engine operated for 321 total hours while using diesel fuel and for 268 total hours while using the 25-75 blend.

The engine running on the diesel control fuel completed the test without requiring a fuel filter change. During 321 hours of engine operation the pressure before the fuel injection pump was maintained between 20-28 kPa according to the manufacturer's specification. A fuel filter change was necessary while operating on the 25-75 blend after 100 hours of testing.

Engine Performance Results

Fig. 1 shows engine output and operating conditions recorded at the beginning and end of the test on diesel control fuel. The power output during the 200 hour test did not show any deterioration. The decrease in peak power at 2200 RPM after testing was insignificant (1.1%). BSFC (break specific fuel consumption) showed slight regression in conjunction with increasing time of engine operation. The change observed at peak torque (1.5%) was similar at 2000 RPM (1.7%). Higher speed provided somewhat improved BSFC and at 2300 RPM was 0.5% lower than at the beginning of the test. Both the loss and gain in BSFC were relatively modest in absolute value. The variation in RB smoke level appears to be consistent with the variation in BSFC. The increase in exhaust temperature was caused by higher air intake temperature and by slightly lower engine performance. The gain was insignificant.

Fig. 2 compares initial engine performance and operating conditions for diesel fuel and the 25-75 blend test. Engine power output over the tested engine speed range was slightly higher for the 25-75 mixture. At rated speed (2300 RPM) the difference was 2.5%. At 1800 RPM the gain in power was 6%. The tendency toward larger differences in power output at lower engine speed was possibly caused by increasing time available for combustion, which appears beneficial for the 25-75 mixture. The RB smoke level of the 25-75 blend compared to diesel fuel, increased at higher engine speed (from 1.0 to 2.2) and decreased at lower engine speed (from 4.0 to 2.6). BSFC was similar for the tested fuels. However, the differences in energy content per unit of mass between the 25-75 blend and diesel fuel would cause the brake specific energy consumption for the 25-75 blend to be lower than for diesel fuel. Greater exhaust temperature for the 25-75 blend was caused by higher intake air temperature during the test. Higher exhaust temperature caused the turbocharger to run faster and produce greater intake manifold pressure.

Fig. 3 shows engine performance and operating conditions recorded at the beginning and end of the 25-75 blend test. The power output after 200 hours of the test (163 hours on the fuel injection nozzles) did not show a significant change with the exception of a 2.5% loss at 2300 RPM. This could indicate the beginning of injection nozzle deterioration. After 200 hours, the BSFC was unexpectedly lower. This may have been caused by a change in fuel system calibration. The RB smoke level was higher after the test. At 2300 RPM the smoke level increased from 2.2 to 3.5 RB smoke units and at 1800 RPM from 2.6 to 3.1 RB smoke units. The exhaust temperature increased but not significantly.

Statistical Analysis of Engine Performance: Linear regression analysis was used to test for a significant relationship between the engine performance variables (power, BSFC, and exhaust temperature) and hours of engine operation. For all tested parameters the hypothesis of dependence was rejected at the 5% level of significance. The standard errors for the test variables indicated a greater standard deviation of the dependent variables for hours of engine operation on the 25-75 mixture compared to diesel fuel. This was caused mainly by erratic deterioration of the fuel injection system.

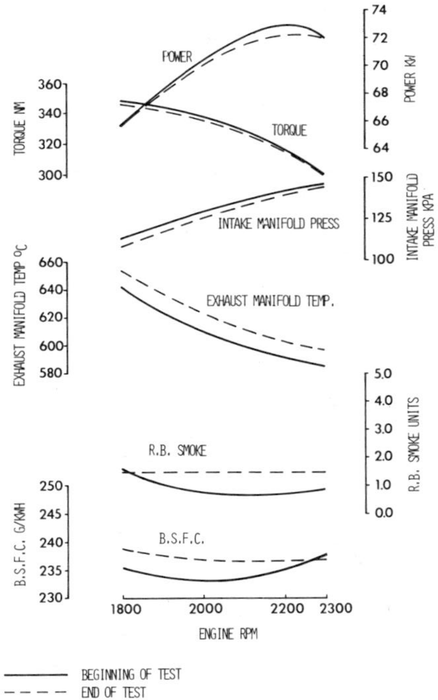

Fig. 1. Engine Performance at Beginning and End of EMA Test Cycle while using diesel fuel.

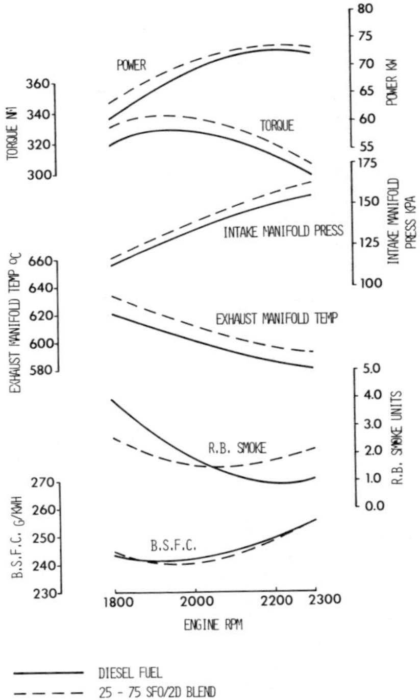

Fig. 2. Comparative Engine Performance between diesel fuel and a 25-75 SFO - 2D blend.

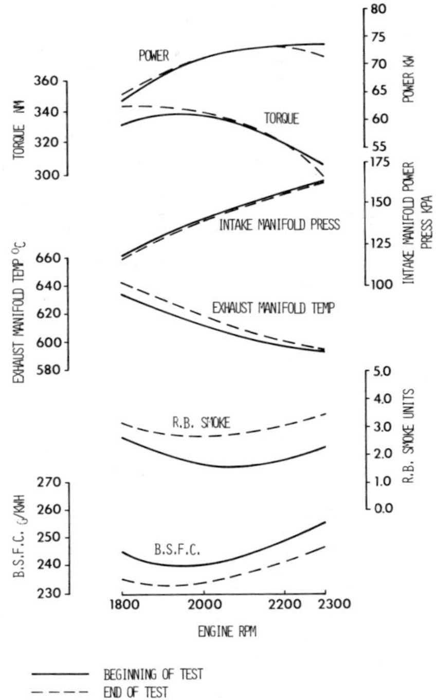

Fig. 3. Engine Performance of Beginning and End of EMA Test Cycle while using a 25-75 SFO - 2D blend.

Injection System

Injection Nozzles: One set of the injection nozzles was used for the 321 hours of engine operation on diesel fuel. After the test all tips showed normal carbon residue. The highest nozzle opening pressure drop was about 10% for nozzle #3. The lowest nozzle opening pressure drop was about 8% for nozzles #1 and #2. Inspection showed a reduction in orifice diameters for nozzles #2 and #3 only. Following the diesel test no leak was observed for all nozzles when tested on the injection nozzle test stand. During 268 hours of engine operation with the blended fuel, the injection nozzles were changed twice. The first set of nozzles were changed at 29 hours of engine operation. The change was made because of problems with the fuel injection pump which may have resulted in injector coking not caused by the test fuel. The second change was made after 105 hours of engine operation (76 hours of operation with the injectors) when a drop in power and significant differences in fuel injection line pressures for cylinders #2 and #3 were experienced. The power output dropped at 2300 RPM from 77.6 kW to 73.4 kW. The BSFC was slightly elevated. The smoke level was unexpectedly low (0.5 RB smoke units). All tips showed abnormal carbon residue. The highest nozzle opening pressure drop was about 18% for nozzles #2 and #4. The remaining two nozzles had a 13% drop in nozzle opening pressure. Inspection showed a reduction in orifice diameter for all tested nozzles. Scuffing was observed on the needles from nozzles #2 and #3. This suggests an abnormal reduction in the needle to housing clearance and difficulty in needle movement. During the injection nozzle stand test no leak was observed for all nozzles. The third set of nozzles performed 163 hours. All tips also showed an abnormal build-up of carbon. The nozzle opening pressure drop was 7% and was uniform for all nozzles. Inspection showed a reduction in orifice diameter for all nozzles. For nozzles #2 and #4 a uniform reduction was observed. Nozzle #1 had one plugged orifice. The second set of nozzle needles did not show excessive wear. During the injection nozzle stand test, no leak was observed for all nozzles.

Injection Line Pressure: Injection line pressure analysis during the 25-75 blend test indicated erratic deterioration of the injection system. During the test sporadic nozzle needle sticking, needle reopening, and secondary injection was experienced. A drop in amplitude of the residual line pressure due to needle sticking and/or poor needle seating with consequent fuel dribble was recognized. (Fig. 4a & 4b). For nozzle #4 after 105 hours of operation, the residual line pressure dropped at 2300 RPM from 6.9 MPa to 4.1 MPa (Point A. Fig. 4a & 4b) and at 1800 RPM from 5.5 MPa to 1.4 MPa. The maximum residual line pressure decreased at 2300 RPM from 20.0 MPa to 13.8 MPa (Point C, Fig. 4a & 4b) and at 1800 RPM from 17.9 MPa to 11.0 MPa. Needle reopening and secondary injection for nozzle #1 was apparent after 163 hours of operation (Fig. 4c & 4d). Reduction in orifice diameters for nozzle #1 due to carbon deposits was significant. One orifice was plugged, two orifices showed a reduction in diameter from 0.32 mm to 0.305 mm, the fourth orifice decreased from 0.32 mm to 0.31 mm. This translated to a total nozzle orifice area reduction of approximately 30% which increased the maximum injection line pressure, (Point B, Fig. 4c & 4d), the residual injection line pressure (Point A, Fig. 4c & 4d) and maximum residual injection line pressure (Point C, Fig. 4c & 4d). For this nozzle, the maximum injection line pressure increased from 50.3 MPa initially to 62.1 MPa after 163 hours of engine operation. Similarly, the residual injection line pressure increased from 5.5 MPa to 9.0 MPa and the maximum residual injection line pressure increased from 21.4 MPa to 29.7 MPa. The injector had an opening pressure of 26.9 MPa initially which decreased to 15.5 MPa after 163 hours of engine operation. Hence, when the injector became coked the maximum residual line pressure was greater than the nozzle opening pressure and secondary injection occurred.

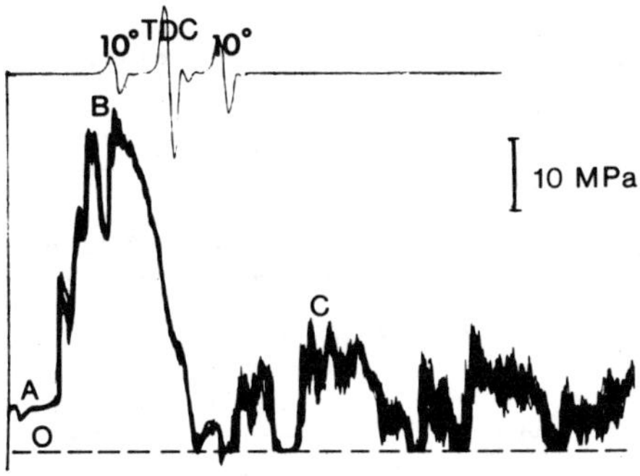

Fig. 4a. Nozzle #4 - New

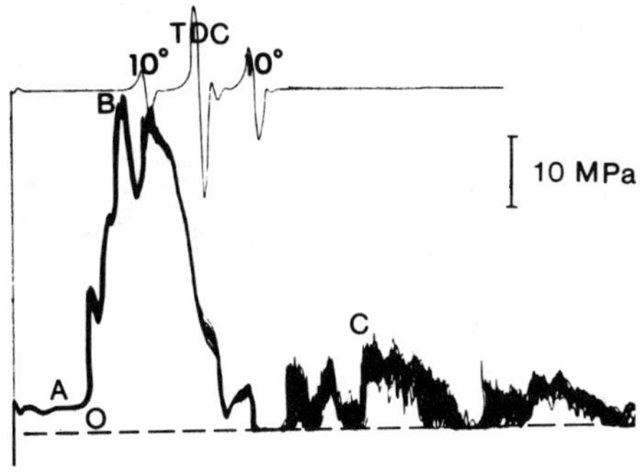

Fig. 4b. Nozzle #4 - 105 hours of use

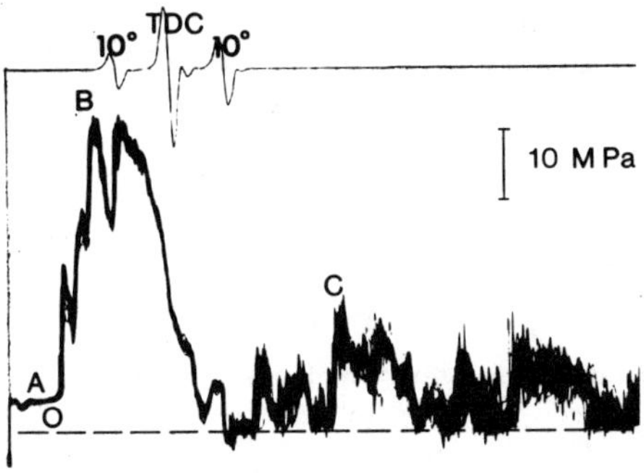

Fig. 4c. Nozzle #1 - New

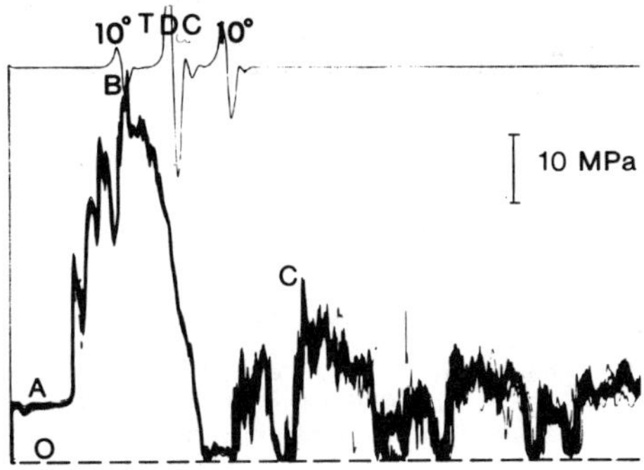

Fig. 4d. Nozzle #1 - 163 hours of use

----- Baseline Pressure (Fuel Vapor Pressure)
A Residual Line Pressure
B Maximum Line Pressure
C Maximum Residual Line Pressure

Fig. 4. Timing Marks and Injection Pressures for 15 Observations at Rated Speed and Load.

Lubricating Oil

During the D-2 diesel fuel test the lubricating oil consumption was 16.04 g/h (Fig. 5a). On the 25-75 blend the lubricating oil consumption from 0 to 105 hours of engine operation was very low (11.61 g/h). Between 105 and 150 hours the oil consumption was 78 g/h even though no outside leaks were observed. The average oil consumption for the whole test was 16.11 g/h (Fig. 5b). For both tests the blowby stayed at a satisfactory level. The blowby for the diesel fuel test was 92 L/min (Fig. 5a). The blowby for the 25-75 blend test was 90 L/min (Fig. 5b).

Oil samples taken every 15 hours were used to determine the change in kinematic viscosity and dispersivity characteristics. The lubricant viscosity was within normal limits during both tests (Fig. 6). For both tested fuels, analysis of "blotter spot" samples did not indicate abnormal changes in lubricant dispersivity characteristics during the tests.

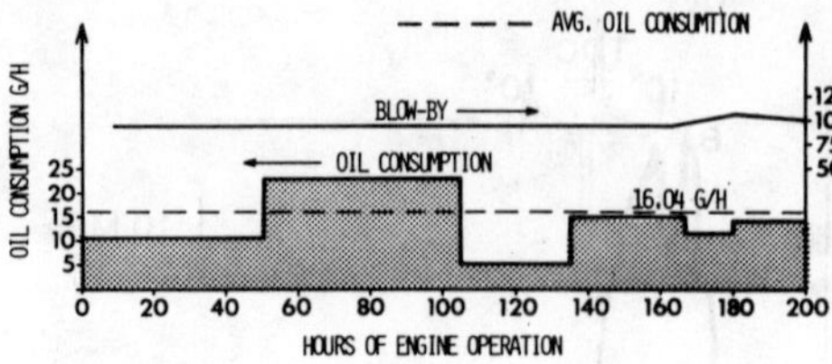

Fig. 5a. Diesel Fuel

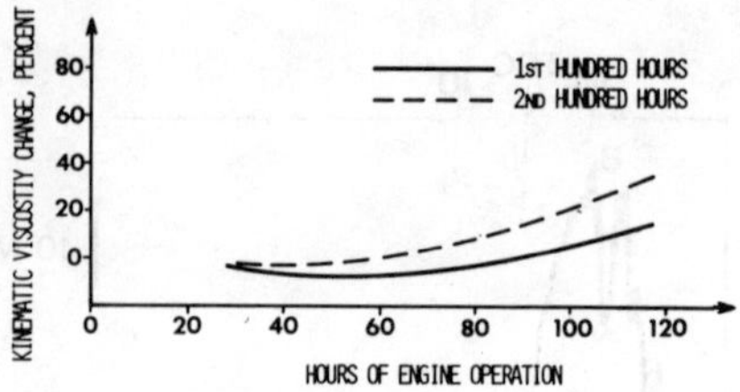

Fig. 6a. Diesel Fuel

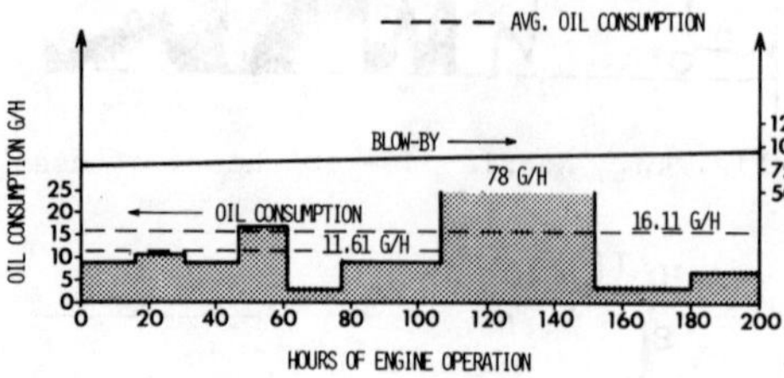

Fig. 5b. 25-75 SFO - D2 Blend

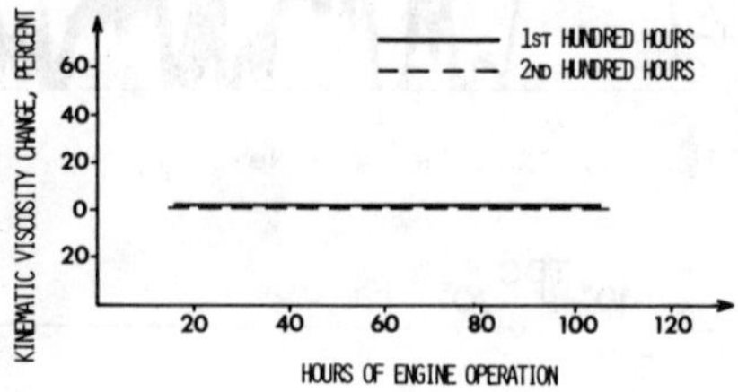

Fig. 6b. 25-75 SFO - D2 Blend

Fig. 5. Oil Consumption and Blowby

Fig. 6. Kinematic Viscosity Change

Engine Teardown and Final Inspection

After each durability test, the carbon, sludge, and varnish deposits were rated using the CRC test procedure (CRC, 1959). The wear of engine parts was determined by direct measurement.

Cylinder Head, Intake and Exhaust Valves: For both tested fuels the combustion area of the cylinder head showed light, uniform, flat carbon build-up. The manifold and combustion side of the exhaust port as well as the manifold side of the intake port appeared clean. The combustion side of the intake port showed soft, oily, uniform carbon build-up. For the engine tested on the 25-75 blend heavy carbon residue was visible underneath the valve seat. The average CRC rating for the intake passage surface was 1.6 for the 25-75 blend test and 1.13 for the diesel fuel test.

No difference in carbon build-up on the intake and exhaust valve tulips was detected for the tested fuels. However, for all exhaust valves after the 25-75 blend test, the valve stems displayed excessive carbon residue. The average CRC rating for exhaust stem deposits was 1.9 for the 25-75 blend test and 0.8 for the diesel fuel test. All valves faces showed light pitting for both tests.

Cylinder Sleeves: Heavy and hard carbon residue was observed on the cylinder sleeves above the ring travel after the 200 hour test with the 25-75 blend. The cylinder sleeves had some scratches throughout the inside surfaces on the major and minor thrust side. Very light random polish tracks were visible. Under high magnification, it could be seen that the hone marks were deeper than the scratches. Some scoring was caused by the piston rather than the rings because the scratches extended below the lowest level of ring travel and metal was smeared on the side of the pistons. Overall, the cylinder sleeves were in very good condition and showed less wear than when the engine was run on diesel fuel. The average CRC rating for carbon deposits above the top ring was 1.1 for the diesel fuel test compared to 1.22 for the test on the 25-75 blend.

Piston and Rings: Pistons for both tests showed heavy carbon build-up right above the ring groove on the top land and normal carbon residue on the second land. For the pistons after the 25-75 blend test the lacquer and carbon residue on the third land was significantly heavier than after the diesel fuel test. All deposits from the test on the 25-75 blend were hard, shiny, and did not flake off as was found with the dry carbon build-up formed during the run on diesel fuel.

Inspection showed differences in volume of ring groove filling with carbon. After the 25-75 blend test, the piston grooves for all compression rings were filled with carbon. The deposits were greater than those seen on the diesel fuel test (Table 4).

TABLE 4. PISTON DEPOSIT RATINGS

D-2 Diesel Control Fuel

	1st ring		2nd ring		3rd ring		oil ring	
	Ave.	SD	Ave.	SD	Ave.	SD	Ave.	SD
Ring Grove Filling, % Vol.	83	3.02	36	7.61	0	0	0	0
Carbon Build-up on side of Ring Groove								
-top	0	0	11	3.7	0	0	0	0
-bottom	0	0	0	0	0	0	0	0
Carbon Build-up on ring								
-top	8	0.6	9	0	0	0	0	0
-bottom	0	0	7	1.2	0	0	0	0
-back	25	0	25	0	0	0	0	0
-front	0	0	4	1.5	0	0	0	0

25-75 Blend

	1st ring		2nd ring		3rd ring		oil ring	
	Ave.	SD	Ave.	SD	Ave.	SD	Ave.	SD
Ring Grove Filling, % Vol.	92	9.46	51	12.99	12	8.3	0	0
Carbon Build-up on side of Ring Groove								
-top	5.5	2.9	20	2.9	0	0	0	0
-bottom	0	0	0	0	0	0	0	0
Carbon Build-up on ring								
-top	0	0	8	1.7	0	0	0	0
-bottom	0	0	0	0	0	0	0	0
-back	14	5.3	25	0	0	0	0	0
-front	0	0	5	3.9	0	0	0	0

Ave. = Average CRC Diesel Engine Rating for four cylinders
SD = Standard Deviation

Piston ring side clearance was measured right after the test with the carbon build-up in place and remeasured after cleaning. For the diesel test, the top side of the first groove appeared clean, and 44% of the second groove top side was covered by light carbon build-up. On the 25-75 blend, 20% of the top side had carbon residue, and 78% of the second groove top side area appeared black.

Inspection of the piston combustion chamber after the test on 25-75 blend revealed evidence of a denser fuel spray core for three cylinders. Carbon build-up on the piston land wall corresponding to points of fuel impact from a multi-hole nozzle was not uniform. The impingement points were surrounded by a larger area where the fuel vapor was reflected from the combustion chamber wall. The appearance and location of the impingment area varied and corresponded to a relative reduction of nozzle orifice diameters. For cylinder #4 no reduction of the nozzle orifice diameters and corresponding increase in injection line pressure was experienced. Regular, light impingement marks were apparent on the piston combustion chamber of cylinder #4.

After the diesel fuel test, the orifice diameters of two injection nozzles were reduced. Visible on the piston head were impingement marks surrounded by carbon build-up and a larger area of fuel vapor reflection. The form of the carbon build-up, fuel and vapor impingement areas were smaller and more uniform for the diesel test as compared to the 25-75 blend test.

Bearings: For both tests the rod and main bearings were in good condition. However, the rod bearings after the run on 25-75 blend showed slightly heavier uniform wear.

Turbocharger: At the conclusion of both tests the turborchargers were in satisfactory condition. The turbine blowers were covered with normal light carbon residue. For the turbocharger tested on the 25-75 blend the carbon build-up had a sticky and oily appearance.

SUMMARY

1. A linear regression analysis tested for dependence of engine performance parameters (power, BSFC, and exhaust temperature) with time. The hypothesis of dependence was rejected at the 5% level of significance for all parameters of both tests. The standard errors of the test variables indicated a greater standard deviation of the performance parameters for the 25-75 mixture compared to diesel fuel. This was possibly caused by erratic deterioration of the fuel injection system.

2. During the test on the 25-75 blend the injection nozzles were changed twice. The first change was made after 29 hours of engine operation due to injection pump problems. The change was made after 105 hours of engine operation when a 5% drop in power and significant differences in fuel injection line pressures were experienced.

    a. All nozzle tips showed abnormal carbon build-up.
    b. The highest nozzle opening pressure drop was 18%.
    c. For all tested nozzles a reduction in orifice diameters was experienced.
    d. The needles showed excessive wear and difficulty in needle movement for two nozzles from the second set of nozzles.

3. Injection line pressure analysis indicated erratic deterioration of the fuel injection system including sporatic nozzle needle sticking, needle reopening, and secondary injection.

4. The lubricating oil consumption for the 25-75 blend test was slightly better compared to the run on diesel fuel.

5. Blowby stayed at a satisfactory level for both tests.

6. The lubrication oil kinematic viscosity did not change for either test.

7. Comparing the engine condition after the 25-75 blend test to the diesel fuel test:

   a. There was a slightly heavier soft carbon build-up in the intake ports.
   b. All four exhaust valve stems had heavy black carbon deposits.
   c. The cylinder sleeves had a heavier carbon build-up above the ring travel but less polish tracks and scratches in the ring travel area.
   d. The pistons had more lacquer and carbon residue on the third land.
   e. The piston grooves for all compression rings were filled with carbon.
   f. The connecting rod bearings showed a slightly heavier
   g. The carbon deposits did not show a tendency to flake off.

8. Inspection of the piston combustion chamber after the test on both fuels revealed evidence of a dense fuel spray.

9. The appearance and location of the impingement areas varied and corresponded to a relative reduction of nozzle orifice diameters. The impingement areas were smaller and more uniform for the diesel test compared to the 25-75 blend test.

ACKNOWLEDGEMENT

The authors express their appreciation to the following people and organizations for their contributions to this work:

Allis Chalmers Corporation - R. Matthews, E. Jones, F. West, P. Patchen, K. Feuerstein, J. Guhl, and M. Daley.
Lubrizal Corporation - R. Tupa, D. Marn, and B. Hillborn.
USDA SEA/AR - E. Pryde, M. Bagby, and W. Fairbank.

Appreciation is also expressed to the following people in the Agricultural Engineering Department at North Dakota State University for their help: G. Pratt, H. Kucera, R. Cossette, G. Savageau, M. Nelson, G. Hanson and C. Knoll.

REFERENCES

1. CRC. 1959. Diesel Engine Rating Manual. Diesel Vehicle Fuel, Lubricant and Equipment Research Committee of the Coordinating Research Council, CRC, Inc. NY, NY, Nov.

2. SAE. 1980. Engine test code-spark ignition and diesel. SAE J816b. SAE Recommended Practice. SAE Handbook.

ENDURANCE TESTS USING SOYBEAN OIL-DIESEL FUEL MIXTURE TO FUEL SMALL PRE-COMBUSTION CHAMBER ENGINES

S. C. Borgelt
Assoc. Member, ASAE

F. D. Harris
Member, ASAE

ABSTRACT

Three Onan diesel engines were fueled with the following fuels: a) 100% diesel fuel, b) 25% soyoil-75% diesel fuel and c) 50% soyoil-50% diesel fuel. The engines were operated under 50-55% load for 1000 hours. Results were grouped into the following five areas: 1) general operation, 2) fuel system, 3) injector characteristics, 4) engine wear and deposits and 5) lubricating oil test analysis. The engines experienced no significant differences in measured wear, drop in power or increase in fuel consumption. As the amount of soyoil in the fuel mixture increased, fuel and injection systems experienced problems, the deposits increased and lubricating oil analysis showed changes in wear patterns.

INTRODUCTION

Petroleum fuels are important in all sections of our society. Because of steadily increasing prices, decreasing reserves and unstable foreign supplies, a major thrust to develop alternate energy sources has become increasingly important. For diesel engines, organic seed oils such as sunflower, soybean, peanut and possibly others hold promise as fuel alternatives.

At the time of this project concept, little had been done to evaluate the long term effects of vegetable oil as fuel in an engine. However, numerous magazine articles and verbal conversations reported individuals and other research personnel introducing vegetable oil in their diesel engines.

In order to gain data that would help provide additional information concerning vegetable oil fuel, engine test objectives were developed. The thrust of this research was to determine long term effects of soybean oil-diesel fuel blends used to fuel compression ignition engines.

TEST DEVELOPMENT AND SETUP

During test development decisions concerning kind and type of fuels to be used were made. Mixtures of soybean oil-diesel fuel were used because short term tests using 100% vegetable oil have shown potential problems. The experimental procedure was to start at a known point and compare with unknown alternatives. Crude degummed soyoil was used in the mixtures because the intent was to use the least amount of refining possible, but yet refined to reasonable limits. These limits are unknown, but as indicated by the American Soybean Association, crude soyoil is questionable as a fuel. "As

The authors are: S. C. Borgelt, Research Assistant, and F. D. Harris, Professor, Agricultural Engineering Dept., University of Missouri-Columbia.

long as the crude soyoil remains dry, the gums stay in solution, but moisture causes the gums to precipitate or come out of solution and settle. This settling, obviously, could cause problems in the fuel system and engine." Historically, fuel system problems have been experienced.

The engines were fueled in the following manner:

Engine A - 100% No. 2 diesel fuel--as a baseline
Engine B - 25% soyoil-75% diesel
Engine C - 50% soyoil-50% diesel

Three Onan (DJA-3.0) pre-combustion chamber, one cylinder, naturally aspirated, air cooled engines with a rated power output of 5 kw (6.7 hp) were used. Attached to the engines were 3 kw, 120/240 volt, one phase, AC, 60 hz generators. When attached to an electrical load the generators act as individual dynamometers. This loading was obtained through "load racks" consisting of four, 660 watt resistance cone heaters and one 300 watt light bulb. The engines were set up inside a building, with the exhaust piped outside, while combustion and cooling air was supplied through an open door and exhaust fan.

Instrumentation included:

Engine hour meters - to monitor run time and schedule maintenance.

Kilowatt-hours meters - to monitor power output.

Fuel charge pressure gauges - used to detect early signs of fuel filter plugging (unsuccessfully).

Volt-amp meter and frequency counter - used periodically to check engine-generator output.

Following a 50 hour break-in period with medium detergent crankcase oil, the engine crankcases were filled with Farmland Industries (COOP) SAE 30 SE-CD diesel engine oil. Each engine was run intermittently; started, run and shut off everyday. Operating data were taken daily. Samples of the lube oil were taken regularly. Crankcase oil was changed every 100 hours with oil filters replaced every 200 hours per manufacturer's specifications.

RESULTS

Results were grouped into five classifications:

1. General operation
2. Fuel system
3. Injector characteristics
4. Engine wear and deposits
5. Lubricating oil test analysis

General Operation

Daily operation of all three engines was basically good with no serious problems observed from any engine. Using the daily data the following values were calculated.

Table 1. Summary of Operating Data

Engine	A	B	C
Fuel used (liters)	623.1	706.85	680.7
(gallons)	164.6	186.7	179.8
Hours of Operation[a]	525.0	518.0	543.1
Kwhrs produced	1351.	1454.	1386.
Load Factor (Kwhr/hr = Kw)	2.6	2.8	2.55
Fuel used (liters/hr)	1.19	1.36	1.25
Specific fuel consumption (liters/Kwhr)	.461	.486	.491
Thermal efficiency (percent)	20.0	19.3	19.5

[a]Only the middle 500 hours were used for the summary because the Kwhr meters were not installed until 275-300 hours into the test and one hour meter was malfunctioning during the last portion of the test.

Fuel System

The most persistent problem during the duration of the engine tests was fuel filtering in Engine C.

The problem was first noticed after 250 hours of operation. (Fuel filter recommended change interval is 600 hours per operator's manual). Engine C was observed to be low in power and engine speed. Upon cleaning of the sediment bowl and filter screen the correct power and engine speed were resumed. Similar characteristics were observed and action taken at the following times.

Table 2. Fuel Filter Change - Engine C

Total Hours	Interval Between Changes (hrs)
1. 250	
2. 315	65
3. 400	85
4. 465[a]	65
5. 583	118
6. 745	162
7. 825	80
8. 860	35
9. 890[b]	30

[a]Injector also cleaned at 465--explanation in next section.

[b]The last reading is not accurate. Engine hour meter was not functioning properly.

Although this problem was only characteristic with Engine C, the plugging of a sediment screen on Engine B at 531 hours did cause some fuel starving of Engine B until the screen was cleaned.

The fuel filters on Engines A and B were changed at the 600 hour interval as recommended in the operator's manual, not because of any apparent need to change filters.

Injector Characteristics

At 465 hours of operation, Engine C began to lose power and speed as if experiencing fuel system problems, as discussed previously. After the replacement of fuel filters and sediment bowl cleaning offered no relief, the injectors were removed.

Injector C had the most carbon buildup on the face and around the pintle plunger. The coke buildup looked like enough to hamper the spray pattern and hurt engine performance.

Injector B had a small amount of carbon buildup on the face, but did not seem to be in the path of the spray pattern enough to hamper engine performance.

Injector A only experienced a carbon film, resembling soot, covering the face of the injector. All three injectors were cleaned and replaced, returning the engines to normal operating characteristics.

Upon conclusion of the test the injectors were again removed and data collected. Injectors A, B and C visually showed more carbon buildup as the amount of soybean oil in the mixture was increased. Figure 1 shows the face of Injector C.

The opening pressure of each injector was measured, spray pattern observed and internal mechanism checked. Specifications call for an opening pressure of 134.5 bar at manufacture to 120.7 bar after several hundred hours of operation.

Engine A - 134.5 bar	Crisp opening and closing of pintle. Good spray pattern.
Engine B - 144.8 bar	The pintle experiences a small delay in closing. Good spray pattern.
Engine C - 131 bar	The pintle seems gummed up. When pressure is applied, the valve action is not crisp. The valve is not "popping" open or closed. The action is hesitant or mushy. The spray pattern observed was very uneven, the spray pattern was not conical, fuel was not atomized correctly, the start and end of injection was not precise, the fuel tends to dribble at the end of injection.

Upon disassembly of the injectors, both B and C experienced deposits of varnish and gums. As an interesting note, the internal gumming experienced in Injector B caused it to be more difficult to disassemble than Injector C.

Engine Wear and Deposits

Upon disassembly, the engine components were measured and observed. Engine measurements are listed by part as follows. The engine wear did not vary significantly and wear was not significantly different from manufacturer's specifications.

Table 3. Summary of Engine Measurements (cm)

	Specifications Min.	Specifications Max.	Engine A	Engine B	Engine C
Rod journal	5.2316	5.2337	5.232	5.232	5.232
Rod bearing			5.237	5.240	5.237
Journal-bearing clearance	.0036	.0132	.005	.008	.005
Cylinder bore (above rings)	8.2423	8.2563	8.255	8.255	8.255
(below rings)	8.2423	8.2563	8.255	8.255[a]	8.255
Top ring (thickness)	.2350	.2375	.236	.235	.235
(end gap)			.015	.020	.048
Valve stems:					
Intake (upper end)	.8649	.8674	.866	.866	.866
(neck)			.859	.859	.853
Exhaust	.8649	.8674	.864	.864	.864

[a]Out of round. .0025 larger front to rear.

Observations concerning deposits and other characteristics:

1. As the amount of soyoil in the fuel mixture increases, the carbon deposited on the head increases slightly. Also the deposits become darker as the amount of soyoil in the fuel increases.

2. The combustion chamber, piston and rings showed no evidence of carbon deposits, sticking or other detrimental effects.

3. During the disassembly of the intake manifolds, serious carbon deposits were found. Engine C experienced a build up of soft carbon deposited throughout the intake manifold, 2-3 mm thick. Hard carbon buildup was also found in intake valve port and on the valve of Engine C. Deposits of this type were not experienced in A or B.

4. Figure 2 shows valves of engines A, B and C. Both the intake valves of A and B exhibited identical carbon buildup. Likewise the exhaust valves of A, B and C exhibited identical carbon deposits. Figure 2 illustrates the large buildup of carbon on Intake C. Upon closer examination of the valve seat of Intake C, it was noted that the valve was not seating properly. However, it is difficult to determine whether the valve was not seating due to carbon deposits, or vice versa, the carbon deposits were caused by the improper seating of the intake valve. Evidence of this type of carbon deposition has been reported by Peters (1981).

5. Bearing C seems to show slightly, but not measurably more wear. This will be discussed further in the next session.

Lubricating Oil Test Analysis

Lubricating oil samples were analyzed at regular intervals throughout the test. During the first portion of the test, samples were analyzed about every 25 hours with this interval extended to 50-hour intervals in the later stages. Analysis was performed by the Oil Test Laboratory at Farmland Industries. Data analysis included:

1. Viscosity @ 40°C
2. Viscosity @ 100°C
3. Viscosity Index
4. Total Acid Number
5. Total Base Number
6. Total Solids

7. Fuel Dilution
8. Metals - in PPM including the following:

a.	Iron	i.	Silicon
b.	Chromium	j.	Sodium
c.	Aluminum	k.	Boron
d.	Copper	l.	Magnesium
e.	Lead	m.	Calcium
f.	Tin	n.	Barium
g.	Silver	o.	Phosphorus
h.	Nickel	p.	Zinc

Summaries of the analysis indicated:

1. Total solids in the engine oil increased as the soybean oil in the fuel mixtures increased. No visual evidence of oxidation or solidification was ever observed.

2. Iron, lead and chromium, indicators of wear, showed greater contamination in Engine B than either of the other engines. This result was surprising. However, Engine B ran at a higher continous power output than either A or C. Engine C showed a lower wear pattern than Engine B but higher than A.

3. Figures 3, 4, 5 and 6 show graphically levels of Total Solids, Iron, Lead and Chromium in the lube oil samples. The oil analysis values caused some concern, however, observed engine measurements as shown in Table 3 proved this to be insignificant at this time.

CONCLUSIONS AND RECOMMENDATIONS

Engine wear was not significantly different in the test engines fueled with diesel fuel or mixtures of diesel-soyoil. Quantity of wear metals is somewhat different in the B engine and wear rates tend to increase in B and C at about 400-500 hours, but none of the wear rates seemed abnormal, according to measurements.

The potentially alarming factor observed was an increase of iron, lead and chromium in the lube oil samples of engines fueled with soyoil mixtures. Further endurance tests would be needed to determine if this is a critical concern.

As the amount of soyoil in the fuel increased the amount of carbon deposited in the engine increased. The carbon buildup in Intake C was significantly greater than Intake A or Intake B, but could have been caused by poor valve seating. This causes some concern for increasing amounts of soyoil in the fuel.

Engine operation using diesel-soyoil mixtures was basically good. Power output ranged from 2.55 to 2.8 kw. Thermal efficiency ranged from 19.3 to 20.0 percent.

Engine B produced the most continuous power. Engine B also had the highest fuel consumption and showed evidence of the most wear, according to oil sample analysis. These characteristics are probably unique to this engine-generator set and it is impossible to conclude that its operation was a result of the type of fuel used without more evidence. However, because of minimal fuel system and injection system problems, it is concluded that the use of small amounts of soybean oil--25% or less--causes negligible differences in engine performance. Soybean oil as a fuel extender looks extremely favorable.

Certain developments are essential to increase the favorability of soyoil as a fuel extender with our present day diesel engine.

- Filtering of fuel before it is introduced in the engine. This helps to increase the maintenance interval of the fuel system and lessens the chances of possible damage to the engine due to fuel impurities.

- Fuel modifications or additives could be used to decrease the viscosity differences of the components, thus helping to eliminate carbon deposits and gumming of the injector components.

Heating of the fuel has been suggested to help eliminate viscosity differences in past references. This is a possible improvement of the system if waste heat can be easily provided. However, the injector is a tremendous heat sink and large increases in fuel temperature at the fuel tank can mean a minimal temperature increase by the time it passes through the injector to the combustion chamber.

REFERENCES

1. Erickson, D. R. and P. Dixon. 1980. Soy oil as diesel fuel: economic and technical perspectives. American Soybean Association, St. Louis, Missouri.

2. Peters, J. F., C. H. Adams, M. C. Rand and M. C. Ziemke. 1981. Test of plant oil (soybean) as a diesel fuel. University of Alabama, Huntsville, (UAH) Report No. 295.

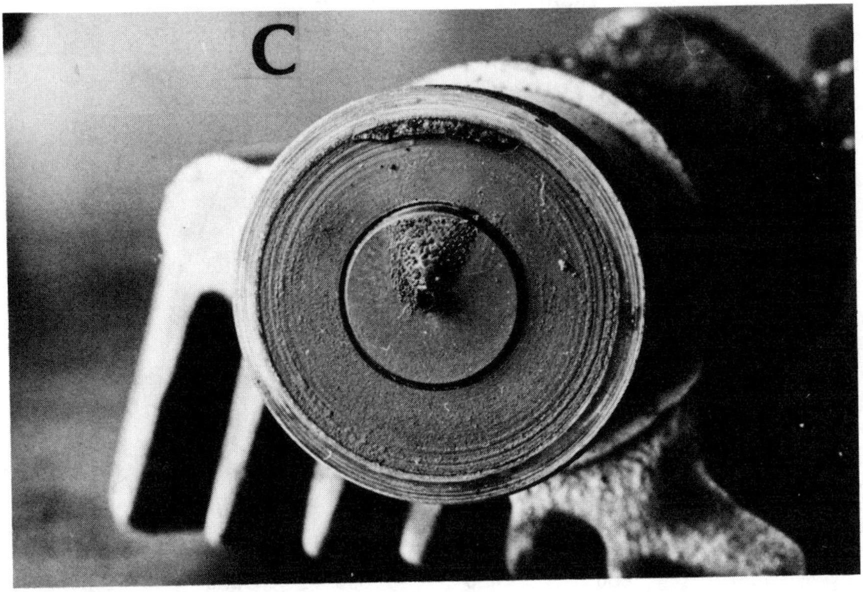

Figure 1. Injection Nozzles Face. Engine C.

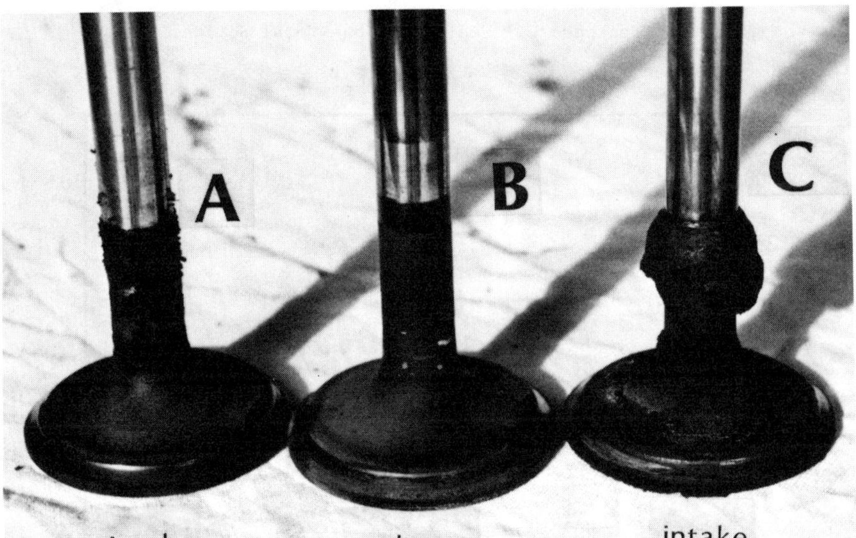

Figure 2. Engine Valves After 1000 Hours.

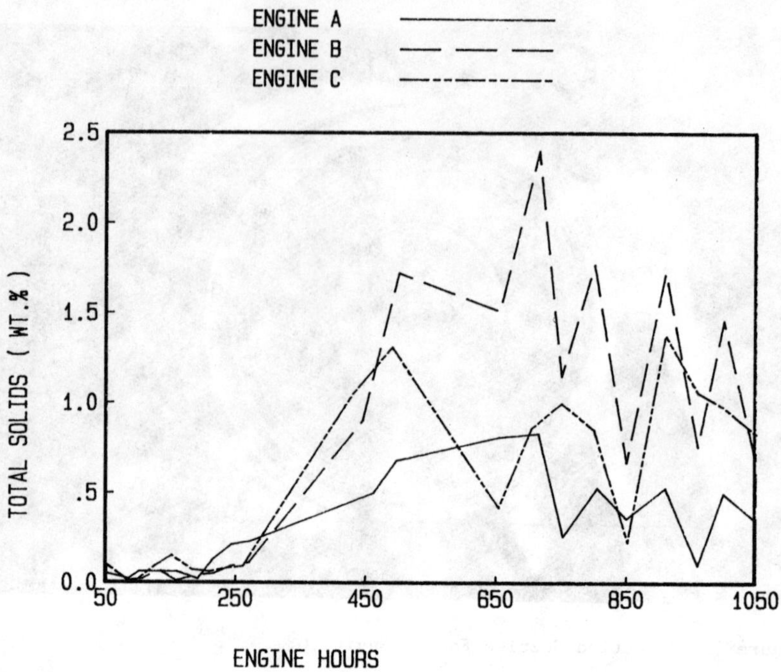

Figure 3. Crankcase Oil Contamination--Total Solids.

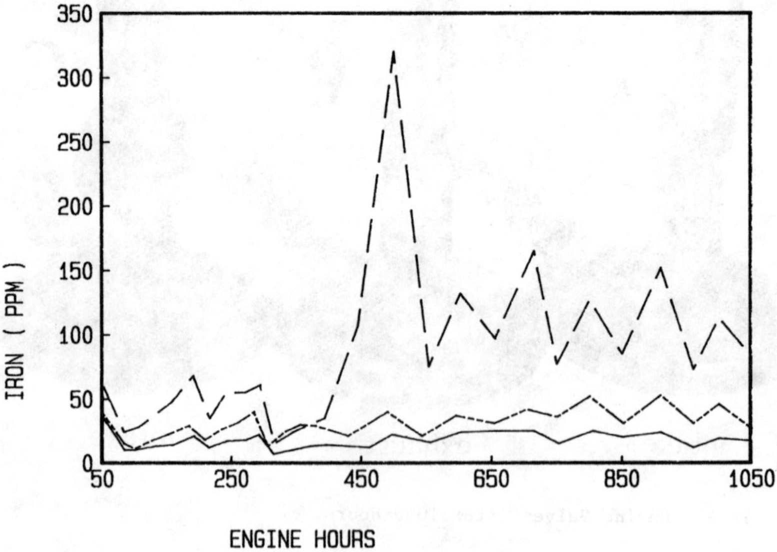

Figure 4. Crankcase Oil Contamination--Iron.

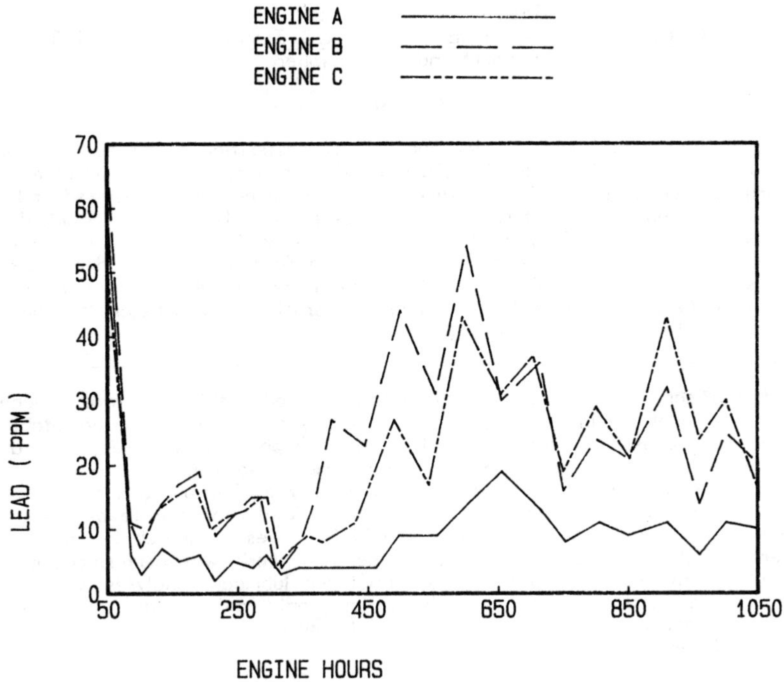

Figure 5. Crankcase Oil Contamination--Lead.

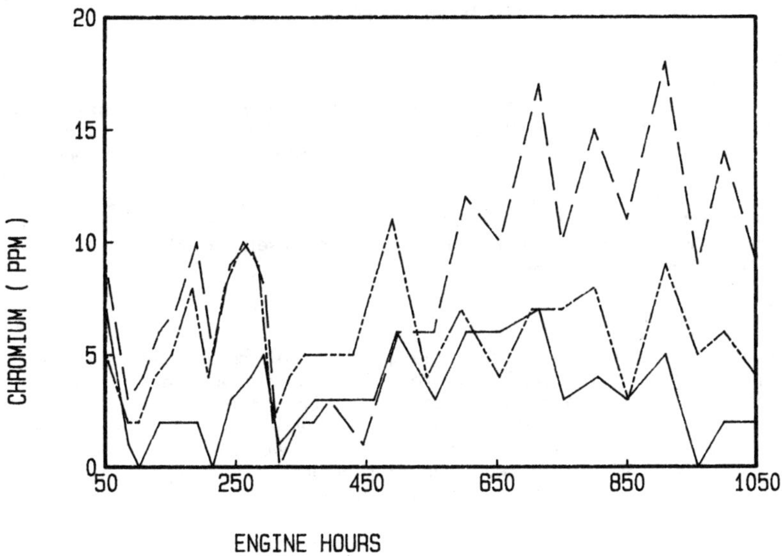

Figure 6. Crankcase Oil Contamination--Chromium.

PERFORMANCE AND DURABILITY OF A TURBOCHARGED DIESEL FUELED WITH COTTONSEED OIL BLENDS

E.F. Fort* and P.N. Blumberg*

ABSTRACT

Experimental fuels made up of cottonseed oil, transesterified cottonseed oil (methyl ester) and No. 2 diesel fuel have been compared to a baseline No. 2 diesel fuel in a turbocharged, open chamber diesel engine. All fuel blends were analyzed and their properties compared to those of typical No. 2 diesel fuels. Test results include data on performance, gaseous emissions and limited (200-hour) cyclic durability. Crankcase oil samples were withdrawn at regular intervals during durability tests and analyzed for indications of fuel-lubricant compatibility. The paper presents data obtained and discusses effects of fuel properties and engine operating conditions on the short and long-term response of existing diesels.

The study that serves as the basis for the data presented in this paper was designed to investigate some of the important technical concerns that emerge when vegetable oils are considered for use in a typical diesel engine. Emphasis was placed on data that reflects durablility effects but information on basic performance and emissions also was obtained to help select fuel blends for durability evaluations and to add to the data base. Support for this project was provided jointly by the Texas Energy and Natural Resources Advisory Council (TENRAC), Southwest Research Institute (SwRI) and International Harvester (IH). All engine test work was performed by SwRI personnel at their San Antonio facilities. Fuel and lubricant analyses were divided between IH and SwRI. Due to space limitations, only major results are presented. Additonal data and discussion will be found in Reference 1** and in a report by SwRI to TENRAC (Reference 2).

FUEL REQUIREMENTS

The unique requirements of the open-chamber, direct-injection diesel combustion process place exacting demands on its fuels. The most important characteristics of a good truck/tractor diesel fuel (grade No. 2) are described below:

Ignition Quality

Satisfactory diesel combustion demands self ignition of the fuel as it is sprayed near top dead center, into the hot, swirling, compressed cylinder gas. Long delays between injection and ignition lead to unacceptable rates of pressure rise (diesel knock) because too much fuel is ready to burn when combustion begins. On the other hand extremely short delays initiate burning quickly and may interfere with proper mixing and combustion of the bulk of the fuel spray. Cetane Number (CN) is the most accepted predictor of ignition delay and, for open chamber diesels, satisfactory fuels normally have values in the range of 40-50.

Viscosity

The open chamber, direct-injection diesel depends heavily on an injection system and its nozzle spray to prepare a fuel-air mixture suitable for combustion. Fuel viscosity can have a significant effect on the ability of an injection system to perform properly which in turn may degrade combustion appreciably. Low viscosity can lead to excessive internal pump leakage, especially when parts become worn, and may cause erratic fuel delivery. High viscosity can increase system pressures to unacceptable levels and also may affect injection duration and spray atomization.

---

* Science and Technology Laboratory, International Harvester

** References are listed at the end of the paper.

Heating Value

Although the diesel combustion process can accept wide variations in heating value, practical systems are most suitable when the heat energy per gallon is high, both to reduce the quantity handled by the injection system and to maximize vehicle or equipment operating range.

Pour Point/Cloud Point

These temperatures are important for cold weather operation. Obviously any substitute for diesel fuel should cloud and pour at temperatures well below freezing or its use either will be restricted to warm climates or seasons, or entail the use of auxiliary heating systems.

Volatility

Although high volatility (such as found in gasolines) is not required for diesel combustion, some degree is necessary to insure easy starting and good fuel-air mixing for smoke-free combustion. Distillation characteristics give some indication of volatility and satisfactory petroleum fuels usually have a boiling range between 180 and 370°C.

Flash Point

From a safety viewpoint, this temperature should be as high as practical. Typical values for commercial No. 2 diesel fuels range from about 50 to 105°C. Naturally, any substitute fuel should not lower the flash point or fire hazards probably will increase.

All of the above properties must be evaluated carefully in considering any material as a component of or a substitute for conventional diesel fuel. Several will be shown to have a measurable effect on engine performance and durability in the test work on cottonseed oil discussed in the following sections.

TEST FUELS AND LUBRICANT

Cottonseed oil was selected for this program since its composition is reasonably typical of other vegetable oils and it is readily available as a commercial product. From analysis of commercially available grades, a "once refined" product appeared to be the most suitable in that its particle content was nearly as low as that of a food grade without the extra cost required in the food quality step. Based on the recently reported work of Bruwer (Reference 3), a transesterified (methyl ester) modification of once refined cottonseed oil also was included as an experimental fuel component. Once-refined cottonseed oil is made up, essentially, of three long chain hydrocarbon molecules connected in a triglyceride structure. Transesterification reacts an alcohol -- methanol in this case -- with cottonseed oil to replace the triglyceride structure with three separate long-chain hydrocarbon methyl esters, each with a molecular weight about one-third that of the once-refined cottonseed oil. Viscosity also is reduced. A more definitive treatment of transesterification, can be found in organic chemical texts such as Reference 4. Phillips' reference diesel fuel blend was used both as a baseline fuel and as the petroleum fraction of other blends with cottonseed oil components. This fuel is commercially available and its properties are controlled. However, for consistency, one batch was purchased and used throughout the tests. As described later, these three basic fuel components - CottonSeed Oil (CSO), TransEsterified CottonSeed Oil (TECSO) and Phillips' diesel fuel (2D) - were run either alone or in various blends with each other for all of the test work reported in this paper. A tabulation of measured fuel properties for typical fuels used appears in the Appendix.

The crankcase lubricating oil used in the project was IH No. 1 SAE 30 Engine Oil, formulated for use in turbocharged diesel engines. This is a CD oil meeting a MIL-L-2104C specification. The normal field drain interval for this oil in IH engines is 200 hours.

ENGINE

All engine test work was run with an in-line, 6-cylinder turbocharged, direct-injection diesel in typical agricultural configuration. Bore is 109 mm and stroke 127 mm for a displacement of 7.1 L. As determined by the injection system and its governor, rated speed is 2500 rpm and peak torque occurs at about 1800 rpm.

TEST PLAN

As noted earlier, the primary emphasis of this test program was on factors related to durability. However performance and emissions data also were measured on a variety of fuel blends at low idle; high idle (no-load, governor-limited; 1800 rpm max. torque; 1800 rpm half-torque; 2500 rpm max. torque; and 2500 rpm half-torque.

It should be noted that torque limits are fixed by the particular injection system/governor installed and do not necessarily represent the limits of the engine itself.

Eight fuels or fuel blends were evaluated: Phillips' reference diesel fuel; 30% CSO, 70% 2D (by volume); 50% CSO, 50% 2D; 65% CSO, 35% 2D; 80% CSO, 20% 2D; 50% CSO, 50% TECSO; 50% TECSO, 50% 2D; and 100% TECSO.

At each speed - load point for each fuel blend, test instrumentation measured: speed; power; fuel consumption; exhaust temperature (each cylinder); boost pressure; cylinder pressure vs. time/crank angle (cylinder No. 6 only); injection line pressure vs. time/crank angle (high pressure line at inlet to No. 6 cylinder only); fuel pressure; gaseous emissions; visible smoke; and inlet air, oil sump and coolant temperatures.

The durability cycle selected follows a proposal of the Engine Manufacturers' Association (EMA) for engine screening work with alternative fuels. Although this cycle is not strenuous in thermal loading, it can stress fuels with properties that contribute to excessive deposit formation in mixed light/heavy load service. The cycle consists of four modes repeated consecutively for fifteen hours followed by a minimum 8-hour shutdown to allow sump oil to cool. A two-hundred hour limit was set for this initial work since it followed EMA recommendations and corresponded to one normal field oil change interval. The four engine operating modes of the cycle are: 1) 30 minutes at low idle; 2) 60 minutes at rated power (2500 rpm in this work); 3) 60 minutes at peak torque speed (1800 rpm); 4) 30 minutes at 80% speed, 25% load (2000 rpm, 27.6 kW*).

During the durability runs engine data was recorded once during each mode. Primary data include engine rpm, power, fuel consumption, coolant temperature, exhaust temperature (each cylinder), boost pressure, oil pressure, fuel pressure and oil temperature. Visible smoke and gaseous emissions were measured periodically. A sample of crankcase oil was taken every 15 hours and retained for later analysis. Following each durability run, the engine was disassembled, inspected, rated for deposits and photographed. Rating procedure followed that described in the CRC Diesel Engine Rating Manual, No. 5, 1958. All parts such as liners, rings, pistons, bearings, etc. likely to be affected by fuel-related factors were replaced before proceeding to subsequent project phases.

RESULTS WITH COTTONSEED OIL/NO. 2 DIESEL FUEL BLENDS (CSO/2D)

Fuel Characteristics

Several important properties of the four experimental CSO/2D blends are compared to those of the baseline diesel fuel in Figure 1. As expected from the characteristics of

* An early version of the cycle specified high idle for this mode and placed it immediately following the low idle mode. This cycle was used during the first durability test with cottonseed oil and was superceded by the version shown above for the transesterified oil run.

vegetable oils, the viscosity and density of the blends goes up with an increased volume fraction of CSO. However, the heating value per pound of CSO is low enough to offset the density trend and results in reduced heat energy per gallon as the CSO content rises. In addition, blending CSO into the baseline 2D lowers the Cetane Number and raises the cloud/pour points significantly. Efforts to assess the volatility of CSO/2D blends through standard atmospheric or vacuum distillation techniques were largely unsuccessful. Usually the 2D portion would distill without any difficulty but as soon as it was exhausted the remainder would smoke or foam sufficiently to require terminating the test.

Engine Performance and Emissions

Performance trends at rated speed and maximum load as a function of the amount of CSO in the fuel blend are shown in Figure 2. As noted in the Test Plan, maximum load is determined by the specific injection system and governor setting used and does not represent a maximum engine limit. Any time that maximum load is indicated, the injection pump control is set to produce a fixed pump stroke which is then held constant for all runs with test fuels. Baseline 2D data are the average of two points. All performance and emission data are reported on an observed basis at local ambient conditions which ranged from 97.6 - 99 kPa and 24 - 34°C with most temperatures in the 27 - 32°C area.

One of the first observations is that, at fixed rpm and pump setting, increased CSO content raises the fuel rate (in gal per hr). A possible explanation is that the more viscous fuel blends reduce normal injection pump leakage enough to make a significant change in the volume discharged per stroke. Although the heating value relationship tends to reduce the specific energy input rate as CSO fraction increases, the net effect is an increase in fuel heat supplied. Calculated brake thermal efficiency indicates the ability of the combustion system to accept the experimental fuel blends. In general, there is little significant change in efficiency with increased concentrations of CSO, over the speed and load range observed. This effect also was found at other speeds and loads. One might expect the relatively high viscosity of the high CSO-content blends to degrade fuel spray characteristics and thus lead to inefficient combustion. However, from the data obtained, any such gross effect is minimal. While not shown in any figures, governed high-idle (no-load) rpm and low idle rpm did not change significantly with fuel blend variation.

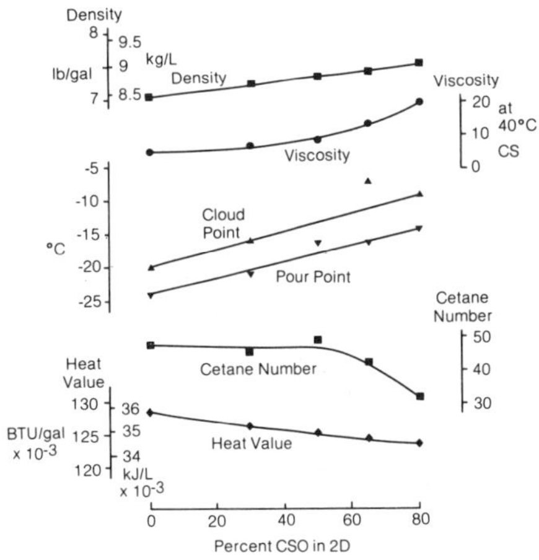

Fig. 1 Fuel Properties of CSO/2D Blends

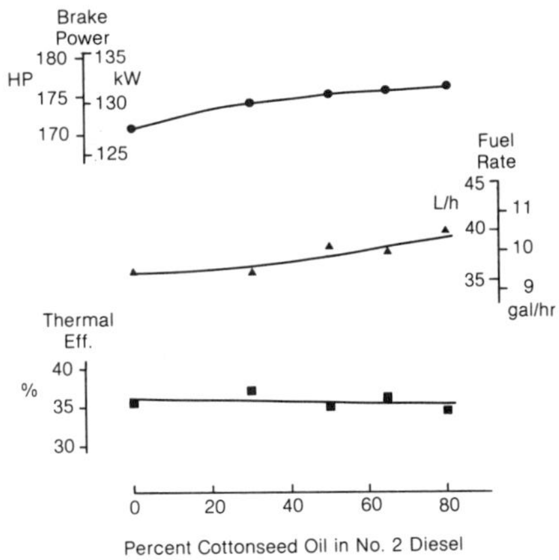

Fig. 2 Performance, CSO/2D Blends, 2500 rpm, Max. Governed Load

As a consequence of the combined effects of fuel rate, heating value and thermal efficiency, the amount of work performed per gallon of fuel used goes down as the concentration of CSO in the fuel blend increases. For this reason, it would be expected that the the effective range or working time of a vehicle in the field would be cut somewhat with CSO substituted for diesel fuel. With 80% CSO the reduction indicated by the above data is about 5%.

Specific emission and visible smoke characteristics as a function of increased CSO fraction are reported in Figure 3. Baseline values are an average of two runs except in the case of BSCO where only one data set is available for use. Trends in these data are not very consistent and in most cases show little significant change as CSO content is varied. Specific carbon monoxide is an exception, with increases observed at 65 and 80% CSO. Since this effect involves CO concentrations below 0.2% and was not found at any other speed/load point, it may be accounted for by minor combustion variations too small to show up in performance or by measurement discrepancies.

Durability

Based on the performance and emissions data, a blend of 50% CSO/50% 2D was selected for a 200-hour cyclic durability test. This particular blend was chosen because: (1) it offers a reasonably high substitution of CSO for diesel fuel; (2) performance and emissions are essentially the same as with the baseline 2D fuel; and (3) cylinder inspection after the performance tests showed no suspicious conditions.

After 183 hours of the cyclic test, it was observed that the engine was noisy at idle and had excessive blowby. Borescope inspection of the cylinders revealed scoring in Nos. 1 and 5. Upon removal of the cylinder head the scoring was found to be severe and serious enough to end the test at that point. Subsequent teardown and inspection showed that the corresponding pistons also were deeply scored with the surfaces torn and the metal rearranged.

The engine was rated after disassembly. The liners from Cylinders 1 and 5 were completely scored in the ring travel area; No. 1 somewhat worse than No. 5. Other liners showed some minor scratches but the honing marks were still visible. For all liners, the surface above the top ring travel had heavy carbon deposits. These deposits

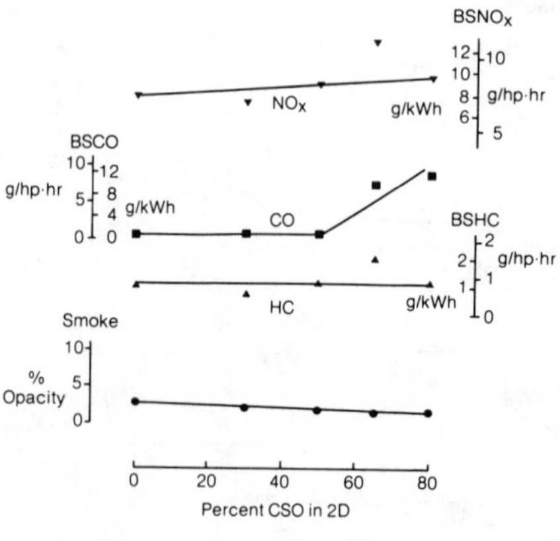

Fig. 3 Specific Emissions, CSO/2D Blends, 2500 rpm, Max. Governed Load

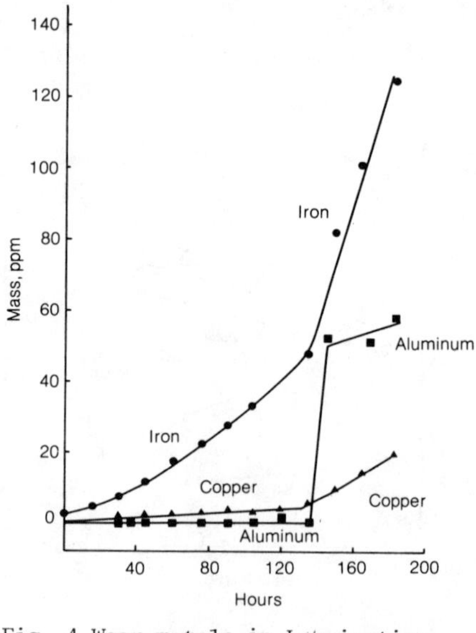

Fig. 4 Wear metals in Lubricating Oil, CSO/2D Durability Test

were extremely hard and required chipping in order to remove the piston upward out of the cylinder. Under normal conditions, using petroleum diesel fuel, these deposits can be removed easily with a knife blade.

All top ring grooves were heavily filled with very hard carbonaceous deposits which interfered with proper ring functions. Apparently, the vegetable oil portion of the fuel blend failed to vaporize and burn completely during the low and high idle modes, ran down the piston lands into the ring grooves and then was pyrolized to hard deposit during the high load modes. Although cylinders No. 1 and 5 failed first, a heavy deposit condition found in the top ring grooves of all other pistons suggests that all cylinders would have failed soon if running were to have continued.

The level of wear-associated metals found in lubricating oil samples withdrawn during the cyclic durability test is presented in Figure 4. For the sake of clarity, only iron, aluminum and copper are shown. However, two additional metals - lead and chromium exhibited similar patterns. A sharp increase in the concentration of all wear metals at about 140 hours indicates that scoring failure involving the cylinder wall, piston, rings and bearings began about 40 hours before its eventual detection. Since the durability cycle is relatively new, there is not yet sufficient baseline data on rates of iron accumulation in crankcase oil for a meaningful comparison between 50% CSO/50% 2D and straight 2D. However, results from one baseline test with 2D fuel, presented in Figure 5, suggest that the wear metal levels noted with 50% CSO/50% 2D, prior to scoring, are not unusually high.

Another potential problem recorded during the 183 hours of cyclic running was a gradual increase in pressure drop across the fuel filters, presumably due to solids in the CSO portion of the fuel. While fuel pressure at the injection pump did not fall below minimum specifications during the test, the value at the end was marginal and indicates that the filter probably would have required replacement at 200 hours. This is much sooner than one would expect with commercial diesel fuels.

Analysis of lubricating oil samples withdrawn periodically showed a marked increase in viscosity as the durability test progressed, from about 110 to about 170 cs (40°C) after 180 hours. Solids also increased from trace to about 7%. The viscosity increase occurred after 150 hours and may be related to excessive fuel passing through the cylinders with scored liners.

RESULTS WITH TRANSESTERIFIED COTTONSEED OIL (TECSO)/NO. 2D BLENDS

Fuel Characteristics

Fuel properties of TECSO and the one experimental blend with No. 2 diesel fuel are compared to the baseline fuel in Figure 6. A significant viscosity reduction is one of the major changes brought about by the transesterification process. Comparison of the viscosity data for TECSO blends with that of the baseline 2D and with the data of CSO blends in Figure 1 illustrates the magnitude of the reduction and indicates that the modified oil's viscosity characteristics are reasonably close to those of some commercial diesel fuels.* As with the base cottonseed oil, TECSO and its blend with 2D have increased density and decreased heat energy compared to the baseline diesel fuel. Interestingly, transesterification raises the Cetane Number of TECSO to about 51, comparable to the upper level of commercial diesel fuels. Unfortunately, the process also changes the pour point to a temperature near freezing which would probably preclude use in most of the U.S. during winter. Distillation of 50% TECSO/50% 2D was unsuccessful due to excessive foaming shortly after the initial point was reached. However, vacuum distillation of 100% TECSO shows that it behaves much like a pure material with a boiling point of about 174°C.

*Reference (5), the Department of Energy publication "Diesel Fuel Oils, 1980" (DOE/BETC/PPS-80/5) notes testing truck/tractor diesel fuels with viscosity at 38°C as high as 4.59 cs.

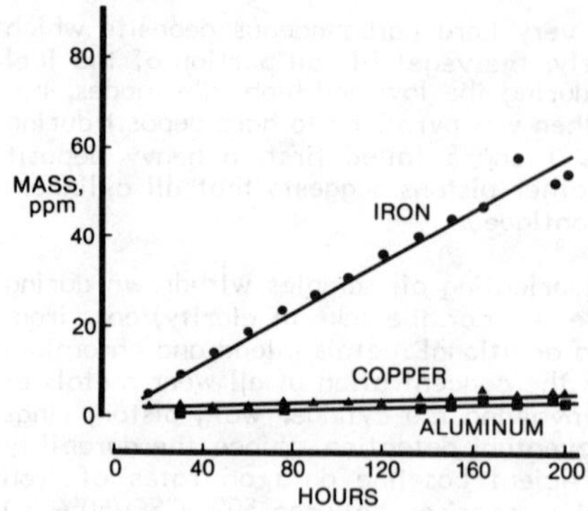

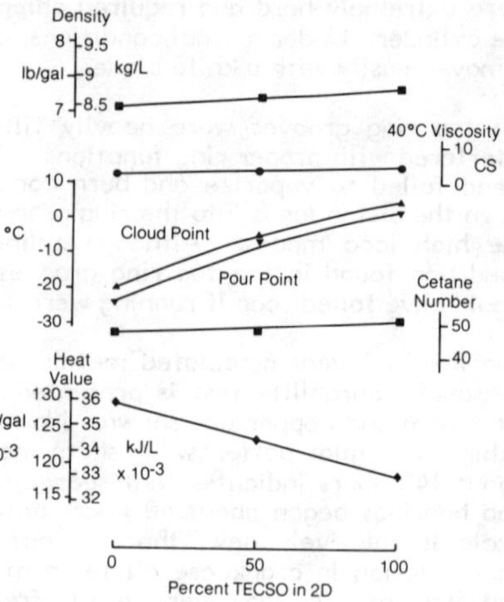

Fig. 5 Wear Metals in Lubricating Oil, 2D Baseline Durability Test

Fig. 6 Fuel Properties of TECSO/2D Blends

Engine Performance

Performance and emission data for TECSO and 50% TECSO/50% 2D, at rated speed and full load are compared with that of the baseline 2D fuel in Figures 7 and 8. Additional results will be found in Reference 1. Since the viscosity of TECSO and its blend with 2D is reasonably close to that of the baseline diesel fuel, injection pump volume delivery varies only slightly over the blend range. However, the decreased heating value of TECSO means less heat input as the TECSO concentration increases and, consequently, the brake horsepower output falls. Since brake thermal efficiency is essentially unaffected by the amount of TECSO in the fuel, all of the power drop is a consequence of the reduction in heat input. The above observations are generally consistent for all speeds and loads tested and indicate that TECSO acts like a petroleum-derived diesel fuel with a reduced heating value over the engines useful operating range when ambient temperatures are around 27 - 32°C. As with CSO blends, substitution of TECSO for diesel fuel will decrease the effective range or working time of a vehicle in the field. Since the heating value of TECSO blends is lower than that of CSO blends and since engine efficiency is about the same, the effect will be more pronounced with TECSO. Additionally the lower power output from TECSO blends will reduce the maximum amount of work that can be performed by a machine in a fixed time period. Consistent with performance under load, governed high-idle (no-load) and low idle engine rpm did not vary signficantly with the substitution of TECSO for 2D.

Specific emissions and visible smoke characteristics shown in Figure 8 indicate that replacement of 2D by TECSO had no significant effect on gaseous emissions. Data at other speeds and loads was similar. In general, these data support the proposition that TECSO behaves much like a petroleum diesel fuel during the combustion process.

Durability

As a result of analyzing the performance and emissions data for TECSO blends, neat TECSO was selected for a second 200-hour cyclic durability test, since: (1) it provides maximum (100%) fuel substitution; (2) with the exception of maximum power, performance and emissions are comparable with results from diesel fuel; and (3) inspection of the cylinders after performance runs did not reveal any potential problems (cylinder walls and piston tops were dry with minimal deposits).

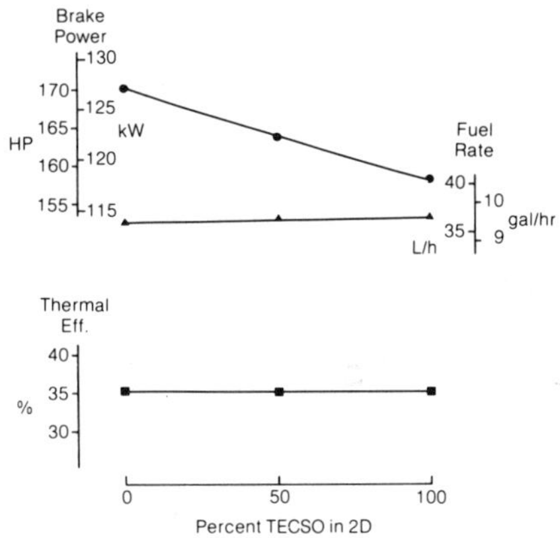

Fig. 7 Performance, TECSO/2D Blends, 2500 rpm, Max. Governed Load

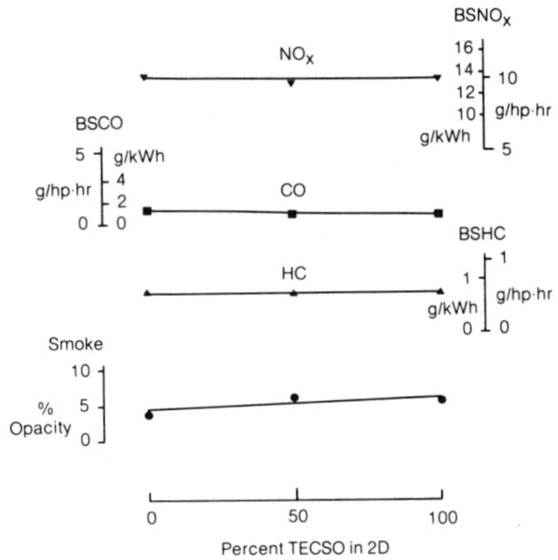

Fig. 8 Specific Emissions, TECSO/2D Blends, 2500 rpm, Max. Governed Load

Results from this 200-hour test tend to confirm the observation that engine performance with TECSO is very much like that with a commercial diesel fuel at ambient temperatures of about 27°C. The test ran the full 200 hours with no noticeable change in engine power or exhaust temperature and without any detectable problems. When the engine was disassembled the general appearance was about the same as if diesel fuel had been used. The engine surfaces were dry except for a small amount of oil in the spray cup of piston No. 5 and a wetness on the parts of the underside of the head. The carbon build-up on theliner above the top ring travel was moderate and soft, not rock hard as it was for the cottonseed oil test. In spite of the general cleanliness of the engine, top ring grooves had extensive deposits. However, unlike the extremely hard material formed with 50% CSO/50% 2D, these deposits were relatively soft and easily removed with a knife blade. Even so, the extensive buildup observed suggests the possibility of scoring or severe ring wear if the test had proceeded further.

The concentrations of three wear metals (iron, aluminum and copper) in lubricating oil samples withdrawn as the durability test progressed are plotted vs. time in Figure 9. Trends for lead and chromium are similar to those of aluminum and copper. In contrast with the data of Figure 4, which represent conditions leading to failure, these data show only increases from liner and bearing metals with no evidence of piston scuffing. As noted earlier in the discussion of CSO/2D durability and referring to Figure 5, the concentration of iron at 200 hours is not considered unusual.

Pressure drop across the fuel filter did not increase during the 200 hour durability test which indicates that the transesterification process also is effective in reducing the filter clogging tendencies of cottonseed oil. Analysis of lubricating oil samples withdrawn every 15 hours showed a decrease in viscosity, from about 107 to about 90 cs (40°C) after 200 hours. Solids increased from a trace to less than 1% by volume. Neither of these trends suggest any potential problem from TECSO contamination of the crankcase lubricating oil.

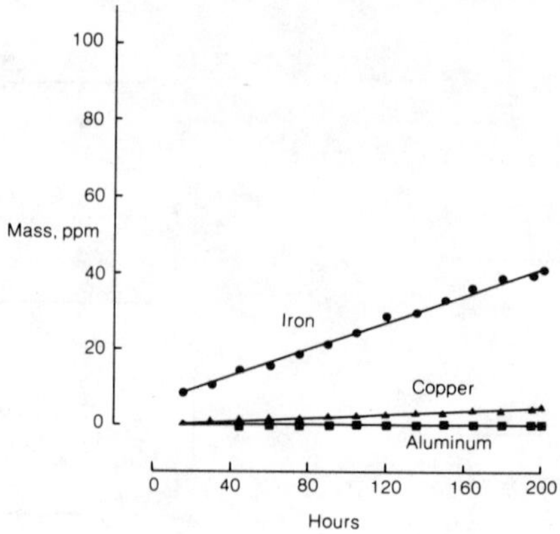

Fig. 9 Wear Metals in Lubricating Oils, TECSO Durability Test

SUMMARY AND CONCLUSIONS

As a result of the work reported in this paper it is apparent that vegetable oils, typified by cottonseed oil, may behave like a petroleum based diesel fuel in short-term performance/emissions tests. All of the experimental fuels appeared reasonably satisfactory when the standards for judgement are power, exhaust temperature, fuel consumption, etc. However, considering deposits, wear or restriction of the fuel system, they are not necessarily suitable for longer term operation, especially under cyclic conditions. Also one should remember that all of the above tests were run at ambient temperatures near 27°C, leaving northern winter operation an open issue.

As in many other instances, durability is the key issue in the use of vegetable oil-based fuels in diesel engines. Since deposit and other durability-related effects often vary among engine design configurations, it will be necessary to run numerous tests for many hours before anything but tentative conclusions can be drawn. The durability run used in this work is very short compared with the 2000 hour tests required for proof-testing engines and related components. For this reason, only failures can be considered highly significant while passing runs merely indicate a potential for eventual success.

Transesterification of vegetable oils such as cottonseed oil appears to have some technical potential for producing a biomass fuel which could substitute for petroluem diesel fuel more successfully than natural oils. However, the cost of such a fuel in volume production is unknown at this time. Low temperature operation appears impractical, due to the high cloud/pour points, unless some curative additive can be developed or means for heating the fuel system is added. Deposit problems may not be completely eliminated by transesterification. Although they are promising, much more work on these types of fuels will be required before they can be considered a commercial product.

ACKNOWLEDGEMENT

The authors wish to acknowledge the valuable contributions of H.E. Staph and J.J. Staudt (SwRI) who directed the engine test program associated with this study.

REFERENCES

1. Fort, E.F., P.N. Blumberg, H.E. Staph and J.J. Staudt. 1982. "Evaluation of Cottonseed Oils as Diesel Fuel". SAE Publication 820317.

2. Staph, H.E. and J.J. Staudt. 1982. "Cottonseed Oil as a Diesel Engine Fuel". Southwest Research Institute, Final Report on TENRAC Project No. 80-B-4-3.

3. Bruwer, J.J., et al. 1982. "Sunflower Seed Oil as an Extender for Diesel Fuel in Agricultural Tractors," Department of Agricultural and Fisheries, Republic of South Africa.

4. Morrison, R.T. and R.N. Boyd. 1959. "Organic Chemistry", 1st Ed., Allyn and Bacon, Boston: 484.

5. Shelton, E.M. 1980. "Diesel Fuel Oils, 1980," DOE/BETC/PPS - 80/5: 18-25.

PROPERTIES OF FUEL BLENDS APPENDIX

		BLEND					
		2D	30% CSO 70% 2D	50% CSO 50% 2D	80% CSO 20% 2D	50% TECSO 50% 2D	TECSO
API Gravity, Deg.		35.2	30.7	28.2	24.5	31.8	28.8
Density,lb/gal		7.07	7.27	7.38	7.57	7.22	7.35
g/mL		.8488	.8722	.8860	.9070	.8665	.8828
Viscosity, cs	40°C	2.43	5.34	9.03	19.95	3.37	4.32
	100°C	1.05	1.96	2.83	8.36	1.41	1.64
Cloud Point	°C	-20	-16	-	-9	-5 Precip.	-5 Precip.
Pour Point	°C	-24	-21	-16	-14	-6	-2
Flash Point	°C	75	78	81	100	87	110
Cetane Number		47.2	45.1	48.6	32.0	48.0	51.2
Gross Heat Value, J/g		45310	43140	42230	40670	42470	40150
Net Heat Value, J/g		42620	40510	39590	38190	39990	37570
BTU/gal		129515	126600	125580	124270	124090	118680
Ash	%	.0022	.001	.0016	.001	.001	.001
Carbon Residue	%	.18	.32	3.79	5.24	.01	.28
Sulfur	%	.37	.11	.54	.61	.07	.001
Copper Corrosion		1a	1a	1a	1a	1a	1a
Color		2.0	2.5	2.5	2.5	2.0	1.0

THE 1981 "FLOWER POWER" FIELD TESTING PROGRAM

J. WALTER P. AAKRE J. DERRY*

ABSTRACT

Twelve typical farm tractors representing three manufacturers were operated on North Dakota farms during the 1981 crop season. The two fuels that were used were blends of 50% sunflower oil - 50% diesel fuel and 25% sunflower oil - 75% diesel fuel. The fuel handling characteristics of the blends and the durability effect on the engines were evaluated. There were no particular problems with fuel handling or engine performance while the tractors accumulated 6,300 total hours. Engine inspection did reveal excess carbon deposits on several engine components. Engines with rectangular style compression rings were more prone to ring sticking problems which accelerated liner wear.

INTRODUCTION

During the 1981 crop season twelve production tractors were operated on 25% - 75% and 50% - 50% blends of sunfower oil and diesel fuel, respectively. The tractors were located on farms throughout the state of North Dakota and operated by individuals selected for the project. The observations and conclusions made during and after completion of the crop season will be presented in this paper.

OBJECTIVES

The objective of this project was to determine the durability and wear patterns of new unmodified production engines utilizing sunflower oil - diesel fuel blends. The work done by the tractor was considered "normal" field work as dictated by the operator. The tractors were subjected to a variety of loads, field conditions, operators, air temperatures, and service procedures. The new tractors were either leased or farmer owned vehicles.

EQUIPMENT

The three manufactures that participated in the project were John Deere, JI Case, and Allis Chalmers. Table 1 lists the tractor identification number, operators, tractor models and blend percentages.

The Flower Power Program is cooperatively funded by the U.S. Economic Development Administration, North Dakota State Mill and Elevator, and North Dakota Sunflower Council.

*J. Walter is a Research Associate and P. Aakre is a Graduate Research Assistant for the Ag. Engineering Dept., North Dakota State University. J. Derry is a fuel injection instructor for the Diesel Dept., North Dakota State School of Science.

Table 1. General Information for 1982 Field Tests

Tractor ID	Operator	Percentage	Tractor	Hours Tractor	Hours Hobbs	Blend Liters	Sunoil Liters
9	Richard Backes	50%	7020 A.C.	393	393	6,814	3,407
8	Ralph Birdsall	25%	7060 A.C.	790	790	10,084	2,521
7	Roger Hanson	50%	7020 A.C.	700	700	10,796	5,398
2	Gene Watne	25%	8550 A.C.	503	503	24,415	6,104
			TOTAL	2,386	2,386	52,110	17,432
4	Don Alexander	25%	2590 Case	350	383	10,039	2,510
3	Tom Alme	50%	2590 Case	616	719	14,892	7,446
1	Everette Dobrinski	25%	2390 Case	470	570	12,306	3,076
10	Ken Tweten	50%	2390 Case	680	732	17,284	8,642
			TOTAL	2,116	2,404	54,521	21,675
5	Willard Pederson	50%	4440 J.D.	324	361	5,337	2,668
12	Morgan Simonson	25%	4440 J.D.	665	686	8,956	2,239
6	Keith Bjerke	50%	4640 J.D.	386	434	7,760	3,880
11	Sev Brekke	25%	4640 J.D.	483	528	7,836	1,959
			TOTAL	1,858	2,009	29,890	10,747
		GRAND	TOTAL	6,360	6,799	136,521	49,854

The tractors had six cylinder in line turbocharged diesel engines with direct injection combustion chambers. Eleven of the twelve tractors were two-wheel drive vehicles in the 120 to 180 PTO horsepower class. The only four-wheel drive vehicle was rated at 253 PTO horsepower.

Electric hour meters were installed on the tractors which had mechanically driven tachometers. The meters were activated by a pressure sensitive switch connected to the oil system.

Each cooperator was supplied with a 1000 gallon above ground bulk tank for fuel storage on the farm. Electric fuel pumps with five micron final filters and volumetric flow meters were also supplied to the farmer. The flow meters were placed on the bulk tank and the farmers field service tank.

Periodic performance checks were made throughout the operating season. Horsepower was measured by A & W and M & W PTO dynamometers. Fuel consumption was determined by an electronic scale and stopwatch. Robert Bosch smoke equipment was used to measure smoke density. API hydrometers were used to determine the specific gravity of the fuel blends. Thermometers were used to measure air and fuel temperatures.

The sunflower oil used in the program was a alkali-refined winterized oil processed by Honeymead Products in Mankato, Minnesota. The fatty acid distribution of the oil is presented in table 2. The oil was delivered by transport to the fuel dealers located in Grand Forks and Velva, North Dakota. The fuel was blended with commercial grades of #2 diesel fuel on a volume basis and stored in the fuel distributors bulk tanks.

Table 2. Fatty Acid Distribution of Two Samples

Palmitic (16:0)	Stearic (18:0)	Oleic (18:1)	Linoleic (18:2)	Linolenic (18:3)	Arachidic (20:0)	Behenic (22:0)
5.78	4.74	18.78	68.92	0.11	0.33	0.72
6.70	4.75	18.56	69.28	0.32	0.26	----

DATA AQUISITION

Daily records were maintained by the cooperators. The field testing team documented the data taken during the performance checks. Lubrication oil was sampled and tested by a reputable company. This data was used to detect potential premature failures.

At the conclusion of the crop season the tractors were disassembled and inspected. Observations were made and documented by photographs. Engineers from the respective manufacturers were able to inspect their engines for their own analysis.

OPERATING PROCEDURES

In North Dakota spring work began about mid-April. At this time all the tractors were on the farms and the cooperators had been informed on the operating procedures for the year. A farm visit was made soon after the start of spring work to establish baseline information on horsepower, fuel consumption, and smoke values with the sunoil blend. The PTO horsepower was measured on a dynamometer under full load using the specified blend. A 18 gallon fuel tank placed on a electronic scale and a stopwatch were used to calculate fuel consumption. The duration of the performance test depended mainly on the quantity of water available for cooling the dynamometer. The minimum time period observed was 15 minutes. Five smoke samples were taken at the midpoint of the test. Other data collected during the test was the wet and dry bulb temperatures. During this visit the flow meters were calibrated within 2% accuracy with a licensed five gallon measuring can. The farmers were instructed to collect oil samples at 25 hour intervals during the first 100 hours and every 50 hours thereafter. Each operator was also supplied with log sheets to keep the daily log of operating conditions, implements, fields, fuel consumption, and any special occurrences. Oil samples and log sheets were collected during successive visits and prepared for analysis. Periodically throughout the season the API gravity of the sunoil blend was determined to confirm the proportions of diesel fuel and sunflower oil in the blend.

At the midpoint of the operating season plans were made to inspect three tractors for polymerized oil and injector appearance and operation.

Throughout the latter part of the season the tractors were retested for performance and injectors were removed before and after the performance test to observe carbon accumulations and injector operation.

After completion of harvest and prior to engine disassembly, the final performance check was made on eight of the twelve tractors. Two of the tractors were disassembled without being tested to observe what effect rated load has on the appearance of the engine components. The two remaining tractors developed mechanical problems prior to obtaining a final performance check.

DISASSEMBLY AND REASSEMBLY

Guidelines were established prior to disassembly to aid in the consistency of the observations. Cylinders number three and four and the related components were used for observation and measurement when applicable. The components used for observation were; cylinder heads, valve and valve guides, injector and injector needle valves, piston and rings, liners, main and rod bearings, valve cover and oil pan.

Care was exercised during disassembly not to disturb any deposits or accumulations built up on the components. Further disassembly was required in some cases when unusual or extreme wear was noticed.

Prior to reassembly the only components that were cleaned were the gasket surfaces to ensure a good fit. New gaskets, seals, lubrication oil and fuel filters were used during assembly. Original rod and main bearings, rings, pistons, and liners were used for reassembly. Following rebuilding the tractors were again checked for performance on the sunoil blends and also on #2 diesel fuel for baseline data for an additional year.

RESULTS AND DISCUSSION

The first three months of the growing season passed with the engines operating very well. There were no unusual starting problems on either blend when the ambient temperature dropped to -5°C. The five micron filter at the bulk tank eliminated most of the particulate matter as only one cooperator changed tractor filters at 172 hours due to a loss of power. Most of the operators changed bulk tank filters only once throughout the operating season. The oil analysis showed no excessive engine wear but there was evidence of break in wear taking place. The operators experienced difficulty in removing spilled fuel if it was not cleaned up immediately.

The mid season inspection of three of the tractors occurred on July 7, 1981. A tractor from each of the manufacturers was selected for inspection. There was no evidence of polymerized oil in tractors 3, 7, and 12. The injectors were checked on a pop-tester for opening pressure, spray pattern and chatter. Most of the injectors had adequate opening pressures and chatter. Spray patterns appeared normal for tractors 7 and 12. The injectors on tractor 3 had five orifice nozzles. Poor atomization was evident from two of the injectors. There was only a moderate build up of carbon on the injector tips. When this tractor was inspected with a bore scope it was found to have several damaged sleeves. The tractor was moved to the local dealer where it was further disassembled. The tractor operated on a 50% blend and had 379 hours on the tachometer at the time of disassembly. Excessive carbon was found on the top ring land of every piston and above the ring travel on the liners. On four pistons the build up was large enough to generate visible wear on the liners. The carbon build up on #1 piston crowded the piston to the rear of the engine and the wear resulted from the heavy carbon deposition rubbing against the cast sleeve. All rings were free in the grooves but carbon was evident behind the rings. Three rod and four main bearings showed contamination wear. The lubrication oil analysis and the wear patterns indicated that assembly contaminates were present during the first few hours of operation. New pistons, rings and sleeves were replaced in cylinders #1, 2, 4, and 5. The seven worn bearings were also replaced. The original American Bosch injectors were replaced with a new set of Robert Bosch injectors supplied by the manufacturer. The heads, valve train components, and cylinders #3 and 6 were not cleaned prior to engine reassembly. After reassembly the tractor was returned to service.

The second failure occurred in the middle of August. The operator noticed excessive oil consumption and smoke through the breather. Tractor 9 was placed on the dynanometer and produced rated power. Periodically the speed and power would drop and excessive smoke was exhausted at the breather. This phenomena lasted ten seconds and would then return back to normal. This condition occurred three times during a 30 minute dynamometer test. The tractor was moved to the dealer where disassembly took place. At this point the tractor had 393 hours operating on a 50% blend. In this case, carbon deposits were evident but much less severe when compared to the preceeding engine. A very slight carbon film was evident on the cylinder head, injector tips, combustion chambers and heads of the pistons. The part numbers were clearly visible on the piston heads. However, on five pistons all the compression rings were stuck. The second compression ring was stuck on piston #3. In cylinders #1 and 5 the oil control ring had shattered and generated severe scoring on the sleeves. On the pistons which had all rings stuck the liner was discolored down to the oil control ring. The material in and between the ring grooves on the pistons had a glossy finish which appeared more like varnish than carbon. The material was very difficult to remove.

This particular operator used an API CC grade motor oil during the first 323 hours. The manufacturer specifications state a minimum of a CD grade lubricant. It was this cooperators operating procedure to use CC grade engine lubricants in all his diesel engines. Since the API classification of engine oils carry different standards concerning additives, this inferior lubricant contributed to the early failure. The engine was rebuilt with new pistons, rings, and sleeves. The cooperator did not feel he could accumulate many more hours so the tractor remained at the dealer in the rebuilt condition.

During the last day of the season the operator of tractor 2 was pulling a chisel plow and noticed excessive smoke from the breather. Soon thereafter the engine started to knock severely and was removed from the field. No performance test was conducted on this tractor because of engine failure. The four wheel drive tractor was moved to the local dealer for disassembly. Inspection revealed #4 rod bearing had rotated on the crankshaft. The resulting bearing and journal wear rapidly increased the connecting rod to journal oil clearance which allowed the piston to strike the valves. This is the condition which caused the knock. Once the bearing rotated the metal particles were carried by the lubrication oil to all the other bearings and caused extreme wear. All the remaining bearings were affected in various degrees. Inspection of the liners revealed stuck compression rings in the pistons. Three of the pistons were removed for observation. On piston #3 the first two compression rings were stuck with the third and oil control ring free. Piston #4 was similar to #3 except the third compression ring was stuck and broken. The most damaged piston was #6 where the land area between the second and fourth ring broke as can be seen in Fig. 1. On this piston the first and second rings were free and stuck respectively. The third compression and oil control rings were broken. The compression rings used in this engine were of the rectangular type. Since the operator had experienced blowby through the breather before the knock developed the rings were starting to stick before the bearing turned. The discoloration of the liners would support the statement. It is possible that ring and liner wear contaminated the lubrication oil causing the #4 bearing to fail. The remaining bearings also showed metal contaminant wear. Therefore, the nature of the failure may have originated from the use of the blended fuel.

Fig. 1. Piston of Tractor 2 with Broken Ring Lands.

There was also abnormal deposits of varnish on the ring lands of the pistons. On the top ring land the deposits had a glossy carbon appearance. This type of build up was common to all engines inspected as will be discussed later. Excessive deposits of carbon were also evident above the ring travel on all the liners. Again, the quantity of deposits was similar among cylinders of any given engine but the excessive build up varied between engines. The variation between engines will be discussed later. The top of the piston and bottom of the head around the injector tip and valves had a light carbon build up similar to tractor 9. The valve cover and oil pan showed no evidence of polymerized oil which was a consistent observation with all the engines. The injectors had slight accumulations of carbon on the tips. The amount was not enough to affect injector operation. Tractor 2 operated the entire season operating under heavy loading conditions. The cleanliness of the combustion chamber and the appearance of the injectors could be the result of the engine operating closer to rated load.

During the performance checks throughout the summer where injectors were inspected before and after the test, it was found that rated load reduced the accumulations of carbon on the injector tips.

As harvest was just about completed the disassembly procedures were finalized with the manufacturers and preparations were made for transporting and housing the vehicles. When the cooperators were finished with the tractors, arrangements were made to transport the vehicles to the local dealer or to the North Dakota State School of Science in Wahpeton, ND. The final power test of the season was performed in Wahpeton with the exception of the J.I. Case tractors. Since two Case engines were to be transported to the manufacturer in an assembled condition, the final power test was performed at the area dealer.

As more hours were accumulated on the tractor fuel filters and the ambient air temperature dropped late in the season, two operators noticed lack of power until the vehicle was warmed up. The problem was solved when the original fuel filters were replaced.

Data was taken throughout the season to monitor the performance of the engines. Since the objective of the program was to determine the durability effects, the performance data was used mainly as a diagnostic tool when failures ensued. Generally, the operators agreed with the performance data when stating the performance of tractors operating on sunoil blends did not deteriorate throughout the operating season. The final power test also indicated that all vehicles performed excellent when compared to the Nebraska standard for that tractor.

Tractor Use

The twelve tractors consumed a total of 136,521 liters of fuel during the operating season. See Table 1. The total volume of sunflower oil used was 49,854 liters while the tractors logged 6,360 total tachometer hours. The total real time hours from the electric meters was approximately 6,800 hours. In some cases the electric hour meter switch failed which caused the hour meter to accumulate hours when the engine was not running or not operate when the engine was running. Therefore, the real time hour figure is approximate. Hence, the tachometer hours will be the values used. Three of the operators logged the most hours pulling a chisel plow. Three others operated a disk for the majority of the time. The others accumulated the most hours operating a row crop planter, field cultivator, plow, auger, dryer, and a rock picker.

Final Engine Inspection

Of the eight remaining engines transported to North Dakota State School of Science for disassembly six were operated on a dynamometer. Tractors 11 and 12 were the two vehicles not operated under load prior to disassembly.

Upon inspection of the components previously selected several observations can be stated. The quantity of carbon seen on the top piston lands and sleeves was not directly related to the blend of fuel used. Tractors 5 and 6 both logged approximately the same hours on a 50% blend and tractor 5 definately had more accumulations present. The loading on both tractors prior to the final dynamometer check was mixed between medium and light loads. Tractor 3 accumulated approximately twice as many hours on a 50% blend as tractor 4 on a 25% blend and there was only a slight difference in appearance of the pistons and liners. Tractor 12 had the greatest build up in the combustion chamber and was operated on a 25% blend. This particular engine was not placed under dynamometer load prior to disassembly. A picture of the head is seen in Fig. 2.

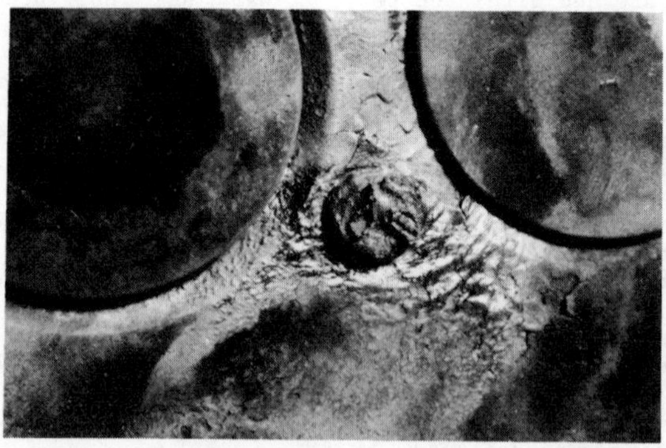

Fig. 2. Carbon on Head of Tractor 12. Injector is Removed.

Once injector #4 was removed from the head the carbon build up around the top remained on the head and the injector appeared as in Fig. 3.

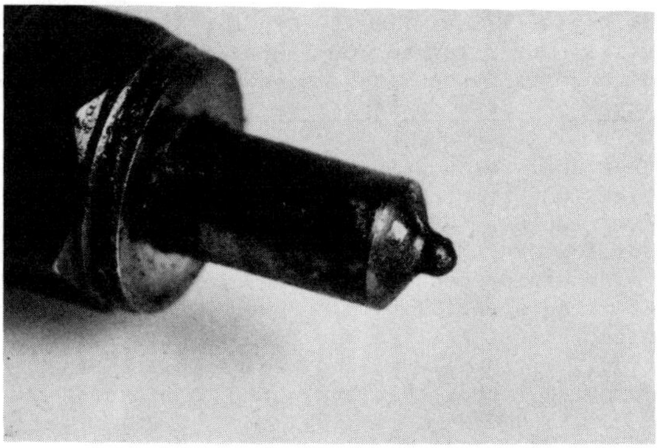

Fig. 3. Injector Tip Corresponding to Figure 2.

The injector spray pattern was altered to the point that excess carbon was present on the piston where the penetrating fuel spray contacted the piston head. None of the other engines inspected demonstrated the extreme build up around the injector tips and on the piston surfaces. Tractor 11 was the other vehicle not operated under load prior to disassembly. Further investigation of tractor 11 indicated that the operator had run the last few hours with #1 diesel fuel. The appearance of the engine components lacked the wet, shiny carbon common to the other engines.

The intake valves and ports of tractors 3 and 4 were components that also exhibited abnormal carbon deposits. An intake valve can be seen in Fig. 4.

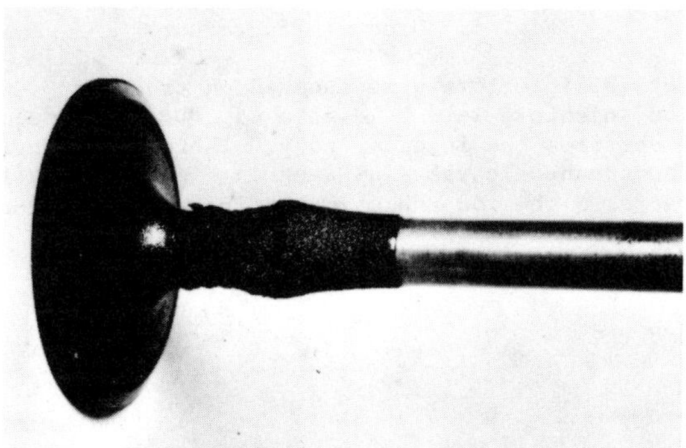

Fig. 4. Carbon Deposit on Intake Valve of Tractor 3.

This deposition can be contributed to the valve timing and the combustion characteristics of the fuel. The manufacturer indicated less but similar deposits of this type on their engines operating on diesel fuel. Tractor 7 while not demonstrating excess carbon on the valve had abnormal accumulations of carbon in the ports. The remaining vehicles did not show such extreme deposits in these areas.

Tractor 7 was the only vehicle to have stuck rings upon disassembly after the final performance check. Since the preselected components showed signs of wear, the remaining pistons and sleeves were removed. The compression rings in the engine were stuck and caused two broken oil rings.

This in turn led to excess material removed from the liner walls. The degree of failure of this engine was similar to tractor 9. Both tractors were of the same model but tractor 7 logged 700 hours of operation while using a lubrication oil of acceptable quality. <u>None of the eight engines which used a keystone style compression ring experienced stuck or broken rings</u>.

The only engine that used rectangular shaped piston rings and did not exhibit ring problems was tractor 8. This tractor accumulated the most hours of all twelve tractors (790) and operated on a 25% blend. This vehicle was used on a dryer for over 300 hours late in the season and appeared the cleanest of all. The engine consistently developed 30 horsepower more than the 160 horsepower rated specification. However, the tractor was not checked for power on straight diesel fuel so any fuel related increase in performance would be speculation. Irregardless, the cleanliness of the engine was an exception to the program rather than the norm. A picture of the head can be seen in Fig. 5.

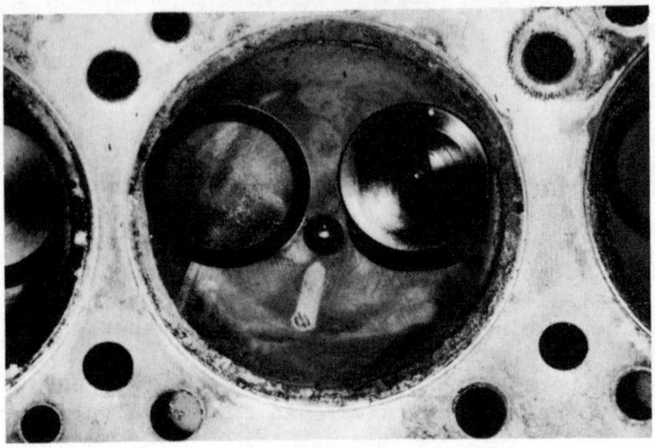

Fig. 5. Head of Tractor 8.

The needle valves of the injectors were checked on tractors 3, 4, 5, 7, 8, and 12. When the injectors were held at a 45° angle the needle valves were able to slide out from the injector body. There was no evidence of varnish or carbon on the needle valve as seen in Fig. 6. In all vehicles except tractor 2 the rod and main bearing wear appeared normal.

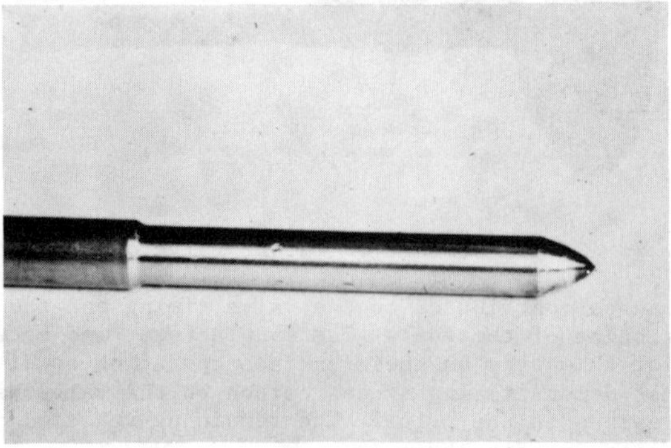

Fig. 6. Injector Needle Valve of Tractor 7.

SUMMARY AND CONCLUSIONS

Twelve unmodified production diesel farm tractors were operated on diesel fuel - sunflower oil blends. The two blends investigated were 25% sunflower oil - 75% diesel fuel and 50% sunflower oil - 50% diesel fuel. The tractors logged 6,360 hours of operation and consummed 136,521 liters of fuel, of which 49,854 liters was 100% alkali-refined winterized sunflower oil purchased from Honeymead Products in Mankato, Minnesota.

During the 1981 crop season the tractors operated successfully with only two vehicles not completing the season due to failure. There was no special equipment necessary for handling the sunflower oil - diesel fuel blends. Minimal fuel filtration problems were encountered. The five micron filter at the bulk tank eliminated potential problems. Care needs to be exercised when handling the fuel to minimize spillage as the cooperators experienced difficulty removing the dried film if not cleaned up immediately. The operators did not notice any lack of power or excessive fuel consumption which can be verified by the performance checks. Normal starting procedures were utilized and found adequate in all cases even when ambient temperature reached -5°C. In some cases near freezing temperatures full power was not available until the tractor and fuel were warmed up properly. The distinct odor of burning sunflower oil was aethetically pleasing to the operators.

Following the conclusion of fall work, the tractor engines were disassembled to check for abnormal wear and engine deposits. Inspection revealed high carbon and varnish deposits on several engine components. Piston ring grooves and ring lands on all engines contained excess deposits. Three of the four engines that had rectangular style rings experienced ring sticking problems which led to broken oil rings and scored cylinder sleeves. Engines with keystone piston rings did not have stuck rings, but the level of excess carbon and varnish was greater than would normally be expected burning straight diesel fuel. The premature failure of the only engine using an API CC engine oil would indicate the importance of using a minimum of an API CD engine oil as recommended by the manufacturer.

The bottom of the engine head including the valve heads and injector tips in most cases appeared relatively free of excess carbon. The load characteristics of the tractor appeared to have a larger influence on engine deposits than the percent of sunflower oil in the fuel blend. The quantity of build up was not proportional to the duration of the test when comparing similar engines. Light loads appeared to increase deposits. One tractor operated under light loads prior to disassembly revealed heavy carbon deposits around the injector tips. Heavy loading prior to disassembly may have reduced the carbon deposits. Some tractors had excess carbon on the intake valve stems and intake ports. These deposits could lead to valve sticking and burned valves under continued use.

Engine wear appeared normal except in the case where sticking piston rings lead to scored sleeves. Engine bearing wear appeared normal in all twelve tractors except tractor 2 where the bearing insert rotated on the crankshaft and contaminated the remaining bearings.

This program indicated that farm diesel tractors can be operated at blends of up to 50% sunflower oil and 50% diesel fuel. However, continued use may cause premature engine problems. Causes of these problems originate from the high viscosity of sunflower oil which causes poorer atomization of the fuel in the combustion chamber. The chemical properties of sunflower oil encourage the creation of varnish like films which can cause ring sticking. Engine modifications, fuel and lubrication oil additives, refinement procedures, and genetic modifications of the sunflower plant should be investigated to improve the use of sunflower oil as a farm fuel.

FIRST RESULTS WITH MERCEDES-BENZ DI DIESEL ENGINES

RUNNING ON MONOESTERS OF VEGETABLE OILS

by Luso M. Ventura

and Aggeo C. Nascimento, Mercedes-Benz do Brasil S.A., São Paulo, Brasil

and Wolfgang Bandel, Daimler-Benz AG, Stuttgart-Untertuerkheim Federal Republic of Germany

ABSTRACT

In their pure form the vegetable oils are not suitable for the use in modern DI diesel engines, due to the excessive carbon deposit on the injection nozzles and in the combustion chamber.
Nevertheless, these oils are promising candidates as raw materials for alternative diesel fuels. Processes are being developed to transform the long vegetable oil molecules into smaller molecules in order to fulfill the fuel requirements of DI diesel engines. Methyl and ethyl esters of fatty acids e.g. obtained by transesterification of vegetable oils through their catalytic reaction with methanol and ethanol, have shown a typical diesel fuel behaviour in conventional DI engines without excessive deposit formation. Problems concerning lubricating oil contamination, and possibile remedial measures to avoid it, are being examined. There are also problems to be solved in relation to white smoke formation and the odor of exhaust gases.

VEGETABLE OILS AS FUELS

At a first glance, conventional fuel characteristics exhibited by products derived from biomass e.g. ethanol and vegetable oils (Fig. 1) suggest that a very good parity exists between the latter and the diesel fuel; some minor deviations in viscosity and cloud point can be easily accomodated by the flexibility of current systems and certainly no problems will arise. The apparent commonality between these fuels and actual engines has been examined on pratical operation basis by various researchers at different times; the results however indicate a quite unexpected pattern: a true incompatibility exists between neat vegetable oils and modern direct injection engine which stems mainly from their physical and chemical properties.
A closer examination of the molecular arrangements of diesel fuel, vegetable oils and common alcohol (Fig. 2) indicates a very distinctive nature for the three fuels where the vegetable oils with long and heavy chains seems to be critical in terms of combustion in the Diesel cycle.
The use of neat vegetable oil seems to be feasible only in diesel engines

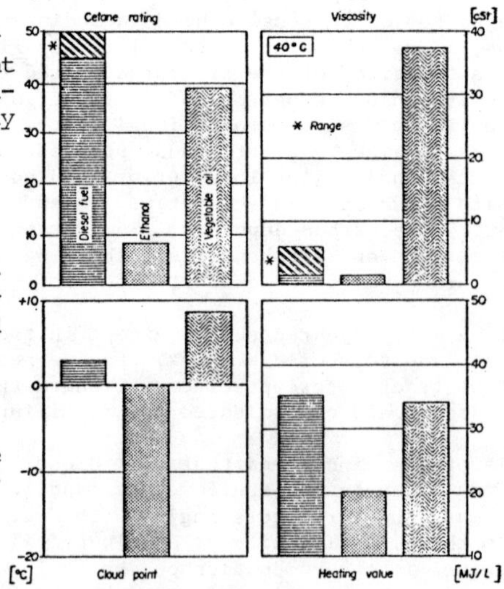

Fig. 1
Properties of vegetable oils and ethanol compared to diesel fuel

with indirect injection. However, it has not been established yet whether this is true with all kinds of oils. At any fuelling rate however even in this case some problems concerning long term deposit formation in combustion chambers, exhaust emissions and lubricating oil contamination remain to be solved.

Further, the use of indirect injection engines involves a penalty in fuel consumption (Fig. 3) which made them unable to compete with the direct injection units introduced more than two decades ago in the commercial vehicle market; possible enforcement on its return should require great expenditure for fleet re-engining and operation.

Some early MB trials with neat vegetable oils in direct injection engines indicated a quite satisfactory engine response with negligible power drop and improved exhaust black smoke levels but moderate endurance runs revealed increased severity of engine oil degradation and unacceptable combustion chamber deposits with a marked performance deterioration. It was clearly understood from this point that engine hardware modifications would hardly provide conditions for better combustion and more substantial improvements could be gained by changing the neat vegetable oil structure.

STRUCTURE MODIFICATION OF THE VEGETABLE OILS

Particular rearrangements of the typical molecular chains of vegetable oil structures in order to approach the natural disposition observed in ordinary diesel fuel are possibile and known as cracking or transesterification via catalytic reaction with methanol or ethanol (Fig.4).

The product obtained by transesterification, the methyl or ethyl monoesters are more favourable and there are clear indications that the process can be realized "in-house" with simplified equipment; the process also adds a bonus in form of pure glycerine, a byproduct which has commercial value in some countries.

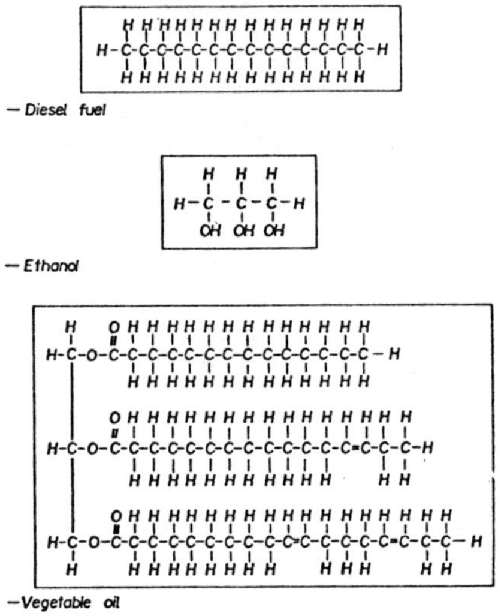

Fig. 2 - Typical structure of diesel oil, ethanol and vegetable oils molecules

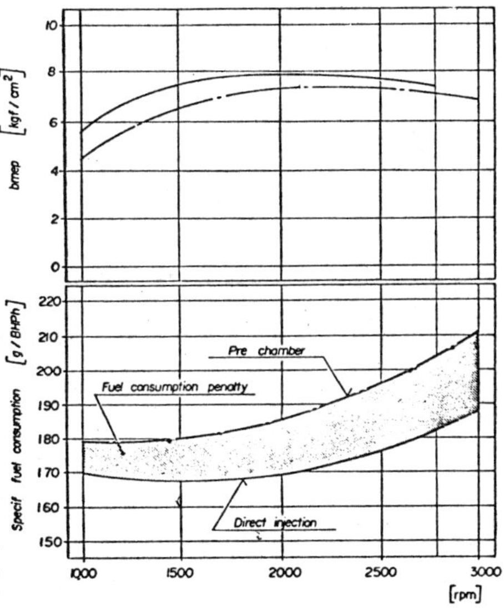

Fig. 3 - Direct injection X prechamber diesel engines Full load performance comparison

The fuel characteristcs of the esters from vegetable oils are closer to those of diesel fuel (Fig. 5) when compared with neat vegetable oils; the esters thus become strong candidates as "full replacement" fuels: Viscosity and cloud point are comparable and the cetane numbers are further improved after processing; heat values are about 10% less than the reference value for diesel fuel. The cloud point however, may pose some problems depending on the type of base oil and certainly will limit fuel utilization in some areas during winter due to reduced cold starting ability. The admixture of diesel fuel can minimize this negative feature.

The first tests were conducted with methyl ester which is relatively easier to produce and certainly will not compromise ethanol programm now under full implementation in Brazil. The base vegetable oil was also carefully selected bearing in mind its immediate availability for processing.

The comparative yields of various proposed liquid fuels from Biomass in Brazil are indicated in figure 6: the apparently reduced output from the first group of oils are due to the yearly character of cropping; the great potential exhibited by palm oil is possible only after 8 years, this being the time needed to attain full production and its widespread as an alternative fuel is hardly foresseeable. The ethanol yields from sugar cane or cassava are more stable. Both are traditional cultures, this being an important point when considering an alternative fuel with abundant sources and uninterrupted supply.

The methyl esters of soybean and peanut were selected as the test fuels; the transesterification has been carried out by expert chemical companies and strict control on fuel quality has been exercised.

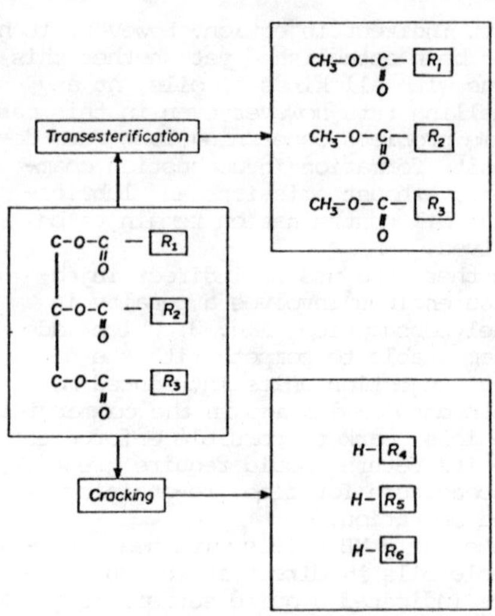

Fig. 4 - Methods for modification of chemical structure of vegetable oils

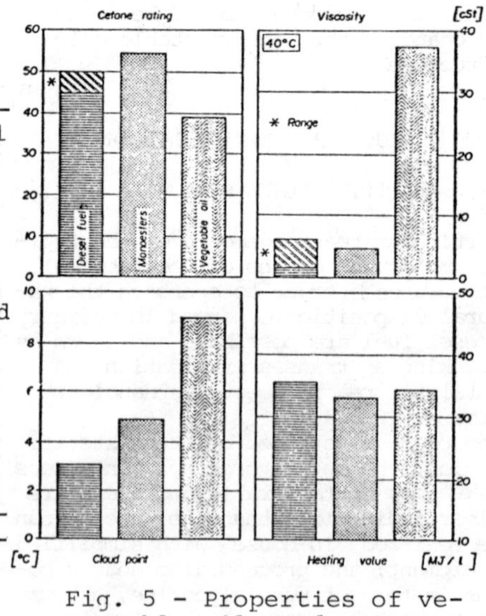

Fig. 5 - Properties of vegetable oils and monoesters compared to diesel fuel

FUEL STANDARIZATION

The desirability for an effective quality control existed at the outset of testing programm and based on actual experience of engine manufacturers and chemical industry the initial parameters were determined; a standard specification sheet has been drawn up (Fig. 7) and is being used to check fuel quality on receipt.

Additionaly, other parameters and associated testing methods which could even better characterize the substance are being realized. The compliance to these initial values and proposed tests will contribute to improve the specifications of fuels; the standard will be up dated accordingly as more knowledge is gained.

ENGINE TESTING

An endurance testing with methyl ester of soybean oil on a production direct injection engine was started aimed at finding clear engine/fuel relationship in short and long term basis; no modifications were made on engine and its systems since previous combustion analysis revealed no discernible differences between the pressure and needle lift traces when compared to diesel fuel and a co-evaluation of lubricating oils was also intended.

Usually recorded engine parameters showed normal values throughout the test. A detailed examination on stripped engine after test completion revealed satisfactory combustion chamber condition with no signs of excessive carbon build-up and nozzle set was also in acceptable conditions without obstructions on injection holes (Fig. 8). The small carbon trumpets observed in nozzle tips did not impair injection characteristcs and combustion was not negatively affected; fuel atomization has been checked after test and showed normal spray patterns with no disturbance being caused by the deposits. Engine lubrication system however was found to be close to full collapse since the whole lube oil charge had started to thicken; oil pump strainer and oil filter suffered a drastic reduction in area by a thick and black paste which also contaminated the remaning parts flushed by the oil. The oil grade used in this phase was a standard product for current Diesels and its condition, monitored during all the test, was worsening at each inspection interval, with increases in viscosity and fast TBN depletion being noteworthy; the cause of severe lubricating oil thickening may be the polymerization of unburnt fuel in contact with engine lube oil. The somewhat coarser spray pattern observed during injection of esters and the poor volatility observed in the distillation curve has a decisive effect on fuel atomization and combustion, both contributing to the presence of unburnt fuel

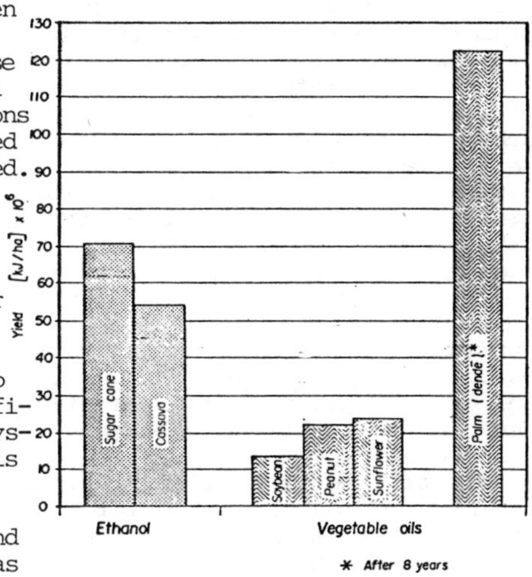

Fig.6 - Yield of liquid fuels from Biomass

ITEM	TEST CONDITION AND/OR UNIT	SPECIFICATION
Acidity	mg KOH/g	⩽ 7
Refraction index	25°C	1,4558 – 1,4548 (M)
		1,4552 – 1,4530 (E)
Peroxide index	mEg/g	⩽ 30
Cloud point	°C	⩽ 6°
Water content	%	⩽ 0,10 %
Viscosity	37,8° C m^2/sec	6 max. at 40°C
Density	25/25° C g/cm^3	0,8817 – 0,8799 (M)
		0,8777 – 0,8756 (E)
Soap	ppm	10 ppm
Free alcohol content	%	< 2%

M = Methyl ester E = Ethyl ester

Fig.7 - Preliminary specification sheet for esters of vegetable oil as fuels

in sump and exhaust.
As a result of the poor performance attained by a standard lubricating oil a test was started to evaluate two newly proposed products of standard lines but with another design of additive package and carried out on two engines; the test was normal up to 200 h; one of the engines showed a marked drop in power output (~ 15%) at 150 h due to a burnt exhaust valve in one cylinder. After repair, the sequence was continued and completed with no further problems.
A probable cause of burnt exhaust valves could be the residual catalyst in fuel (2). However, there is no sufficient data available yet to support these assumption.

Fig. 8 - Nozzle tip condition after 150 h with methyl ester of soybean oil

The lube oils performed satisfactorily without signs of "breakage" even under sustained fuel dilution detected by continous analysis during the test which showed a reduction in viscosity. An assesment of engine conditions revealed satisfactory overall appearance indicating reasonable behaviour of lube oil as a whole.
In order to gain further knowledge on the possible influence of the molecule double bonds on the behaviour of lubricating oils a test was performed with methyl ester of peanut oil, which has a less insaturated nature, with iodine value around 80.
The test run normally up to 100 h, with no noteworthy ocurrences and engine condition remained satisfactory, without deposits on combustion chambers or oil degradation; in fact, oil analyses revealed a trend for improved behaviour when compared with tests on ester of soybean oil, which has an iodine value of ~120.
These findings will be confirmed by testing the ester of Palm oil which has an even lower iodine value (~ 50).
Some earlier testing of Diesel/Ester blend indicated a clear relationship between lube oil dilution rate and ester fraction in the fuel; to reduce or avoid the ocurrence of this phenomenon, improvements in the fuel/air mixture preparation must be studied and this will involve modification of injection and combustion parameters and some benefits should come also from a proper formulation of lubricating oils. On engines' side, the following subjects and their relationship could contribute to improve the conditions for a better combustion:

- Increased coolant temperature
- Injection timing advance
- Proper matching of nozzle geometry
- Increased compression ratio

FIELD TESTING

The field testing followed the evaluation on test bed and currently comprises two vehicles runing on 100% of soybean oil ester; the fleet and test bed times are presented with cumulative value, in fig. 9. The peculiar problems encountered in field, apart from lube oil contamination, are related to white smoke emission in cold conditions with associated pungent odors (3) in exhaust and cold starting ability. Due to unfavourable cloud point, the esters will solidify at low temperatures and the consequent fuel filter blockage is indeed a critical item; this area has been studied recently with the development and testing of flow improver additives (1) for shale oil derived fuels and a similar reaction with esters are likely to occur. Another possible solution consists in allowing the free alcohol content in the esters to increase further to a set limit (max.8% vol) which will tend to depress the cloud point.

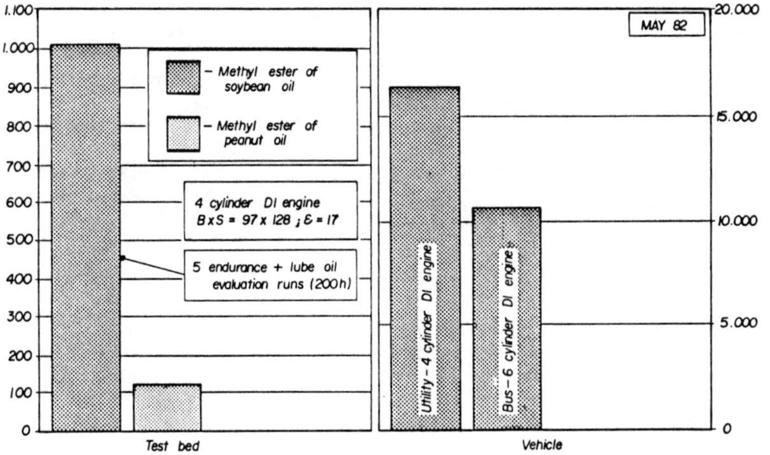

Fig. 9 - Endurance tests carried out with monoesters of vegetable oils

EXHAUST EMISSIONS

One of the main concerns with exhaust emissions previously evaluated on engines burning neat vegetable oils was the presence of acrolein, whose order of magnitude was considered potentially dangerous; however, a simple change over to esters eliminated the possibility of ocurrence of this compound in exhaust. The white smoke observed is typical on diesel engines runing cold, especially during winter; however, in the case of esters, the amount of smoke is greater and it takes a reasonable time until it disappears, being solely influenced by coolant temperature, whose level increases in a slow rate during engine warm-up period. The exhaust smell intensity is worst in this phase but still remains even after engine reaches its normal working temperature. (3)

Extensive field testing aimed at finding some means to reduce or alleviate white smoke and odor emissions were realized under actual conditions and the results presented in fig. 10; the smoke intensity was measured by an USPHS opacimeter and engine operated at idling speed, following a night soaking period before starting. As can be seen, the static injection timing exerts a positive influence on white smoke formation with substantial reductions being attained with injection advances.

The benefits brought about by the injection advance on cold starting are however restricted in practical sense: it surely helps to provide better conditions for combustion, possibly also reducing the rate of lube oil dilution but tends to increase peak cylinder pressures and noise levels. Some mechanical problems related to erosion of piston top caused by a too early

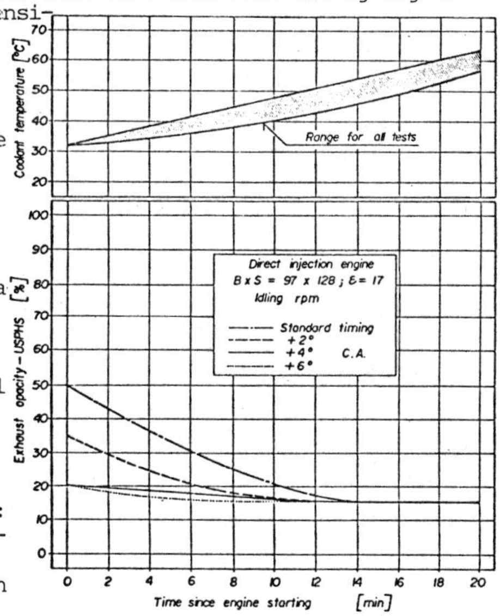

Fig. 10 - Influence of static injection timing on white smoke emission at cold starting

injection begining have already been observed; a proper matching of all influencing factors will need a substantial research effort, already suggested at the end of chapter 4.
The exhaust smell intensity is unbearable during cold starting and seems to be less influenced by timing advance than the white smoke.
A basic concern on the regulated exhaust emissions with esters is not applicable at this stage since the combustion problems encountered must be tackled properly before a true appraisal on the impact of these fuels on environment can be made; the drastic reduction (~ 50%) on black smoke levels at full load with esters can be regarded as an attractive feature of these fuels.

CONCLUSION

It has been demonstrated that current direct injection diesel engines will react more favourably with esters of vegetable oils than with the neat product; particulary, combustion chambers deposits are comparable when operating on Diesel fuel. Few engine modifications and settings together with a strict control on these new fuels properties and proper lube oil formulation could solve the main problems.
The exhaust smell intensity must be observed after the proposed modifications and settings and should the problem persits, new research work will be carried out aimed at finding a true solution.
The blending of esters with diesel fuel could by-pass the majority of reported problems but these action will tend to mask the problems, retarding their solution.
The first consistent works on thermal and catalytic cracking of vegetable oils are showing that these products could be a solution to the engine technical problems as has been demonstrated by lab test on first samples; this research effort however are being retarded by the unavailability of proper laboratory facilities but the difficulties can be overcome by the full engagement of oil refinery in these efforts.
The costs involved in the implementation of a vegetable derived fuel programm are directly related to agricultural primary costs and further aggravated by the necessary after-treatment of neat vegetable oils to adjust their properties.
The case of transesterification via methanol must be studied further in light of the growing availability of ethanol and a change over to the latter must be considered as the methanol production in Brazil is still based on petroleum feed stocks; also the impact of a high glycerine surplus in the world markets is easily predictable and should therefore, receive attention.
In the case of cracking, the utilization of existing refining plants for petroleum processing are very attractive, depending on the development of suitable catalysts and a full assessment of the process efficiency.

REFERENCES

1. Frankenfeld, J.W. & Taylor, W.F. - "Improving low temperature properties of synthetic Diesel fuels derived from oil shale"
U.S.Department of Energy, report nr. DOE/CS/50019-1, November, 1980

2. Hawkins, G.S. - "Alcohol/Sunflower oil ester: fuel performance test in compression ignition engines"
Proceedings of the V ISAFT, Vol. II, Aukland, May, 1982.

3. Ventura, L.M. - "Current research status on vegetable fuel oils by Mercedes-Benz do Brasil S.A.", Seminar II, Northern Agricultural Energy Center, Peoria, Il. - October 21/22 , 1981

ACN/MGGL/BG